Ecological Studies, Vol. 145

Analysis and Synthesis

Ecological Studies

Volumes published since 1994 are listed at the end of this book.

Springer

Berlin
Heidelberg
New York
Barcelona
Hong Kong
London
Milan
Paris
Singapore
Tokyo

G. Bauer K. Wächtler (Eds.)

Ecology and Evolution of the Freshwater Mussels Unionoida

With 156 Figures and 39 Tables

Springer

Professor Dr. Gerhard Bauer
Universität Freiburg
Institut für Biologie I
Hauptstr. 1
79104 Freiburg
Germany

Professor Dr. Klaus Wächtler
Institut für Zoologie
Tierärztliche Hochschule Hannover
Bünteweg 17
30559 Hannover
Germany

ISSN 0070-8356
ISBN 3-540-67268-0 Springer-Verlag Berlin Heidelberg New York

Library of Congress Cataloging-in-Publication Data

Ecology and evolution of the freshwater mussels Unionoida / G. Bauer, K. Wächtler (eds.).
p. cm. – (Ecological studies; v. 145)
Includes bibliographical references (p. ••).
ISBN 3540672680 (alk. paper)
1. Unionoida – Ecology. 2. Unionoida – Evolution. I. Bauer, G. (Gerhard), 1953 – II. Wächtler, K. (Klaus), 1938 – III. Series.
QL430.6.E36 2000
594'.4 – dc21

00-044062

Springer-Verlag Berlin Heidelberg New York
a member of BertelsmannSpringer Science + Business Media GmbH

Production: PRO EDIT GmbH, Heidelberg, Germany
Cover design: design & production GmbH, Heidelberg, Germany
Typesetting: Best-set Typesetter Ltd., Hong Kong

Printed on acid free paper SPIN 10860533 31/3111/Di 5 4 3 2 1

Preface

All those who think that bivalves are boring are in the best company. Karl von Frisch is reported to have turned the pages more quickly in texts where bivalves were treated because, according to him, they literally lack any behaviour.

The fact that they can filtrate huge amounts of water, burrow into the sediment, actively swim, drill holes into rocks and boats or detect shadows with the aid of pretty blue eyes located on the rim of their mantle obviously left v. Frisch unimpressed.

Why, then, a book on the large freshwater mussels (Naiads or Unionoida), which on first sight are much less spectacular than the marine ones?

The main reason is that they are keepers of secrets which they reveal only on close and careful inspection. This is not only true for the pearls some species produce and which over centuries have contributed to the treasures of bishops and kings, but particularly for their ecology: their life cycles are linked with those of fishes, some can occur in incredible densities and some can live for more than 100 years. Thus, the presence or absence of naiads in a lake or stream has manifold implications.

Naiads produce larvae which have to contact a host fish, where they pass through a stage as ectoparasites over a period from a few days to up to 10 months before they are dropped to start life as filter feeders on the sediment. The spectrum of suitable host fish may be wide or narrow. Some can develop on more than 15 species, and some are specialized on very few closely related species. There are host-parasite systems, and this is unique, where the generation time of the parasite exceeds that of the host by a factor of 10. Here, the host should win the race for defensive adaptations; however, these systems have been stable for at least 60 million years.

The maximum life span may vary, with only 5 years in some species, whereas others belong to the longest-lived invertebrates on earth. Fertility is very high; some can produce up to 4 million larvae every year over several decades. Although there are interesting mechanisms to increase the chance of host contact, the mortality of larvae is enormous.

The life histories resulting from these characters and their interactions with the hosts, their constraints, their success, their risks and their mode of evolution will be described in this book.

In addition, aspects of nature conservation will also be covered, as many naiad species are seriously threatened. Some have been used successfully as sensitive pollutant indicators in habitat monitoring and as "archives" for environmental changes, incorporating the chemical history of the habitat for many decades.

Biologists from many geographical regions are agreed that it is time to give a survey on recent insights in naiad biology.

Finally, we should add that we are very grateful to C. Vogel-Bauer, A. Bogan and H. Zwölfer for their great help in preparing this book.

August 2000

G. Bauer and K. Wächtler

Contents

Part III Populations and Communities

Contributors

Rafael Araujo

Museo Nacional de Ciencias Naturales, José Gutiérrez Abascal 2,
28006 Madrid, Spain

Guido Badino

Dipartimento di Biologia Animale e dell'Uomo, Università di Torino,
Via Accademia Albertina, 17, 10123 Torino, Italy

Gerhard Bauer

Institut für Biologie I, Hauptstr. 1, 79104 Freiburg, Germany

Victor Beletsky

Institute of Developmental Biology of Russian Academy of Sciences,
117808, Vavilov st. 26, Moscow, Russia

Arthur E. Bogan

Curator of Aquatic Invertebrates, Research Laboratory,
North Carolina State Museum of Natural Sciences, 4301 Reedy Creek Road,
Raleigh, North Carolina 27607, USA

Maria Byrne

Department of Anatomy and Histology, University of Sydney,
New South Wales 2006, Australia

Peter J. Cosgrove

Cairngorm Partnership, 14 The Square, Granton-on-Spey,
Morayshire PH26 3HG, Scotland

J. Mark Hanson

Invertebrate Fisheries, Science Branch, Gulf Fisheries Centre, P.O. Box 5030
(343 University Avenue) Moncton, New Brunswick, E1C 9B6 Canada

Lee C. Hastie

Culterty field Station, University of Aberdeen, Newburgh, Ellon, Aberdeenshire AB41 6AA, Scotland

William H. Heard

Department of Biological Sciences, Florida State University, Tallahassee, Florida 32306, USA

Christopher W. Hickey

National Institute of Water and Atmospheric Research, P.O. Box 11-115, Hamilton, New Zealand

Susanne Hochwald

Lehrstuhl Biogeographie, Gebäude Geowissenschaften, Universität Bayreuth, Universitätsstr. 30, 95447 Bayreuth, Germany

Walter R. Hoeh

Department of Biological Sciences, Kent State University, Kent, Ohio 44242, USA

Daniel J. Hornbach

Department of Biology, Macalester College, 1600 Grand Avenue, St. Paul, Minnesota 55105, USA

Wolfgang Jansen

Department of fisheries & Oceans, Freshwater Institute, 501 University Cres., Winnipeg, MB, R3T 2N6 Canada

Sviatoslav Kaliuzhin

Institute of Developmental Biology of Russian Academy of Sciences, 117808, Vavilov st. 26, Moscow, Russia

Maria Cristina Dreher-Mansur

Museu de Ciẽncias e Tecnologia, Pontifícia Universidade Católica do Rio Grande do Sul, Av. Spiranga 6681, 90619-900 Porto Alegre – RS – Brasil

Doris Müller

Institute of Zoology, University of Salzburg, Hellbrunnerstr. 34, 5020 Salzburg, Austria

Harry Mutvei

Swedish Museum of Natural History, Department of Palaeozoology, Svante Arrhenius väg 9-11, Box 50007, SE-19405 Stockholm, Sweden

KARL-OTTO NAGEL

Dipartimento di Biologia Animale e dell'Uomo, Università di Torino, Via Accademia Albertina, 17, 10123 Torino, Italy
Present address: Schwarzwaldstr. 15, 79189 Bad Krozingen, Germany

ROBERT A. PATZNER

Institute of Zoology, University of Salzburg, Hellbrunnerstr. 34, 5020 Salzburg, Austria

EKATERINA POPKOVITCH

Institute of Developmental Biology of Russian Academy of Sciences, 117808, Vavilov st. 26, Moscow, Russia

MARTIN PUSCH

Institut für Gewässerökologie und Binnenfischerei, Müggelseedamm 310, 12561 Berlin, Germany

MARÍA ANGELES RAMOS

Museo Nacional de Ciencias Naturales, José Gutiérrez Abascal 2, 28006 Madrid, Spain

TORSTEN RICHTER

Institut für Zoologie der Tierärztlichen Hochschule Hannover, Bünteweg 17, 30545 Hannover, Germany

DAVID S. ROPER

Mighty River Power Limited, P.O. Box 445, Hamilton, New Zealand

JÖRG SIEFERT

Institut für G Gewässerökologie und Binnenfischerei, Müggelseedamm 310, 12561 Berlin, Germany

DOUGLAS G. SMITH

Dept. of Biology and Graduate Program in Organismic and Evolutionary Biology, University of Massachusetts, Amherst, MA 01003, USA

KLAUS WÄCHTLER

Institut für Zoologie der Tierärztlichen Hochschule Hannover, Bünteweg 17, 30559 Hannover, Germany

KEITH F. WALKER

Cooperative Research Centre for Freshwater Ecology, Department of Environmental Biology, University of Adelaide, South Australia 5005, Australia

NORBERT WALZ

Institut für Gewässerökologie und Binnenfischerei, Müggelseedamm 310, 12561 Berlin, Germany

G. THOMAS WATTERS

Ohio Biological Survey and Aquatic Ecology Laboratory, Ohio State University, 1315 Kinnear Road, Columbus, Ohio 43212, USA

TORBJÖRN WESTERMARK

Royal Institute of Technology, Department of Chemistry, SE-10044 Stockholm, Sweden

MARK R. YOUNG

Culterty field Station, University of Aberdeen, Newburgh, Ellon, Aberdeenshire AB41 6AA, Scotland

ELKE ZAHNER-MEIKE

Blumenstraße 29, 85586 Poing, Germany

VALERY ZIUGANOV

Institute of Developmental Biology of Russian Academy of Sciences, 117808, Vavilov st. 26, Moscow, Russia

Part I
Systematics and Distribution

1 Characterization of the Unionoida (= Naiads)

GERHARD BAUER

1.1 Introduction

This chapter will be a short introduction for a non-specialist to the order Unionoida (= naiads). These large mussels live exclusively in freshwaters all over the earth (Chaps. 2, 10, 15, this Vol.). A problem for the people dealing with this group is its systematic classification (Chaps. 3, 4, 14, this Vol.). This problem is due to the low number of adult characters which may be used for systematics (Chaps. 4, 14, 15, this Vol.). Though there are some 1000 species, naiads have radiated very little in basic structure. Furthermore, a high degree of plasticity (Chaps. 5, 7, this Vol.) renders characters like shell morphology or size unreliable for classification. New biochemical methods (e.g. immunoelectrophoretic or DNA sequence analyses) may be a way out of this dilemma, and most contributions in this book (e.g. Chaps. 5, 13, 15, this Vol.) therefore follow the classification proposed in Chapter 14 (this Vol.). The order Unionoida comprises two superfamilies: Etherioidea (with the families Mycetopodidae and Iridinidae) and Unionoidea (with the families Hyriidae, Unionidae and Margaritiferidae). Very little is known about the Etherioidea, which are confined to the tropics of America and Africa (Chap. 6, this Vol.). However, the data so far suggest that their life history is quite similar to that of the Unionoidea.

1.2 Phylogeny

Naiads evolved from an as yet unidentified marine group by at least the Triassic (Chap. 15, this Vol.). Due to different types of larvae (Chaps. 6, 14, this Vol.) there have been speculations that the group is polyphyletic (Chap. 15, this Vol.). However, a parasitic mode of larval development common to all naiads and new biochemical data (Chap. 14, this Vol.) argues for a monophyletic origin, but an early divergence between Unionoidea and Etherioidea.

1.3 The Adult Stage

Naiads are characterized by a semi-infaunal mode of life. They dig with their foot into the substrate such that they are usually buried to the posterior edge of the shell.

The soft part groundplan is very uniform. The foot, a tongue-like organ in the middle of the anterior part of the animal, is used for digging. On each side of the foot there are two large and morphologically complex gills which have some important functions besides oxygen uptake. All naiads are filter feeders. Cilia on the gills and on the mantle create a water current and transport mucus to the mouth in front of the

Ecological Studies, Vol. 145
G. Bauer and K. Wächtler (eds.) Ecology and Evolution of the Freshwater Mussels Unionoida

animal. The mucus, together with small particles trapped from the water, is eaten. Furthermore, the gills are used as marsupia for the developing eggs and larvae. The different gill and marsupial morphologies are treated in Chapters 2, 3, 5, 14, (this Vol.). Two symmetrical mantle flaps enclose the whole body. They secrete the two-piece shell and at the posterior end form an inhalent and exhalent aperture for the water current, but never a syphon as in many marine taxa.

Mussels have lost their head and therefore most of the sense organs have withdrawn completely from the anterior end. The margin of the mantle, particularly at the posterior end, is now the site of contact with the environment and carries light-sensitive spots and tactile organs. Within the animal, paired osphradia near the attachment of the gills perceive chemical stimuli.

Naiad shells consist of two principal aragonitic layers: outer prismatic and inner nacreous. On the outside, the shells are covered by periostracum, a protein-like substance preventing dissolution of the shell (Chaps. 5, 21, this Vol.). Due to cessation of growth during winter, in temperate climates the periostracum usually shows concentric winter lines (annuli), which may be used to estimate the age of the animal (Chaps. 2, 5, 7, 13, 18, 21, this Vol.).

The earliest part of the shell is the beak or umbo. The hinge, that portion of one shell that is connected to the other, may show a variety of teeth which interlock the shells when closed by the two adductor muscles (Chaps. 3, 8, 15, this Vol.).

1.4 Larvae and Larval Development

There are a variety of sexual and brooding strategies, as described in Chapters 2, 5, 8, 13 (this Vol.). Apart from very few exceptions (Chap. 6, this Vol.) naiad larvae have to pass through a parasitic phase on a fish host to metamorphose into the young mussel. The larvae are produced in large numbers. They are small (0.05–0.4 mm), short-lived and not able to swim actively (Chaps. 6, 11, this Vol.). There are three different types of larvae: the glochidium among the Unionoidea (Hyriidae, Unionidae, Margaritiferidae), the lasidium among the Mycetopodidae and the haustorium among the Iridinidae. The tremendous differences in the morphology of these types and in their development are treated extensively in Chapter 6 (this Vol.).

1.5 Summary

The large freshwater mussels Unionoida (= naiads) evolved by at least the Triassic. Presumably, they are a monophyletic group. The adults are semi-burrowed filter feeders whereas the larvae have to contact a fish host on which they pass through a parasitic phase.

2 Freshwater Mussels (Hyriidae) of Australasia

Keith F. Walker[1], Maria Byrne[2], Christopher W. Hickey[3], and David S. Roper[4]

2.1 Introduction

In Australia, New Guinea, Solomon Islands and New Zealand, freshwater mussels are represented by the families Hyriidae and Unionidae. The Hyriidae includes 27 species in 8 genera, in the subfamilies Cucumerunioninae, Hyridellinae, Lortiellinae and Velesunioninae (a fifth subfamily, the Hyriinae, is endemic to South America). Two other species (one genus) from New Guinea are placed tentatively in the Unionidae.

This perspective is based on taxonomic revisions made four decades ago (e.g. McMichael and Hiscock 1958; Parodiz and Bonetto 1963). It is uncertain, because taxa below family level are defined by morphological features that vary in response to local environmental conditions. Without a more robust framework, intriguing problems in ecology, evolutionary biology and biogeography have languished. For example, the disjunct regional distributions of *Cucumerunio*, *Hyridella*, *Westralunio* and other hyriid genera present puzzles for biogeographers (Walker 1981a).

With the advent of new molecular techniques in phylogeny (e.g. Lydeard et al. 1996), these problems are reopened for study. Work is not far advanced, and in some respects a review is premature. Nevertheless, the present framework is a valid hypothesis, within the constraints of morphological taxonomy, and even if changes are indicated, the older literature will remain as a rich source of questions. In this spirit, we offer an outline of present knowledge of the Australasian fauna.

2.2 Phylogeny and Biogeography

The order Unionoida (Unionacea) nominally includes two superfamilies, the Unionoidea and Etherioidea, distinguished by larval forms (Parodiz and Bonetto 1963; Haas 1969; Heard and Guckert 1970; Davis and Fuller 1981; Boss 1982). The Unionoidea, with *glochidia* larvae, includes the Hyriidae (Australasia, South America), Margaritiferidae (Eurasia, North America) and Unionidae (Africa, Eurasia, India, North America). The Etherioidea (Muteloidea), with *lasidia* larvae, includes the Etheriidae (= Mycetopodidae fide Heard and Vail 1976) (Africa, South

[1] Cooperative Research Centre for Freshwater Ecology, Department of Environmental Biology, Adelaide University, South Australia 5005, Australia
[2] Department of Anatomy and Histology, University of Sydney, New South Wales 2006, Australia
[3] National Institute of Water and Atmospheric Research, P.O. Box 11-115, Hamilton, New Zealand
[4] Mighty River Power Limited, P.O. Box 445, Hamilton, New Zealand

Ecological Studies, Vol. 145
G. Bauer and K. Wächtler (eds.) Ecology and Evolution of the Freshwater Mussels Unionoida

America) and Iridinidae (Mutelidae) (Africa). The perceived groups rest upon Parodiz and Bonetto's (1963: p. 185) argument that "the two different types of larvae ... cannot be considered to be derived one from the other or from any hypothetical direct ancestry".

Recent studies challenge this view. Preliminary mitochondrial gene trees suggest that the Unionoida is monophyletic (Hoeh et al. 1998; Graf and Ó Foighil 2000), and that one of the larval types is indeed ancestral. The Hyriidae shares derived characters with the Etherioidea, namely a marsupium formed by the inner demibranchs, anterior attachment of the inner demibranchs near (or in contact with) the labial palps and fusion of the mantle between the incurrent and excurrent apertures (Ortmann 1912, 1921; McMichael and Hiscock 1958). Thus, the hyriids may share a more recent ancestor with the Etherioidea than with any of the Unionoidea (Graf 2000).

This hypothesis is partly corroborated by historical biogeography. The Unionoida date from at least the Triassic (Haas 1969), and their fossil history includes iridinids (Morris and Williamson 1988) and hyriids (Henderson 1935; Parodiz and Bonetto 1963) from the Mesozoic of North America. The hyriid fossils were validated by the presence of "radial" beak sculpture, a feature of modern South American Hyriinae that is less developed or absent in Australasian species (Parodiz and Bonetto 1963). This feature, however, is not clearly distinguished from the beak sculpture of distantly related forms (e.g. the Indian *Parreysia*), and may prove to be unreliable. For the present, the fossil evidence is pivotal: the modern Etherioidea and Hyriidae suggest a Gondwanan origin, but the fossils indicate an earlier, Pangaean origin.

Ideas of continental drift were not widely accepted, of course, at the time of McMichael and Hiscock (1958) or Parodiz and Bonetto (1963). McMichael and Hiscock postulated – from scant evidence – that an ancestral form invaded Australia from SE Asia, but soon afterward McMichael and Iredale (1959) acknowledged that dispersal via an Antarctic land bridge was feasible.

Preliminary 12S mitochondrial data suggest that the Velesunioninae, and perhaps all hyriids, are monophyletic (K.J. Roe, University of Alabama, pers. comm.). According to McMichael and Hiscock, members of the Velesunioninae are primitive, the Lortiellinae and Hyridellinae are derived from the Velesunioninae and the Cucumerunioninae is derived from the Hyridellinae. This speculation rests upon a single synapomorphy (a perforate diaphragm dividing the infra- and suprabranchial chambers). The fossil evidence includes a probable velesunionid, *Prohyria eyrensis*, from the Australian Triassic, and more recent fossil species of *Alathyria*, *Hyridella* and *Velesunio* (McMichael 1957; McMichael and Hiscock 1958). Incidentally, Ludbrook (1961) described Triassic "unionid" fossils from Australia, but without regard for contemporary taxonomy (other, more recent, literature in archaeology and geology mistakenly refers to the presence of "unionids" in Australia).

2.3 Taxonomy

Smith (1881) ventured the first taxonomic synthesis of Australasian species, although the fauna was regarded as part of the Unionidae until after 1900. Simpson

(1914) assigned the Australian species to *Diplodon*, in a new subfamily, the Hyriinae, and referred all but one species to the subgenus *Hyridella*. Ortmann (1909, 1912, 1920) transferred the Hyriinae to the Mutelidae, with *Diplodon* species from South America. Cotton and Gabriel (1932) elevated *Hyridella* to generic status, displacing the name *Diplodon* in Australia. Iredale (1934, 1943) discarded the Hyriinae, defined the family Propehyridellidae and admitted new genera and species.

This conveys a false impression of orderly progress; in fact, the literature is replete with nomenclatural errors and descriptions and type material generally are inadequate. The confusion is illustrated by Hiscock (1950, 1951a,b), who referred to *Velesunio ambiguus* as *Hyridella australis* (cf. also Ortmann 1912, 1920) and did not distinguish that species from *Alathyria jacksoni.*

Dell's (1953) revision of New Zealand species was the first to apply a more rigorous approach, and was followed by McMichael and Hiscock's (1958) revision of the entire regional fauna. Supplementary data were provided by McMichael (1956, 1957, 1958, 1967), McMichael and Iredale (1959), Hiscock (1960) and, more recently, Smith (1992, 1998) and Walker (1998). McMichael and Hiscock (1958) rationalized earlier diagnoses of genera and species (for example, the morphologically variable *V. ambiguus* was formed from 25 synonyms). They regarded the dominant Australasian mussels as Mutelidae, later corrected to "Hyriidae" by Parodiz and Bonetto (1963). Thus, Unionoidea are distinguished by glochidia larvae, and the glochidia of most Hyriidae (those that are parasitic) typically have S-shaped, pointed teeth (Fig. 2.1).

Two questions arising from the 1958 review are unresolved. First, the status of two New Guinea species thought to be unionids (*Haasodonta fannyae, H. vanheurni*) remains unclear. McMichael and Hiscock had access only to shells and their assignment of *Haasodonta* to the Unionidae (Rectidentinae) was provisional, pending anatomical and larval data that are still wanting. Second, their inclusion among the Hyriidae of the Lortiellinae (*Lortiella froggatti, L. rugata*), from NW Australia, was tentative. The doubts are undermined by later, informal observations of internal anatomy, suggesting an affinity with the Velesunioninae (Hiscock in McMichael 1967), and by the apparent similarity of the glochidia to those of Hyridellinae (Bonetto et al. 1987; Fig. 2.1).

McMichael and Hiscock's review remains the regional standard, but diagnoses of species, genera and even subfamilies are equivocal given the traditional emphasis on shell characters and the morphological variability of most taxa. Gross internal anatomy is scarcely differentiated below family level, and information about glochidia was (and still is) too fragmentary to evaluate their taxonomic potential. McMichael and Hiscock (1958, p. 384) stated that "the species concept adopted here is one of widespread, variable species, with some genetically divergent populations; these, however, are not recognized nomenclaturally unless they occur over a considerable geographic area and occupy a range more or less distinct from that of the parent population". Two unfortunate consequences were that the geographic ranges of taxa became *de facto* criteria for identification, and the review did not venture a key to species (but see McMichael 1967; Walker 1999). Until new criteria and diagnoses emerge, questions of hyriid phylogeny and biogeography must be framed cautiously.

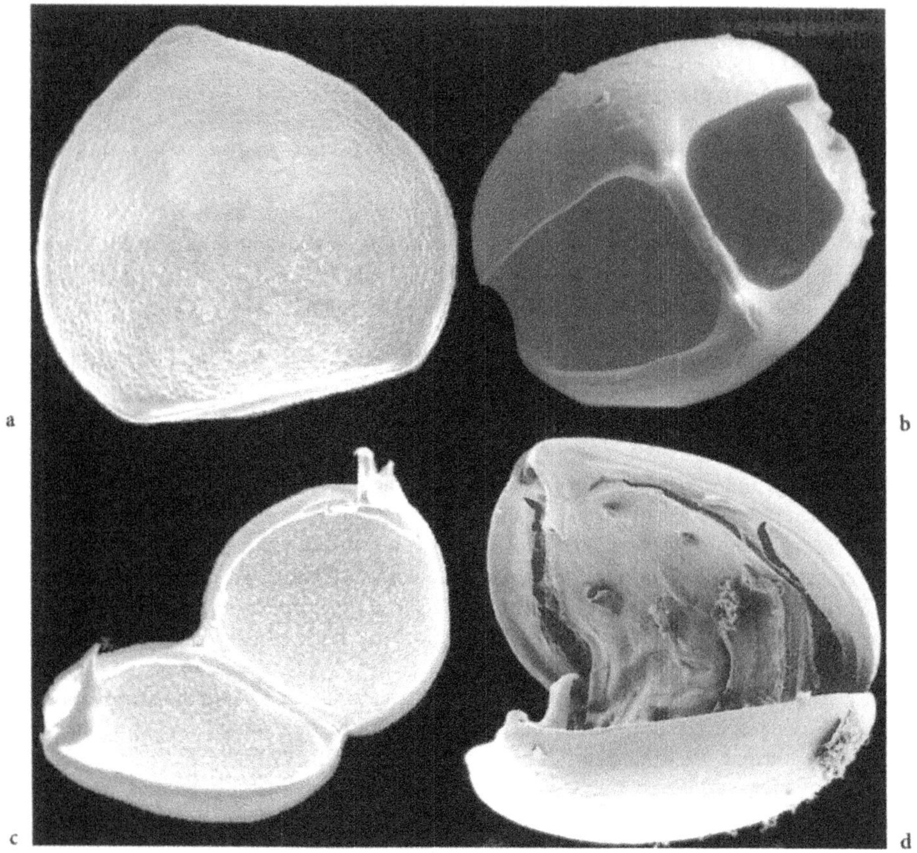

Fig. 2.1a–d. Glochidial shells of Hyriidae. **a** *Velesunio ambiguus* (length 250 μm). **b** *Lortiella froggatti* (190 μm). **c** *Alathyria jacksoni* (270 μm). **d** *Hyridella depressa* (215 μm). (Walker 1981b; Jupiter and Byrne 1997; H.A. Jones and M. Byrne unpubl.)

2.4 Regional Distribution

Table 2.1 lists species and indicates their distributions. Australia has 17 species in 6 genera. Among the genera only *Lortiella* is endemic, but 13 of the 15 remaining species are endemic. New Guinea has nine hyriid species in six genera, including four genera and two species shared with Australia, one genus shared with Solomon Islands and New Zealand, and one species shared with the Solomons. New Guinea also has two endemic species of the presumed unionid, *Haasodonta*. The Solomons fauna consists of *Hyridella guppyi*, shared with New Guinea. New Zealand has three endemic species in two genera shared with Australia, and one genus shared with New Guinea and the Solomons.

The subfamily Velesunioninae is widespread in Australia and New Guinea, but absent from New Zealand. The Lortiellinae is confined to NW Australia. The Hyridellinae and Cucumerunioninae occur in coastal SE Australia, with

Table 2.1. Freshwater mussels of Australasia and their broad distributions. Includes range extensions for *H. narracanensis*, *V. angasi* and *V. wilsonii* (K. F. Walker unpubl.) (NSW New South Wales, NT Northern Territory, Q Queensland, SA South Australia, T Tasmania, V Victoria, WA Western Australia)

Family	Subfamily	Species	Distribution	Notes
Hyriidae				
	Cucumerunioninae			
		Cucumerunio novaehollandiae (Gray)	Australia	E coast
		Cucumerunio websteri (Simpson)	New Zealand	N Island
		Virgus beccarianus (Tapparone Canefri)	New Guinea	Fly, Kikori, Merauke rivers
	Hyridellinae			
		Hyridella aucklandica (Gray)	New Zealand	N, S Islands
		Hyridella australis (Lamarck)	Australia	SE coast
		Hyridella depressa (Lamarck)	Australia	SE coast
		Hyridella drapeta (Iredale)	Australia	SE coast
		Hyridella glenelgensis (Dennant)	Australia	SW Vic (status uncertain)
		Hyridella guppyi (Smith)	New Guinea, Solomons	
		Hyridella menziesi (Gray)	New Zealand	N, S Islands
		Hyridella misoolensis (Schepman)	New Guinea	Mainland; island of Misool
		Hyridella narracanensis (Cotton & Gabriel)	Australia	SE SA, E Vic, N Tas
	Lortiellinae			
		Lortiella froggatti Iredale	Australia	NW WA, NE NT
		Lortiella rugata (Sowerby)	Australia	NW WA, NE NT
	Velesunioninae			
		Alathyria condola Iredale	Australia	Murray-Darling Basin (part); inland
		Alathyria jacksoni Iredale	Australia	Murray-Darling Basin; inland
		Alathyria pertexta Iredale	Australia	E coast; Lake Eyre Basin (part)
		Alathyria profuga (Gould)	Australia	E NSW (part)
		Microdontia anodontaeformis Tapparone Canefri	New Guinea	S coast
		Velesunio ambiguus (Philippi)	Australia	E coast and inland
		Velesunio angasi (Sowerby)	Australia	N Australia, W WA; coastal
		Velesunio moretonicus (Reeve)	Australia	N Tas (syn. *V. legrandi* (Petterd))
		Velesunio sentaniensis (Haas)	New Guinea	N coast
		Velesunio wilsonii (Lea)	Australia, New Guinea	E to W coast, mainly inland
		Westralunio albertisi Clench	New Guinea	Restricted; ecotype of *W. flyensis*?
		Westralunio carteri Iredale	Australia	SW WA
		Westralunio flyensis (Tapparone Canefri)	New Guinea	S coast
Unionidae				
	Rectidentinae			
		Haasodonta fannyae (Johnson)	New Guinea	Restricted: Bian, Merauke rivers
		Haasodonta vanheurni (McMichael & Hiscock)	New Guinea	Restricted: Bian, Merauke rivers

outliers in New Guinea and New Zealand, but are absent from inland Australia (Walker 1981a).

Although the global distribution of hyriids (and other unionoids) clearly accords with ideas of plate tectonics, their regional distribution invites speculations in vicariance and dispersal biogeography. Thus, the distributions of some genera and species might be interpreted as remnants of once-extensive ranges fragmented by sea-level changes, modified perhaps by post-Pleistocene aridity. This is a credible explanation for the presence of *Alathyria* and *Velesunio* species in continental Australia, Tasmania and New Guinea, and *Hyridella narracanensis* in Tasmania, Victoria and South Australia (Walker 1981a, unpubl.). The origins of the New Zealand and Solomons species, however, are more difficult to explain in terms of land connections.

Dispersal of adult mussels via transoceanic rafting (cf. Walker 1969), passage on migratory birds (Cotton 1934; McMichael 1958) and even transport by humans is feasible, but is rarely confirmed. Glochidia are best equipped for adventitious dispersal, by virtue of their parasitic association with fish. Metamorphosis allows significant time for dispersal as it occurs over days or weeks, depending on temperature and other factors (see Sect. 2.6). Anadromous fish in particular could be implicated in dispersal of glochidia across marine barriers. For example, the glochidia of *Hyridella menziesi* infect eels (Hine 1978), although metamorphosis has not been confirmed. The telling question may be whether encysted glochidia are able to survive transit through seawater.

H. menziesi is widespread in New Zealand. It is common in both the North and South Islands, in habitats ranging from small, fast-flowing streams to lakes. In contrast, *H. aucklandica* and *Cucumerunio websteri* are restricted to the northern North Island, near Auckland, and are little known (e.g. Powell 1979).

In Australia, the distributions of lotic animals have been related to faunal assemblages or *fluvifaunulae* associated with major drainage basins (Iredale and Whitley 1938; McMichael and Hiscock 1958). Although there is some correspondence with the putative fluvifaunulae, most of the major basins have lacked sufficient integrity to maintain distinctive assemblages (Walker 1981a). Thus, some species extend over several basins (e.g. *A. pertexta*, *V. ambiguus*, *V. angasi*, *V. wilsonii*).

Dispersal and vicariance mechanisms are not exclusive, and both may have contributed to the regional, if not global, distribution of Hyriidae. These ideas are problematic for biogeographers, however, because they are not falsifiable. Further, the perceived distributions of species need to be confirmed as real, and not artefacts of taxonomy. Again, this highlights the need to secure the taxonomic and phylogenetic basis for research in biogeography.

2.5 Phenotypic Variation

Most Australasian Unionoidea show strong phenotypic variation, particularly *Hyridella menziesi* (Dell 1953; Roper and Hickey 1994), *Velesunio ambiguus* and *V. angasi* (McMichael and Hiscock 1958; Walker 1981b; Humphrey 1984; Wirtado 1994). Variation is most apparent in features of the shell, but may also extend to differences in physiology and behaviour. Thus, *H. menziesi* and *V. ambiguus* vary in glycogen levels, oxygen consumption, heart rate and patterns of valve movements (Hiscock 1950; Walker 1981b; Hickey et al. 1995; C.W. Hickey, unpubl.).

Shells vary in size and shape, beak and surface sculpture, hinge dentition, growth lines, erosion of the periostracum and the colour and thickness of the nacre. The differences between taxa may be subtle. For example, *H. australis* is ovoid (height to length ratio 60%), with a prominent posterior ridge and glossy black periostracum. *H. depressa* is elongate (ratio 50%), with a flattened posterior ridge and dull periostracum. *H. drapeta* has more delicate beak sculpture than either of its congeners. All three occur in coastal SE Australia.

In some variable species, notably *H. menziesi*, interactions of physical, chemical and biological factors are implicated (Roper and Hickey 1994). In others, the causal factors are less complex. Thus, distinctive river and lake forms occur in some species (e.g. Dell 1953; McMichael and Hiscock 1958). Lotic populations of *V. ambiguus* have heavy, elongate shells (Fig. 2.2), and some lentic populations have rounded or swollen shells (Walker 1981b). The "creek" form of the tropical *V. angasi* is more swollen then the "billabong" form, perhaps related to its capacity for aestivation and drought tolerance (Humphrey 1984). Other variants are associated with water velocity. For example, *Alathyria jacksoni*, an obligate river-dweller in the Murray-Darling Basin, tends toward one of two growth forms (Fig. 2.2): in moderate currents the dorsal margin of the shell is produced as a blade or "wing", whereas in strong currents the dorsum is arched and there is a corresponding ventral inflection. Electrophoretic data suggest that the two forms are one species (Balla and Walker 1991). Internally, the arched shells have comparatively large adductor muscles and more elongate posterior adductors, and the anterior and posterior adductors are displaced ventrally. It appears that the arch confers a wider shell-

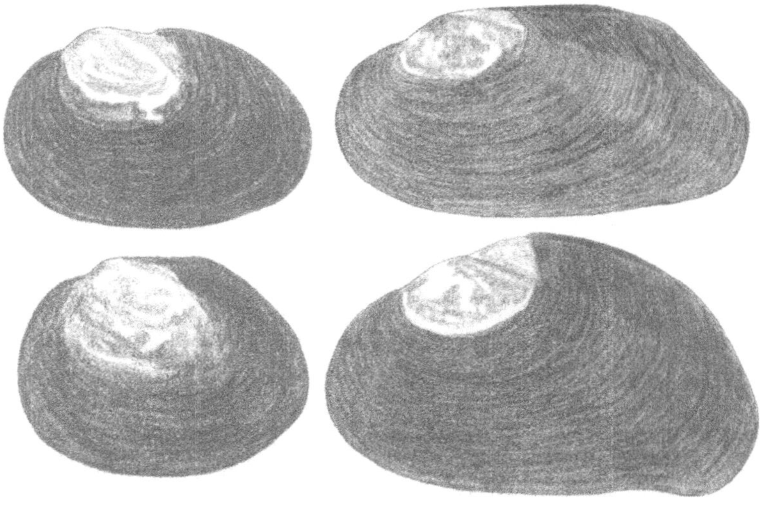

Fig. 2.2. Hyriidae from the Murray-Darling Basin. Specimens of *Velesunio ambiguus* at *left* are from river (*upper*) and lake populations (*lower*), and those of *Alathyria jacksoni* at *right* are the winged (upper) and arched forms (lower) typical of moderate and strong currents, respectively. Scales are similar; the shell at *lower right* is 140 mm long

gape, more foot extension and stronger anchorage. Similar variation occurs in *V. angasi* (Humphrey 1984).

2.6 Reproduction

There are no external features to distinguish male and female hyriids, but gill morphology is diagnostic. In mature females the interlamellar septa of the inner demibranchs are thickened as marsupia, more prominent than the septa of males or young females (McMichael and Hiscock 1958; Humphrey 1984; Jones et al. 1986; Jupiter and Byrne 1997). This allows for rapid identification of mature females, including non-brooding individuals. The reproductive status of individuals may be obscured, however, by hermaphroditism and parasitic infestation.

A balanced sex ratio is typical of populations of *Cucumerunio novaehollandiae*, *Hyridella australis*, *H. depressa* (Jones et al. 1986) and, allowing for some local variations, *Alathyria jacksoni* and *Velesunio ambiguus* (Walker 1981b). There is an imbalance in favour of males in *V. angasi*, perhaps due to seasonal mortality of females in hypoxic environments (Humphrey 1984). A female imbalance occurs in low-density populations of *H. depressa*, although some "females" are cryptic microhermaphrodites with gonadal acini that are a mosaic of oogenic and spermatogenic tissues (Byrne 1998). Some of these "females" appear to be facultative hermaphrodites capable of self-fertilization (cf. Bauer 1987), and so may boost recruitment when populations are too sparse to be maintained by outcrossing. Hermaphrodites are reported also in *V. ambiguus* and *V. angasi* (Humphrey 1984; Wirtado 1994).

Hyriids often harbour parasitic trematodes (e.g. Angel 1961), sometimes in densities enough to castrate the host. Wirtado (1994) recorded more than 80% infection in tropical populations of *V. ambiguus*, and comparable levels occur in *V. ambiguus* and *A. jacksoni* in the River Murray, South Australia (K.F. Walker, unpubl.). Indeed, hermaphroditism may be an adaptive response to the effects of parasites (cf. Kat 1983).

Size at sexual maturity varies among species. In *V. angasi*, males attain maturity (30 mm length) in 1 year and females (40 mm) in 1.5 years, suggesting protandry. *Hyridella* species are mature in 2–3 years (40 mm) and *C. novaehollandiae* matures at 3–4 years (60 mm) (Jones et al. 1986; Byrne 1998). *A. jacksoni* and *V. ambiguus* mature at 3–4 years in the River Murray (Walker 1981b), and the latter species probably matures earlier in the tropics (cf. Wirtado 1994).

Glochidial release may be seasonal and opportunistic. In temperate Australia, *A. profuga* and *Hyridella* species brood through the warmer months (Fig. 2.3), whereas *C. novaehollandiae* broods in autumn and winter (Atkins 1979; Jones et al. 1986; Byrne 1998). In the Murray-Darling species *A. jacksoni*, spawning is seasonal (the females brood in spring and spawn in summer) but may be interrupted by flood or drought (Walker 1981b, unpubl.). The Murray-Darling wetland species *V. ambiguus* also is seasonal (embryos are incubated over winter and glochidia are spawned in spring-summer), but it breeds over a longer period and is thereby more adaptable to changeable conditions. In tropical Australia, *V. ambiguus* and *V. angasi* brood year-round (Humphrey 1984; Wirtado 1994).

Gametogenesis appears to be continuous, or potentially so, in all species (Humphrey 1984; Jones et al. 1986; Wirtado 1994; Byrne 1998). In seasonal spawners like *H. depressa* a peak occurs just before spawning, when the gonads are filled with

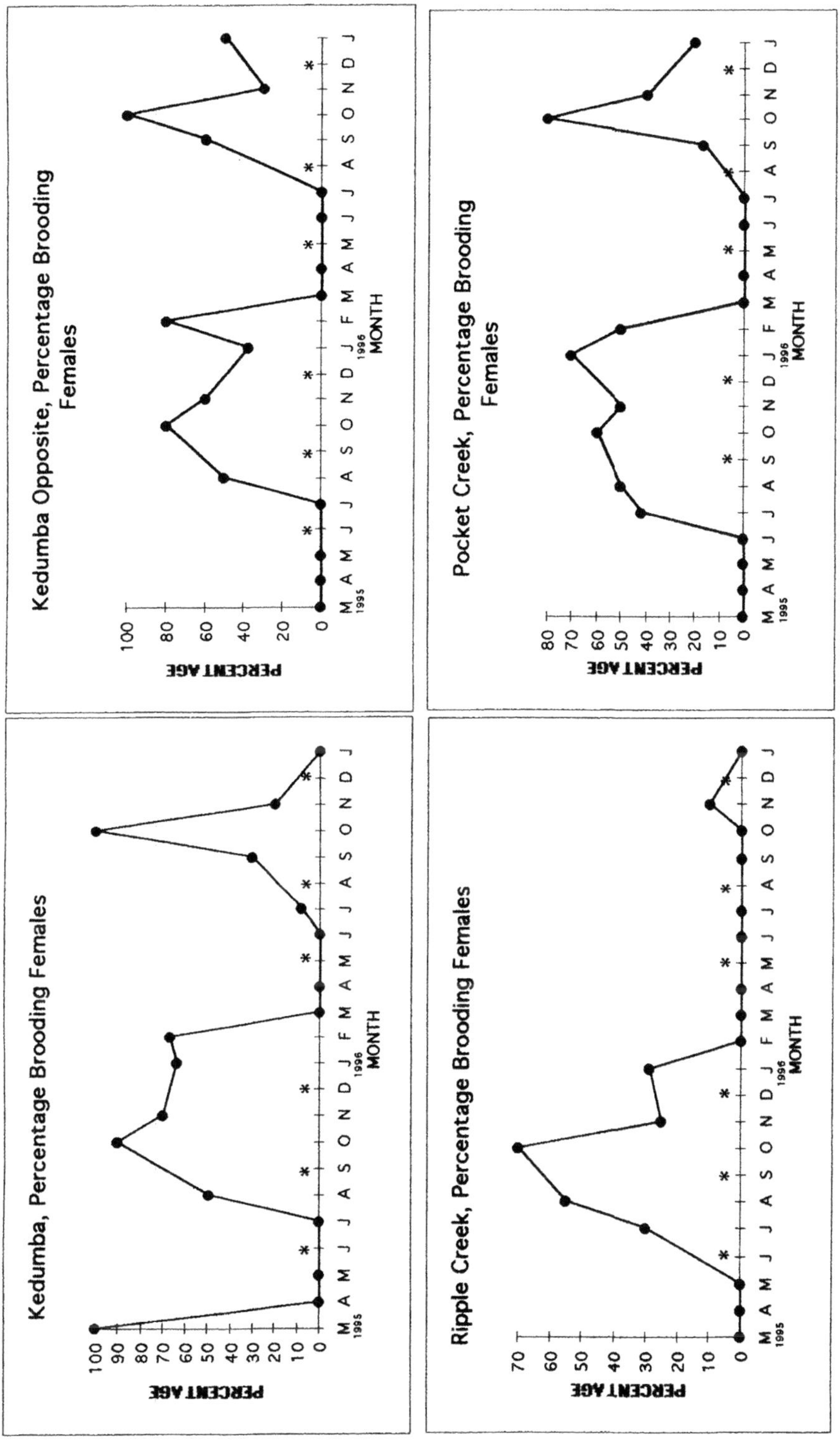

Fig. 2.3. Proportions of brooding females in samples ($n = 7$–10) of *Hyridella depressa* at sites in Lake Burragorang, New South Wales. *Asterisks* denote months not sampled. (From Byrne 1998)

advanced gametes and interacinal spaces are minimal; afterward, the gametes are dispersed and the acini are less crowded (Byrne 1998). Enhanced food increases fecundity and extends gametogenesis, gamete storage and spawning, so that mussels at eutrophic sites maintain advanced gametes throughout the year. At oligotrophic sites gametogenesis is more seasonal and may fail in some years. The value of maintaining advanced gametes outside the spawning season is unclear, but it could provide for opportunistic breeding in unstable environments.

The marsupium in female hyriids is confined to the middle third of the innermost demibranchs (Fig. 2.4), except in *C. novaehollandiae*, where it occupies most of the inner demibranchs (H. A. Jones, NSW Department of Land and Water Conservation, pers. comm.). Developmental stages are synchronous within females, but different females in a population may contain embryos at different stages. Marsupia laden with embryos are cream, and progressively assume the pale brown colour of mature glochidia.

In tropical Australia, embryogenesis in *V. angasi* takes 12 days in summer and up to 40 days in winter (Humphrey 1984). In temperate areas, *H. australis* requires about 8 weeks in summer (Jones et al. 1986). Depending on female condition, glochidial release is promptly followed by the release of new eggs, ensuring serial clutches through the breeding season. The spawning interval is influenced by floods and changes in water temperature (Jones et al. 1986; Byrne 1998), and access to host fish also may be limiting. Spring and summer incubation effectively shortens the embryonic period. The viability and rates of metamorphosis

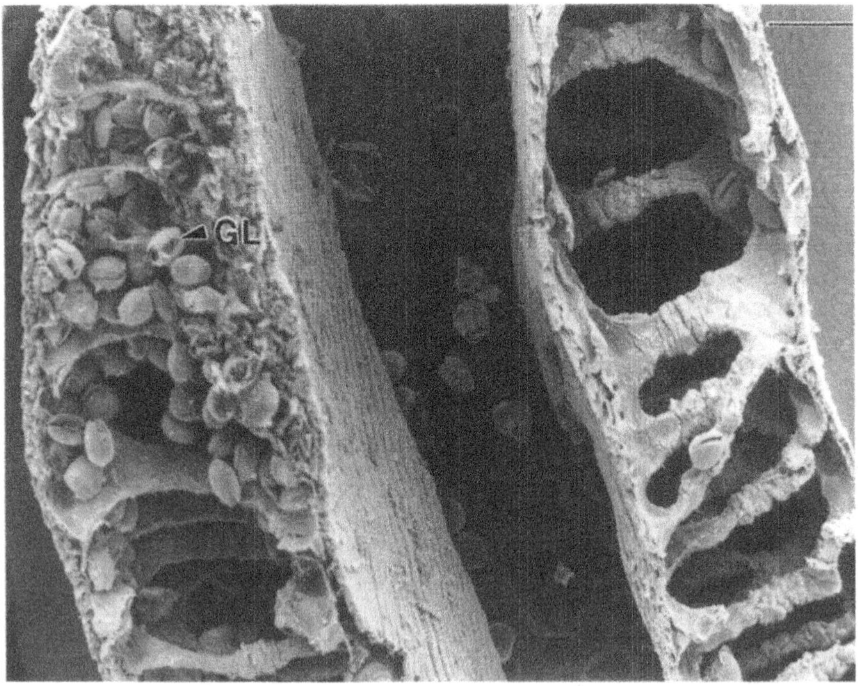

Fig. 2.4. *Hyridella depressa*: inner demibranch containing glochidia (*GL*) and outer demibranch (*right*). *Bar* 500 μm (From Byrne 1998)

of the glochidia of *A. jacksoni* and *V. ambiguus* are temperature-dependent (Walker 1981b), and temperature also appears to cue glochidial release in *H. drapeta* (Atkins 1979).

Clutch size generally increases with body size and is strongly influenced by habitat trophic status (Fig. 2.5). Female *H. depressa* may carry 10000–100000 embryos, and the annual output by large females is conservatively estimated as 400 000 glochidia (Byrne 1998). This is considerably less than in the females of Northern Hemisphere species, which fully utilize their demibranchs for brooding (Kat 1984; Bauer 1992). Although data are few, the most fecund hyriid may be the tropical *V. angasi*, in which gametogenesis is rapid, sexual maturity is attained quickly and breeding continues year-round (Humphrey 1984). A challenger may be *C. novaehollandiae*, from temperate eastern Australia, which has large marsupial demibranchs (see above) and produces two broods in some years (Jones et al. 1986).

Hyriid glochidia typically have a straight hinge and subtriangular valves, each with one or two recurved teeth (Fig. 2.1). There is sufficient diversity in the size and shape of the shell and teeth to suggest that they may have taxonomic value. The glochidia generally are released in straw-coloured, worm-like conglutinates that must

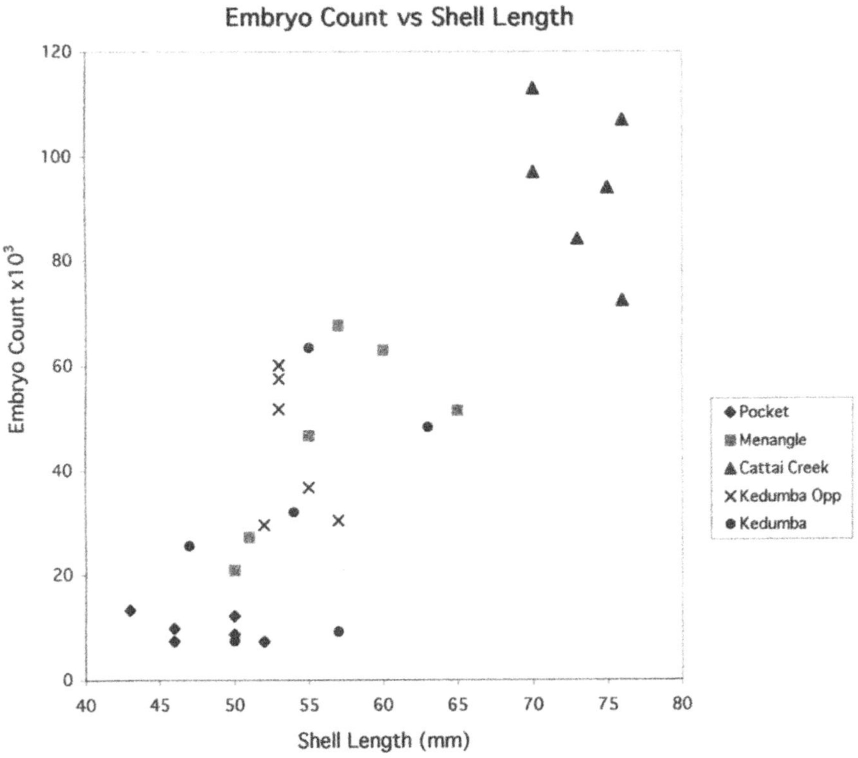

Fig. 2.5. Relation of clutch size to shell length for six female *Hyridella depressa* from each of three sites in Lake Burragorang (Kedumba, Kedumba Opposite, Pocket Creek) and two sites in the Hawkesbury-Nepean river system (Menangle, Cattai Creek), New South Wales. (See further Byrne 1998)

attract fish, and they attach to the fins, body surface and sometimes the gills of their host (cf. Atkins 1979). Hyriids appear to have few, if any, obligate hosts. In Australia, numerous native and introduced fish are hosts for *A. jacksoni*, *V. ambiguus*, *V. angasi* and species of *Hyridella* (e.g. Atkins 1979; Walker 1981b; Humphrey 1984). In New Zealand, *H. menziesi* glochidia are reported from eels (*Anguilla* spp.) (Hine 1978), koaro (*Galaxias brevipinnis*), giant bully (*Gobiomorphus gobioides*) (Percival 1931) and common bullies (*G. cotidianus*) (D. S. Roper and C. W. Hickey, unpubl.). There is inconclusive evidence that cyprinid fish, including common carp (*Cyprinus carpio*) and goldfish (*Carassius auratus*), are unsuitable hosts for some species (M. Byrne and K.F. Walker, unpubl.). Finally, the glochidia of *V. ambiguus* (and probably other species) can be experimentally induced to metamorphose on tadpoles (Walker 1981b), and this may enhance recruitment in some circumstances.

Recruitment in hyriid populations is often patchy in time and space, indicated by unimodal age profiles, a predominance of large individuals and few or no juveniles (Grimmond 1968; Walker 1981b; James 1985; Roper and Hickey 1994). For example, populations of *V. ambiguus* in Lake Alexandrina, near the mouth of the River Murray, fail to spawn in seasons when strong winds and waves strand many shells, temporarily, along the shore (Walker 1981b, unpubl.). In other cases there may be a wide range of shell sizes and the substratum often contains juveniles (M. Byrne, unpubl.). Observations of juvenile *V. ambiguus* in situ are recorded by Atkins (1980).

2.7 Abundance, Growth and Longevity

Population densities of *Hyridella menziesi* in New Zealand lakes average 6 m^{-2}, but may attain 814 m^{-2} (Table 2.2). Densities of 0.0–18.2 m^{-2} are reported for

Table 2.2. Population density and biomass data for *Hyridella menziesi*

Lake	North/South Island	Mean density (m^{-2})	Mean DW (g)	Reference
Taupo	N	5.6	2.8	James (1985)
Rotokawau	N	160[a] (max 814)	0.9	James (1987)
Rotokakahi	N	13	–	Forsyth (1978)
Ngapouri	N	5	–	Forsyth (1978)
Karapiro	N	<1	–	Roper and Hickey (1994)
Tuakitoto	S	5.5	2.5[b]	Ogilvie and Mitchell (1995)
Rotoroa	S	9.7	1	Timms (1980)
Rotoiti	S	4.3	1.4	Timms (1980)
Grassmere	S	7	1.4	Timms (1983)
Pearson	S	5	0.4	Timms (1983)
Marymere	S	7	1.5	Timms (1983)
Sarah	S	2	2.5	Timms (1983)
Mean		*6*	*1.6*	

DW dry weight. AFDW ash-free dry weight.
[a] Data excluded from mean.
[b] AFDW to DW conversion factor 1.17.

Velesunio angasi in billabongs of Magela Creek, northern Australia (Humphrey 1984). In the River Murray mussels are patchily distributed; records include shoals of 2.7 m^{-2} *V. ambiguus* in Lake Alexandrina and 1.2 *A. jacksoni* m^{-2} and 3.7 *V. ambiguus* m^{-2} in the river near Overland Corner (Walker 1981b). There is, however, one record of 303 *A. jacksoni* m^{-2} in the Murray near Yarrawonga (Walker 1981b). Similar aggregations are known for *H. depressa* near Sydney (Byrne 1998).

Hyriids are relatively long-lived among bivalves. In New Zealand, *H. menziesi* attains over 100 mm length (Ogilvie 1993), and reported maximum ages are 13 years (61 mm) in Lake Taupo (James 1985) and 33 years (84 mm) in lakes on the Waikato River (Roper and Hickey 1994). Reported mean ages include 7 years (46 mm) in Lake Taupo (James 1985), 15 years in Lake Tuakitoto (Ogilvie 1993) and 20–25 years in Lake Waipori (Grimmond 1968).

The shells of *A. jacksoni* and *V. ambiguus* show apparently annual growth-interruption lines (Walker 1981b), although growth may be arrested by irregular disturbances like flood and drought. Use of these annuli to determine age profiles is questionable (e.g. Kesler and Downing 1997) and should be validated for each population. With this caveat, however, growth patterns in riverine populations of the two Murray species appear to be similar, but are markedly different from lacustrine populations of *V. ambiguus* (Walker 1981b). *A. jacksoni* attains a length of 170 mm (H.A. Jones, pers. comm.) and *V. ambiguus* attains 108 mm (Walker 1981b). The largest shells of *A. jacksoni* occur in coarse sandy sediments along the middle river, where a 140-mm specimen may have an age of nearly 30 years. The largest *V. ambiguus*, generally from riverine habitats, indicate an age of 11 years, but lacustrine specimens of similar age may be only 75 mm long. Estimates for *V. angasi* in northern Australia, based on growth lines that almost certainly were true annuli, indicate a 11–35-years life span (Humphrey 1984).

Growth and activity are influenced by temperature. In the tropics, *V. angasi* is active most of the year, but may aestivate in the dry season. In temperate regions, growth is slower and there is little or no activity in winter. When water temperatures fall to 10–12 °C, entire *H. depressa* populations burrow into the sediment and remain quiescent, and the same is true of *A. jacksoni* and *V. ambiguus* in the Murray-Darling Basin. Further, the sizes and growth rates of shells in the Murray appear to be temperature-dependent. If 12 °C is taken as a lower limit for activity and growth (Walker 1981b; Sheldon and Walker 1989), the effective season for growth along the Murray (2560 km) increases from 7 months near Albury, 350 km from the source, to 11 months near Murray Bridge, in the lower reaches.

Growth rates of *V. angasi* in Magela Creek billabongs are correlated with the availability of oxygen and food (see Sect. 2.9), both determined by monsoonal floods (Humphrey 1984). The size and growth rates of shells generally increase with depth, except at depths prone to seasonal anoxia. Production, excluding gametic and larval production, is reported as 0.39–1.75 $g\,m^{-2}$ $year^{-1}$ shell-free dry weight, but turnover ratios (production/biomass) may be only 10%.

Growth in *Hyridella* species also is sensitive to eutrophication (Roper and Hickey 1994; Byrne 1998). In an experiment in New South Wales (M. Julli and M. Byrne, unpubl.), *H. depressa* translocated from oligotrophic to eutrophic sites affected by

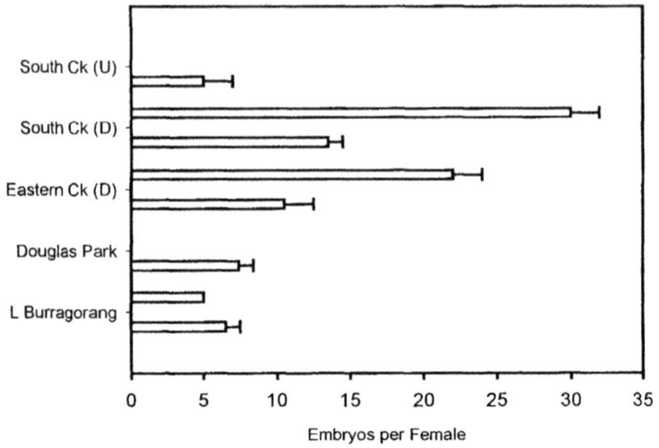

Fig. 2.6. Fecundity in samples of *Hyridella depressa* ($n = 30$–40) 6–9 months after translocation from an oligotrophic lake (Lake Burragorang, New South Wales) to an oligotrophic river (Douglas Park) and two creeks upstream (*U*) and downstream (*D*) of sewage plants. Two replicate samples are shown for each site except Douglas Park and South Creek (upstream). All mussels were from one population and had shell lengths of 40–55 mm before deployment. Data are means and standard deviations. (M. Julli and M. Byrne, unpubl.)

sewage effluent for 6–9 months increased in shell length, weight and fecundity. The clutch size in particular increased to about twice that of the parent population (Fig. 2.6).

2.8 Environmental Tolerances

Hyriids are widespread in Australian inland waters, except in ephemeral and saline waters. They are absent from streams in southern Tasmania, possibly for reasons related to water chemistry (cf. Hart and McKelvie 1986). Their broad geographic distributions (Table 2.1) hint at their ecological and physiological limitations. Dispersal is one constraint; others are imposed by the tolerances of adult and juvenile mussels, glochidia and their hosts.

No molluscs occur in Lake Tikitapu, New Zealand, where the calcium concentration is $0.7\,mg\,l^{-1}$, but nearby Lake Rotokakahi has $1.9\,mg\,l^{-1}$ calcium and supports *Hyridella menziesi* and other molluscs (Forsyth 1978; Timperley 1987). These data imply a calcium requirement of at least $1\,mg\,l^{-1}$.

Five *Hyridella* species occur in coastal SE Australia, but not in the adjacent Murray-Darling Basin. The intervening mountains are not a barrier to waterfowl, and the possibility of obligate host fish requirements is remote. In Victoria, *H. drapeta* occurs only in streams of salinity $<300\,mg\,l^{-1}$, suggesting that it may be excluded from inland rivers (which generally are more saline) for that reason (Dean 1968). This would not explain its absence from northern Queensland, where stream salinities are like those in the south. Temperature, velocity, substratum type and water chemistry may be implicated.

The salinity tolerances of the adults and glochidia of *A. jacksoni* and *V. ambiguus* are described by Walker (1981b, 1998). These species, and mussels in general, are restricted to salinities $<3\,g\,l^{-1}$, and there are reports of mussel populations disappearing from salinized streams (e.g. *Westralunio carteri*: Kendrick 1976).

The occurrence of *V. ambiguus* and *V. wilsonii* in temporary floodplain billabongs and episodic streams in remote inland areas indicates tolerance of hypoxia, elevated temperatures and drought (Walker 1981b; Sheldon and Walker 1989). Indeed, *V. ambiguus* tolerates a year or more removed from water and is capable of anaerobic metabolism (Ch'ng-Tan 1968). There is anecdotal evidence that *V. wilsonii* survives out of water for 3 years (McMichael and Iredale 1959), and this species presumably has physiological features like those of *V. ambiguus*. The ability of individual mussels to tolerate drought is maximized if the valves fit tightly and the shell is encased in mud or otherwise protected from exposure and, given appropriate conditions, it is conceivable that *V. ambiguus* also could survive drought for 3 years. The riverine *A. jacksoni* lacks a tight valve seal and succumbs to 20% weight loss due to dehydration, whereas *V. ambiguus* forms a close seal and tolerates 40% weight loss. *A. jacksoni* is also less tolerant of hypoxia and exposure to air, and this may explain its absence from floodplain billabongs. Whilst *V. ambiguus* is physiologically equipped for diverse environments, it does not occur in big-river habitats, except in impounded or slow-flowing reaches, perhaps because it is unable to maintain an anchorage in strongly flowing water.

2.9 Food and Feeding

Freshwater mussels occur in habitats of diverse trophic status, organic enrichment, suspensoid levels and other features that may modify their diet and feeding behaviour. Dense mussel populations may even modify algal composition and microbial productivity.

Few Australasian studies have investigated the diets of mussels, but it is clear that the primary food source is material suspended in the water. Thus, tissue levels of heavy metal contaminants in *Hyridella menziesi* may not be correlated with local sediment concentrations (Hickey et al. 1995). Stable isotope data for *Velesunio ambiguus* in billabongs suggest that most dietary carbon is derived from algae rather than benthic detritus or vegetation (Bunn and Boon 1993). Mussels may counter algal blooms, as in control of the dinoflagellate *Gymnodinium* by *H. menziesi* (Ogilvie and Mitchell 1995), and are sometimes used to filter water in aquaria and aquaculture tanks.

In tropical Australia, *V. angasi* filters unicellular algae and plant detritus and apparently is not selective (Humphrey 1984). Feeding is inhibited in seasonally turbid or hypoxic waters. Strong correlations are reported for shell and somatic growth with phytoplankton abundance, but not with organic detritus. Food availability and mussel condition peak in the mid-dry season, when Magela Creek billabongs become eutrophic, and there is a lesser peak in the early wet season, when nutrient inputs increase with the first rain.

Table 2.3 summarizes field and laboratory data for filtration by *H. menziesi* in trophic environments differing in planktonic algal biomass (to $98\,\mu g\,l^{-1}$) and suspensoids ($1–35\,mg\,l^{-1}$). Mean filtration rates are $0.096–1.71\,g^{-1}h^{-1}$ in mussels with mean dry tissue weights of 0.39–2.5 g, consistent with other unionoids but eight fold lower than for the unrelated zebra mussel (*Dreissena polymorpha*) (Kryger and Riisgård

Table 2.3. Filtration data for resident, caged and laboratory *Hyridella menziesi*

	Chlorophyll a ($\mu g l^{-1}$)	Suspended Sediment ($mg l^{-1}$)	Condition[a] ($mg DW g SW^{-1}$)	Mean DW (g)	Filtration[b] ($l h^{-1} g^{-1}$)	Filtration method	Reference
Resident							
Waikato River, Hamilton	98[c]	8.0[c]	-	-	0.2–0.55	-	Nobes (1980[d])
Lake Rotokawau	0.15–1.8	0.5–2.5[e]	-	0.7	0.34	1	James (1987)
Lake Tuakitoto	25–35	-	-	2.5[b]	1.6	2	Ogilvie and Mitchell (1995)
Caged[e]							
Lake Taupo	12[c]	1.1[c]	99	0.57	0.35	3	Hickey et al. (1995)
Lake Aratiatia	8[c]	2.1[c]	106	0.61	0.40	3	Hickey et al. (1995)
Lake Maraetai	49[c]	2.2[c]	111	0.64	0.38	3	Hickey et al. (1995)
Tarawera River, Kawerau	-	7.3[f]	76	0.49	0.51	4	Hickey et al. (1993)
Tarawera River, SH30[g]	-	14.9[f]	121	0.88	0.096	4	Hickey et al. (1993)
Tarawera River, Matata[g]	-	17.7[f]	105	0.65	0.34	4	Hickey et al. (1993)
Pond 22[g]	-	-	80	-	0.38	4	Burggraaf et al. (1996)
Laboratory[i]							
Low food	0	<1	59	0.39	1.7	4	Roper and Hickey (1995)
Medium food	0	5	76	0.39	1.1	4	Roper and Hickey (1995)
High food	0	35	70	0.39	0.97	4	Roper and Hickey (1995)

Filtration methods: 1 natural lake-water phytoplankton at 18 °C; 2 algal culture (*Choricystis coccoides*) at 18–21 °C; 3 neutral red dye at 20 °C; 4 yeast cells at 20 °C (– no data).

[a] Calculated from dry weight (DW) and shell weight (SW).

[b] Converted to DW.

[c] Huser (1991).

[d] See James (1987).

[e] From particulate C assuming 50% C.

[f] C. W. Hickey, unpubl.

[g] Site receives pulp and paper kraft mill effluent.

[i] From L. Rotokawau, kept at 20 °C for 3 weeks, fed milk solids and kaolin to simulate silt; pooled data for food with 0–35 $mg l^{-1}$ inorganic solids.

1988). The data show an inverse relationship between (log) filtration rate and mussel dry weight (DW) ($r = -0.96$, $n = 10$, $P < 0.0001$, excluding Lake Tuakitoto), indicating a 91% reduction in filtration for an 0.5–1.0g increase in DW. In contrast, measurements of particulate nitrogen removal in Lake Rotoakawau show a 40% increase in filtration for a comparable DW increase (James 1987). Mussel health may have been poor in some environments, as there was a negative relationship between (log) filtration rate and mussel condition (tissue weight/shell weight) ($r = -0.87$, $n = 10$, $P < 0.001$).

Both food quantity and quality affect filtration and respiration. Laboratory studies indicate a 43% reduction in filtration when food increases from <1 to 35 $mg\,l^{-1}$ (Roper and Hickey 1995; Table 2.3). Silt (non-volatile suspended solids) levels up to 35 $mg\,l^{-1}$ do not affect filtration or condition after 3 weeks' exposure, but mussels exposed to high silt produce copious pseudofaeces, indicating some metabolic cost (cf. James 1987). Respiration increases when food concentrations are low and silt levels are high (Roper and Hickey 1995), but mussel health may not be affected in the short term. Food-silt interactions may have influenced the environmental filtration rates cited above.

The effects of mussel filtration on plankton depend on the maximum filtration rate, filtering time, densities of mussels and plankton, water depth, current velocity and other factors. The riverine species *A. jacksoni* increases its filtration rate as current velocity increases, whereas the wetland species *V. ambiguus* filters more slowly (Humphreys 1986). *H. menziesi* populations may filter more than 90% of the time (Ogilvie and Mitchell 1995), but patterns of activity among individuals are highly variable (C. W. Hickey unpubl.; cf. Hiscock 1950). In Lake Tuakitoto (mean depth 0.7 m in summer) (Tables 2.1, 2.3), the *H. menziesi* population may filter a volume of water equal to that of the lake every 32 h (Ogilvie and Mitchell 1995). Under these conditions, mussel filtration may directly affect algal abundance.

2.10 Biomonitoring

Freshwater mussels store toxicants in their tissues and pseudofaeces, and may play a key role in maintaining water quality (James 1987; Strayer et al. 1994). They exhibit most characteristics of "sentinel organisms" (Phillips and Rainbow 1992), and potentially are biomonitors of ecosystem health (Jeffree et al. 1995a). In New Zealand, for example, *Hyridella menziesi* has proven effective as a sentinel for bacteria and viruses in waters exposed to faecal pollution (Donnison and Ross 1999). Reproductive responses also may have biomonitoring applications (e.g. Fig. 2.6): the propensity of females to abort their young when disturbed also has been used to test their sensitivity to pollutants (e.g. Humphrey et al. 1990). In recent years, freshwater mussels have been included in "mussel watch" programs, with their marine counterparts. Protocols for monitoring elements in marine mussels need to be reassessed, however, given the physiological implications of low calcium availability in freshwater (Metcalfe-Smith et al. 1996).

Some early work in Australia concerned metal accumulation in tissue homogenates of *V. ambiguus* (Jones and Walker 1979; Millington and Walker 1983; cf. Walker 1981b). Although *V. ambiguus* readily accumulates metals, the degree of inter-individual variability appeared to preclude its use as a short-term monitor (Walker 1986). In research prompted by uranium mining, Jeffree and colleagues investigated the uptake kinetics of various elements in species of *Velesunio* and *Hyridella* (Jeffree and Simpson 1984; Jeffree and Brown 1992; Jeffree et al. 1993). They developed the calcium-mimic hypothesis, suggesting that mussels absorb trace metals as metabolic analogues of calcium (Jeffree and Brown 1992; Jeffree et al. 1993; Markich and Jeffree 1994). Thus, most of the interindividual variance in metal loads could be explained by tissue calcium flux. By normalizing metal levels relative to calcium in whole-tissue homogenates, variability was reduced to acceptable levels and mussels were reinstated as potential monitors (Jeffree et al. 1993). Calcium usually is bound to various forms of phosphate (Brown et al. 1996). As many elements follow the Ca metabolic pathway, environmental calcium may play a key role in reducing bioavailability and metal toxicity in mussels.

Like other freshwater mussels, hyriids sequester calcium and other elements in small (0.5–1.5 μm) granules that may account for half of the dry tissue weight (Jeffree and Simpson 1984; Silverman et al. 1985; Pynnönen et al. 1987; Byrne and Vesk 1996, 2000; Vesk and Byrne 1999; Byrne 2000). In addition to Ca granules, *H. depressa* possesses metallothionein-like S granules (Fig. 2.7). Research on these granules has reinvigorated studies of metal pollution and biomonitoring. Postulated functions include calcium homeostasis, detoxification of harmful elements and calcium storage for larval growth (Silverman et al. 1985, 1987a,b; Mason and Jenkins 1995). Both Ca and S granules might be used in biomonitoring investigations and are advantageous, especially for rare species, because tissues can be assayed without killing the mussel (e.g. Berg et al. 1995). Work on *H. depressa* indicates a link between granule composition and the ambient water (Byrne and

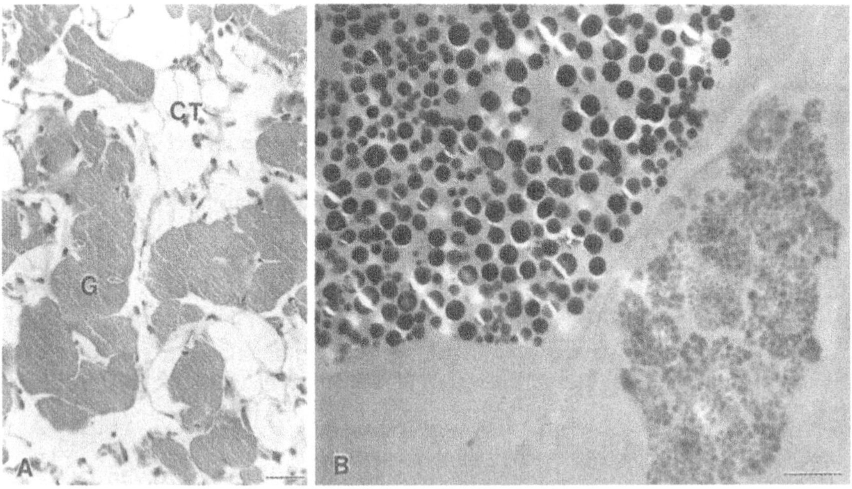

Fig. 2.7A,B. Light **A** and Hscanning transmission electron **B** micrographs showing granules (*G*) in the mantle tissue of *Hyridella depressa*. In **B** the Ca-type granules (*left*) are stained more heavily than the S-type granules (*right*); *CT* connective tissue. Bars: **A** 70 μm; **B** 5 μm. (After Vesk and Byrne 1999)

Vesk 2000). For common elements, granule composition reflects catchment lithology, site trophic status and exogenous inputs. In coastal SE Australian streams, for example, iron is a major component of the granules, reflecting high concentrations in the water (Adams et al. 1997; Vesk and Byrne 1999). For less common elements, the relationship with water chemistry is obscured by complex elemental substitutions.

Studies on organic contaminants involve species of *Hyridella* and *Westralunio* (Ryan et al. 1972; Storey and Edward 1989; Burggraaf et al. 1996; Hickey et al. 1997; Gifford et al. 1995, 1996; Wilkins et al. 2000). In the Waikato River, New Zealand, *H. menziesi* accumulates contaminants from agricultural and atmospheric sources, but the tissue pesticide concentrations would not affect human consumers (Hickey et al. 1997). On the other hand, blooms of neurotoxic Cyanobacteria are common in impoundments and rivers in Australia, and *Alathyria condola* is known to accumulate paralytic shellfish poisons (PSP) (Humpage et al. 1994; Negri and Jones 1995). PSP levels conceivably could jeopardize the health of predators including birds, fish and mammals.

2.11 Parasites and Predators

At several sites in New Zealand, the shells of *Hyridella menziesi* have abnormalities caused by chironomid larvae (Diptera: *Xenochironomus canterburyensis*) (Roper and Hickey 1994). The larvae occupy the mantle cavity and the space beneath the periostracal membrane at the valve margins (Forsyth and McCallum 1978a,b), and dead individuals become embedded in the nacre. The larvae also appear to stimulate growth of the periostracum at the posterior shell margin.

Parasitic trematodes are common (see Sect. 2.6). Adult and juvenile mites (Acari: Unionicolidae) occur on the siphons and gills and in the mantle cavity of *Alathyria*, *Hyridella* and *Velesunio* species, and the eggs and embryos form conspicuous cysts on the gills (Gavey 1979; Viets 1980a,b; Jupiter and Byrne 1997).

Most information about predators is anecdotal, but there are references to fish, turtles, water rats (e.g. Thomson 1958; Woollard et al. 1978), several bird species (e.g. Vestjens 1973; Van Tets 1994) and crayfish (Parastacidae: *Paranephrops planifrons*) (C.W. Hickey, unpubl.). One intriguing report refers to white-winged choughs (Corcoracidae: *Corcorax melanorhamphus*) using sticks to prise open shells (Hobbs 1971).

2.12 Fisheries

Freshwater mussel shells are associated with the earliest-known sites of human occupation in Australia (e.g. Bowler et al. 1970). They were an important source of food and utensils for indigenous peoples in Australia (e.g. Walker 1981b; Humphrey 1984; Clark and Hope 1985; Allison and Simpson 1989) and New Zealand (Hiroa 1921), but are no longer exploited on a large scale. There is occasional interest in using them for pearl shell, but they are unlikely to realize the commercial value of Northern Hemisphere species, as populations generally are too sparse and the shells too thin and discoloured to endow them with fishery potential. Natural

Fig. 2.8. An endangered, possibly extinct species: *Hyridella glenelgensis* from the Glenelg-Wannon river system, Victoria. The shell is 42 mm long

densities also limit their potential as human food, and their propensity to accumulate pollutants might prejudice the health of consumers, unless the mussels are maintained in aquaculture.

2.13 Conservation

Mussels, indeed most aquatic invertebrates, have a low priority in conservation issues throughout Australasia. This may change as government agencies become more aware of the decline of molluscs in the Murray-Darling Basin and elsewhere (e.g. Sheldon and Walker 1997). Whilst Australia abounds with exotic animals, there are fortuitously no problems with exotic freshwater bivalves (cf. *Anodonta, Corbicula, Dreissena* in North America: Bogan 1993; Ricciardi et al. 1998; Watters 1997). Instead, the main threats to mussels are habitat changes through flow regulation and urban and agricultural pollution. The distributions of riverine and wetland species have changed (Walker 1992) and declines and local extinctions are reported for some rivers (Kendrick 1976; Byrne 1998). Larval and juvenile mussels (and host fish) probably are less tolerant of stress than adults. In changed environments, populations of large mussels are likely to persist for some years, despite failing recruitment, and gradual declines will go undetected in the short term. If inertia prevails, the Australasian fauna may emulate the perilous situation of freshwater mussels in the Northern Hemisphere (Bauer 1991; Williams et al. 1993). Clearly, investigations of population biology are needed, as a strategic response.

A more immediate response is needed for species with restricted (e.g. *Cucumerunio websteri, Haasodonta* spp., *Hyridella aucklandica, Lortiella* spp.) or disjunct distributions (e.g. *H. narracanensis*). One species in need of urgent action is *Hyridella glenelgensis* (Fig. 2.8), restricted to the Glenelg-Wannon river system in SW Victoria, where it is threatened by flow regulation and salinization associated with irrigated agriculture. *H. glenelgensis* is a small (40-mm) species that resembles *H. narracanensis* but for its strong shell sculpture. Museum collections are sparse. After intensive searches, live specimens were found in one small, isolated population in 1980 and again in 1991 (K.F. Walker, unpubl.), but subsequent searches have failed. If *H. glenelgensis* does survive, it is on the verge of extinction. Its plight has been registered in the IUCN *Invertebrate Red Data Book* for over 15 years (Wells et al. 1983).

2.14 Conclusion

This chapter began with an explicit statement that studies of Australasian Unionoidea have languished, pending the need for a phylogenetic review. Whilst this may not wholly preclude investigations of population biology and biogeography, it is a major impediment. With a new framework, or perhaps confirmation of the present one, the way is open for a vigorous expansion of research. In turn, this will require a stronger commitment to studies unrelated to commercial interests. Priority should go to a review of the distribution and conservation status of the Hyriidae, and in that regard *Hyridella glenelgensis* is a beacon for what we hope may be a new beginning.

2.15 Summary

In Australia, New Guinea, Solomon Islands and New Zealand, freshwater mussels are represented by 27 species of Hyriidae (two other species from New Guinea are regarded as Unionidae). Taxonomic diagnoses and phylogenetic interpretations are confounded by the inherent variability of shells, although some variation is associated with particular habitats. Glochidia also may prove to have taxonomic value. Research in molecular phylogeny may open the way for new studies in biogeography and ecology. The family shows a Gondwanan distribution (Australasia, South America), but Triassic fossils suggest Pangaean origins. Regional distributions may have been influenced by sea-level changes and aridity, rafting and dispersal by fish, birds and humans. Hyriids are widespread except in ephemeral, saline or low-calcium waters, and some species tolerate hypoxia, high temperatures and drought. Populations often are sparse, although dense aggregations may occur. Activity is year-round or seasonal, depending on food, suspensoids, oxygen, temperatures and drought. Dense populations may modify algal composition and microbial productivity. Females are distinguished by inner demibranchs modified as marsupia. Reproductive functions in both sexes may be impaired by hermaphroditism and parasitic trematodes. Maturity is attained in 1–4 years, depending on species and climate. Gametogenesis is continuous, but glochidial release is seasonal and opportunistic. Clutch size increases with body size, and is influenced by habitat trophic status. All species appear to have several potential host fish species. Hyriids readily store elements in granules within the tissues. Early work on trace-metal accumulation suggested that interindividual variability should preclude their use as biomonitors, but later work showed that variability is reduced to acceptable levels by adjusting for tissue calcium concentrations. Mussels were used extensively by indigenous peoples, but their low densities and propensity to accumulate pollutants limit their fishery potential. Declines are reported for some species and, as long-lived individuals (30–50 years) may persist despite patchy recruitment, incipient declines may be undetected. Investigations of population biology are needed to determine the conservation status of Hyriidae. At least one species is endangered, if not extinct.

Acknowledgements. We are grateful to Dan Graf, University of Michigan, Hugh Jones, Department of Land and Water Conservation, Sydney, and Kevin Roe, University of Alabama, for helpful comments on a draft manuscript.

References

Adams SM, Shorey CD, Byrne M (1997) An ultrastructural and microanalytical study of metal-ion content in granular concretions of the freshwater mussel *Hyridella depressa*. Micron 28:1–11

Allison HE, Simpson RD (1989) Element concentrations in the freshwater mussel *Velesunio angasi* in the Alligator Rivers region. Supervising Scientist Alligator Rivers Region, Tech Memo 25. Australian Government Publishing Service, Canberra

Angel LM (1961) Larval trematodes from Australian freshwater molluscs, XV. *Cercaria velesunionis* n. sp. Trans R Soc S Aust 84:63–70

Atkins L (1979) Observations on the glochidial stage of the freshwater mussel *Hyridella drapeta* Iredale (Mollusca: Pelecypoda). Aust J Mar Freshwater Res 30:411–416

Atkins L (1980) The use of freshwater mussels (Mollusca: Pelecypoda) as indicators of heavy metals in inland waters. MAgSc Thesis, LaTrobe University, Australia (unpublished)

Balla SA, Walker KF (1991) Shape variation in the Australian freshwater mussel *Alathyria jacksoni* Iredale (Bivalvia, Hyriidae). Hydrobiologia 220:89–98

Bauer G (1987) Reproductive strategy of the freshwater pearl mussel *Margaritifera margaritifera*. J Anim Ecol 56:691–704

Bauer G (1991) Plasticity in life history traits of the freshwater pearl mussel – consequences for the danger of extinction and for conservation measures. In: Seitz A, Loeschicke V (eds) Species conservation: a population-biological approach. Birkhäuser, Basel, pp 103–120

Bauer G (1992) Variation in the life span and size of the freshwater pearl mussel. J Anim Ecol 61:425–436

Berg DJ , Haag WG, Guttman SI, Sickel JB (1995) Mantle biopsy – a technique for non-destructive tissue-sampling of freshwater mussels. J N Am Benthol Soc 14:577–581

Bogan AE (1993) Freshwater bivalve extinctions (Mollusca: Unionoida): a search for causes. Am Zool 33:599–609

Bonetto AA, Tassara MP, Rumi A (1987) *australis* new-subgenus of *Diplodon* Spix (Bivalvia Unionacea) and its possible relationships with Australian Hyriidae. Bol Soc Biol Concepcion 57:55–62

Boss KJ (1982) Mollusca. In: Parker SP (ed) Synopsis and classification of living organisms, vol 1. McGraw-Hill, New York, pp 945–1116

Bowler JM, Jones R, Allen H, Thorne AG (1970) Pleistocene human remains from Australia: a living site and human cremation from Lake Mungo, New South Wales. World Archaeol 2:39–60

Brown PL, Jeffree RA, Markich SJ (1996) Kinetics of Ca-45, Co-60, Pb-210, Mn-54 and Cd-109 in the tissue of the freshwater bivalve *Velesunio angasi*: further development of a predictive and mechanistic model of metal bioaccumulation. Sci Total Environ 188: 139–166

Bunn SE, Boon PI (1993) What sources of organic carbon drive food webs in billabongs? A study based on stable isotope analysis. Oecologia 96:85–94

Burggraaf S, Langdon AG, Wilkins AL, Roper DS (1996) Accumulation and depuration of resin acids and fichtelite by the freshwater mussel *Hyridella menziesi*. Environ Toxicol Chem 15:369–375

Byrne M (1998) Reproduction of river and lake populations of *Hyridella depressa* (Unionacea: Hyriidae) in New South Wales: implications for their conservation. Hydrobiologia 389:29–43

Byrne M (2000) Calcium concretions in the interstitial tissues of the Australian freshwater mussel *Hyridella depressa* (Hyriidae). In: Harper EM, Taylor JD, Crame JA (eds) The evolutionary biology of the Bivalvia. Geological Society of London, London, pp 227–329

Byrne M, Vesk PA (1996) Microanalysis of elements in granules in *Hyridella depressa* (Bivalvia): multivariate analysis and biomonitoring potential. Asian J Ecotoxicol 2:91–97

Byrne M, Vesk PA (2000) Elemental composition of mantle tissue granules in *Hyridella depressa* (Unionoida) from the Hawkesbury-Nepean River system, Australia: inferences from catchment chemistry. Mar Freshwat Res 51:183–192

Ch'ng-Tan SK (1968) Some aspects of renal and excretory physiology in the Australian freshwater mussel, *Velesunio ambiguus* Philippi (Mollusca, Bivalvia). PhD Thesis, Monash University, Australia (unpublished)
Clark P, Hope J (1985) Aboriginal burials and shell middens at Snaggy Bend and other sites on the central Murray River. Aust Archaeol 20:68–89
Cotton BC (1934) A freshwater mussel attached to a duck's foot. S Aust Nat 15:113
Cotton BC, Gabriel CJ (1932) Australian Unionidae. Proc R Soc Vict 44:155–161
Davis GM, Fuller SLH (1981) Genetic relationships among Recent Unionacea (Bivalvia) of North America. Malacologia 20:217–253
Dean JC (1968) A study of the distribution of freshwater mussels (Mollusca: Pelecypoda) in the coastal streams of Victoria. BSc(Hons) Thesis, Department of Zoology, Monash University, Australia (unpublished)
Dell RK (1953) The fresh-water Mollusca of New Zealand, 1. The genus *Hyridella*. Trans R Soc NZ 81:221–237
Donnison AM, Ross CM (1999) Animal and human faecal pollution in New Zealand rivers. NZ J Mar Freshwat Res 33:119–128
Forsyth DJ (1978) Benthic macroinvertebrates in seven New Zealand lakes. NZ J Mar Freshwat Res 12:41–49
Forsyth DJ, McCallum JD (1978a) *Xenochironomus canterburyensis* (Diptera: Chironomidae): an insectan inquiline commensal of *Hyridella menziesi* (Mollusca: Lamellibranchia). J Zool (Lond) 186:331–334
Forsyth DJ, McCallum JD (1978b) *Xenochironomus canterburyensis* (Diptera: Chironomidae), a commensal of *Hyridella menziesi* (Lamellibranchia) in Lake Taupo: features of pre-adult life history. NZ J Zool 5:795–800
Gavey LC (1979) Population regulation of water mites (Unionicolidae) parasitic on the freshwater mussel *Velesunio ambiguus* (Philippi) in Lake Alexandrina, South Australia. BSc(Hons) Thesis, Department of Zoology, University of Adelaide, Australia (unpublished)
Gifford JS, Judd MC, McFarlane PN, Anderson SM (1995) Pentachlorophenol (PCP) in the New Zealand environment: assessment near contaminated sites and remote freshwater lakes. Toxicol Environ Chem 48:69–82
Gifford JS, Buckland SJ, Judd MC, McFarlane PN, Anderson SM (1996) Pentachlorophenol (PCP), PCDD, PCDF and pesticide concentrations in a freshwater lake catchment. Chemosphere 32:2097–2113
Graf DL (2000) The Etherioidea revisited: a phylogenetic analysis of hyriid relationships (Mollusca: Bivalvia: Paleohetrodonta: Unionoida). Occasion Pap Univ Mich Mus Zool 729:1–21
Graf DL, ÓFoighil D (2000) The evolution of reproductive characters among the freshwater pearly mussels (Mollusca: Bivalvia: Unionoidea) of North America. J Molluscan Stud (in press)
Grimmond NM (1968) Observations on growth and age of *Hyridella menziesi* Gray (Mollusca: Bivalvia) in a freshwater tidal lake. MSc Thesis, University of Otago, New Zealand (unpublished)
Haas F (1969) Superfamily Unionacea Fleming 1828. In: Moore RC (ed) Treatise of invertebrate paleontology, Part N Mollusca 6, vol 1. Bivalvia. Geological Society of America, Boulder, Colorado; University of Kansas Press, Lawrence, Kansas, pp N411–N414
Heard WH, Guckert RH (1970) A re-evaluation of the Recent Unionacea (Pelecypoda) of North America. Malacologia 10:333–335
Heard WH, Vail VA (1976) Anatomical systematics of *Etheria elliptica* (Pelecypoda: Mycetopodidae). Malacol Rev 9:15–24
Henderson J (1935) Fossil non-marine Mollusca of North America. Spec Pap Geol Soc Am 3, 313 pp
Hickey CW, Roper DS, Martin ML (1993) Mussel biomonitoring in the Tarawera River:1993. Bay of Plenty Regional Council NZ, Rep BPR215, 23 pp
Hickey CW, Roper DS, Buckland S (1995) Metal concentrations of indigenous and transplanted freshwater mussels *Hyridella menziesi* (Unionacea: Hyriidae) and sediments in the Waikato River, New Zealand. Sci Tot Environ 175:163–177

Hickey CW, Buckland SJ, Hannah DJ, Roper DS, Stuben K (1997) Polychlorinated biphenyls and organochlorine pesticides in the freshwater mussel *Hyridella menziesi* from the Waikato River, New Zealand. Bull Environ Contam Toxicol 59:106–112

Hine PM (1978) Distribution of some parasites of freshwater eels in New Zealand. NZ J Mar Freshwat Res 12:179–187

Hiroa TR (1921) Maori food-supplies of Lake Rotorua, with methods of obtaining them, and usages and customs appertaining thereto. Trans R Soc NZ 53:433–451

Hiscock ID (1950) Shell movements of the freshwater mussel, *Hyridella australis* Lam. (Lamellibranchiata) Aust J Mar Freshwater Res 1:259–268

Hiscock ID (1951a) A contribution to the knowledge of the physiology of the Australian freshwater mussel *Hyridella australis* Lamarck. PhD Thesis, University of Adelaide, Australia (unpublished)

Hiscock ID (1951b) A note on the life history of the Australian freshwater mussel, *Hyridella australis* Lam. Trans R Soc S Aust 74:146–148

Hiscock ID (1960) Supplementary data on freshwater mussels (Mollusca: Pelecypoda) of the Australian region. Aust J Mar Freshwater Res 11:123–126

Hobbs JN (1971) Use of tools by the white-winged chough. Emu 71:84–85

Hoeh WR, Bogan AE, Cummings KS, Guttman SI (1998) Evolutionary relationships and morphological character state transitions in unionoid bivalves (Bivalvia: Unionoida): a molecular systematic perspective. Abstr World Congress of Malacology, Washington DC, 156 pp

Humpage AR, Rositano J, Bretag AH, Brown R, Baker PD, Nicholson BC, Steffensen DA (1994) Paralytic shellfish poisons from Australian cyanobacterial blooms. Aust J Mar Freshwater Res 45:761–771

Humphrey CL (1984) Biology and ecology of the freshwater mussel *Velesunio angasi* (Bivalvia: Hyriidae) in the Magela Creek, Alligator Rivers region, Northern Territory. PhD Thesis, University of New England, Australia (unpublished)

Humphrey CL, Bishop KA, Brown VM (1990) Use of biological monitoring in the assessment of effects of mining wastes on aquatic ecosystems of the Alligator Rivers region, tropical northern Australia. Environ Monit Assess 14:139–181

Humphreys I (1986) Effects of water velocity on the filtration rates of two freshwater mussel species (Unionacea: Hyriidae) of the River Murray, South Australia. BSc(Hons) Thesis, Department of Zoology, University of Adelaide, Australia (unpublished)

Huser B (1991) Waikato river water quality monitoring programme. Waikato Regional Council NZ, 100 pp

Iredale T (1934) The freshwater mussels of Australia. Aust Zool 8:57–78

Iredale T (1943) Guide to the freshwater shells of New South Wales. Aust Nat 11:85–95, 113–127

Iredale T, Whitley GP (1938) The fluvifaunulae of Australia. S Aust Nat 18:64–68

James MR (1985) Distribution, biomass and production of the freshwater mussel, *Hyridella menziesi* (Gray), in Lake Taupo, New Zealand. Freshwater Biol 15:307–314

James MR (1987) Ecology of the freshwater mussel *Hyridella menziesi* (Gray) in a small oligotrophic lake. Arch Hydrobiol 108:337–348

Jeffree RA, Brown PL (1992) A mechanistic and predictive model of metal accumulation by the tissue of the Australian freshwater mussel *Velesunio angasi*. Sci Total Environ 125:85–95

Jeffree RA, Simpson RD (1984) Radium-226 is accumulated in calcium granules in the tissue of the freshwater mussel *Velesunio angasi*: support for a metabolic analogue hypothesis? Comp Biochem Physiol 79A:61–72

Jeffree RA, Markich SJ, Brown PL (1993) Comparative accumulation of alkaline earth metals by two freshwater mussel species from the Nepean River, Australia – consistencies and a resolved paradox. Aust J Mar Freshwater Res 44:609–634

Jeffree RA, Markich SJ, Brown PL (1995a) Australian freshwater bivalves: their applications in metal pollution studies. Asian J Ecotoxicol 1:33–41

Jeffree RA, Markich SJ, Lefebvre F, Thellier M, Ripoll C (1995b) Shell microlaminations of the freshwater bivalve *Hyridella depressa* as an archival monitor of manganese water concentration - experimental investigation by depth profiling using Secondary Ion Mass Spectrometry (SIMS). Experientia 51:838–848

Jones HA, Simpson RD, Humphrey CL (1986) The reproductive cycles and glochidia of freshwater mussels (Bivalvia: Hyriidae) of the Macleay River, northern New South Wales, Australia. Malacologia 27:185–202

Jones WG, Walker KF (1979) Accumulation of iron, manganese, zinc and cadmium by the Australian freshwater mussel *Velesunio ambiguus* (Philippi) and its potential as a biological monitor. Aust J Mar Freshwater Res 30:741–751

Jupiter SD, Byrne M (1997) Light and scanning electron microscopy of the embryos and glochidia larvae of the Australian freshwater bivalve *Hyridella depressa* (Hyriidae). Invert Reprod Dev 32:177–186

Kat PW (1983) Sexual selection and simultaneous hermaphroditism among the Unionidae (Bivalvia: Mollusca). J Zool (Lond) 201:395–416

Kat PW (1984) Parasitism and the Unionacea (Bivalvia). Biol Rev 59:189–207

Kendrick GW (1976) The Avon: faunal and other notes on a dying river in southwestern Australia. W Aust Nat 13:97–114

Kesler DH, Downing JA (1997) Internal shell annuli yield inaccurate growth estimates in the freshwater mussels *Elliptio complanata* and *Lampsilis radiata*. Freshwater Biol 37:325–332

Kryger J, Riisgard HU (1988) Filtration rate capacities in six species of European freshwater bivalves. Oecologia 77:34–38

Ludbrook NH (1961) Mesozoic non-marine Mollusca (Pelecypoda: Unionidae) from the north of South Australia. Trans R Soc S Aust 84:139–147

Lydeard C, Mulvey M, Davis GM (1996) Molecular systematics and evolution of reproductive traits of North American freshwater unionacean mussels (Mollusca: Bivalvia) as inferred from 16S rRNA gene sequences. Trans R Soc Lond B 351:1593–1603

Markich SJ, Jeffree RA (1994) Absorption of divalent trace metals as analogues of calcium by Australian freshwater bivalves - an explanation of how water hardness reduces metal toxicity [review]. Aquat Toxicol 29:257–290

Mason AZ, Jenkins KD (1995) Metal detoxification in aquatic organisms. In: Tessier A, Turner DR (eds) Metal speciation and bioavailability in aquatic systems. Wiley, Chichester, pp 479–583

McMichael DF (1956) Notes on the freshwater mussels of New Guinea. Nautilus 70:38–48

McMichael DF (1957) A review of the fossil freshwater mussels (Mollusca, Pelecypoda) of Australasia. Proc Linn Soc NSW 81:222–244

McMichael DF (1958) The nature and origin of the New Zealand freshwater mussel fauna. Trans R Soc NZ 85:427–432

McMichael DF (1967) Australian freshwater Mollusca and their probable evolutionary relationships: a summary of present knowledge. In: Weatherley AH (ed) Australian inland waters and their fauna: eleven studies. Australian National University Press, Canberra, pp 123–149

McMichael DF, Hiscock ID (1958) A monograph of the freshwater mussels (Mollusca: Pelecypoda) of the Australian Region. Aust J Mar Freshwater Res 9:372–507

McMichael DF, Iredale T (1959) The land and freshwater Mollusca of Australia. In: Keast A, Crocker RL, Christian CS (eds) Biogeography and ecology in Australia. Dr W Junk, The Hague, pp 224–245

Metcalfe-Smith JL, Green RH, Grapentine LC (1996) Influence of biological factors on concentrations of metals in the tissues of freshwater mussels (*Elliptio complanata* and *Lampsilis radiata radiata*) from the St Lawrence River. Can J Fish Aquat Sci 53:205–219

Millington PJ, Walker KF (1983) The Australian freshwater mussel *Velesunio ambiguus* (Philippi) and its potential as a biomonitor for zinc, iron and manganese. Aust J Mar Freshwater Res 34:873–892

Morris PJ, Williamson WG (1988) *Pleiodon* (Conrad) (Bivalvia: Mutelidae: Pleiodoninae) from the Late Cretaceous of Montana - a first North American record for the Mutelidae. J Paleontol 62:758–765

Negri A, Jones GJ (1995) Bioaccumulation of paralytic shellfish poisoning (PSP) toxins from the cyanobacterium *Anabaena circinalis* by the freshwater mussel *Alathyria condola*. Toxicon 33:667–678

Nobes RG (1980) Energetics of the freshwater mussel *Hyridella menziesi* (Gray) MSc Thesis, University of Waikato, New Zealand (unpublished)

Ogilvie SC (1993) The effects of the freshwater mussel *Hyridella menziesi* on the phytoplankton of a shallow Otago lake. MSc Thesis, University of Otago, New Zealand (unpublished)

Ogilvie SC, Mitchell SF (1995) A model of mussel filtration in a shallow New Zealand lake, with reference to eutrophication control. Arch Hydrobiol 133:471–482

Ortmann AE (1909) A new system of the Unionidae. Nautilus 23:114–120

Ortmann AE (1912) The anatomy of the najad *Hyridella australis* (Lamarck) (= *Diplodon australis*). Nautilus 25:100–103

Ortmann AE (1920) Über die australische Muschelgattung *Hyridella*. Arch Molluskenkol 52:119–121

Ortmann AE (1921) South American naiades: a contribution to the knowledge of the freshwater mussels of South America. Mem Carnegie Mus 8:451–670

Parodiz JJ, Bonetto AA (1963) Taxonomy and zoogeographic relationships of the South American naiades (Pelecypoda: Unionacea and Mutelacea). Malacologia 1:179–213

Percival E (1931) A note on the life history of *Diplodon lutulentus* Gould. Trans NZ Inst 62:86–91

Phillips DJ, Rainbow PS (1992) Biomonitoring of trace aquatic contaminants. Elsevier, London

Powell AWB (1979) New Zealand Mollusca: marine, land and freshwater shells. Collins, Auckland

Pynnönen K, Holwerda DA, Zandee DI (1987) Occurrence of calcium concretions in various tissues of freshwater mussels, and their capacity for cadmium sequestration. Aquat Toxicol 10:101–114

Ricciardi A, Neves RJ, Rasmussen JB (1998) Impending extinctions of North American freshwater mussels (Unionoida) following the zebra mussel (*Dreissena polymorpha*) invasion. J Anim Ecol 67:613–619

Roper DS, Hickey CW (1994) Population structure, shell morphology, age and condition of the freshwater mussel *Hyridella menziesi* (Unionacea, Hyriidae) from seven lake and river sites in the Waikato River system. Hydrobiologia 284:205–217

Roper DS, Hickey CW (1995) Effects of food and silt on filtration, respiration and condition of the freshwater mussel *Hyridella menziesi* (Unionacea: Hyriidae): implications for bioaccumulation. Hydrobiologia 312:17–25

Ryan S, Bacher GJ, Martin AA (1972) The mussel *Hyridella australis* as a biological monitor of the pesticide Endrin in fresh water. Search 3:446–447

Sheldon F, Walker KF (1989) Effects of hypoxia on oxygen consumption by two species of freshwater mussel (Unionacea: Hyriidae) from the River Murray. Aust J Mar Freshwater Res 40:491–499

Sheldon F, Walker KF (1997) Changes in biofilms induced by flow regulation could explain extinctions of aquatic snails in the lower River Murray, Australia. Hydrobiologia 347:97–108

Silverman H, Steffens WL, Dietz TH (1985) Calcium from extracellular concretions in the gills of a freshwater unionid mussel is mobilized during reproduction. J Exp Zool 236:137–147

Silverman H, Kays WT, Dietz TH (1987a) Maternal calcium contribution to glochidial shells in freshwater mussels (Eulamellibranchia: Unionidae). J Exp Zool 242:137–146

Silverman H, McNeil JW, Dietz TH (1987b) Interaction of trace metals Zn, Cd and Mn, with Ca concretions in the gills of freshwater unionid mussels. Can J Zool 65:828–832

Simpson CT (1914) A descriptive catalogue of the naiades or pearly freshwater mussels. Bryant Walker, Ann Arbor

Smith BJ (1992) Non-marine Mollusca, vol 8. In: Houston WWK (ed) Zoological catalogue of Australia. Australian Government Publishing Service, Canberra, 399 pp

Smith BJ (1998) Order Unionoida. In: Beesley PL, Ross GJB, Wells A (eds) Mollusca: the southern synthesis. Fauna of Australia, vol 5. CSIRO, Melbourne, pp 296–298

Smith EA (1881) On the freshwater shells of Australia. J Linn Soc Lond 16:255–316

Storey AW, Edward DHD (1989) The freshwater mussel *Westralunio carteri* Iredale as a biological monitor of organochlorine pesticides. Aust J Mar Freshwater Res 40:587–593

Strayer DL, Hunter DC, Smith LC, Borg CK (1994) Distribution, abundance, and roles of freshwater clams (Bivalvia, Unionidae) in the freshwater tidal Hudson River. Freshwater Biol 31:239–248

Thomson WP (1958) Shellfish and rats. Bull Conchol Sect Auckland Mus NZ 14:24–26

Timms BV (1980) The macrobenthos of Lakes Rotoroa and Rotoiti, South Island, New Zealand, with special reference to the influence of allochthonous organic detritus. Arch Hydrobiol 90:182–196

Timms BV (1983) Benthic macroinvertebrates of seven lakes near Cass, Canterbury High Country, New Zealand. NZ J Mar Freshwater Res 17:37–49

Timperley MH (1987) Regional influences on lake water chemistry. In: Viner AB (ed) Inland waters of New Zealand. DSIR Bull Wellington NZ 241:97–111

Van Tets GM (1994) Do cormorants eat freshwater mussels – if not what does? Emu 94:127–128

Vesk PA, Byrne M (1999) Metal levels in tissue granules of the freshwater bivalve *Hyridella depressa* (Unionoida) for biomonitoring: the importance of cryopreservation. Sci Total Environ 225:219–229

Vestjens WJM (1973) Feeding of white ibis on freshwater mussels. Emu 72:71–72

Viets KO (1980a) New Unionicolidae (Acari, Hydrachnellae) from Australia. Trans R Soc S Aust 104:27–40

Viets KO (1980b) Weitere Wassermilben (Hydrachnellae, Acari) aus Australien. Gewässer Abwasser 66/67:143–169

Walker KF (1969) The ecology and distribution of *Halicarcinus lacustris* (Brachyura: Hymenosomatidae) in Australian inland waters. Aust J Mar Freshwater Res 20:163–173

Walker KF (1981a) The distribution of freshwater mussels (Mollusca: Pelecypoda) in the Australian zoogeographic region. In: Keast A (ed) Ecological biogeography in Australia. Dr W Junk, The Hague, pp 1233–1249

Walker KF (1981b) Ecology of freshwater mussels in the River Murray. Aust Water Resour Counc Tech Pap 63, 119 pp

Walker KF (1986) The freshwater mussel *Velesunio ambiguus* as a biomonitor of heavy metals associated with particulate matter. In: Hart BT (ed) Water quality management: the role of particulate matter in the transport and fate of pollutants. Chisholm Institute of Technology, Victoria, pp 175–185

Walker KF (1992) A semi-arid lowland river: the River Murray, Australia. In: Calow PA, Petts GE (eds) The rivers handbook, vol 1. Blackwell, Oxford, pp 472–492

Walker KF (1998) Molluscs of inland waters. In: Beesley PL, Ross GJB, Wells A (eds) Mollusca: the southern synthesis. Fauna of Australia, vol 5. CSIRO, Melbourne, pp 59–65

Walker KF (1999) A provisional guide to the freshwater mussels (Hyriidae, Unionidae) of Australasia. In: Gunn B, Cranston PS, Dimitriadis S, Trueman JWH (eds) Interactive CD-ROM guide to Australian aquatic invertebrates, 2nd edn. CSIRO, Melbourne

Watters GT (1997) A synthesis and review of the expanding range of the Asian freshwater mussel *Anodonta woodiana* (Lea 1834) (Bivalvia: Unionidae). Veliger 40:152–156

Wells SM, Pyle RM, Collins NM (1983) IUCN invertebrate red data book. International Union for the Conservation of Nature, Lausanne, Switzerland

Wilkins AL, Hickey CW, Langdon AG, Roper DS, Singh-Thandi M, Simpson CD (2000) Accumulation of resin acids and fichtelite in the freshwater mussel *Hyridella menziesi* in the Tarawera River, New Zealand. Bull Environ Contam Toxicol (in press)

Williams JD, Warren ML Jr, Cummins KS, Harris JL, Neves RJ (1993) Conservation status of freshwater mussels of the United States and Canada. Fisheries 18:6–22

Wirtado TH (1994) Aspects of the biology of *Velesunio ambiguus* Philippi from a tropical freshwater environment, Ross River, Townsville, Australia. MTropSci Thesis, James Cook University of North Queensland, Australia (unpublished)

Woollard P, Vestjens WJM, MacLean L (1978) The ecology of the eastern water rat *Hydromys chrysogaster* at Griffith, NSW: food and feeding habits. Aust Wildl Res 5:59–73

3 Systematics and Distribution of the Recent Margaritiferidae

Douglas G. Smith

3.1 Introduction

3.1.1 The Margaritiferidae Defined

A study of the morphologically conservative and allegedly most primitive family of freshwater mussels, the Margaritiferidae, can provide much information in understanding the evolution and adaptive radiation of the only large uniquely freshwater bivalve group, the Unionoida. Although containing only about 5% of the described and recognized unionoid species occurring in Holarctic regions, the Margaritiferidae is wide-ranging in its distribution. Recent species are known from the North American continent, Europe including northernmost Africa, the Middle East, and throughout much of southern and eastern Asia. Margaritiferids, colloquially known as pearl mussels, have received intense study only in the areas of life history, generally confined to European populations, and in earlier literature the general pattern of distribution. Very little work has been done on the relationships among the known species of the family and the characteristics used to assign species to the family. In order to provide a foundation for the ensuing discussion, a brief description of the Margaritiferidae and a definition of the morphological and anatomical traits that diagnose the family from other unionoid groups is presented.

The Margaritiferidae probably contains no more than 15 Recent species worldwide; the number of recognized species varies as much as the number of people actually working on the group. Species of margaritiferids reach a large size, with adults in some species reaching 200 mm in shell length. The entire group (excluding alleged fossil forms) is confined to streams and rivers, there being no living species known to occur permanently in standing water. The muscular foot of pearl mussels is greatly extensible and is used as a powerful digging and anchoring device allowing the animal to borrow well into the substrate, a behavior comparatively well developed in margaritiferids among unionoids in general. Several margaritiferid species contain populations that live in alkaline environments but most are found in neutral to weakly acidic waters. As with all unionoid mussels, the Margaritiferidae have an intermittent parasitic larva that must attach to a suitable host fish in order to complete metamorphosis to the adult form. Most "northern" species utilize a salmonid fish as a host; however, the "southern" species (generally below 40°N latitude) infest some other kinds of fish.

Ecological Studies, Vol. 145
G. Bauer and K. Wächtler (eds.) Ecology and Evolution of the Freshwater Mussels Unionoida

Fig. 3.1. Hinge teeth of the left valve of a juvenile *Cumberlandia monodonta*. *AC* Anterior cusp (pseudocardinal); *LT* lateral tooth; *PC* posterior cusp (pseudocardinal). *Bar* 3 mm

3.1.2 Diagnosis of the Margaritiferidae

Shell elongate, compressed, rhomboid and often arcuate, particularly in older individuals. Hinge teeth always present in the juvenile stage (Fig. 3.1), often reduced in the adult. Small mantle attachment scars present on inner surface of shell extending diagonally from beak cavity ventrally and posteriorly and limited by the pallial border. Shell conchiolin consists of one differentiated layer; if homogeneous, unlike thin homogeneous layers of the Unionidae. Periostracum almost always without rays or pigment patterns during any life stage, as a rule light brown or greenish brown in young animals, ranging to black in oldest individuals. Gill demibranchs equipped with separate solid vascularized interlamellar junctions, becoming variously coalesced or fused to form incomplete septa, arranged in oblique linear rows. All four demibranchs marsupial; however, junctions or septa not modified for marsupial function. Gills not fused with mantle posteriorly; however, diaphragm complete, formed by coupling of gills and diaphragmatic septa on mantle lobes to functionally separate inhalant and exhalant currents. Anal (exhalant) aperture long and crenulate along mantle margin, extending dorsally to a point slightly anterior to anus; no fusion of mantle lobes, thus no supraanal opening formed separate from anal. Branchial (inhalant) apertural margin of mantle containing bifid or branched papillae which continue a short distance anteriorly along the ventral border.

All species normally short-term spring, summer, or fall brooders, normally dioecious, with occasional individual or populational simultaneous hermaphroditism reported in a few species (*Margaritifera falcata, M. auricularia*). Neither parasitic hydracaroid unionicoline mites nor aspidogastrid (trematode) flukes known from any species.

3.1.3 Review of Classifications

When Ortmann (1910, 1911) first proposed and subsequently elevated to family rank the Margaritanidae (= Margaritiferidae fide Kennard et al. 1925) there was mixed reaction. Some accepted his view (e.g., Hannibal 1912) while others questioned the new status (e.g., Leferve and Curtis 1912; Simpson 1914). The salient anatomical features discussed by Ortmann (1910, 1912a) eventually convinced most students of the unique position of margaritiferids among the Unionoida. In subsequent years, various interpretations of the Margaritiferidae have appeared, however. Haas (1940), a major student of the group, placed all Palearctic species in the single genus *Margaritifera*, then extending the genus to include all described species in his later monograph (Haas 1969). Modell (1964), while accepting the familial status of the group, rejected all of Ortmann's original reasons for establishing the family and using shell characters alone expanded the scope of the family to include several additional taxa. Heard and Guckert's (1970) treatment, although restricted to North American species, restored Ortmann's emphasis on anatomical characters in their arrangement. However, like Modell (1964), Heard and Guckert (1970) established subfamilies within the Margaritiferidae. Subsequently, Smith (1980) proposed a conservative arrangement within the family that included only two genera without subfamily groups. Thereafter, Davis and Fuller (1981), proposed reducing the Margaritiferidae to subfamilial status and suggested that only a single genus, *Margaritifera*, be recognized. Smith and Wall (1984) refuted Davis and Fuller's (1981) conclusions and reestablished the Margaritiferidae as a family.

Studies subsequent to Smith and Wall (1984) have analyzed shell structure (Kat 1983; Chesney et al. 1993) and DNA sequences (Lydeard et al. 1996) with results supporting Smith and Wall's (1984) position. Ziuganov et al. (1994) reviewed a few published classifications and an unpublished treatment, all of which, including Ziuganov et al. (1994), recognize the family. The most recent classification (Starobogatov 1995) reverts somewhat to Modell's (1964) conchological system, adds several species traditionally placed in the Unionidae, splits populations into species, and resurrects several genus names.

The present account, though somewhat brief, attempts to organize the family into a natural system utilizing morphological (primarily conchological) features in a systematic arrangement with emphasis on examination of as much historical material, including types, as possible.

3.2 Materials and Methods

Space does not allow for the inclusion of all museum specimens, including types, of Margaritiferidae that have been examined in the preparation of this chapter. A complete listing of museums and holdings of margaritiferid specimens is being compiled and will be discussed elsewhere (D.G. Smith and J. Cordeiro, in prep.). In this study, over 140 specimens and color photographs representing 27 described taxa (Table 3.1) from 23 institutions were examined. Also, the large holdings of margaritiferid species at Harvard University and the University of Massachusetts were examined. Both conchological and available fluid preserved material was examined.

3.3 Systematics

3.3.1 Characters of the Margaritiferidae

New species of Margaritiferidae have been diagnosed and described by use of conchological characters exclusively. Characters most often considered include the development and expression of hinge teeth and shape and relative dimensions of the shell. Tooth development varies considerably, ranging from the near obliteration of both pseudocardinal and lateral teeth in the adult of a few species to robust, cuspate

Table 3.1. List of taxa in their original combinations examined in this study[a]

Alasmodon falcata Gould, 1850 *
Dahurinaia kurilensis Zatravkin and Starobogatov, 1984 *
Dahurinaia shigini Zatravkin and Bogatov, 1987 *
Dahurinaia sujfunensis Moskvicheva, 1973 *
Margaritana dernaica Pallary, 1928 *
Margaritana alleni Castro (Locard, 1889)
Margaritana jetschini Westerlund, 1890 *
Margaritana marocana Pallary, 1918 *
Margaritana middendorffi Rosen, 1926 *
Margaritana pyrenaica Bourguignat, 1889
Margaritana sachalinensis Zhadin, 1938 *
Margaritana syriaca Pallary, 1929 *
Margaritifera durrovensis Phillips, 1928
Margaritifera marrianae Johnson, 1983 *
Margaritinopsis woodthorpi Godwin-Austen, 1919 *
Mya margaritifera Linneaus, 1758
Ptychorhynchus laevis Haas, 1910 *
Unio auricularius Spengler, 1793 !
Unio dahuricus Middendorff, 1850 *
Unio decumbans Lea, 1861 *
Unio elongata Lamarck, 1819 !
Unio hembeli Conrad, 1838
Unio homsensis Lea, 1864 *
Unio laosensis Lea, 1863 *
Unio monodonta Say, 1829
Unio margaritifer var. *minor* Rossmaessler, 1835
Unio sinuata Lamarck, 1818 !

[a] Types examined indicated by *, photographs of types indicated by !.

pseudocardinal and distinctly lamellate lateral teeth in adults of those species living in the southern part of the family's distribution.

Used alone, relative dimensions of the shell, however, are the least reliable morphological traits upon which to base a description, due to the well-documented notorious ability of environment to affect shell growth, hence relative dimensions. Another restriction on the use of shell dimensions is the difficulty in establishing consistent reference points for measurements of shell dimensions.

A little-used but unique trait of the Margaritiferidae is the mantle attachment scars on the inner surface of the valves formed by modified epithelial cells (Smith 1983). All species of the Margaritiferidae have such scars whether or not their anatomy is known (Fig. 3.2). Specific patterns of distribution and abundance of the scars can be identified among taxa. It is proposed that their arrangement and density has taxonomic value, especially among margaritiferid species occurring in the Mediterranean region.

Anatomical characters have been rarely used in the diagnosis of margaritiferid species. Ortmann (1912a) investigated the general anatomy of several species and drew attention to the difference in gill structure among a few of the species he exam-

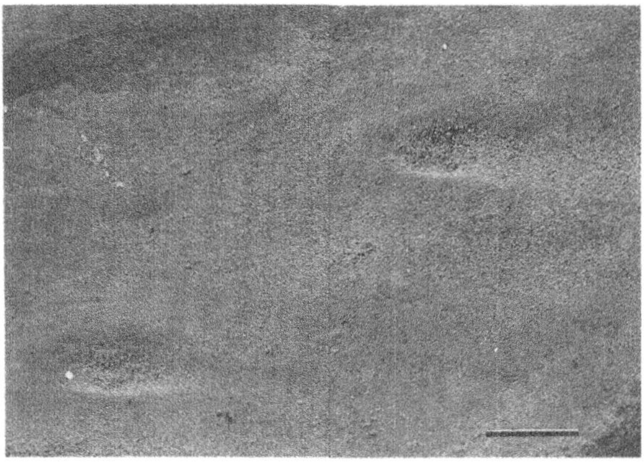

Fig. 3.2. Mantle attachment scars (SEM) on a specimen of *Margaritifera laosensis* (ANSP 41288). *Bar* 1 mm

ined. An anatomical study of *Margaritifera durrovensis* Phillips 1928 was carried out by Bloomer (1928) and later Chesney et al. (1993). Haas (1910) provided a general description of the anatomy of *Margaritifera auricularia* and Smith (1980, 1986, 1988) has investigated the anatomy of several North American species.

The morphology of the larva or glochidium of species should provide information on species relationships and taxonomy. Unfortunately, the only published investigations using both conventional histology and electron microscopy are on *M. margaritifera* (Nezlin et al. 1994; Ziuganov et al. 1994; Pekkarinen and Valovirta 1996) and *M. auricularia* (Araujo and Ramos 1998).

Studies analyzing species differences among species at the subcellular level may provide important data that support or challenge existing conclusions from conchological and anatomical data. Davis and Fuller (1981) analyzed immunological characteristics and Lydeard et al. (1996) examined sequence data retrieved from ribosomal DNA. Although both of these studies were concerned with the higher relationships among North American unionaceans in general, they included several species of North American margaritiferids. Although more general in nature, their results substantiated the conchological and anatomical basis for species recognition among certain North American margaritiferid taxa. Ideally, studies that combine morphological, anatomical, and biochemical characters, using the investigation by Chesney et al. (1993) as a model, will clarify the relationships among species and populations of the Margaritiferidae.

3.3.2 On Certain Species of the Margaritiferidae

Among the described taxa of the Margaritiferidae there are several forms which remain very poorly known. Three of these occur in Morocco, *Margaritifera marocana*, *Margaritifera redomica*, and *Margaritifera dernaica*. There are very few specimens of two of these taxa and none has been located of *M. redomica*. The available specimens of *M. marocana* and *M. dernaica* reveal no substantial differences other than size. The *M. marocana* specimens are less than 90 mm shell length while the few specimens of *M. dernaica* are well over 100 mm. In this study I accept Haas' (1940) synonymy of the two later-named taxa with *M. marocana*.

Margaritifera marocana and the small number of extant *M. dernaica* specimens contain conchological features distinct from *Margaritifera auricularia*. The posterior cusp of the pseudocardinal tooth of the right valve is not tubercular as in *M. auricularia*. In *M. marocana* sensu lato the cusp is flattened and forms a sharp angle with the sulcus. The ventral edge of the cusp is continuous with and parallel to the hinge. Whereas the cusp in *M. auricularia* contains a distinct, oblique serration and tubercles (Fig. 3.3), the serration pattern in *M. marocana* is weak and parallel to the hinge (Fig. 3.4). Additionally, the interior surfaces of the shells of the Moroccan taxa contain abundant mantle-attachment scars, with a distinct and diagnostic crescent of such scars along the upper (dorsal) midsurface. In *M. auricularia* there are few scars and occasional specimens lack them all together. On the basis of these findings it is suggested that *M. marocana* sensu lato represents a distinct species separated geographically from *M. auricularia* by a narrow marine barrier (Straits of Gibraltar).

Margaritana syriaca, and its races ponderosa, minor, and quadratus, occur in eastern Syria. According to Falkner (1994), Pallary placed his *M. syriaca* in *Margaritana* mostly because it resembled *M. marocana* and other related forms described by him from northwestern Africa (Pallary 1927, 1928). Pallary (1920) knew that *M. marocana* was similar to *M. auricularia* of southern Europe and, although possibly not

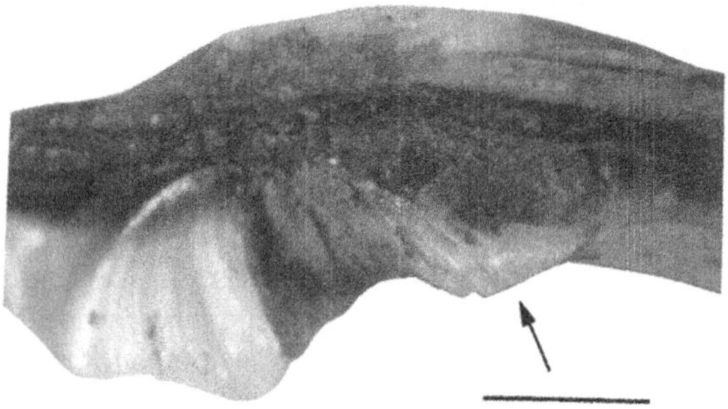

3

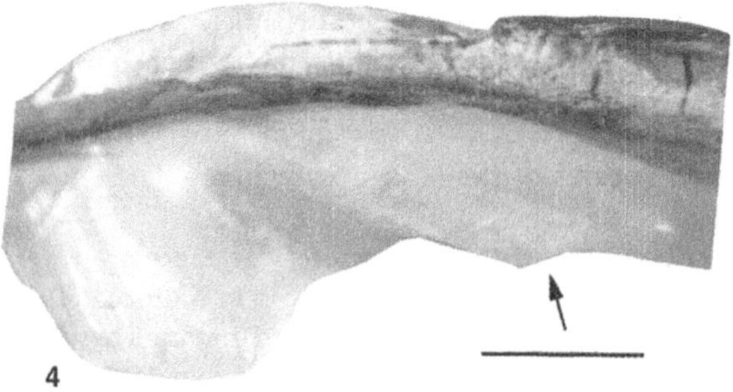

4

Fig. 3.3. Pseudocardinal teeth of the right valve of *Margaritifera auricularia*. *Arrow* denotes posterior cusp. *Bar* 5 mm

Fig. 3.4. Pseudocardinal teeth of the right valve of *Margaritifera marocana*. *Arrow* denotes posterior cusp. *Bar* 5 mm

aware of Haas' (1910) anatomical observations on *M. auricularia*, made a point of the presence of mantle attachment scars in *M. marocana*, a distinct margaritiferid feature. Nevertheless, *M. syriaca*, along with a few other taxa, were synonymized with *Unio homsensis* (Haas 1940), itself considered a subspecies of *Potomida littoralis* (Lamarck 1801). The type specimens of *Unio homsensis* were collected in the Orontes (= Asi) River system somewhere near Homs, Syria. The proximity of *U. homsensis* to other taxa including *M. syriaca*, *Unio episcopalis* Tristam 1865, and *Unio barroisi* Drouët 1893, and an overall similarity amongst their shells probably guided Haas' (1940) synonymy of these forms. Several subsequent studies (Modell 1964; Haas 1969; Schütt 1983; Kinzelbach and Roth 1984; Kinzelbach 1987; Schütt 1987; Nesemann 1993; Falkner 1994) have variously considered *U. homsensis* to be a *Potomida* (Unionidae), *Pseudunio*, *Margaritana*, or *Margaritifera* (Margaritiferidae).

An important but previously overlooked feature of *M. homsensis* is the presence of mantle-attachment scars on the inner surface of the shell. All extant specimens assigned to this taxon contain such scars and the scars are identical in morphology to those of other margaritiferid species. They generally radiate away from the umbo region in typical margaritiferid fashion, do not form any specific clusters, and are not numerous. In the absence of anatomical data, the presence of mantle attachment scars indicates that *M. homsensis* sensu lato is properly assigned to the Margaritiferidae. The remaining forms currently in synonymy with *M. homsensis*, *U. episcopalis* and *U. barroisi*, have not been seen. Johnson's (1971: Plate 2, Fig. 1) figure of *U. episcopalis* shows it to be very similar to *M. homsensis*.

Compared with *M. auricularia*, *M. homsensis* is a much smaller species. The shell length of 150 mm published by Kinzelbach and Roth (1984) and a "large, thick-shelled freshwater mussel" referred to by Schütt (1987) and the identical figures published in each paper are probably of some other species, possibly a *Unio*. All material examined in this study and the type of *U. episcopalis* examined by Johnson (1971) are under 90 mm shell length. The ventral margin is not gently curved as in Fig. 6 of Kinzelbach and Roth (1984), but straight to slightly concave in older specimens. The posterior slope of the shell of *M. homsensis* sensu lato has a moderately well-developed ridge and above the ridge in the corselet and extending a bit to the flank below are well-defined corrugations or folds, especially developed in the younger shells but also somewhat persistent in the older shell. Based on the contrasting conchological characters, *M. homsensis* is considered to be a species distinct from *M. auricularia*. The discovery and study of a living population of *M. homsensis* would greatly enhance our understanding of this poorly known and rare species.

Observations on the anatomy of Asiatic margaritiferids are limited to *M. laevis* from Japan (Taylor and Uyeno 1965; Awakura 1968; Smith 1983, unpubl.). The anatomy of all mainland Russian Margaritiferidae remains unknown. A description of the anatomy of *Dahurinaia dahurica* (Middendorff 1850) is to be presented elsewhere (D.G. Smith, accepted); however, characteristics of the gill are necessarily discussed below. Zhadin (1965) was aware of the anatomical characteristics that distinguish the Margaritiferidae from the Unionidae and placed several Russian species within the Margaritiferidae; however, he did not stipulate to seeing the species anatomy nor did he provide reference to any previous anatomical investigation of species he placed in the family. Zatravkin and Bogatov's (1987) review of the Margaritiferidae of eastern Russia provided no anatomical data on any of the six taxa they recognized including one newly described species. Shells of each described taxon

(6) listed by Zatravkin and Bogatov (1987) have been examined in this study and all contain mantle-attachment scars.

Ziuganov et al. (1994), like many workers since Haas (1940), have not recognized any Russian species described after 1926, leaving only three; *M. dahurica* in the Amur River basin, *M. middendorffi* on the Kamchatka Peninsula, and *M. laevis* on the Kuriles, Sachalin, and Japan. This study follows Ziuganov et al. (1994).

Included by Taylor and Uyeno (1965), Smith (1980), and Taylor (1988) in the Asian "cluster" of species is *Margaritifera falcata*, which appears to be most similar to *M. laevis*. Both species utilize salmonid fish species (*Oncorhynchus* spp.) as hosts for the glochidium larva (Murphy 1942; Awakura 1968, see also Watters 1994). Both species show reduction of the anterior cusp of the left pseudocardinal tooth (Smith 1980; Taylor 1988). In both species the lateral teeth are poorly developed, if developed at all, but a difference exists in the degree of expression. In *M. laevis* the lateral teeth in the adult are developed enough to fully contact one another on valve closure and in the majority of specimens the teeth interlock. The lateral teeth in adult *M. falcata* are reduced to a point such that only simple contact, or more commonly noncontact, occurs in closed valves. Additionally, the nacre of *M. falcata* tends to be bluish purple to deep purple in adults, less frequently a light iridescent blue, while the nacre of *M. laevis* is typically pearly white to a light iridescent salmon or blue.

Anatomically, the gill distinguishes the two species (Fig. 3.5). In *M. laevis*, the free portion of the outer gill is deep and forms a single angle along the midventral margin, and the interlamellar junctions are generally fused only in the anterior part of the gill. In *M. falcata*, the free portion of the outer gill is shallow, biangulate along the ventral margin, and the interlamellar junctions are generally fused into oblique bands.

3.3.3 Revised Classification of the Margaritiferidae

Within the Margaritiferidae, three groups can be identified using conchological and, to a lesser extent, anatomical data. Without doubt, the proposed arrangement will be tested by new data, especially of an anatomical and biochemical nature. The establishment of the three genera system includes the resurrection of two genera originally proposed by Haas, *Pseudunio* (Haas 1910) and *Margaritinopsis* (Haas 1912). *Pseudunio* was proposed as a new genus for *M. auricularia*, with *M. sinuatus* (= *auricularia)* listed as the type species. Although dormant, the genus was recently resurrected to include *P. homsensis* (Falkner 1994). The five species included in the genus *Pseudunio* are given with complete citation in Table 3.2. Besides the persistence of lateral teeth in the adult, the principal feature used by Haas (1910) to establish the genus, the included species also show a fully developed anterior cusp of the left pseudocardinal tooth and the presence of sculpturing (Fig. 3.6) described as corrugate plication (Johnson 1983), on the dorso-posterior flank or corselet of each valve. Plication is persistent in the adult of the North American *P. hembeli* and *P. marrianae*, obsolete to obliterated in the adult of *P. auricularia*, *P. marocana*, and *P. homsensis*. In those species in which the anatomy is at least partly known (*P. auricularia, P. hembeli, P. marrianae*) the gills contain separate, oblique interlamellar junctions with no tendency towards fusion.

The remaining species have undergone reduction or loss of well-developed lateral hinge teeth in the adult (except *M. laosensis*) and lack plication on the shell surface.

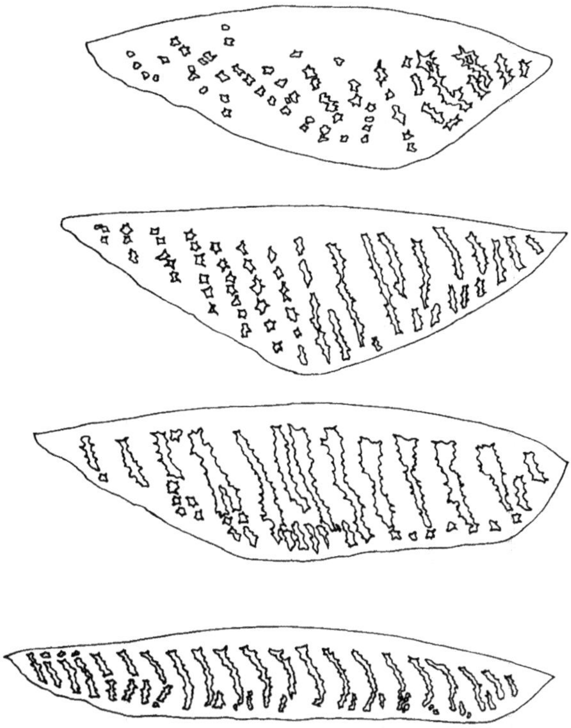

Fig. 3.5. Variation in the interlamellar junctions of the right outer gill of *Margaritinopsis*. Not to scale. *Top to bottom M. laevis* (Hokkaido, Japan), *M. laevis* (Iturup Island, Kurile Islands), *M. falcata* (Montana, USA), *M. monodonta* (Missouri, USA)

These species can be separated into two genera on the basis of the development of the anterior cusp of the left pseudocardinal tooth and the presence or absence of two lateral teeth in the left valve in the juvenile stage. The genus *Margaritifera* retains the fully developed cusps of the left pseudocardinal tooth (Smith 1980). Lateral hinge teeth are absent in the adult. In the juvenile stage, a single short lateral tooth is present in each valve. The gills contain separate, obliquely oriented interlamellar junctions without indications towards fusion (Ortmann 1912a, 1919; Smith 1976a, 1979). Only one species is recognized, the type species *M. margaritifera*, occurring on both sides of the northern Atlantic Ocean. Several named populations and subspecies of this species have been placed in synonymy with *M. margaritifera* (Simpson 1914; Haas 1940; Chesney et al. 1993), and the suggestion to separate the North American and European populations taxonomically (Starobogatov 1995) is not supported by biological evidence. Thus, where previously encompassing several species from throughout the world, the genus *Margaritifera* is now monotypic.

The Asiatic species and the North American *M. falcata* and *C. monodonta* are placed in *Margaritinopsis*, introduced by Haas (1912) for *M. laosensis*, the type species by monotypy. The margaritinopsids (Table 3.2) have in common a single conchological character, reduction of the left pseudocardinal tooth, especially the anterior cusp. This conchological trait was originally observed by Smith (1980) and later by Taylor

Table 3.2. Proposed classification and species of the Margaritiferidae

- Margaritiferidae
  - (Type species: *Mya margaritifera* Linneaus, 1758)
  - *Pseudunio* Haas, 1910
    - (Type species: *Unio sinuata* Lamarck 1819 (= *Unio auricularius* Spengler, 1793)]
    - *Pseudunio auricularia* (Spengler, 1793)
    - *Pseudunio hembeli* (Conrad, 1835)
    - *Pseudunio homsensis* (Lea, 1864)
    - *Pseudunio marocana* (Pallary, 1918)
    - *Pseudunio marrianae* (Johnson, 1983)
  - *Margaritifera* Schumacher, 1816
    - (Type species: *Mya margaritifera* Linneaus, 1758)
    - *Margaritifera margaritifera* (Linnaeus, 1758)
  - *Margaritinopsis* Haas, 1912
    - (Type species: *Unio laosensis* Lea, 1863)
    - *Margaritinopsis dahurica* (Middendorff, 1850)
    - *Margaritinopsis falcata* (Gould, 1850)
    - *Margaritinopsis laevis* (Haas, 1910)
    - *Margaritinopsis laosensis* (Lea, 1863)
    - *Margaritinopsis middendorffi* (Rosen, 1926)
    - *Margaritinopsis monodonta* (Say, 1829)

Fig. 3.6. Examples of plicate shell sculpture in *Pseudunio*; *top P. auricularia*, 73 mm, enlargement of corselet to right; *bottom left P. marrianae*, 111 mm; *bottom right P. homsensis*, holotype (ANSP 125718), 62 mm

(1988) in species of western North American and Asiatic margaritiferids. Smith (1980) proposed recognition of *Margaritinopsis* as a subgenus to encompass all Asiatic species and *M. falcata* but not *C. monodonta* of the central North American interior.

In a few of the included species, *M. falcata* and *M. laevis*, the reduction of the left anterior cusp is slight to moderate, whereas in other forms, e.g., *M. dahurica*, *M. monodonta*, and *M. laosensis*, the reduction is extensive and includes the complete loss of the cusp. Lateral teeth are present and well-formed in the juvenile stage of those species examined, two in the left valve, one in the right. In examined adult *M. laevis* the teeth are somewhat persistent in that they form partial occlusion while in *M. laosensis* they are developed, although very short and possibly neotenic. In all other species of *Margaritinopsis* lateral teeth are absent in the adult. The anatomy remains unknown in several species of *Margaritinopsis*; however, the shells of all species have a very typical margaritiferid type of arcuate, reniform shell and contain well-marked mantle-attachment scars (Fig. 3.2; see also Smith 1983). In species with known anatomy the gills contain separate and obliquely oriented interlamellar junctions; however, there is a clear tendency towards fusion, especially in the anterior portion of the gill (Fig. 3.5).

The system presented here eliminates two established genera, *Cumberlandia* and *Dahurinaia*. *Cumberlandia* was originally established by Ortmann (1912b) on anatomical grounds; specifically the fusion of the interlamellar junctions into oblique, complete rows of septa-like junctions. However, there is a clear west to east transition in species of *Margaritinopsis* (e.g., *M. dahurica*, *M. laevis* of Japan and the Kurile Islands, *M. falcata*, *M. monodonta*) of fusion of the interlamellar junctions and gill shape (Fig. 3.5). The greatest degree of fusion is exhibited by *M. monodonta*. Also, the shell shape and dentition of *M. monodonta* is clearly convergent with that of large, old *M. laosensis* specimens. Although *M. laosensis* exhibits lateral teeth in the adult shell, the pseudocardinals of *M. laosensis* are reduced, extensively so in *M. "woodthorpi"* (= *M. laosensis*). This study supports Davis and Fuller (1981, p. 237), who concluded that *Cumberlandia* was not monophyletic.

The genus *Dahurinaia* (Starobogatov 1970) was defined entirely by subtle features of the shell (shape, depth), all encompassed by other species in the genus *Margaritinopsis* (see Table 3.2). It is not recognized as a valid genus by Smith (1980), Taylor (1988), or Ziuganov et al. (1994).

The bizarre *Unio decumbans* Lea 1861, allegedly from somewhere in Alabama and known only from the type specimen, was placed in *Margaritana* (= *Margaritifera* sensu lato) by Simpson (1900, 1914), where it has remained since. Simpson (1914) correctly observed a few mantle attachment scars on the inner valve surface. However, the presence of weak color rays on the periostracum, the aberrant hinge, and the extreme angulation of the shell margins suggest that *M. decumbans* is a aberration of some other possibly unionid species exhibiting an atavistic trait of mantle shell attachment.

3.4 Distribution

The expansive distribution of the Margaritiferidae precludes the use of a single map showing the ranges of all species. The distribution of *M. margaritifera* in Europe and eastern North America is well-documented and the species range in Europe has been discussed with detailed maps elsewhere (Jackson 1925; Kearney 1975; Jungbluth and

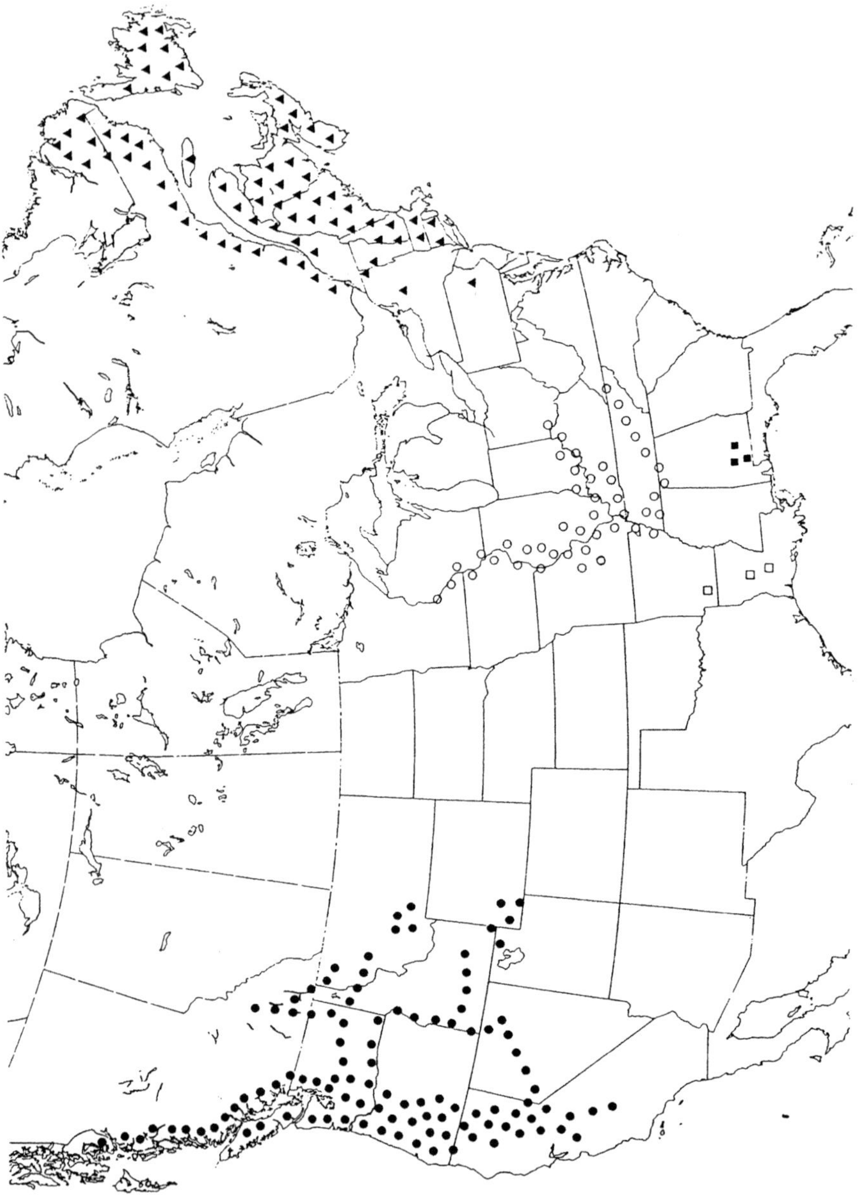

Fig. 3.7. Distribution of North American Margaritiferidae, *Margaritifera margaritifera* (closed triangles), *Margaritinopsis monodonta* (open circles), *M. falcata* (closed circles), *Pseudunio hembeli* (open squares), and *P. marrianae* (closed squares)

Lehmann 1976; Bauer 1986). The North American distribution is shown in Fig. 3.7. In the south of its range, the Iberian Peninsula of Europe and south of New England in eastern North America, *M. margaritifera* becomes fragmented into isolated populations (Ortmann 1919; Bauer 1986), and these are most likely remnants of late Pleistocene distributions. Although declining throughout its range (IUCN 1983), new populations continue to be reported (Ziuganov et al. 1994) and still others may yet be discovered (Bauer 1986).

The distribution of *Pseudunio* is also transatlantic with two species in the Gulf Coast region of North America (Fig. 3.7) and three species in the Mediterranean region (Fig. 3.8). Currently, the range of only a few of the included species is well understood. The ranges of *P. marocana* and *P. homsensis* are imperfectly known, a problem further complicated by the lack of any direct knowledge of their habitats or biology; actually almost nothing is known about the north African and Middle Eastern margaritiferid fauna.

The genus *Margaritinopsis* occurs along a broad south to north belt from 33°N latitude to 56°N latitude in western North America (Fig. 3.7) and from 15°N latitude discontinuously to 55°N latitude in eastern most Asia (Fig. 3.9). Oddly, nearly all species are limited to coastal areas abutting the Pacific Ocean, the only inland species being *M. monodonta* of the upper Mississippi River system of North America. The greatest south-north discontinuity exists between *M. laosensis* of the upper Salween, Mekong, and Koi (Rouge) River systems of southeast Asia and *M. dahurica*, an Amur River system endemic. There is no reason to exclude the possibility that undiscovered species occur in eastern China.

There have been a few models proposed explaining the world distribution of the Margaritiferidae, the earliest involving land bridges (e.g., Scharff 1911) and more

Fig. 3.8. Distribution of *Pseudunio* in the Mediterranean and southern European region, *P. auricularia* (*closed circles, question marks* indicate insufficient data), *P. marocana* (*open squares*), and *P. homsensis* (*closed triangles*)

Fig. 3.9. Distribution of *Margaritinopsis* in Asia, *M. middendorffi* (*crosses*), *M. laevis* (*closed triangles*), *M. dahurica* (*open circles*), and *M. laosensis* (*closed squares*)

recently plate tectonics (Smith 1976b). Evolution, dispersal, and extinction within the Margaritiferidae was most likely affected by tectonic history (Smith 1976b). Rising land and the formation of mountain ranges, especially the Hercynian and Uralide orogenies, created hydrological pathways for dispersing fish and their larval margaritiferid passengers. It is suggested that before the breakup of Pangaea, the family had dispersed throughout the northern mountainous parts of the supercontinent. Subsequent splitting of Pangaea into what is now the North American and Eurasian continents during the Cretaceous split all three genera, leaving vast oceanic gaps between distributions. With the rise of the Unionidae, margaritiferids were increasingly forced into mountainous refugia, especially along the east Asian and western North American coasts where young mountain ranges resulting from an expanding Pacific plate provided abundant habitats. Therefore, the vast discontinuity in some ranges, especially that of *Margaritinopsis* in western Siberia and Europe and eastern North America, can be explained only as extinction along a former broad west-east distribution.

3.5 Summary

The Margaritiferidae is defined using anatomical, conchological, biological, and ecological characters. The described taxa of the Margaritiferidae are reassessed on the basis of examination of both non-type and type material from several collections. Several poorly known margaritiferid taxa are discussed. A revised classification is proposed recognizing three genera, *Pseudunio* with five species, *Margaritifera* with one species, and *Margaritinopsis* with six species.

Acknowledgments. I would like to extend my sincerest gratitude to Dr. Alexander Martynov of the Zoological Institute RAS, St. Petersburg, Russia, and Ms. Elena Sayenko of the Institute of Biology and Soil Science, Vladivostok, Russia, for providing specimens, including many type specimens, of eastern Russian species. Without their efforts, this study would be impossible. Mr. Jay Cordeiro of the American Museum of Natural History has been a valuable source of information on margaritiferid collections and Mr. Richard Johnson of the Museum of Comparative Zoology, Harvard University, has been a continuous source of support and encouragement. Several curators, too many to list here, have graciously allowed inspection of specimens under their care. To all I extend my sincerest thanks.

References

Araujo R, Ramos MA (1998) Description of the glochidium of *Margaritifera auricularia* (Spengler 1793) (Bivalvia, Unionoidea). Philos Trans R Soc Lond Ser B 353:1553–1559

Awakura T (1968) The ecology of parasitic glochidia of the fresh-water pearl mussel, *Margaritifera laevis* (Haas). Sci Rep Hokkaido Salmon Hatchery 23:1–21 (in Japanese)

Bauer G (1986) The status of the freshwater pearl mussel *Margaritifera margaritifera* L. in the south of its European range. Biol Conserv 38:1–9

Bloomer HH (1928) *Margaritfera durrovensis* – anatomical notes. Proc Malacol Soc Lond 18:74–76

Chesney HCG, Oliver PG, Davis GM (1993) *Margaritifera durrovensis* Phillips, 1928: taxonomic status, ecology and conservation. J Conchol 34:267–299

Davis GM, Fuller SLH (1981) Genetic relationships among Recent Unionacea (Bivalvia) of North America. Malacologia 20:217–253

Falkner G (1994) Systematik vorderorientalischer Najaden als Vorstudie zur Bearbeitung archäologischer Funde. In: Kokabi M, Wahl J (eds) Sonderdruck aus Beiträge zur Archäozoologie und Prähistorischen Anthropologie. Forsch Ber Vor Frühgeschichte Baden-Württemberg 53:135–162

Haas F (1910) *Pseudunio*, neues Genus für *Unio sinuatus* Lam. Nachrichtsbl Dtsch Malakozool Ges 42:181–183

Haas F (1912) 5. Genus *Margaritinopsis* Haas 1912. Martini und Chemnitz, Systematisches Conchyliencabinet IX. 2(11):121–123, Taf 12 (1,2). Baver and Raspe, Nürenburg

Haas F (1940) A tentative classification of the Palearctic unionids. Field Mus Nat Hist Zool Ser 24:115–141

Haas F (1969) Das Tierreich. Lieferung 88 – Superfamilia Unionacea. de Gruyter, Berlin

Hannibal H (1912) A synopsis of the Recent and Tertiary freshwater Mollusca of the Californian Province, based upon an ontogenetic classification. Proc Malacol Soc Lond 10:112–165
Heard WH, Guckert RH (1970) A re-evaluation of the Recent Unionacea (Pelecypoda) of North America. Malacologia 10:333–355
IUCN (1983) The IUCN invertebrate red date book. International Union of Conservation of Nature. Gland, Switzerland, 632 pp
Jackson JW (1925) The distribution of *Margaritifera margaritifera* in the British Isles. Part I. J Conchol 17:195–204
Johnson RI (1971) The types and figured specimens of Unionacea (Mollusca: Bivalvia) in the British Museum (Natural History). Bull Br Mus Nat Hist Zool 20:75–108
Johnson RI (1983) *Margaritifera marrianae*, a new species of Unionacea (Bivalvia: Margaritiferidae) from the Mobile-Alabama-Coosa and Escambia River systems, Alabama. Occas Pap Mollusks Mus Comp Zool Harv Univ 4:299–304
Jungbluth J, Lehmann G (1976) Studies on the distribution, morphology and ecology of the *Margaritifera* populations in the atypical biotopes of the young Tertiary basalt area of the Vogelsberg/Upper Hesse (Mollusca: Bivalvia). Arch Hydrobiol 78:165–212 (in German)
Kat PW (1983) Conchiolin layers among the Unionidae and Margaritiferidae (Bivalvia): microsculptural characteristics and taxonomic implications. Malacologia 24:298–311
Kearney MP (1975) European distribution maps of *Pomatias elegans* (Muller), *Discus ruderatus* (Ferussac), *Eobania vermiculata* (Muller) and *Margaritifera margaritifera* (Linne). Arch Molluskenkd 106:243–249
Kennard AS, Salisbury AE, Woodward BB (1925) Notes on the British Post-Pliocene Unionidae with more especial regard to the means of identification of fossil fragments. Proc Malacol Soc Lond 16:267–290
Kinzelbach R (1987) Faunal history of freshwater invertebrates of the northern Levant. In: Krupp F, Schneider W, Kinzelbach R (eds) Proceedings of the symposium on the fauna and zoogeography of the Middle East. Dr Ludwig Reichert, Wiesbaden, pp 115–128
Kinzelbach RK, Roth G (1984) Patterns of distribution of some freshwater molluscs of the Levant region. Foliae Hist-Nat Mus Matrensis 9:115–128
Lefevre G, Curtis WC (1912) Studies on the reproduction and artificial propogation of freshwater mussels. Bull Bur Fish 30:105–201
Lydeard C, Mulvey M, Davis GM (1996) Molecular systematics and evolution of reproductive traits of North American freshwater unionacean mussels (Mollusca: Bivalvia) as inferred from 16S rRNA gene sequences. Philos Trans R Soc Lond Ser B 351:1593–1603
Modell H (1964) Das natürliche System der Najaden. 3. Arch Molluskenkd 93:71–126
Murphy G (1942) Relationship of the fresh-water mussel to trout in the Truckee River. Calif Fish Game 28:89–102
Nesemann H (1993) Zoogeographie und Taxonomie der Muschel-Gattungen Unio Philipsson 1788, *Pseudanodonta* Bourguignat 1877 und *Pseudunio* Hass 1910 im oberen und mittleren Donausystem (Bivalvia: Unionidae, Margaritiferidae). Nachrichtenbl Ersten Vorarlberger Malakol Ges 1:20–40
Nezlin LP, Cunjak RA, Zotin AA, Ziuganov VV (1994) Glochidium morphology of the freshwater pearl mussel (*Margaritifera margaritifera*) and glochidiosis of Atlantic salmon (*Salmo salar*): a study by scanning electron microscopy. Can J Zool 72:15–21
Ortmann AE (1910) A new system of the Unionidae. Nautilus 23:114–120
Ortmann AE (1911) The anatomical structure of certain exotic naiades compared with that of the North American forms. Nautilus 24:127–131
Ortmann AE (1912a) Notes upon the families and genera of naiades. Ann Carnegie Mus 8:222–365
Ortmann AE (1912b) *Cumberlandia*, a new genus of naiades. Nautilus 26:13–14
Ortmann AE (1919) A monograph of the naiades of Pennsylvania. Part III. Systematic account of the genera and species. Mem Carnegie Mus 8:1–384
Pallary P (1920) Récoltes malacologiques du Captaine Paul Martel dans la Partie Septenrionale du Maroc. J Conchyliol 65:131–160

Pallary P (1927) Complément a la fauna Malacologique de la Berbérie. J Conchyliol 71:197–277
Pallary P (1928) Notice sur seize mollusques nouveaux du Maroc decouverts en 1926 et 1927. J Conchyliol 72:1–24
Pekkarinen M, Valovirta I (1996) Anatomy of the glochidia of the freshwater pearl mussel, *Margaritifera margaritifera* (L.). Arch Hydrobiol 137:411–423
Scharff RF (1911) Distribution and origin of life in America. Constable, London
Schütt H (1983) Die Molluskenfauna der Sü ßwässer im Einzugsgebiet des Orontes unter Berücksichtigung benachbarter Flußsysteme. Arch Molluskenkd 113:17–91
Schütt H (1987) The molluscs of the Oasis Palmyra. In: Krupp F, Schneider W, Kinzelbach R (eds) Proceedings of the symposium on the fauna and zoogeography of the Middle East. Dr Ludwig Reichert, Wiesbaden, pp 62–72
Simpson CT (1900) Synopsis of the Naiades, or pearly fresh-water mussels. Proc U S Nat Mus 22:501–1044
Simpson CT (1914) A descriptive catalogue of the Naiades or pearly freshwater mussels. Bryant Walker, Detroit, Michigan
Smith DG (1976a) Notes on the biology of *Margaritifera margaritifera margaritifera* (Lin.) in central Massachusetts. Am Midl Nat 96:252–256
Smith DG (1976b) The distribution of the Margaritiferidae: a review and a new synthesis. Am Malacol Union Inc Bull 1976:42 (Abstr)
Smith DG (1979) Marsupial anatomy of the demibranch of *Margaritifera margaritifera* (Linn.) in northeastern North America (Pelecypoda: Unionacea). J Molluscan Stud 45:39–44
Smith DG (1980) Anatomical studies on *Margaritifera margaritifera* and *Cumberlandia monodonta* (Mollusca: Pelecypoda: Margaritiferidae). Zool J Linn Soc 69:257–270
Smith DG (1983) On the so-called mantle muscle scars on shells of the Margaritiferidae (Mollusca, Pelecypoda), with observations on mantle-shell attachment in the Unionoida and Trigonioida. Zool Scr 12:67–71
Smith DG (1986) The stomach anatomy of some eastern North American Margaritiferidae (Unionoida: Unionacea). Am Malacol Bull 4:13–19
Smith DG (1988) Notes on the biology and morphology of *Margaritifera hembeli* (Conrad, 1838) (Unionacea: Margaritiferidae). Nautilus 102:159–163
Smith DG (Accepted) Observations on the morphology and anatomy of *Dahurinaia dahurica* (Middendorff, 1850) (Unionoida: Margaritiferidae). J Conchol
Smith DG, Wall WP (1984) The Margaritiferidae reinstated: a reply to Davis and Fulller (1981), Genetic relationships among Recent Unionacea (Bivalvia) of North America. Occas Pap Mollusks Mus Comp Zool Harv Univ 4:321–330
Starobogatov Y (1970) Molluscan fauna and zoogeography of the continental water bodies. Nauka, Leningrad (in Russian)
Starobogatov Y (1995) The pearly freshwater mussels (Mollusca, Unionoida, Margaritiferidae of Russia. In: Valovirta I, Harding PT, Kime D (eds) Proc 9th Int Colloq Eur Invertebrate Survey, Helsinki, 3–4 September 1993. WWF Finland Rep (7):109–123
Taylor DW (1988) Aspects of freshwater mollusc ecological biogeography. Palaeogeogr Palaeoclimatol Palaeoecol 62:511–576
Taylor DW, Uyeno T (1965) Evolution of host specificity of freshwater salmonid fishes and mussels in the North Pacific region. Venus 24:199–209 (in Japanese)
Watters GT (1994) An annotated bibliography of the reproduction and propagation of the Unionoidea (primarily of North America). Ohio Biol Surv Misc Contrib 1
Zatravkin MN, Bogatov VV (1987) Large bivalve freshwater and brackish water mollusks of the Far East of the U.S.S.R. Academy Sciences, USSR Vladivostok (in Russian)
Zhadin VI (1965) Mollusks of fresh and brackish waters of the U.S.S.R. Israel Program Scientific Translations, Jerusalem. (Originally published in Russian as Keys to the fauna of the U.S.S.R. No 46. Academy Sciences, USSR, Moscow, 1952)
Ziuganov V, Zotin A, Nezlin L, Tretiakov V (1994) The freshwater pearl mussels and their relationships with salmonid fish. (English edition) VNIRO Publishing House, Moscow

4 Population Genetics and Systematics of European Unionoidea

KARL-OTTO NAGEL and GUIDO BADINO

4.1 Introduction

The taxonomy and systematics of European freshwater mussels (naiades: Unionidae and Margaritiferidae) have suffered from an overestimation of shell characters. Since ecological factors have a strong influence on shell form (Ortmann 1920; Agrell 1948; Kodolova and Logvinenko 1973; Watters 1994), the need to study other characters became obvious. Among others, Davis (1983, 1994) advocated comparing all available information from as many independent sources as possible to arrive at a stable and comprehensive scheme of phylogenetic relationships. Nagel (1988; see also Nagel et al. 1998) proposed a system of European naiades based on anatomical (gills, stomach), microstructural (shell) and morphological (larvae) characters (see Table 4.1). This formed the background for our study of population genetics.

Enzyme variability is one of the characters that are less determined by environmental factors. The genetic basis of this variability was demonstrated, for example, by Badino (1980: PGI) and Badino and Celebrano (1978: LAP; 1981: MDH and PGM). With adequate models of population structure, it is possible to infer the genetic composition of populations and species from the analysis of allozyme variability. Moreover, these data are an important source of information for detecting phylogenetic units at various systematic levels (Badino and Celebrano 1978). Thus, molecular techniques have made a significant contribution to the understanding of relationships among populations, species and genera of European naiades (Badino et al. 1991; Nagel et al. 1996, 1998).

Kodolova and Logvinenko (1973) were the first to use molecular variability to resolve systematic problems of European Unionoidea. They analyzed banding patterns of muscle proteins from mussels from the European part of the former USSR. They concurred with Haas (1969) on the number of species in the genus *Unio*, but they showed the existence of two *Anodonta* species. When *Margaritifera margaritifera* was compared to other species, the degrees of similarity were consistent with a differentiation on the family level. The authors also put forward phylogenetic hypotheses (Logvinenko and Kodolova 1979). Their results will be referred to later on.

In our studies we were mainly concerned with relationships among the Anodontinae and within the genus *Unio*. Preliminary results were also obtained for *Potomida* and *Microcondylaea*.

Ecological Studies, Vol. 145
G. Bauer and K. Wächtler (eds.) Ecology and Evolution of the Freshwater Mussels Unionoida

Table 4.1. Systematics of European naiades (superfamily Unionoidea)

Family: Margaritiferidae Haas 1940
  Genus: *Margaritifera* Schuhmacher 1816
    Species: *M. margaritifera* (Linnaeus 1758)
      M. auricularia (Spengler 1793)
Family: Unionidae Rafinesque 1820 (= Fleming 1828)
  Subfamily: Ambleminae Rafinesque 1820 *sensu* Davis & Fuller 1981
   Tribe: Gonideini Ortmann 1916 *sensu* Davis & Fuller 1981
    Genus: *Potomida* Swainson 1840
     Species: *P. littoralis* (Cuvier 1798)
   Tribe: Pleurobemini Hannibal 1912 *sensu* Davis & Fuller 1981
    Genus: *Microcondylaea* von Vest 1866
     Species: *M. bonellii* (A. Férrusac 1827)
  Subfamily: Unioninae Rafinesque 1820 *sensu* Heard & Guckert 1970
    Genus: *Unio* Philipsson 1788
     Species: *U. pictorum* (Linnaeus 1758)
      U. mancus (Lamarck 1819)[a]
      U. crassus Philipsson 1788
      U. tumidus Philipsson 1788[b]
  Subfamily: Anodontinae Rafinesque 1820 *sensu* Ortmann 1910
    Genus: *Anodonta* Lamarck 1799
     Species: *A. cygnea* (Linaeus 1758)
      A. anatina (Linnaeus 1758)
      Anodonta woodiana (Lea 1834) (introduced) (plus two other taxa identified in Italy that may deserve specific rank)
    Genus: *Pseudanodonta* Bourguignat 1877
     Species: *P. complanata* (Rossmässler 1835)

[a] In this chapter we argue for it to be treated as a subspecies of *U. pictorum*.
[b] Probably to be placed into a separate genus (Nagel et al. 1998, and this study).
(After Nagel 1988; Nagel et al. 1996; Nagel et al. 1998, with nomenclature according to Haas 1969; Heard and Guckert 1970; Davis and Fuller 1981; Falkner 1990a,b and in prep.: Nomenclature of *Microcondylaea*).

4.2 Methods

4.2.1 Material

Most of our populations were sampled and analyzed between 1986 and 1991. Some additional samples of Italian *Anodonta* (Table 4.2) were taken in 1998. The mussels were sampled at depths of at most 2 m. Only *Unio pictorum* from the Zürichsee were collected at about 3 m depth by SCUBA divers (H.A. Arter, pers. comm.). All animals were transported alive to the laboratory and maintained for a few days in aquaria before analysis.

4.2.2 Electrophoresis

Standard techniques were used for the separation of proteins through electrophoresis (on cellulose acetate strips and polyacrylamide gels) and for the staining of various

Table 4.2. Collection localities and sample sizes for genetic analyses of additional samples of European Anodontinae

N(pop)[a]	Name of collection site	N(sp)[b]	River or lake	Drainage system
Italian *Anodonta* group I				
1	Poirino	11	Fish farming pond	Po
2	Ghirla	11	Lake of Ghirla	Po
Italian *Anodonta* group II				
3	Candia	21	Lake of Candia	Po
4	Annone di Brianza	7	Lake of Annone	Po
5	Alserio	10	Lake of Alserio	Po
6	Comabbio	2	Lake of Comabbio	Po
7	Bosisio	2	Lake of Pusiano	Po
8	Poirino	1	Fish farming pond	Po
9	Ghirla	1	Lake of Ghirla	Po
Anodonta anatina				
10	Freiburg	10	Lake Flückiger	(Rhine)
11	Trebur	8	Landgraben	Rhine
12	Lieblos	9	Kinzig	Rhine
13	Kassel	10	Fulda	Weser
Anodonta cygnea				
14	Freiburg	10	Lake Flückiger	(Rhine)
15	Trebur	1	Landgraben	Rhine
16	Lieblos	1	Kinzig	Rhine

[a] Number of population.
[b] Number of specimens.

enzymes. Badino et al. (1991) and Nagel et al. (1996) report details and occasional variations of standard methods.

4.2.3 Analysis of Data

4.2.3.1 Genetic Variability

Allele frequencies (Nagel et al. 1998) and various measures of genetic variability (of Anodontinae only: Nagel et al. 1996) were published previously. They were calculated using BIOSYS-1 (Swofford and Selander 1989). In all cases, the expected heterozygosity is the unbiased estimate, based on conditional expectations (Levene 1949; Nei 1978). Expected genotype frequencies were corrected for small sample sizes (Levene 1949). Genotypes were pooled in χ^2-tests for deviation from Hardy-Weinberg equilibrium.

4.2.3.2 Cluster Analysis of Genetic Distances

NTSYS-pc (Rohlf 1988) was used to calculate genetic distances (Nei 1972) and to perform cluster analysis of distance data with the UPGMA algorithm. We tested the robustness of the results in different ways. First, we calculated genetic distances between repeated samples (allele frequencies: Table 4.3, genetic variability and distances: Table 4.4) and between neighbouring populations (Table 4.5) for which exchange of gametes or larvae is assumed to be possible. We then indicated the range of these "spurious" distances on the dendrograms, which show the results of the cluster analysis of the genetic distances for the full set of populations of Anodontinae and for the *Unio pictorum/U. mancus* complex. Second, we performed consensus analysis on bootstrapped allele frequencies using PHYLIP (Felsenstein 1993). The consensus trees were arbitrarily rooted by the population indicated with every dendrogram. The results are shown together with the dendrogram of the original data whenever possible, otherwise they are discussed in the text.

Note: All previously published dendrograms of European Unionidae (Badino et al. 1991: 268; Nagel et al. 1996: 351; Nagel et al. 1998: 100, 101) use biased distance (Nei 1972) and not, as was written, unbiased distance (Nei 1978). This is due to an erroneous description in the manual of NTSYS-pc, version 1.50 (F.J. Rohlf, pers. comm.). However, mean genetic distances reported in Nagel et al. (1996: 349) were calculated with BIOSYS-1 using the Nei (1978) distance.

4.2.3.3 Gene Flow Estimates

Gene flow was estimated from the variance of allele frequencies [parameters F(sg), F(gt), and F(st)]. The use and interpretation of these values in taxonomy and systematics follows Porter (1990) as outlined in Nagel et al. (1996). The basic relationship is: $Nm(st) \approx [1/F(st) - 1]/4$, which relates gene flow to genetic differentiation. Following abbreviations are used – F: variance in gene frequencies; N: effective population size; m: effective proportion of migrants between populations. Letters in parentheses indicate levels of hierarchy, s represents (sub)populations, g groups of populations and t the total set of populations. Two different types of hierarchical analysis of variance (Anova) were performed. Gene flow between neighbouring populations of *Unio pictorum/U. mancus* was determined by a two-way Anova, using the program of Weir (1990). This program calculates estimators theta(st) of the parameter F(st) (Weir and Cockerham 1984). A three-way Anova yielded F(gt)-values for the estimation of gene flow between regions, subspecies or species of Anodontinae and the *Unio pictorum/U. mancus*-complex when the total set of populations was subdivided into regional population groups. Three-way Anova results were obtained from BIOSYS-1. In this case, standard errors were obtained after jackknifing over polymorphic loci only using Tukey's method (Hinkley 1982). Nm values and their confidence limits were directly calculated from the basic relationship. Because of a number of small sample sizes we used Hardy-Weinberg heterozygosities rather than observed heterozygote frequencies. Panmixis is assumed whenever F-values are negative.

4.2.4 Limits to Population Genetics Methods

Population models used to analyze genetic variability idealize reality in one way or another. We will briefly list some of the difficulties in complying with the assump-

Table 4.3. Allele frequencies for repeated samples. Locus abbreviations in text; sample sizes in parentheses

Anodonta anatina

Population	Kassel		Ginsheim	
Locus and allele	year 1987	1998	1986	1987
Mdh-1	(7)	(10)	(37)	(9)
100	0.143			0.111
109	0.857	1.000	1.000	0.889
Est	(11)	(10)	(35)	(12)
100	1.000	1.000	0.914	0.833
109			0.086	0.167
Lap	(9)	(10)	(37)	(12)
95	0.667	1.000	1.000	1.000
100	0.333			

All populations were monomorphic for Pgi: 100, Pgm: 100 and for the same allele at the following loci: Mdh-2, Idh-1, Idh-2, Got-1, Got-2.

Unio pictorum pictorum

Population	Kassel		Ginsheim	
Locus and allele	year 1986	1987	1986	1987
Pgm-2	(32)	(40)	(25)	(36)
88	0.875	0.825	0.900	0.889
94	0.125	0.175	0.100	0.111
Mdh-1	(38)	(40)	(25)	(40)
100	0.947	0.975	1.000	0.937
130	0.053	0.025		0.063
Est-1	(29)	(37)	(25)	(39)
100	0.466	0.135	0.880	0.500
107		0.324		0.500
115	0.534	0.541	0.120	
Lap-1	(37)	(40)	(25)	(40)
96	0.135	0.188	0.040	0.050
100	0.865	0.812	0.960	0.950
Lap-2	(37)	(40)	(24)	(40)
93			0.021	
100	0.297	0.250	0.229	0.363
106	0.311	0.338	0.583	0.612
112	0.392	0.412	0.167	0.025

All populations were monomorphic for Pgi: 100 and for the same allele at the following loci: Pgm-1, Pgm-3, Mdh-2, Est-2, Idh-1, Idh-2, Got-1, Got-2, Me-1, Me-2, Me-3.

Table 4.4. Genetic variability measures of repeated samples

Population	Year	N[a]	n[b]	n(av)[c]	%P[d]	H(o)[e]	H(e)[f]	χ^2 (df)[g]	D[h]
Anodonta anatina									
Kassel	1987	11	10.4 (0.4)	1.2 (0.1)	20	0.000 (0.000)	0.073 (0.051)		
	1998	10	10.0	1.0	0	0.000 (0.000)	0.000 (0.000)	10.95 (2)**	0.013
Ginsheim	1986	37	36.8 (0.2)	1.1 (0.1)	10	0.011 (0.011)	0.016 (0.016)		
	1987	12	11.7 (0.3)	1.2 (0.2)	20	0.000 (0.000)	0.050 (0.034)	9.64 (2)**	0.002
Unio pictorum pictorum									
Kassel	1986	38	37.0 (0.6)	1.4 (0.1)	29	0.077 (0.033)	0.102 (0.049)		
	1987	40	39.8 (0.2)	1.4 (0.2)	29	0.066 (0.030)	0.112 (0.053)	34.18 (7)***	0.008
Ginsheim	1986	25	24.9 (0.1)	1.4 (0.2)	24	0.065 (0.037)	0.063 (0.037)		
	1987	40	39.7 (0.2)	1.4 (0.1)	29	0.077 (0.037)	0.084 (0.041)	55.63 (8)***	0.015

[a] Number of animals sampled.
[b] Mean sample size per locus.
[c] Average number of alleles.
[d] Percentage of polymorphic loci.
[e] Observed heterozygosity.
[f] Expected heterozygosity: standard error in parentheses.
[g] χ^2 values of contingency test for homogeneity of allelic frequencies between populations (degrees of freedom).
[h] Genetic distance (Nei 1972).
** $p < 0.01$, *** $p < 0.001$.

Table 4.5. Homogeneity and genetic distances between neighbouring populations[a]

Population	Distance (km)	χ^2 (df)	D
Anodonta anatina			
Ginsheim – Trebur	6	2.78 (2) n.s.	0.001
Commercy – St. Mihiel	18	18.09 (2) ***	0.034
St. Mihiel – Dieue	20	21.61 (4) ***	0.025
Montfrin – Comps	4	29.55 (3) ***	0.049
Brèhèmont – Candes-St. Martin	14	2.38 (2) n.s.	0.015
Unio pictorum			
Lays-sur-le-Doubs – Verjux	33	51.56 (16) ***	0.012
Montfrin – Comps	4	48.33 (11) ***	0.029
Brèhèmont – Candes-St. Martin	14	32.51 (12) **	0.005
Cloyes-Vitry-le-François	15	7.62 (8) n.s.	0.004
Capriva del Friuli – Gabria	35	39.37 (8) ***	0.019

[a] Abbreviations as in Table 4.4.
n.s. – not significant.

tions of the models that are frequently encountered in population genetics studies of unionid mussels.

Large ("Infinite") Population Size. Unionids typically have a patchy distribution pattern even in apparently homogeneous habitats. Thus population size is hard to assess without specific studies, especially when collecting unionids from rivers or lakes (e.g. Miller and Payne 1988). Only in some cases do we have estimates of the total number of mussels of a particular population (*Anodonta anatina*: Ederbringhausen < 1000; *Unio pictorum*: Ederbringhausen < 1000; *Unio crassus*: Oberpfreimd < 10000, Rennertehausen < 1000, Neuenschmidten < 4000).

Random Mating. This assumption is violated at least in populations from running waters where the dispersal of spermatozoa is determined by the direction of water flow. There is also evidence that some unionid species aggregate. This might counteract the effects of random mating (Burla 1971, 1972).

Adequate Sampling Strategy. The evaluation of gene flow and genetic distances is sensitive to the sampling strategy (Porter 1990; Hoeh and Gordon 1996, and references therein). Numbers of populations, sample sizes and numbers of loci should be sufficiently large in studies attempting to resolve taxonomic and systematic questions. When estimating genetic distances, the ideal number of loci should be 30 or more since by this factor the standard error of genetic distance is influenced the most (Nei 1978). Additionally, the effects introduced by heterogeneous population size may also be an important source of error. While the first two conditions are approximately met in our studies of the Anodontinae and *Unio pictorum*, the number of loci, however, is small. We analyzed eight enzymes in *Unio* spp. and *Potomida littoralis* which yielded 17 and 13 loci, respectively. Seven enzymes with ten loci could be analyzed in the Anodontinae species, while *Microcondylaea bonellii* showed seven loci from five enzymes. The genetic distances should therefore be interpreted with some caution. In other studies the numbers of enzymes and loci are as follows: Baagøe et al. (1985) – 12 enzymes, 14 loci originally given, 13 loci used in this study; Purser (1985) – 4 enzymes resolved, 5 loci; Hvilsom and Pedersen (1988) – 13 enzymes, 18 loci; Hoeh

(1990) – 20 enzymes (from his Table 2), 24 loci; Chesney et al. (1993) – 18 enzymes, 24 loci.

Complete Detection of Molecular Variability. Standard electrophoretic procedures used for screening large numbers of specimens underestimate genetic variability (Davis et al. 1981). Badino and Celebrano (1978), for example, showed an additional polymorphism after the application of isoelectric focusing to apparently uniform bands in the Lap-2 locus of *Unio tumidus* from Sweden.

Selective Neutrality of Alleles. This condition may not be fulfilled in reality. Badino (1980) found a correlation between environmental characters and the frequency of certain alleles in locus Pgi for *Unio mancus*. In eutrophicated waters which underwent periodic hypoxia, the frequency of the most common allele was distinctly higher than in running water habitats or a panmictic lake. Selection for certain alleles in a highly eutrophic environment may also explain the lack of polymorphism found in population 11* (Trebur; see Table 4.5) of *Anodonta anatina*.

Origin of Mussel Populations. In Central Europe the stocking of ponds and lakes with fish dates back at least to the Middle Ages, in Southern Europe probably back to Roman times. It cannot always be excluded that mussels from standing waters and their outlets are not the descendents of local stock. We consider this problem to be most relevant for *Anodonta*. Four samples come from man-made reservoirs which have a restricted faunal exchange with nearby natural waters (*A. anatina*: Malaucène, Freiburg; *A. cygnea*: Virelles; *A.* sp. I: Poirino). Mussels were introduced into these water bodies, whether intentionally or unintentionally, through (re)stocking with fishes infected with glochidia. The *A. anatina* population from Lieblos, although inhabitingy a small river, is likely to have originated from restocking as well, since all native unionids became extinct in the 1970s because of water pollution. Badino et al. (1984) calculated a high gene flow for *Unio elongatulus* (= *U. mancus*) from Piedmont and Lombardia (northern Italy). On the basis of this rather unexpected result, they speculated about the role of repopulation measures and intentional release of fish infected with mussel larvae that might obscure the expected medium level of gene flow.

4.3 Margaritiferidae

We refer here to two studies dedicated to the genetics of European populations of *Margaritifera margaritifera*. Purser (1985) analyzed British and German populations while Chesney et al. (1993) compared four Irish populations to one from Pennsylvania, USA. The results were almost identical: a very low number of polymorphic loci and very low values for mean heterozygosity [P = 4% and H(o) = 0.007 – 0.023 in Irish populations]. Part of this genetic variability even seems to be under environmental control (Purser 1985). The mean genetic distance for Irish populations is 0.005, while the range of distance values between Irish populations and a population from Pennsylvania is in the range of 0.099–0.119. The data of Chesney et al. (1993) and Purser (1985) suggest that all populations belong to a single species. Chesney et al. (1993) hypothesize that some biological or ecological factors must prevent further genetic divergence. One of these factors may be the ability of *Margaritifera margaritifera* to become self-fertilizing hermaphrodites when in small populations (Bauer 1987). This strategy intensifies the founder effect of a small number of animals colonizing new habitats by avoiding outcrossing. The same is true for the effect of population bottlenecks. If the above strategy were common to all Margaritiferidae, it would at least in part explain the small species diversity of the family despite its presumed old geological age (Chesney et al. 1993).

4.4 Unionidae: Anodontinae

The controversies about the number of *Anodonta* species in Europe offer an example for the inappropriate use of shell characters like contour line, thickness, obesity or coloring. The role of environmental factors acting upon these characters was often greatly underestimated. Thus, the proposed number of species in Europe ranged from 1 to over 400. Today, the existence of two *Anodonta* species and one species of the genus *Pseudanodonta* in central and northern Europe is widely accepted, but there are uncertainties about the status of the Anodontinae of southern Europe.

Nagel et al. (1996) studied the genetic differentiation of European Anodontinae in 31 populations [*A. anatina*: 18, *A. cygnea*: 3, *A. woodiana* (a species introduced from East Asia): 1, two Italian taxa of *Anodonta*: 7, *P. complanata*: 2] and the probability of gene flow. Since then some additional populations and samples were studied in our laboratory (*A. anatina*: 4, *A. cygnea*: 3, two Italian taxa of *Anodonta*: 9; Table 4.2: collection localities; Table 4.6: allele frequencies). The results of the recent analysis are discussed together with our previous results.

4.4.1 Molecular Variability and Systematics

Molecular variability is generally low and frequently polymorphic loci are not in Hardy-Weinberg equilibrium (Nagel et al. 1996). In some cases, the low variability may be explained by population bottlenecks (see Sect. 4.2.4). An increased number of diagnostic loci (four) together with relatively high values of mean genetic distances support the classification of *Pseudanodonta complanata* into a separate genus. This concurs with morphological and anatomical findings that had previously been put forward in favour of this interpretation. Within the genus *Anodonta*, both *A. cygnea* and *A. woodiana* show two diagnostic loci. Comparing *A. anatina* with *A. cygnea* yielded the same number.

All Italian *Anodonta* studied so far present a unique Est allele. Furthermore, there were individuals sharing Lap alleles with the central European *A. anatina* (group I) and others showing a new allele (group II). There were no hybrids between these two types. By analogy to the situation found in central Europe we interpreted these results to be indicative of two distinct taxa of Anodontinae in Italy.

4.4.2 Genetic Differentiation

The cluster analysis of genetic distances (Fig. 4.1; see also Nagel et al. 1996, 1998) produced some distinct branches that correspond to different species and to geographically defined subgroups within species. The *Anodonta anatina* populations from the drainage basins of the rivers Seine, Rhine and from affluents of the North Sea all separate at $D = 0.110$ from the group containing the Loire and Danube

Table 4.6. Allele frequencies at four polymorphic loci for additional samples of European Anodontinae

Locus abbreviations in text; sample sizes in parentheses.

Locus and allele	Italian *Anodonta*, group I		Italian *Anodonta*, group II						
	N(pop)[a]								
	1	2	3	4	5	6	7	8	9
Pgm	(11)	(11)	(21)	(7)	(10)	(2)	(2)	(1)	(1)
100	1.000	1.000							
108			1.000	1.000	1.000	1.000	1.000	1.000	1.000
Mdh-1	(11)	(11)	(21)	(7)	(10)	(2)	(2)	(1)	(1)
100			1.000			1.000			1.000
109	1.000	1.000		1.000	1.000		1.000	1.000	
Lap	(11)	(11)	(21)	(7)	(10)	(2)	(2)	(1)	(1)
95	1.000	1.000							
124			1.000	1.000	1.000	1.000	1.000	1.000	1.000
Idh-1	(11)	(11)	(21)	(7)	(10)	(2)	(2)	(1)	(1)
76		1.000					0.500		1.000
100	1.000		1.000	1.000	1.000	1.000	0.500	1.000	

Locus and allele	*Anodonta anatina* N(pop)				*Anodonta cygnea*		
	10	11	12	13	14	15	16
Pgm	(10)	(8)	(9)	(10)	(10)	(1)	(1)
100	1.000	1.000	1.000	1.000			
105					1.000	1.000	1.000
Mdh-1	(10)	(8)	(9)	(10)	(10)	(1)	(1)
100					1.000	1.000	1.000
109	1.000	1.000	1.000	1.000			
Est	(10)	(8)	(9)	(10)	(10)	(1)	(1)
100	1.000	1.000	1.000	1.000			
107					1.000	1.000	1.000
Lap	(10)	(8)	(9)	(10)	(10)	(1)	(1)
95	1.000	1.000	1.000	1.000			
119					1.000	1.000	1.000

All populations were monomorphic for the allele Pgi^{-100} and for the same allele at the loci Idh-2, Got-1, Got-2, Me and Cat (Catalase); Italian *Anodonta* were monomorphic for $Est\text{-}2^{-102}$; *Anodonta anatina* and *A. cygnea* were monomorphic for $Idh\text{-}1^{-100}$.
[a] Number of population from Table 4.2.

Fig. 4.1. European Anodontinae: phenogram derived from genetic distances (Nei 1972). *First column* Population numbers; (*) with asterisk: from Table 4.2; other numbers: Nagel et al. (1998);[a] two samples united for analysis. *Second column* Drainage system of sampling locality. *Third column* Species or taxon: *Aa Anodonta anatina*; *A-I* Italian *Anodonta* group I; *A-II* Italian *Anodonta* group II; *Ac A. cygnea*; *Aw A. woodiana*; *Pc Pseudanodonta complanata*; *Numbers above branches of the phenogram* number of trees in which the branch occurred, out of 300 trees (consensus analysis); root population: 30 (*P. complanata*: Granges-sur-Aube); *white arrow* maximum genetic distance between repeated samples of *A. anatina* (see Table 4.4); *black arrow* maximum genetic distance between neighbouring populations of *A. anatina* (see Table 4.5)

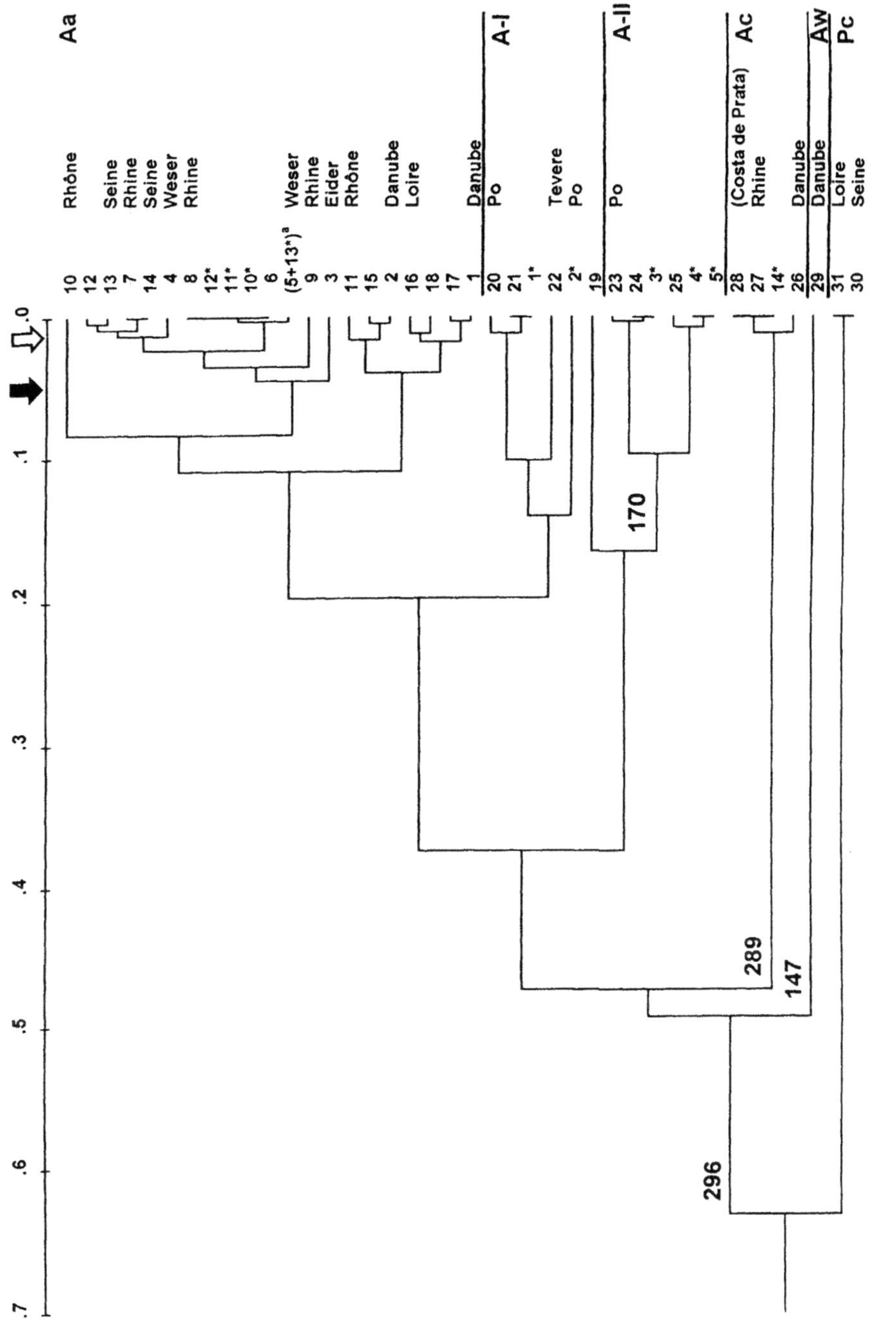
Aa
A-I
A-II
Ac
Aw
Pc
Rhône
Seine
Rhine
Seine
Weser
Rhine
Weser
Rhine
Eider
Rhône
Danube
Loire
Danube
Po
Tevere
Po
Po
(Costa de Prata)
Rhine
Danube
Danube
Loire
Seine
10 12 13 7 14 4 8 12* 11* 10* 6 (5+13*)[a] 9 3 11 15 2 16 18 17 1 20 21 1* 22 2* 19 23 24 3* 25 4* 5* 28 27 14* 26 29 31 30
170
289
147
296
.0
.1
.2
.3
.4
.5
.6
.7

populations. The Rhone populations are distributed among these clusters. The Italian *Anodonta* group I populations are distantly linked to *A. anatina* at D = 0.198. The Italian *Anodonta* group II forms a distinct branch at D = 0.373 and the population 19 (Viverone; see Nagel et al. 1998) of group I also clusters here. *A. woodiana*, *A. cygnea* and *Pseudanodonta complanata* cluster into separate groups with *P. complanata* separating at D = 0.632 from the rest, while the branch containing *A. cygnea* and *A. woodiana* differs at D = 0.473 from *A. anatina* and the Italian *Anodonta*. The relationships between the two Italian taxa may be more complicated than was previously suspected. Both taxa consist of subgroups that separate at D values > 0.05. Marked genetic differentiations correspond with a large geographical distance [population 22 (Passignano; see Nagel et al. 1998) of group I, central vs. northern Italy] or with monomorphism for a rare allele [population 2* (Ghirla; see Table 4.5) of group I, Idh-1^{-76}]. The Viverone sample of group I agrees with all group II populations in being monomorphic for the allele Pgm^{-108}. Relationships within group II are determined by the alleles present in the Mdh-1 locus which cause populations from Piedmont and Lombardia to cluster into separate groups. Consensus analysis shows that the clusters containing *Pseudanodonta complanata* and *Anodonta cygnea* populations, respectively, are highly confirmed. Among the *A. anatina*/*A.* group I/*A.* group II complex only the group II cluster appears in more than half of the bootstrapped trees, while the *A. woodiana* population is just under this value. Genetic distances are of the same order of magnitude as previously reported (Nagel et al. 1996). Our estimates of the times of divergence between taxa remain the same and point to the middle and late Pleistocene as the main periods of differentiation of genera and species.

4.4.3 Gene Flow

The estimated level of gene flow (Table 4.7) between population groups from different drainage basins of the Central European *A. anatina* [Nm(gt) = 0.754] indicates a good probability of recent or actual genetic exchange (Porter 1990). This value is of the same order of magnitude between *A. anatina* and the Italian *Anodonta* group I [Nm(gt) = 0.861]. There is no indication of gene flow between *A. anatina* from Central Europe and *A. cygnea* [Nm(gt) = 0.290]. The same is true for the two Italian *Anodonta* groups [Nm(gt) = 0.322]. These results agree with the previously published data of Nagel et al. (1996).

4.4.4 Conclusions

Diagnostic loci, genetic distances and gene flow analysis show that (1) the *A. anatina* populations, coming from a wide range of drainage basins from Western and Central Europe, belong to a single species, (2) *A. anatina* and *A. cygnea* are good species, (3) Italian *Anodonta* can be separated into two taxa (*Anodonta* group I and II), which may live sympatrically, (4) the introduced species *A. woodiana* should be retained in the genus *Anodonta* and (5) *Pseudanodonta complanata* is correctly placed into a separate genus.

What is the taxonomic rank for the two Italian *Anodonta* groups? Studies on North American *Anodonta* suggest that the evolution of conchological characters and

Table 4.7. Gene flow estimates between groups of populations including additional populations of *Anodonta anatina*, *A. cygnea* and Italian *Anodonta* groups I and II

Comparison	Level of comparison	No. of loci	No. of alleles	F(gt)[a]	Standard error	Nm(gt)[a]	95% confidence limit	
							Lower	Upper
(1)	Drainage basins	5	11	0.249	0.0739	0.754	0.275	3.519
(1) and (2)	? Subspecies	6	17	0.225	0.1573	0.861	0.110	Panmictic
(2) and (3)	? Species	5	13	0.437	0.2353	0.322	Panmictic	Panmictic
(1), (2) and (3)	? Subspecies/species	6	18	0.470	0.0722	0.282	0.114	0.623
(1) and (4)	Species	5	14	0.466	0.1057	0.286	0.084	1.316

[a] See text for explanation.
(1) *Anodonta anatina*, western and central European populations.
(2) Italian *Anodonta* group I.
(3) Italian *Anodonta* group II.
(4) *Anodonta cygnea*.

allozyme diversity is poorly correlated. Two fixed allelic differences may be sufficient to separate species that show little genetic divergence (Hoeh 1990). Our genetic data are still insufficient to propose a certain taxonomic rank for the Italian taxa. Studies are underway examining anatomical and morphological differences. Recent results suggest that the two Italian taxa may belong together closely and probably have a common ancestor.

4.5 Unionidae: Unioninae

Most malacologists agree on the presence of four species of the genus *Unio* in Europe. A study of allozyme diversity showed that *U. tumidus* occupies a position isolated from the other species. Furthermore, *U. crassus*, *U. pictorum* and *U. mancus* group together into a mixed cluster (Nagel et al. 1998). Since conchological and anatomical characters leave no doubt that *U. crassus* is a distinct species this result must be regarded as a shortcoming of the method. Nonetheless, allozyme characters have proven valuable elucidating the relationship between *U. mancus* and *U. pictorum*. In our opinion the traditional characters used to discriminate between species of the genus *Unio* are very unreliable in this case. Biochemical characters thus offer a different source of information in solving taxonomic and phylogenetic questions.

Results from Hvilsom and Pedersen (1988) agree well with our findings. They analyzed two populations of *Unio pictorum* and *U. tumidus*, respectively, and one population of *U. crassus* from Denmark. Based on the distribution of alleles they inferred a close relationship between *U. pictorum* and *U. tumidus*. Previously, Logvinenko and Kodolova (1979) had arrived at the same conclusion by comparing electrophoretic patterns from muscle proteins of the same species taken from the European parts of the former USSR. Contrary to the interpretation given by Hvilsom and Pedersen (1988), the clustering of Nei (1972) genetic distances calculated from their (modified) allele frequencies (Table 4.8) shows the close affinity of *U. crassus* to *U. pictorum* and the rather isolated position of *U. tumidus*.

4.5.1 *Unio pictorum*

In a previous study (Badino et al. 1991) we addressed the problems of taxonomic status and an appropriate phylogenetic model for the species traditionally recognized as *Unio pictorum* and *U. mancus*. The former lives in central, northern and eastern Europe, while the latter has a mainly circummediterranean distribution. We argued that central European *U. pictorum* and Italian *U. mancus* are members of a Rassenkreis (Rensch 1929) or, in modern terms, of a polytypic species. We have tested this idea by adding French populations to our analysis. These additional populations are of special interest because Haas (1969) supposed that *U. pictorum* and *U. mancus* live there sympatrically and that they eventually hybridize.

4.5.1.1 Molecular Variability and Systematics

Table 4.8 shows that molecular variability is particularly low in populations from the Italian peninsula, Sardinia and Corsica. In other regions there are no obvious patterns of change in these parameters. There are no true diagnostic loci but some major

Table 4.8. Genetic variability of European Unionidae (see Nagel et al. 1996 for additional data of Anodontinae)

N(pop)-1998[a,b]	Locality	River or lake	N	n	n(av)	%P	H(o)	H(e)	Remarks[c]
Anodonta anatina									
	Lake Tystrup outlet?	Suså	71	42.8 (6.7)	1.3 (0.2)	23	(*)[d]	0.066 (0.042)	(1)
Anodonta cygnea									
	Lake Tystrup outlet?	Suså	79	55.5 (7.9)	1.2 (0.2)	8	(*)	0.023 (0.023)	(1)
Unio pictorum mancus – French populations									
1	Lays-sur-le-Doubs	Doubs	33	30.5 (1.3)	1.9 (0.4)	35	0.074 (0.034)	0.139 (0.055)	
2	Verjux	Saône	32	31.4 (0.5)	1.7 (0.4)	29	0.069 (0.030)	0.133 (0.059)	
3	Châtillon	Lac du Bourget	27	25.5 (0.9)	1.5 (0.2)	29	0.055 (0.031)	0.099 (0.045)	
4	Comps	Gardon	43	41.1 (1.0)	1.6 (0.3)	29	0.085 (0.039)	0.126 (0.057)	
5	Montfrin	Gardon	11	10.4 (0.3)	1.2 (0.1)	18	0.018 (0.010)	0.049 (0.032)	
6	Clérey	Seine	37	37.0 (0.0)	1.5 (0.3)	24	0.067 (0.031)	0.101 (0.051)	
7	Grange-sur-Aube	Aube	8	7.8 (0.2)	1.3 (0.1)	24	0.083 (0.041)	0.114 (0.054)	
8	Migennes	Yonne	34	33.3 (0.4)	1.5 (0.3)	29	0.057 (0.030)	0.090 (0.046)	
9	Cloyes	Marne	13	13.0 (0.0)	1.4 (0.2)	24	0.077 (0.042)	0.087 (0.048)	
10	Vitry-le-François	Saulx	13	13.0 (0.0)	1.4 (0.2)	29	0.036 (0.018)	0.098 (0.049)	
11	Tourny	Lavaux	28	27.2 (0.6)	1.5 (0.3)	24	0.039 (0.023)	0.084 (0.051)	
12	Chazeuil	Allier	14	13.6 (0.2)	1.2 (0.1)	18	0.049 (0.030)	0.062 (0.034)	
13	Brèhèmont	Indre	35	31.4 (1.7)	1.6 (0.4)	35	0.072 (0.036)	0.106 (0.054)	
14	Candes-St. Martin	Vienne	37	36.4 (0.3)	1.5 (0.4)	24	0.061 (0.035)	0.102 (0.052)	
Unio pictorum mancus – Italian populations									
15	Barauda	Po	88	78.9 (3.6)	1.8 (0.4)	35	(*)	0.132 (0.047)	(2)
16	Angera	Lago Maggiore	55	54.7 (0.1)	2.0 (0.4)	35	(*)	0.167 (0.059)	
17	Asti	Tanaro	24	23.9 (0.1)	1.6 (0.3)	29	0.103 (0.050)	0.143 (0.062)	
18	Ravara	Taro	40	39.2 (0.4)	1.5 (0.3)	24	(*)	0.099 (0.045)	(3)
19	Scardovari	Po	49	48.4 (0.4)	1.8 (0.4)	29	(*)	0.132 (0.052)	(4)
20	Poggibonsi	Staggia	94	88.5 (2.8)	1.2 (0.1)	18	0.026 (0.022)	0.040 (0.030)	
21	Acquapendente	Paglia	52	51.7 (0.2)	1.1 (0.1)	12	0.006 (0.005)	0.010 (0.009)	
22	Benevento	Calore	48	45.6 (1.5)	1.1 (0.1)	12	0.005 (0.003)	0.032 (0.029)	
23	Montescaglioso	Bradano	18	17.9 (0.1)	1.1 (0.1)	12	0.071 (0.049)	0.056 (0.039)	

Table 4.8. *(Continued)*

N(pop)-1998[a,b]	Locality	River or lake	N	n	n(av)	%P	H(o)	H(e)	Remarks[e]
Unio pictorum mancus – populations from Corsica and Sardinia									
24	Aleria	Tavignano	41	39.2 (1.1)	1.1 (0.1)	12	0.029 (0.027)	0.027 (0.026)	
25	Asuni	Araxisi	28	28.0 (0.0)	1.1 (0.1)	12	0.017 (0.015)	0.018 (0.016)	
26	Orosei	Cedrino	30	28.9 (1.1)	1.2 (0.1)	12	0.020 (0.018)	0.021 (0.016)	
Unio pictorum mancus – population from Spain									
27	Ciudad Real	Embalse de Gasset	17	16.5 (0.4)	1.2 (0.1)	18	0.052 (0.032)	0.072 (0.040)	
Unio pictorum pictorum – central European populations									
28	Zeitlarn	Regen	30	29.8 (0.1)	1.7 (0.4)	24	0.087 (0.041)	0.127 (0.061)	
29	Felden	Chiemsee	19	18.6 (0.2)	1.6 (0.3)	29	0.092 (0.046)	0.148 (0.064)	
30	Hainburg	Donau	29	28.8 (0.1)	1.8 (0.4)	24	0.099 (0.047)	0.144 (0.069)	
31	Fonyód	Pogány-völgyi-viz	24	23.6 (0.3)	1.7 (0.4)	24	0.081 (0.044)	0.150 (0.069)	
32	Dziekanow	lake at Dziekanow	4	4.0 (0.0)	1.4 (0.1)	29	0.176 (0.074)	0.145 (0.060)	
	Næsbybro	Suså	48	47.7 (0.3)	1.7 (0.2)	50	(*)	0.124 (0.044)	(5)
		lake Tissø	46	45.9 (0.1)	1.9 (0.2)	50	(*)	0.140 (0.044)	(5)
33	Depenau	Kührener Au	9	8.9 (0.1)	1.2 (0.1)	18	0.029 (0.023)	0.073 (0.040)	
34	Bollingstedt	Bollingstedter Au	11	10.6 (0.1)	1.4 (0.1)	35	0.106 (0.042)	0.142 (0.053)	
35	Kayhude	Alster	17	16.6 (0.4)	1.2 (0.1)	24	0.075 (0.037)	0.089 (0.041)	
36	Ederbringhausen	Orke	19	18.6 (0.3)	1.3 (0.1)	24	0.030 (0.022)	0.074 (0.037)	
37	Gombeth	Schwalm	30	29.9 (0.1)	1.5 (0.2)	29	0.053 (0.025)	0.086 (0.044)	
38	Kassel	Fulda	78	76.8 (0.8)	1.4 (0.2)	29	0.072 (0.031)	0.110 (0.052)	
39	Zürich	Zürichsee	21	21.0 (0.0)	1.1 (0.1)	6	0.017 (0.017)	0.015 (0.015)	
40	Ginsheim	Rhine	65	64.6 (0.2)	1.4 (0.2)	29	0.071 (0.035)	0.081 (0.041)	
41	Bas-Oha	Meuse	16	15.8 (0.2)	1.3 (0.2)	18	0.067 (0.036)	0.081 (0.045)	

Unio pictorum pictorum – Italian populations									
42	Capriva del Friuli	Versa	17	16.2 (0.4)	1.4 (0.2)	24	0.059 (0.032)	0.073 (0.041)	
43	Gabria	Vipacco	33	32.9 (0.1)	1.5 (0.2)	24	0.099 (0.051)	0.089 (0.046)	
Unio pictorum pictorum – population from Portugal									
44	Vilanova de Cerveira	Minho	52	51.7 (0.2)	1.1 (0.1)	12	0.015 (0.012)	0.046 (0.032)	
Unio crassus									
	Holløse Molle	Suså	9	9.0 (0.0)	1.2 (0.1)	17	(*)	0.046 (0.028)	(5), cr
46	Kayhude	Alster	25	24.4 (0.4)	1.4 (0.2)	25	0.053 (0.029)	0.112 (0.053)	cr
47	Rennertehausen	Eder	20	19.9 (0.1)	1.3 (0.1)	19	0.025 (0.019)	0.039 (0.025)	cr
48	Neuenschmidten	Bracht	24	23.8 (0.1)	1.2 (0.1)	12	0.019 (0.014)	0.026 (0.021)	na
	Dave	Meuse	4	4.0 (0.0)	1.1 (0.1)	7	0.016 (0.16)	0.016 (0.016)	(6), na
49	Clérey	Seine	33	32.4 (0.4)	1.5 (0.2)	35	(*)	0.110 (0.051)	na
50	Candes-St. Martin	Vienne	25	24.4 (0.6)	1.3 (0.2)	18	(*)	0.080 (0.044)	na
45	Oberpfreimd	Pfreimd	28	28.0 (0.0)	1.2 (0.1)	12	0.051 (0.036)	0.061 (0.042)	cy
Unio tumidus									
	Næsbybro	Suså	56	56.0 (0.0)	1.4 (0.2)	22	(*)	0.048 (0.025)	(5)
		Lake Tissø	54	54.0 (0.0)	1.3 (0.2)	22	(*)	0.043 (0.024)	(5)
55	Bollingstedt	Bollingstedter Au	15	12.4 (0.4)	1.1 (0.1)	12	0.038 (0.028)	0.062 (0.043)	
54	Beringstedt	Mühlenbek	49	49.0 (0.0)	1.1 (0.1)	12	0.040 (0.030)	0.041 (0.031)	
56	Kassel	Fulda	39	39.0 (0.0)	1.1 (0.1)	12	0.044 (0.030)	0.040 (0.027)	
53	Dziekanow	Lake at Dziekanow	13	12.8 (0.2)	1.1 (0.1)	6	0.023 (0.023)	0.024 (0.024)	
51	Hainburg	Danube	23	21.8 (0.8)	1.0 (0.0)	0	0.000 (0.000)	0.000 (0.000)	
52	Fonyód	Pogány-völgyi-viz	15	14.9 (0.1)	1.1 (0.1)	6	0.008 (0.008)	0.030 (0.030)	(7)

Table 4.8. *(Continued)*

N(pop)-1998[a,b]	Locality	River or lake	N	n	n(av)	%P	H(o)	H(e)	Remarks[c]
Potomida littoralis[e]									
	Vilanova de Cerveira	Minho	19	19.0 (0.0)	1.3 (0.1)	25	0.066 (0.035)	0.088 (0.049)	
	Montfrin	Gardon	17	17.0 (0.0)	1.2 (0.1)	17	0.078 (0.053)	0.071 (0.049)	
	Candes-St. Martin	Vienne	20	19.7 (0.2)	1.3 (0.1)	25	(*)	0.073 (0.042)	
	Brèhèmont	Indre	18	17.7 (0.2)	1.3 (0.1)	25	(*)	0.073 (0.041)	
	Clérey	Seine	33	32.8 (0.3)	1.3 (0.1)	25	(*)	0.081 (0.051)	

[a] Population number from Nagel et al. [1998, pp 86–87 (Table 2)].
[b] Other abbreviations as in Table 4.4.
[c] Remarks
[d] Frequency data only available, no observed heterozygosity.
[e] Same sampling localities as for corresponding *Unio* populations.
(1) Allele frequency data from Baagøe et al. (1985) with modifications: omission of data for ME, GOT and locus UV-Est-III because of inconsistent interpretation; recoding of loci Odh and Icd into two loci (Odh-1, -2 and Icd-1, -2, respectively) with monomorphism for one of two alleles in all cases.
(2) Data modified with respect to Nagel et al. (1998): Est-1: sample size 59.
(3) Data modified with respect to Nagel et al. (1998): Lap-1: sample size 35.
(4) Data modified with respect to Nagel et al. (1998): Lap-1^{-96}: 0.035, Lap-1^{-100}: 0.795, Lap-1^{-102}: 0.045.
(5) Allele frequency data from Hvilsom and Pedersen (1988) with modifications: omission of locus CA1-Ib; recoding of locus Odh into two loci (Odh-1 and Odh-2) with monomorphism for one of two alleles in all cases; omission of uninterpretable cases (the "?" alleles) with other frequencies recalculated.
(6) Sampling locality: 5 km south of Namur, Belgium.
(7) Data modified with respect to Nagel et al. (1998): sample size: 15 except for Lap-1 and Lap-2: 14; allele frequencies: Pgi$^{-A'}$: 0.533, Pgi^{-A}: 0.467, all other loci monomorphic (as before).
cr, na, cy – subspecies of *Unio crassus*: cr – *U. c. crassus*; na – *U. c. nanus* Lamarck 1819; cy – *U. c. cytherea* Küster 1836.

groups can be identified within and among the two taxa by unshared alleles (Badino et al. 1991; Nagel et al. 1998). Pgm-2^{-94} is present only in Central European populations and it is replaced by the allele Pgm-2^{-100} in French and Italian populations. Therefore we propose that the latter belong to the same taxon. Only one heterozygous individual was found in population 1 (Lays-sur-le-Doubs; see Table 4.8) with the hybrid genotype 94/100. Italian populations are characterized by the lack of Est-1^{-115}. Two populations from north-eastern Italy belong to the Central European group on the basis of their allele Pgm-2^{-94}.

4.5.1.2 Genetic Differentiation

The cluster analysis of the Nei (1972) genetic distances (Fig. 4.2) shows a major subdivision at D = 0.150, while the Spanish population already separates at D = 0.165 from all others. One branch consists of *U. pictorum* only (two populations from the Isonzo basin, northeastern Italy, one population from Portugal and all Danube populations) while the other is a mixed one. The latter is further subdivided with *U. mancus* from Italy, Corsica, Sardinia and French populations from the Rhone and Loire drainage diverging at D = 0.094 from all the rest.

Only some population groups are highly confirmed by consensus analysis. They come from (in order of times of confirmation): (1) Po (without Tanaro), (2) Danube, (3) Isonzo and (4) affluents of the North and Baltic Sea, while the Rhine group appears in less than half of the bootstrapped trees.

According to Davis (1983), a distance value of $D > 0.139$ indicates with good probability the presence of two allopatric species, provided that this value is accompanied by substantial differences in allele frequencies. The majority of populations of the two presumed species *U. pictorum* and *U. mancus* separate at values below this threshold. Since neither diagnostic loci nor striking differences in allele frequencies are involved, higher distance values alone are no sufficient argument to put the populations from the Iberian peninsula, the northern Adriatic region and the Danube drainage system into another species.

With Nei's (1972) D = 0.1 corresponding to ca. 100000 years of separation of different clades (see Nei 1971 and Kat 1983 for underlying assumptions), the distance data point to the late Pleistocene as the main period of diversification in *U. pictorum* and *U. mancus*.

4.5.1.3 Gene Flow

Table 4.9a–c gives Nm values for a variety of comparisons among groups of populations. The Iberian populations were excluded from this analysis because we consider their number to be too small to give an adequate picture of the genetic variability in this area. Parameter values of gene flow within population groups [Nm(sg), Table 4.9a] are all well above 0.5, which indicates present or past gene flow to be an important factor for the genetic cohesiveness of each group.

When gene flow between all population groups is considered, the parameter value is greater than 1 for French and Italian mussels. The corresponding value for Central European populations is close to 0.5. Pairwise comparisons reveal that variance in allele frequencies is especially large between the Rhine and Danube groups of populations [F(gt) = 0.300]. This value is even greater than any obtained from pairwise

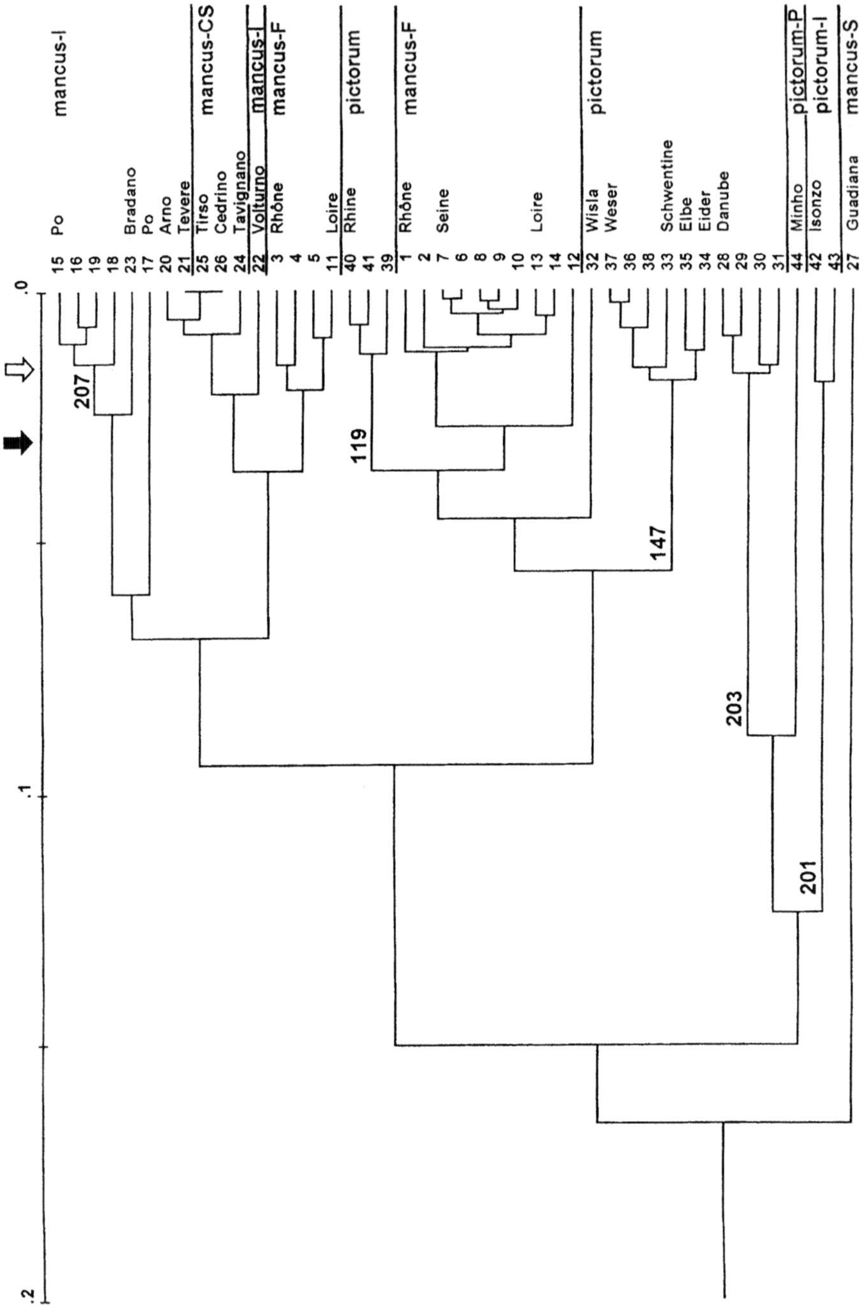
.0
.1
.2
207
119
147
203
201
15 Po mancus-I
16
19
18
23 Bradano
17 Po
20 Arno
21 Tevere
25 Tirso mancus-CS
26 Cedrino
24 Tavignano
22 Volturno mancus-I
3 Rhône mancus-F
4
5
11 Loire
40 Rhine pictorum
41
39
1 Rhône mancus-F
2
7 Seine
6
8
9
10
13 Loire
14
12
32 Wisla pictorum
37 Weser
36
38
33 Schwentine
35 Elbe
34 Eider
28 Danube
29
30
31
44 Minho pictorum-P
42 Isonzo pictorum-I
43
27 Guadiana mancus-S

comparisons between neighbouring population groups belonging to different species (Table 4.9b). Gene flow estimates across presumed taxa boundaries all indicate considerable past or present gene flow [Nm(gt) > 0.8]. The hypothesis of genetic isolation between *U. pictorum* and *U. mancus* is not supported. This is especially true when French population groups are compared to more eastern ones [Nm(gt) > 1.2] while the comparison between neighbouring population groups of these two taxa in Italy yield slightly smaller estimates of gene flow [Nm(gt) < 0.9].

Population differentiation is small or even non-existing for pairs of populations living relatively close together (Table 4.9c). The estimated gene flow [theta(st)] between four pairs of French populations decreases with increasing distance. It is smallest but still important for the maintenance of genetic similarity in the presence of a river dam separating two populations. There are, however, exceptions to this rule. Estimated gene flow between two populations separated by seven dams is in the range of a panmictic population.

With the exception of the Italian populations, the confidence intervals of Nm(sg) values are relatively close to the parameter values. Confidence intervals of Nm(gt) values are much greater, possibly ranging from negative values (panmixis) to values indicating the absence of gene flow. This result may, in part, be an effect of only a few polymorphic loci. Therefore the interpretation of Nm values in this study should be made with some caution.

4.5.1.4 Conclusions

The distribution of alleles among populations suggests that the presumed species *U. pictorum* and *U. mancus* are very closely related taxa. The genetic distances and estimates of gene flow corroborate this view. There is no diagnostic locus but the presence of diagnostic alleles at the Pgm-2 locus permits the separation of two distinct groups of populations. The distribution of these groups roughly coincides with the core regions of typical *U. pictorum* and *U. mancus*. There is no evidence from our genetic data that these groups live sympatrically. We disagree with Haas (1969) who, on the basis of shell characters, recognized overlapping distribution areas and hybrids. We consider these characters to be largely shaped by environmental factors. Nesemann and Nagel (1989) give an example from the Loire drainage system. One individual with a hybrid genotype of the Doubs population did not differ from the rest of the population in shell morphology. The database at hand supports the view that in an evolutionary context both taxa might best be treated as belonging to the polytypic species (Badino et al. 1991), *Unio pictorum*. We propose that two subspecies of *U. pictorum* should be recognized, *U. p. pictorum* and *U. p. mancus*. Their

Fig. 4.2. *Unio pictorum*: phenogram derived from genetic distances (Nei 1972). *First column* N(pop) - 1998 from Table 4.8. *Second column* Drainage system of sampling locality; *Third column* subspecies or regional group; *CS* from Corsica and Sardinia; *F* from France; *I* from Italy; *P* from Portugal; *S* from Spain. *Numbers above branches of the phenogram* number of trees in which the branch occurred, out of 250 trees (consensus analysis); root population: 27 (*U. p. mancus*-S: Ciudad Real); *white arrow* maximum genetic distance between repeated samples of *U. p. pictorum* (see Table 4.4); *black arrow* maximum genetic distance between neighbouring populations of *U. p. mancus*-F and *U. p. pictorum*-I (see Table 4.5)

Table 4.9. Population differentiation and gene flow estimates in *Unio pictorum*

a) Within subspecies Subspecies and population groups	n(l)[a]	n(a)[b]	F(sg)	F(gt)	SE[c]	Nm	Lower	Upper	N(pop) – 1998[e]
							95% CI[d,f]		
Unio pictorum mancus	6	31	0.192		0.0309	1.052	–6.908	1.812	1–26
				0.170	0.0274	1.221	–0.662	–0.585	
France	6	25	0.166		0.0269	1.256	0.806	2.284	1–14
				0.029	0.0250	8.371	–8.069	2.337	
Rhône – Seine	6	25		0.041	0.0309	5.848	–7.746	1.744	1–5/6–10
Rhône – Loire	6	24		–0.023	0.0156	–11.120	–4.056	16.765	1–5/11–14
Seine – Loire	6	20		0.014	0.0251	17.607	–5.651	2.776	6–10/11–14
Italy	6	27	0.231		0.0967	0.832	–72.495	0.256	15–23
				0.113	0.0343	1.962	0.993	9.813	15–19/20–23
Sardinia and Corsica	2	6	0.003		0.0025	83.083	–8.790	7.044	24–26
				0.110	0.0589	2.023	–0.647	0.041	
Unio pictorum pictorum	6	26	0.171		0.0193	1.212	0.867	1.753	28–43
				0.314	0.0738	0.546	0.249	1.801	
Rhine – North and Baltic Sea	6	17		0.052	0.0390	4.558	–5.911	1.347	32–38/39–41
Rhine – Danube	6	24		0.300	0.1432	0.583	–3.424	0.130	28–31/39–41
North and Baltic Sea – Danube	6	26		0.200	0.1007	1.000	–4.026	0.304	28–31/32–38
Danube – Adria	5	23		0.217	0.0540	0.902	0.430	3.441	28–31/42–43

b) Between neighbouring populations of different subspecies

Population groups	n(l)	n(a)	F(gt)	SE	Nm	95% CI Lower	95% CI Upper	N(pop) - 1998
Rhine - upper Rhône	6	23	0.061	0.0377	3.848	−7.556	1.317	1–2/39–41
Danube - upper Rhône	6	27	0.169	0.1104	1.229	−2.309	0.310	1–2/28–31
Rhine - Seine	5	19	0.094	0.0558	2.410	−4.443	0.747	6–10/39–41
Po - Adria	6	28	0.224	0.1217	0.866	−3.067	0.216	15–19/42–43
central Italy - lower Rhône	5	19	0.054	0.0771	4.380	−2.062	0.610	3–5/20–22
Sardinia and Corsica - lower Rhône	5	18	0.129	0.1061	1.688	−2.246	0.288	3–5/24–26
Sardinia and Corsica - central Italy	4	10	−0.089	0.0327	−3.059	−1.510	25.140	20–22/24–26

c) Between neighbouring populations

Subspecies	n(l)	n(a)	theta(st)	SE	Nm	95% CI Lower	95% CI Upper	N(pop) - 1998	Distance (km)	Barriers
U. p. mancus	6	18	0.024	0.0118	9.954	−37.563	2.840	13–14	14	None
	5	13	−0.019	0.0088	−13.548	−6.245	356.893	9–10	15	None
	6	22	0.046	0.0103	5.208	3.010	192.058	1–2	33	None
	5	17	0.157	0.0157	1.340	1.013	5.220	4–5	4	1 dam
U. p. pictorum	4	12	0.150	0.0188	1.417	1.075	5.908	42–43	35	None
	5	13	−0.002	0.0121	−119.298	−9.022	8.139	36–37	65	7 dams

Note:
Wright (1931): Nm > 0.5: gene flow as major source of genetic differentiation; good probability of gene flow across group boundaries; 0.5 > Nm > 0.0: genetic drift as major source of genetic differentiation; poor probability of gene flow across group boundaries.
Porter (1990): Nm > 1.0: gene flow is an important factor for between-group similarity; 1.0 > Nm > 0.5: gene flow is weak but eventually permits the exchange of favourable genes; 0.5 > Nm > 0.0: gene flow is almost or fully absent and unimportant relative to genetic drift; Nm < 0: panmictic populations.
[a] Number of polymorphic loci.
[b] Number of alleles in polymorphic loci.
[c] Standard error.
[d] Confidence interval.
[e] See Table 4.8.
[f] Other abbreviations explained in the text.

respective distribution areas concur best with the descriptions given by Modell (1951) and Falkner (1994) although neither author suspected the presence of *U. pictorum* (our subspecies *U. p. pictorum*) in the northern Adriatic region.

4.5.2 *Unio crassus*

For *Unio crassus* a number of subspecies with local forms of uncertain taxonomic rank are widely accepted (cf. Haas 1969; Falkner 1990a; Nesemann 1993). We were able to analyze allozyme variability of a few populations belonging to three different subspecies. The focus of our study was to compare genetic relationships with the traditional concept of subspecific division.

4.5.2.1 Molecular Variability and Systematics

On the average, genetic variability in *U. crassus* populations is lower than in polytypic *U. pictorum*. With the exception of locus Pgm-2 there are no conspicuous differences in allele frequencies. Pgm-2 alleles allow us to define two groups of populations (see Table 4.8 for localities and further references): (1) in the rivers Pfreimd, Seine and Vienne Pgm-2^{-100} is the most abundant allele (three mussels from the river Aube and seven from the river Lavaux also belong to this group; sampling localities as of *U. pictorum mancus*); (2) Pgm-2^{-102} is the dominating allele in populations from the rivers Alster, Eder, Meuse and Bracht. The first group is made up of populations from the drainage systems of the rivers Seine, Loire and Danube while the second group consists of populations from the Rhine River basin and from two affluents of the North Sea. In this way the populations belonging to the western subspecies *U. crassus nanus* are distributed between two biochemically defined groups. However, the database at hand seems too small to draw conclusions about the validity of the proposed subspecies from a genetic point of view.

4.5.2.2 Genetic Differentiation

The phenogram (Fig. 4.3) shows two main branches which separate at D = 0.086. One branch with populations from affluents of the Rhine River and the North Sea is more strongly confirmed than the other. Making the same assumptions about time of diver-

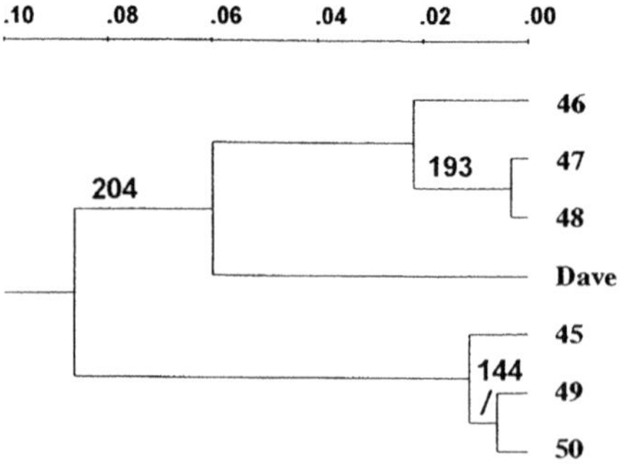

Fig. 4.3. *Unio crassus*: phenogram derived from genetic distances (Nei 1972). *Numbers and name* N(pop) – 1998 and locality name from Table 4.8; *numbers above branches of the phenogram* number of trees in which the branch occurred, out of 300 trees (consensus analysis); root population: 45 (Oberpfreimd)

gence as for Anodontinae and *Unio pictorum*, Central and Western European *Unio crassus* may have differentiated in the late Pleistocene.

4.5.3 *Unio tumidus*

Since *Unio tumidus* has a rather uniform appearance throughout its European range there have been few speculations about the recognition of subspecies (Haas 1969). At best, populations from the affluents of the Black Sea might deserve subspecific rank, but in more recent publications (e.g. Falkner 1990a) such a differentiation is no longer proposed. We studied genetic diversity for a number of populations coming from affluents of the North and Baltic Sea and from the middle part of the Danube basin.

4.5.3.1 Molecular Variability and Systematics

Unio tumidus, too, shows a relatively low genetic variability (see Table 4.8) which seems to decrease from north to south. In population 51 (Hainburg; see Table 4.8) no polymorphism could be detected at all though it is probably quite numerous and there are no indications of a population bottleneck. The sampling locality lies about 50 km downstream from Vienna (Austria) which marks the uppermost occurance of the species in the river Danube (Falkner 1990a). The small database does not allow us to draw conclusions about a genetic differentiation of the species. The low genetic variability, however, concurs with the hypothesis that the species is uniform within its Central European range.

4.5.3.2 Genetic Differentiation

Genetic distances are small, although the populations were sampled at distant localities in central Europe. In the phenogram (Fig. 4.4a) the Danube basin populations are distributed among the populations from affluents of the North and Baltic Sea. The consensus tree, however, bears evidence that the Danube populations belong together more closely (Fig. 4.4b).

4.6 Unionidae: Ambleminae

Only two species of this subfamily occur in Europe. While *Potomida littoralis* has a distribution area which comprises river systems in western Europe, in western North Africa and in some parts of the eastern mediterranean area, *Microcondylaea bonellii* is restricted to rivers and lakes in northern Italy and to parts of the eastern Adriatic region.

4.6.1 *Microcondylaea bonellii*

Still little is known about this species. Nagel (1988; see also Nagel et al. 1998) showed it to be a member of the Ambleminae. Recently, Cencetti (1994) demonstrated that *M. bonellii* is a short-term breeder with glochidia produced and released in spring

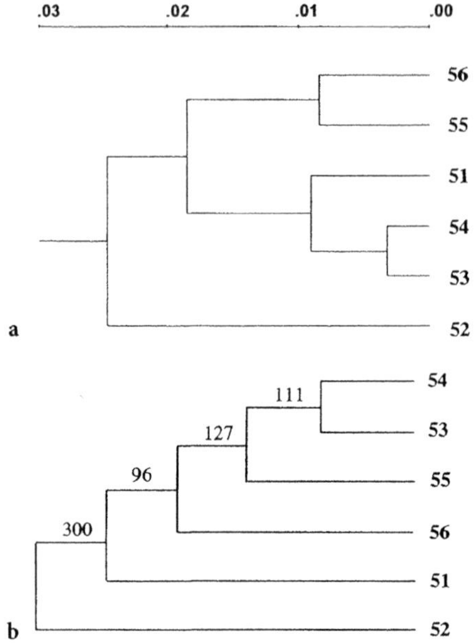

Fig. 4.4. **a** *Unio tumidus*: phenogram derived from genetic distances (Nei 1972). **b** *Unio tumidus*: consensus analysis of genetic distances. *Numbers* N(pop) – 1998 from Table 4.8; *numbers above branches*: number of trees in which the branch occurred, out of 300 trees (consensus analysis); root population: 52 (Fonyód)

and summer. Other details of the reproductive cycle are still unknown. The species' appearance is rather homogenous throughout its range so that no further subdivision of the species has been proposed in recent publications (Haas 1969; Falkner 1990a).

Twelve specimens were collected from the Torrente Versa (province of Gorizia, northeastern Italy) in 1990. Five enzymes (PGI, PGM, MDH, EST and LAP) were studied that showed a pattern of seven bands with two bands appearing in MDH and EST, respectively. We interpret these results as showing a single allele at each of the loci Pgi, Pgm, Mdh-1, Mdh-2, Est-1, Est-2 and Lap. Thus, the sampled population was monomorphic at all loci.

The electrophoretic pattern of *Microcondylaea bonellii* was compared to that of *Potomida littoralis* (population Montfrin; see Table 4.8) because of the close phylogenetic relationships between the two species (Nagel et al. 1998). Only one allele (EST-2) was found in common between the two species.

The species has disappeared from many water bodies in northern Italy and the same is probably true for the rest of its limited distribution area. Extant populations, though locally quite numerous [e.g. estimated population size more than 1000 in the Torrente Versa; sampling site as of population 42 of Italian *U. p. pictorum* (see Table 4.8)], are likely to be isolated. This may account for the absence of genetic variability of the sample through population bottlenecks or random genetic drift leading to the fixation of alleles.

4.6.2 *Potomida littoralis*

Our populations come from France and the northern Iberian peninsula. According to Haas (1969) they all belong to the nominal subspecies *littoralis*. This author rec-

Table 4.10. Allele frequencies at three polymorphic loci of *Potomida littoralis*[a]

Locus abbreviations in text; sample sizes in parentheses

Locus and allele	N(pop)[b] 1	2	3	4	5
Pgi	(19)	(17)	(18)	(18)	(33)
100	0.658	0.500	0.833	0.889	0.970
175	0.342	0.500	0.167	0.111	0.030
Mdh-1	(19)	(17)	(20)	(16)	(33)
78	0.105	0.206	0.700	0.313	0.561
100	0.895	0.794	0.300	0.687	0.439
Lap	(19)	(17)	(18)	(16)	(30)
100	0.737	1.000	0.917	0.875	0.717
112	0.263		0.083	0.125	0.283

[a] All populations were monomorphic for the same allele at the following loci: Pgm-1, Pgm-2, Mdh-2, Est, Idh-1, Idh-2, Got-1, Got-2, Me-1 and Me-2.
[b] Population number (with sampling locality and name of river in parentheses; F – France, P – Portugal).
1 – Vilanova de Cerveira (Minho, P).
2 – Montfrin (Gardon, F).
3 – Candes-St. Martin (Vienne, F).
4 – Brèhèmont (Indre, F).
5 – Clérey (Seine, F).

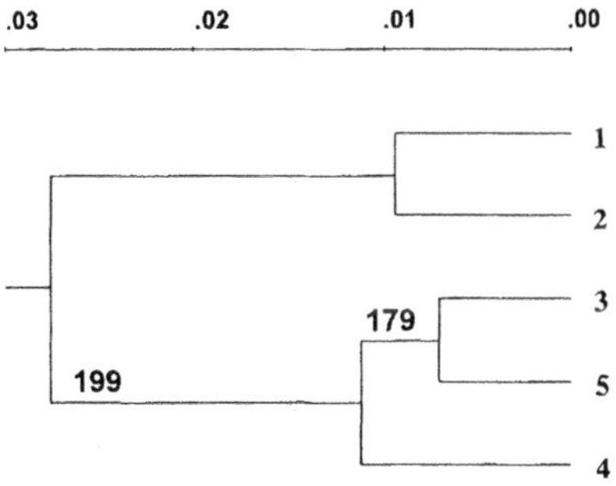

Fig. 4.5. *Potomida littoralis*: phenogram derived from genetic distances (Nei 1972). *Numbers* N(pop) from Table 4.10; *numbers above branches of the phenogram* number of trees in which the branch occurred, out of 300 trees (consensus analysis); root population: 1 (Vilanova de Cerveira)

ognized seven other subspecies. From his descriptions it appears that they are all quite similar in their shell features.

Table 4.10 gives the allele frequencies of three polymorphic loci. The various measures of genetic variability of *P. littoralis* are in good agreement with data from *Unio* species and from *Anodonta anatina*. (Table 4.8; Nagel et al. 1996). Genetic distances among populations are small. They are within the range of values found for *Unio pictorum* populations coming from the same drainage basin. The dendrogram (Fig. 4.5) has two branches which separate at D = 0.028. The consensus analysis confirms these branches quite well, but since neighbouring populations are not nearest in the dendrogram (Vienne, Indre) it can be doubted that the relationships are more than accidental. First, because the genetic distances involved are small, second, because populations are not strictly arranged according to geographical distances or stream relationships.

4.7 Summary

Electrophoretic studies of enzyme variability have contributed new aspects to understanding the relationships among populations of European freshwater mussels (families Margaritiferidae and Unionidae). In particular, molecular variability and allozyme frequencies were used to estimate genetic differentiation and gene flow. The results bear consequences for the systematics of unionid taxa. All populations of *Margaritifera margaritifera* showed a very low genetic variability, probably as a consequence of the species' reproductive strategy which includes facultative hermaphroditism. The European Anodontinae comprise two genera with three autochthonous and one introduced species. Two groups of Italian *Anodonta* could not yet be assigned to a definite taxonomic rank. Three species of freshwater mussels belong to the Unioninae. *Unio tumidus* seems to be only distantly related to *U. crassus* and *U. pictorum*. Genetic variability is small in both *U. tumidus* and *U. crassus*. The database for these species is as yet too small to draw more detailed conclusions. *U. pictorum* is a polytypic species and comprises populations from the mediterranean area known as *U. mancus* or *U. elongatulus*. The subfamily Ambleminae is represented by two species, both of which could be sampled and analyzed in small numbers only. One population of *Microcondylaea bonellii* showed no genetic variability at all. This might be an effect of isolation of extant populations of this rare species. *Potomida littoralis* from France and Portugal was as variable as *Unio* spp. but genetic distances among populations are very low.

Acknowledgements. We are most grateful to the following persons for their contributions to this work: E. Leinardi and E. Greppi collected additional samples of Italian Anodonta and E. Greppi analyzed them. G. Celebrano ran the laboratory and helped with the enzyme analysis. M. Young made literature available to us. H. Weigel (computing center of the University of Freiburg) helped run the computer programs, while R. Senger and B. Schlüter provided additional computer facilities. J. Plant revised the English and the comments of an anonymous reviewer further improved the text.

References

Agrell I (1948) The shell morphology of some Swedish unionides as affected by ecological conditions. Ark Zool 41 A:1–30

Baagøe P, Hvilsom MM, Pedersen BV (1985) The species rank of *Anodonta anatina* (L.) and *A. cygnea* (L.), with remarks on *Pseudanodonta complanata* (Rossmässler) (Bivalvia, Unionidae). Vidensk Medd Dan Naturhist Foren 146:75–83

Badino G (1980) Genetic variability in freshwater bivalves. 2. PGI polymorphism and its ecological significance in *Unio elongatulus* Pfeiffer (Bivalvia Eulamellibranchia). Monit Zool Ital (NS) 14:199–210

Badino G, Celebrano G (1978) Leucine aminopeptidase variation in two species of the freshwater clam *Unio* (Bivalvia Eulamellibranchia). Monit Zool Ital (NS) 12:219–228

Badino G, Celebrano G (1981) Variabilità genetica nei Bivalvi d'acqua dolce. 3. – Polimorfismi enzimatici e loro impiego nella definizione sistematica delle Unio nord-italiane. Boll Mus Zool Univ Torino 3:25–48
Badino G, Sella G, Celebrano G (1984) Sistemi gene-enzima nello studio della differenziazione genetica di popolazioni naturali di molluschi a diversa vagilità. Lavori SIM 22:241–258 (Atti I. Congr Soc Ital Malacol)
Badino G, Celebrano G, Nagel K-O (1991) *Unio elongatulus* and *Unio pictorum* (Bivalvia: Unionidae): molecular genetics and relationships of Italian and central European populations. Boll Mus Reg Sci Nat Torino 9:261–274
Bauer G (1987) Reproductive strategy of the freshwater pearl mussel *Margaritifera margaritifera*. J Anim Ecol 56:691–704
Burla H (1971) Gerichtete Ortsveränderungen bei Muscheln der Gattung *Anodonta* im Zürichsee. Vjschr Naturforsch Ges Zürich 116:181–194
Burla H (1972) Die Abundanz von *Anodonta*, *Unio pictorum*, *Viviparus ater*, *Lymnaea auricularia* und *Lymnaea ovata* im Zürichsee, in Abhängigkeit von der Wassertiefe und zu verschiedenen Jahreszeiten. Vjschr Naturforsch Ges Zürich 117:129–151
Cencetti E (1994) Ecologia e ciclo riproduttivo di *Microcondylaea compressa* Menke, 1828 (Bivalvia: Unionidae), una specie poco conosciuta. Thesis, University of Siena (Italy), 47 pp (unpublished)
Chesney HCG, Oliver PG, Davis GM (1993) *Margaritifera durrovensis* Phillips, 1928: taxonomic status, ecology and conservation. J Conchol Lond 34:267–299
Davis GM (1983) Relative roles of molecular genetics, anatomy, morphometrics and ecology in assessing relationships among North American Unionidae (Bivalvia). In: Oxford GS, Rollinson D (eds) Protein polymorphism: adaptive and taxonomic significance. Academic Press, London, pp 193–222
Davis GM (1994) Molecular genetics and taxonomic discrimination. Nautilus Suppl 2:3–23
Davis GM, Fuller SLH (1981) Genetic relationships among Recent Unionacea (Bivalvia) of North America. Malacologia 20:217–253
Davis GM, Heard WH, Fuller SLH, Hesterman C (1981) Molecular genetics and speciation in Elliptio and its relationships to other taxa of North American Unionidae (Bivalvia). Biol J Linn Soc 15:131–150
Falkner G (1990a) Binnenmollusken. In: Steinbach G (ed) Weichtiere. Europäische Meeres- und Binnenmollusken (Steinbachs Naturführer 10). Mosaik Verlag, München, pp 112–280
Falkner G (1990b [1991]) Vorschlag für eine Neufassung der Roten Liste der in Bayern vorkommenden Mollusken (Weichtiere). Mit einem revidierten systematischen Verzeichnis der in Bayern nachgewiesenen Molluskenarten. Schriftenreihe Bayerisches Landesamt für Umweltschutz, Heft 97 [= Beiträge zum Artenschutz 10]:61–112
Falkner G (1994) Systematik vorderorientalischer Najaden als Vorstudie zur Bearbeitung archäologischer Funde. Forschungen und Berichte zur Vor- und Frühgeschichte in Baden-Württemberg. In: Kokabi M, Wahl J (eds) Beiträge zur Archäozoologie und Prähistorischen Anthropolgie, vol 53. K Theiss, Stuttgart, pp 135–162
Felsenstein J (1993) PHYLIP (Phylogeny inference package), version 3.5c. Distributed by the author. Department of Genetics, University of Washington, Seattle
Haas F (1969) Superfamilia Unionacea. Tierreich 88:I–X, 1–663
Heard WH, Guckert RH (1970) A re-evaluation of the Recent Unionacea (Pelecypoda) of North America. Malacologia 10:333–355
Hinkley D (1982) Jackknife. In: Kotz S, Johnson NL (eds) Encyclopedia of statistical sciences, vol 4. John Wiley, New York, pp 280–287
Hoeh WR (1990) Phylogenetic relationships among eastern North American *Anodonta* (Bivalvia: Unionidae). Malacol Rev 23:63–82
Hoeh WR, Gordon ME (1996) Criteria for the determination of taxonomic boundaries in freshwater unionoids (Bivalvia: Unionoida): comments on Stiven and Alderman (1992). Malacologia 38:223–227

Hvilsom MM, Pedersen BV (1988) The species rank of *Unio pictorum* (L.), *U. tumidus* Philipsson, and *U. crassus* Philipsson (Bivalvia, Unionidae). Vidensk Medd Dan Naturhist Foren 147:37–46

Kat PW (1983) Morphologic divergence, genetics, and speciation among *Lampsilis* (Bivalvia: Unionidae). J Molluscan Stud 49:133–145

Kodolova OP, Logvinenko BM (1973) Comparison of different populations of bivalves *Unio pictorum* and *U. tumidus* (Unionidae) by systems of myogens and shell morphology. Zool Zh 52:988–999 (in Russian)

Levene H (1949) On a matching problem arising in genetics. Ann Math Stat 20:91–94

Logvinenko BM, Kodolova OP (1979) Comparison of myogen systems of some mollusk species of the superfamily Unionacea. Vestnik Moskovskogo Universiteta. Biologiya 34:65–69 (original in Russian, cited from an English translation)

Miller AC, Payne BS (1988) The need for quantitative sampling to characterize size demography and density of freshwater mussel communities. Am Malacol Bull 6:49–54

Modell H (1951) Die Najaden Vorderasiens. Rev Fac Sci Univ Istanbul B 16:351–366

Nagel K-O (1988) Anatomische, morphologische und biochemische Untersuchungen zur Taxonomie und Systematik der europäischen Unionacea (Mollusca: Bivalvia). PhD Thesis, University of Kassel, pp 1–100

Nagel K-O, Badino G, Alessandria B (1996) Population genetics of European Anodontinae (Mollusca: Bivalvia: Unionidae). J Molluscan Stud 62:343–357

Nagel K-O, Badino G, Celebrano G (1998) Systematics of European naiades (Bivalvia: Margaritiferidae and Unionidae): a review and some new aspects. Malacol Rev Suppl 7 (Bivalvia I):83–104

Nei M (1971) Interspecific gene differences and evolutionary time estimated from electrophoretic data on protein identity. Am Nat 105:385–398

Nei M (1972) Genetic distance between populations. Am Nat 106:283–292

Nei M (1978) Estimation of average heterozygosity and genetic distance from a small number of individuals. Genetics 89:583–590

Nesemann H (1993) Zoogeographie und Taxonomie der Muschel-Gattungen *Unio* Philipsson 1788, *Pseudanodonta* Bourguignat 1877 und *Pseudunio* Haas 1910 im oberen und mittleren Donausystem (Bivalvia: Unionidae, Margaritiferidae) (mit Beschreibung von *Unio pictorum tisianus* n. ssp.). Nachrichtenblatt Ersten Vorarlberger Malakol Ges 1:20–40

Nesemann H, Nagel K-O (1989) Die Flußmuscheln (Bivalvia: Unionacea) im Einzugsgebiet der Loire (Zentralfrankreich) – eine erste Bestandserfassung. Mitt Dtsch Malakozool Ges 44/45:1–15

Ortmann AE (1920) Correlation of shape and station in fresh-water mussels (naiades). Proc Am Philos Soc 59:269–312 (cited from Watters 1994)

Porter AH (1990) Testing nominal species boundaries using gene flow statistics: the taxonomy of two hybridizing admiral butterflies (*Limenitis*: Nymphalidae). Syst Zool 39:131–147

Purser GJ (1985) Factors affecting the distribution of the freshwater pearl mussel, *Margaritifera margaritifera* (L.) in Britain. Thesis, University of Aberdeen, Scotland (unpublished)

Rensch B (1929) Das Prinzip geographischer Rassenkreise und das Problem der Artbildung. Bornträger, Berlin

Rohlf FJ (1988) NTSYS-pc, Numerical taxonomy and multivariate analysis system (Manual). Exeter, New York

Swofford DL, Selander RB (1989) BIOSYS-1, Release 1.7 (Manual). Ill Nat Hist Surv v + 41 pp

Watters GT (1994) Form and function of unionoidean shell sculpture and shape (Bivalvia). Am Malacol Bull 11:1–20

Weir BS (1990) Intraspecific differentiation. In: Hillis D, Moritz C (eds) Molecular systematics. Sinauer, Sunderland, pp 373–410

Weir BS, Cockerham CC (1984) Estimating F-statistics for the analysis of population structure. Evolution 38:1358–1370

Wright S (1931) Evolution in Mendelian populations. Genetics 16:97–159

Part II
Life-History Strategies

5 Life-History Variation on Different Taxonomic Levels of Naiads

Gerhard Bauer

5.1 Introduction

A conservative groundplan (e.g. two pairs of gills, two adductor muscles etc.), the freshwater habitat, a semi-infaunal way of life, the mode of filter feeding and a larva which has to pass through a parasitic phase are features which seem to have never been altered during evolution (except for some few cases where larvae develop directly; see below). This uniformity (Chap. 1, this Vol.) seems to be a contradiction to the nearly global distribution of naiads (Chaps. 1, 2, 3, 4, 6, this Vol.) and to their high species numbers (Chaps. 12, 15, this Vol.), which, as Watters (this Vol.) points out, is not only a recent phenomenon, but occurred similarly during the Mesozoic. Naiads are able to adapt to the conditions of nearly all freshwater habitats. They occur in arctic rivers extremely poor in nutrients (Chaps. 19, 20, this Vol.), in highly eutrophic streams and lakes (Chaps. 17, 18, this Vol.) and even in temporary floodplain billabongs and episodic streams in the tropics (Chap. 2, this Vol.). Occupying so many different niches requires adaptability and thus variability of life-history traits.

This chapter will demonstrate this variability on various taxonomic levels, and it will become evident that the degree of variation shown by some traits is, in fact, enormous. A crude overview is presented in Fig. 5.1. It is mainly based on published information. More detailed data are shown only for some cases where published information is rare or lacking.

5.2 Life Span, Size and Shell Morphology

Already on the species level these life-history traits are variable to such a degree (Fig. 5.1) that a large number of species had been described erraneously (reviewed, e.g. in Chap. 4, this Vol.). For example, the maximum life span attained by Swedish freshwater pearl mussels (*Margaritifera margaritifera*) is reported to be more than 200 years (Chap. 21, this Vol.), whereas in other populations they live for hardly 40 years and the maximum size varies between 80 and 145 mm shell length (Bauer 1992). In German *Unio crassus* populations, the maximum life span is between 8 (Chap. 7, this Vol.) and nearly 40 years (Lechner 1997) and maximum size between ca. 50 mm (Chap. 7, this Vol.) and ca. 80 mm (Engel 1990). The growth process may be influenced by a number of factors, like temperature (Chaps. 7, 13, this Vol.), water turbulence (Bailey and Green 1987; Chap. 2, this Vol.) or food availability (Chap. 18, this Vol.). Also water acidity may affect shell growth, as demonstrated by Machado et al. (1988) for *Anodonta cygnea*. Similarly, in *U. crassus* there is a strong negative corre-

Ecological Studies, Vol. 145
G. Bauer and K. Wächtler (eds.) Ecology and Evolution of the Freshwater Mussels Unionoida

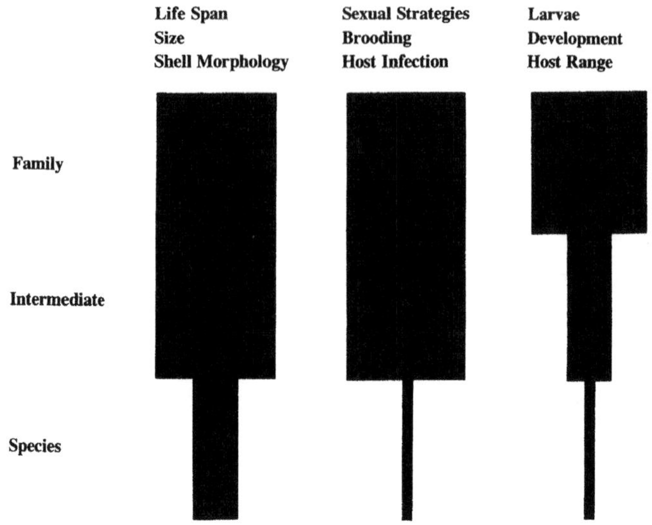

Fig. 5.1. Degree of variation of life-history traits on different (*Species, Intermediate, Family*) taxonomic levels. Systematics according to Chapters 1, 14 (this Vol.). The *thickness of the bars* indicates low, (if any), moderate and high degree of variation

lation between (standardized) shell weight and pH, such that specimens living in limestone areas are thin-shelled compared to those from primary rocks (Hochwald 1997).

This intraspecific variability contributes to tremendous variation in age, size and shell morphology above the species level. The largest (native) naiad in Europe is *Margaritifera auricularia*, with a maximum shell length of ca. 200 mm (Chap. 8, this Vol.) and in America it is *Megalonaias nervosa*, with nearly 300 mm as compared with *Toxolasma lividus*, reaching not more than 25 mm shell length (Watters 1995). Within the Margaritiferidae the freshwater pearl mussel is among the longest-lived invertebrates on earth, whereas *Cumberlandia monodonta*, another margaritiferid, is comparatively short-lived, attaining a life span of hardly more than 20 years (pers. observ.). Among the Unionidae *Anodonta anatina* in the Rhine river near Freiburg does not live longer than 4 years (pers. observ.), whereas for *Unio crassus* in Estonia a life span of more than 70 years is reported (Timm and Mutvei 1993).

Different shell morphologies evolved dependent on life history and environment. In long-lived species of soft waters the periostracum is very thick and many conchiolin layers are deposited in the shell in order to prevent dissolution (Fig. 5.2). *M. margaritifera* allocates more than 30% of its organic substance to the shells (Bauer 1989), such that they last for decades. A characteristic pattern can be observed along the length of a river. Headwater taxa living in highly turbulent water are frequently characterized by unsculptured, streamlined shells. Sculptured shells are found among large river taxa (the sculpture is an aid to burrowing), and taxa living in soft substrata usually have thin, alate and complanate shells (Stone et al. 1982; Watters 1994).

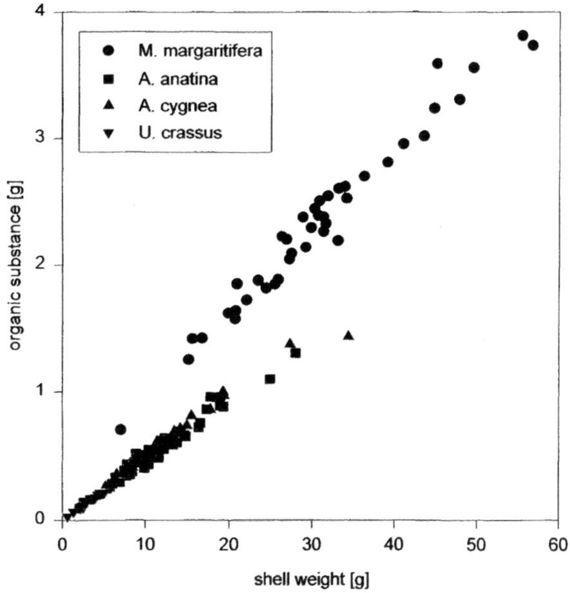

Fig. 5.2. Amount of organic substance invested in the shell by the long-lived *M. margaritifera*, compared to some other European naiad species. The dried shells were weighed, then heated to 450 °C for 2 h and then weighed again, allowing an estimate of their organic content. (Joergensen 1976)

5.3 Sexual Strategies, Brooding and Host Infection

Interspecific comparisons exhibit a wide variety of different sexual strategies even within genera. For example, *Anodonta anatina* was found to be of separate sex with very few hermaphrodites in some populations (perhaps due to infection of the gonads with trematodes, Kat 1983), whereas *A. cygnea* populations consist of only a few females and a large proportion of hermaphrodites (Franke 1993; Teutsch 1997). Similar variations in American *Anodonta* species are described by Heard (1975). Downing et al. (1993) found 80% hermaphrodites in a population of *Elliptio complanata*. *M. margaritifera* is of separate sex in dense populations, whereas the females become hermaphroditic (with proportionally much more ovarian than testicular tissue) in sparse populations such that fertilization of eggs is ensured (Bauer 1987a). The same mechanism might exist in *M. auricularia* since Araujo (this Vol.) found only hermaphrodites in a small population. Such "cryptic microhermaphrodites" are also described for *M. falcata* and some hyriid species, whereas others are of separate sex (Chap. 2, this Vol.). *Unio crassus*, however, is strictly dioecious, with a male-female relationship of 1:1 (Engel 1990; Hochwald 1997; Lechner 1997; Lang 1998), like *Alasmidonta heterodon* (Michaelson and Neves 1995), *Cyclonaias tuberculata* (Haggerty et al. 1994) and *Mutela bourguignati* (Herrmann 1995) and presumably most Lampsilids (Davis and Fuller 1981). These strategies seem to be species-specific, i.e. there is hardly any evidence that there are different sexual strategies within a species (Fig. 5.1). Strictly dioecious species like *U. crassus* or *A. anatina* have always been found to be dioecious, hermaphroditic species have always found to be hermaphroditic and species with a plastic strategy have always been found to be plastic.

Sexual dimorphism seems to be absent in hyriids (Chap. 2, this Vol.) and common in dioecious unionids like *A. anatina* (Franke 1993; Teutsch 1997), but apparently it is often difficult to detect. For example, in some populations of *U. crassus* females are thicker than males of the same length allowing more space for the marsupia, whereas in other populations of *U. crassus* no difference between males and females could be proven (Hochwald 1988). Sexual dimorphism is common and conspicuous among the Lampsilidae (Walker 1917).

Brooding of larvae occurs exclusively in gill chambers. However, the number of gills serving as marsupia and their degree of development is highly variable above the species level. This was a main character used for systematics by many workers (e.g. Heard and Guckert 1971). Morphologically simple are the marsupia in the Margaritiferidae, where all four gills incubate eggs (Chap. 3, this Vol.), a character termed tetragenous. Among the Mutelidae only the inner gills are marsupial (Chap. 14, this Vol.). Marsupia confined to discrete areas of the inner gills are found among the Hyriidae (Chap. 2, this Vol.). Within the Unionidae the Anodontinae show outer gills which are highly developed for marsupial purpose (Lefevre and Curtis 1911; Davis and Fuller 1981; Tankersley and Dimock 1993), whereas among the Ambleminae two or four gills may serve as marsupia (Davis and Fuller 1981). Highly developed marsupia which are confined to discrete areas of the outer gills are found among the Lampsilinae (Davis and Fuller 1981; Richard et al. 1991). Thus, this trait shows high variability at high and intermediate (here particularly within the Unionidae) level, but is canalized within species.

Short-term breeders, like margaritiferids or *Unio*, keep eggs and glochidia for only a few weeks within the marsupia, whereas long-term breeders, like most *Anodonta* species, retain them for many months and supply them with nutrients (Wood 1974).

Different strategies to improve host infection have been developed above the species level: (1) High fertility due to small glochidia which are just released into the water. This strategy, associated with very high glochidial mortality (Chap. 11, this Vol.) is common in lotic habitats, for example among the Margaritiferidae and to a lesser extent (lower fertility) among *Unio*. (2) Synchronization of the timing of reproduction to correspond to availability of the host (Davenport and Warmuth 1965). (3) Glochidia are released in small numbers when the presence of host fish is encountered olfactorically (Teutsch 1997; Trübsbach 1998; Welte 1999; Chap. 13, this Vol.). So far, this strategy is known only from some Anodontinae living frequently in lentic habitats. (4) Attracting fish by releasing glochidia within conglutinates which mimic fish food items like worms and leeches (Kat 1984) or dipteran larvae (Hartfield and Hartfield 1996). This strategy occurs among Unionidae and Hyriidae (Chap. 2, this Vol.). Among the Lampsilinae some species attract fish with mantle flaps resembling prey fish (Kat 1984).

5.4 Larvae, Larval Development and Host Range

It is only at the highest taxonomic level that we observe different types of larvae, namely the lasidium in the Mycetopodidae, the haustorium in the Mutelidae and the glochidium in the remaining groups (Hyriidae, Margaritiferidae, Unionidae; Chaps.

1, 6, this Vol.). The tremendous differences in the developmental mode of these larval types are described in Wächtler et al. (this Vol.).

For comparisons at lower levels only data on glochidia are available.

At intermediate levels the size of glochidia may vary moderately, for example from 0.07 mm shell length in *M. margaritifera* to 0.134 mm in *M. auricularia* (Chap. 8, this Vol.) or from 0.23 mm in *Anodonta imbecellis* to 0.41 mm in *A. grandis* (Fuller 1974).

Large glochidia, such as in the Anodontinae, frequently attach to the exterior surfaces of their hosts (i.e. fins and epidermis). They are adapted for gripping strength by a hook at each shell margin, a large adductor muscle and depressed shells. Hookless glochidia usually are gill parasites. According to the position of the adductor muscle, they have long resistance arms that are adapted for a large area of sweep (Hoggarth and Gaunt 1988).

Since there is a tradeoff between size and number of glochidia, fertility is highest in those species with the smallest glochidia (Bauer 1994); however, large glochidia are morphologically more accomplished (Harms 1908, 1909).

While the carbon content of the larvae generally increases during parasitism (Wächtler et al., this Vol.), shell growth is rare, seems to be associated with low initial size (i.e. small glochidia) and evolved independently in different taxonomic groups. *M. margaritifera* glochidia grow from 0.07 to 0.4–0.5 mm shell length and *M. auricularia* from 0.14 to 0.19 mm (Buddensiek 1991; Bauer 1994; Chap. 8, this Vol.), *Pseudodon omiensis* (a unionid) glochidia grow from 0.06 to 0.18 mm (Kondo 1987) and the Lampsilinae *Proptera laevissima* and *Plagiola donaciformis* from 0.06 or 0.1 mm, respectively, to 0.3–0.35 mm (Surber 1912).

The available data on the host range suggest that it is very narrow in the Margaritiferidae (Table 5.1), comprising only a few closely related fish species (Chap. 8, this Vol.; Bauer 1997). Host specificity is lower in most Unionidae (Bauer 1994), where many *Anodonta* species are able to utilize a very broad host range (Trdan and Hoeh 1982; Franke 1993) comprising even exotic fish species like *Danio rerio* (Teutsch

Table 5.1. Example of high (*Margaritifera*), intermediate (*Unio crassus*) and low host specifity (*Anodonta anatina*) among European naiads. In particular, the data for *A. anatina* must be considered minimum values since many fish species have not yet been tested for their suitability. This mussel species is even able to metamorphose on tadpoles and *Axolotl* (Pekkarinen and Hanstén 1998; Terberger 1999). (Data from Nagel 1997; Bauer 1987b; Maaß 1987; Hochwald 1988; Engel 1990; Franke 1993; Niemeyer 1993; Lechner 1997; Teutsch 1997; Pekkarinen and Hanstén 1998; chap. 8, this Vol.)

	No. of fish species in the host range	No. of fish families
M. margaritifera	4	1
M. auricularia	?	1
U. crassus	12	5
A. anatina	15	7

1997), or *Lepomis gibbosus* (Terberger 1999). Metamorphosis on amphibians is reported from some *Anodonta* (Pekkarinen and Hansten 1998; Terberger 1999) and hyriid species (Chap. 2, this Vol.). The host range of the Unioninae seems to be less broad (Table 5.1; Nagel 1997).

On the intraspecific level there is hardly any evidence for variation in these life-history traits (for example see Bauer 1997).

Meanwhile there can be no doubt that there are some species or some populations of a species which have secondarily evolved away from the parasitic mode of reproduction, i.e. glochidial development occurs without parasitism within the marsupia of the adult. These cases are limited to a few (less than ten) North American long-term brooders and a number of South American Hyriidae (Chap. 6, this Vol.). Under what circumstances this mode of development occurs and how glochidia are supplied with food remains completely unknown. Howard (1915) and Wood (1974) postulate that components of the mucus in which the glochidia are embedded can serve as nourishment.

5.5 Conclusions

This description of life-history variation, though very crude, renders it possible to identify major lines along which evolution proceeds among naiads.

Some traits, like the general soft tissue ground plan, the mode of filter feeding etc. (see sect. 5.1) are essentially identical in all Unionoida. Apparently they are inherited from the common ancestor. Such traits, which are fixed up to the highest taxonomic level, offer no genetic variation on which evolution could work, and thus do not represent traits involved in naiad evolution.

Another category are traits which are fixed at low (species), moderately variable at intermediate and highly variable at high taxonomic level. Perhaps the most striking example among naiads is the different types of larvae, namely glochidium, lasidium and haustorium (Chap. 6, this Vol.). As shown in Chapter 5.4, glochidial size, size of the host range and mode of development on the host are moderately variable at intermediate levels. These are traits which are closely related to the host relationship. They represent important lines of evolution and show that a variety of different parasitic strategies exists. Apparently, these traits were optimized during evolution and then fixed, such that they are now canalized within the species.

The reproductive traits (sexual strategy, brooding, host infection) are fixed on the species level, but are highly variable already at intermediate and higher levels. They represent important lines of evolution adopted during Recent radiation. Examples are the different sexual strategies of *A. anatina* and *A. cygnea* (Sect. 5.3), the different modes of brooding within the subfamily Ambleminae (Davis and Fuller 1981; Chap. 14, this Vol.), or the different strategies to improve host infection among the Lampsilinae (Kat 1984). Apparently, these reproductive traits are a main playing ground for naiad evolution.

The interpretation of variability already on the species level of naiads is difficult. This variability might be due to either microevolution, or simply to phenotypic plasticity, or both. Only measuring heritability of plasticity could solve this problem. However, due to the long life of most naiads, such experiments are diffcult and to my knowledge have not yet been conducted. The results presented so far on intraspecific variability of age and size suggest that at least a great deal of this variability must be attributed to passive phenotypic plasticity (Bauer 1992; Chaps. 7, 13, this Vol.).

5.6 Summary

Naiad life-history variation is analyzed on low (species), intermediate and high (family) taxonomic level.

The growth process, i.e. life span, maximum size and shell morphology is variable on all levels. A great deal of this variability in age and size must be attributed to phenotypic plasticity.

Reproductive traits appear to be canalized within species, but are highly variable above, sometimes even within, genera. Apparently, they are a main goal of naiad evolution.

Host-related traits are canalized within species, moderately variable at intermediate and highly variable at high level. They represent important lines of early naiad evolution during which they were optimized and then fixed.

References

Bailey RC, Green RH (1987) Within-basin variation in the shell morphology and growth rate of a freshwater mussel. Can J Zool 66:1704–1708

Bauer G (1987a) Reproductive strategy of the freshwater pearl mussel *Margaritifera margaritifera*. J Anim Ecol 56:691–704

Bauer G (1987b) The parasitic stage of the freshwater pearl mussel. III. Host relationships. Arch Hydrobiol Suppl 76/4:413–423

Bauer G (1989) Die bionomische Strategie der Flußperlmuschel. Biol Unserer Zeit 19,3:69–75

Bauer G (1992) Variation in the life span and size of the freshwater pearl mussel. J Anim Ecol 61:425–436

Bauer G (1994) The adaptive value of offspring size among freshwater mussels (Bivalvia; Unionoidea). J Anim Ecol 63/4:933–944

Bauer G (1997) Host relationships at reversed generation times. Margaritifera and salmonids. In: Dettner K, Bauer G, Völkl W (eds) Vertical food web interactions: evolutionary patterns and driving forces. Ecological Studies, vol 130, Springer, Berlin Heidelberg New York, pp 69–79

Buddensiek V (1991) Untersuchungen zu den Aufwuchsbedingungen der Flußperlmuschel *Margaritifera margaritifera* LINNAEUS (Bivalvia) in der frühen postparasitären Phase. PhD Thesis, University Hannover, 288 pp

Davenport D, Warmuth M (1965) Notes on the relationship between the freshwater mussel *Anodonta implicata* Say and the alewife *Pomolobus pseudoharengus* (Wilson). Limnoe Oceanogr 10:R74–R78

Davis GM, Fuller SLH (1981) Genetic relationships among Recent Unionacea (Bivalvia) of North America. Malacologia 20(2):217–253

Downing J, Rochon Y, Pérusse M, Harvey H (1993) Spatial aggregation, body size, and reproductive success in the freshwater mussel *Elliptio complanata*. J N Am Benthol Soc 12(2): 148–156

Engel H (1990) Untersuchungen zur Autökologie von *Unio crassus* (Philipsson) in Norddeutschland. PhD Thesis, University Hannover, 213 pp

Franke G (1993) Zur Populationsökologie und Geschlechtsbiologie der Teichmuscheln *A. anatina* L. und *A. cygnea* L. MS Thesis, University Bayreuth, 84 pp

Fuller SLH (1974) Clams and mussels. In: Hart CW, Fuller SLH (eds) Pollution ecology of freshwater invertebrates. Academic Press, New York, pp 215–273

Haggerty TM, Garner JT, Patterson GH, Jones LC (1995) A quantitative assessment of the reproductive biology of *Cyclonaias tuberculata* (Bivalvia: Unionidae). Can J Zool 73:83–88

Harms W (1908) Die postembryonale Entwicklung von *Unio pictorum* und *Unio tumidus*. Zool Anz 32:693–703

Harms W (1909) Postembryonale Entwicklungsgeschichte der Unioniden. Zool Jahrb 28: 325–386

Hartfield P, Hartfield E (1996) Observations on the conglutinates of *Ptychobranchus greeni*. Am Midl Nat 135:370–375

Heard WH (1975) Sexuality and other aspects of reproduction in *Anodonta* (Pelecypoda: Unionidae). Malacologia 15(1):81–103

Heard WH, Guckert RH (1971) A re-evaluation of the recent Unionacea of North America. Malacologia 10:333–353

Herrmann F (1995) Untersuchungen zum Lebenszyklus der afrikanischen Süßwassermuschel *Mutela bourguignati* Bourguignat 1885. PhD Thesis, Tierärztliche Hochschule Hannover, 96 pp

Hochwald S (1997) Das Beziehungsgefüge innerhalb der Größenwachstums- und Fortpflanzungsparameter bayerischer Bachmuschelpopulationen (*Unio crassus* Phil.1788) und dessen Abhängigkeit von Umweltfaktoren. PhD Thesis, University Bayreuth, 172 pp

Hochwald S (1988) Untersuchungen zur Populationsökologie und Fortpflanzungsbiologie der Bachmuschel *Unio crassus* (Phil.) 1788. MS Thesis, University Bayreuth, 86 pp

Hoggarth MA, Gaunt AS (1988) Mechanics of glochidial attachment (Mollusca: Bivalvia: Unionidae). J Morphoe 198:71–81

Howard AD (1915) Some exceptional cases of breeding among the Unionidae. Nautilus 29:4–11

Joergensen CB (1976) Growth efficiencies and factors controlling size in some mytilid bivalves. Ophelia 15:175–192

Kat PW (1983) Sexual selection and simultaneous hermaphroditism among the Unionidae (Bivalvia: Mollusca). J Zool Lond 201:395–416

Kat PW (1984) Parasitism and the Unionacea (Bivalvia). Biol Rev 59:189–207

Kondo T (1987) Breeding seasons of seven species of unionid mussels (Bivalvia: Unionidae) in a small creek. Venus 46,4:227–236

Lang K (1998) Die Bachmuschel *Unio crassus* (Philipsson 1788) im Niederbach: Bestandssituation, Fortpflanzungsbiologie und Wanderverhalten. MS Thesis, University Freiburg, 76 pp

Lechner S (1997) Populationsökologische Untersuchungen an der Bachmuschel *Unio crassus* (Philipsson 1788) im Einzugsgebiet der hessischen Kinzig. MS Thesis, University Freiburg, 91 pp

Lefevre G, Curtis WC (1911) Metamorphosis without parasitism in the Unionidae. Science 33,857:863–865

Maaß S (1987) Untersuchungen zur Fortpflanzungsbiologie einheimischer Süßwassermuscheln der Gattung Unio. PhD Thesis, Tierärztliche Hochschule Hannover, 107 pp

Machado J, Coimbra J, Sà C, Cardoso I (1988) Shell thickening in *Anodonta cygnea* by induced acidosis. Comp Biochem Physiol 91A,4:645–654

Michaelson DL, Neves RJ (1995) Life history and habitat of the endangered dwarf wedgemussel *Alasmidonta heterodon* (Bivalvia: Unionidae). J N Am Benthol Soc 14(2):324–340

Nagel K-O (1997) Flußmuscheln: Bestandsentwicklung und Schutzmöglichkeiten in Hessen. Mitteilungen aus dem Auenzentrum Hessen 1:61–72

Niemeyer B (1993) Vergleichende Untersuchungen zur bionomischen Strategie der Teichmuschelarten *Anodonta cygnea* L. und *Anodonta anatina* L. PhD Thesis, University Hannover, 143 pp

Pekkarinen M, Hanstén C (1998) Success of metamorphosis of the lake mussel (*Anodonta anatina*) Glochidia in different hosts and *in vitro*. Proc 50th Annu Meet Scan Soc Electron Micross 115–116

Richard PE, Dietz TH, Silverman H (1991) Structure of the gill during reproduction in the unionids *Anodonta grandis*, *Ligumia subrostrata*, and *Carunculina parva texasensis*. Can J Zool 69:1744–1754

Stone NM, Earll R, Hodgson A, Mather JG, Parker J, Woodward FR (1982) The distributions of three sympatric mussel species (Bivalvia: Unionidae) in Budworth Mere, Cheshire. J Molluscan Stud 48:266–274
Surber T (1912) Identification of the glochidia of freshwater mussels. Bur Fish Doc 771:3–13
Tankersley RA, Dimock RV (1993) The effect of larval brooding on the filtration rate and particle-retention efficiency of *Pyganodon cataracta* (Bivalvia: Unionidae) Can J Zool 71:1934–1944
Terberger J (1999) Wirtsbeziehungen der Teichmuschel *Anodonta*. Ms Thesis, University Freiburg
Teutsch S (1997) Populationsbiologische Untersuchung an *Anodonta*: Fortpflanzungsstrategien und Charakterisierung von Teilpopulationen innerhalb eines Stillgewässers. MS Thesis, University Freiburg, 82 pp
Timm H, Mutvei H (1993) Shell growth of the freshwater unionid *Unio crassus* from Estonian rivers. Proc Estonian Acad Sci Biol 42:55–67
Trdan RJ, Hoeh WR (1982) Eurytopic host use by two congeneric species of freshwater mussels. Am Midl Nat 108:381–388
Trübsbach J (1998) Fortpflanzungsbiologie und Mortalität der Gemeinen Teichmuschel *Anodonta anatina* (LINNAEUS 1758). MS Thesis, University Freiburg, 58 pp
Walker B (1917) The method of evolution in the unionidae. Occ Pap Mus Zool Univ Mich 45:1–10
Watters GT (1994) Form and function of unionoidean shell sculpture and shape (Bivalvia). Am Malacol Bull 11(1):1–20
Watters GT (1995) A guide to the freshwater mussels of Ohio. The Division of Wildlife, Columbus, 121 pp
Welte R (1999) Wirtserkennung bei der Teichmuschel *Anodonta*. Ms Thesis, University Freiburg
Wood E (1974) Development and morphology of the glochidium larva of *Anodonta cygnea*. J Zool Lond 173:1–13

6 Larval Types and Early Postlarval Biology in Naiads (Unionoida)*

KLAUS WÄCHTLER, MARIA C. DREHER-MANSUR, and TORSTEN RICHTER

6.1 Introduction

In freshwater bivalves development takes place in three ways (Fig. 6.1).

1. By producing veliger larvae as in *Dreissena polymorpha*, which is particularly successful wherever it first appears.
2. By releasing fully developed young mussels from brood pouches (viviparity) as in Sphaeriidae.
3. By passing early development as a parasitic stage on a host as in naiads (Unionoida).

In this chapter the third of these developmental strategies will be described. Ectoparasitic larvae are found worldwide: (Fig. 6.2) glochidia in Unionoidea and lasidia or haustoria in Etherioidae = [Mycetopodidae / Mutelidae (=Iridinidae) (see Hoeh et al., this Vol.).

While the development from glochidia to young mussels is relatively uniform, that from lasidia to young mussels shows interesting modifications. In both cases, however, the ectoparasitic development leads to juvenile mussels almost identical in size and internal structure.

6.2 The Glochidium: the Most Common Larva in Naiads

6.2.1 Early Observations

Glochidia were described first as bivalve parasites (glochidium parasiticum) by Rathke (1797; Fig. 6.10) long before they were recognized as larval stages (Carus 1832; Leydig 1866).

6.2.2 Formation, Storage and Release of Glochidia

Glochidia form from fertilized eggs and are stored within pouches (marsupia) of the modified gills. In the formation of the marsupia one or both pairs of gills may be involved (Mackie 1984): the outer pair in Unionidae, the inner pair in Hyriidae and both pairs in Margaritiferidae (Walker et al., this Vol.). Filled marsupia can be

* Dedicated to the memory of Dr. W. D. Bischoff (1918–1999) general practitioner and naturalist who, over many years, drew our attention to the biology and cultural history of *Margaritifera margaritifera* in northern Germany.

Ecological Studies, Vol. 145
G. Bauer and K. Wächtler (eds.) Ecology and Evolution of the Freshwater Mussels Unionoida

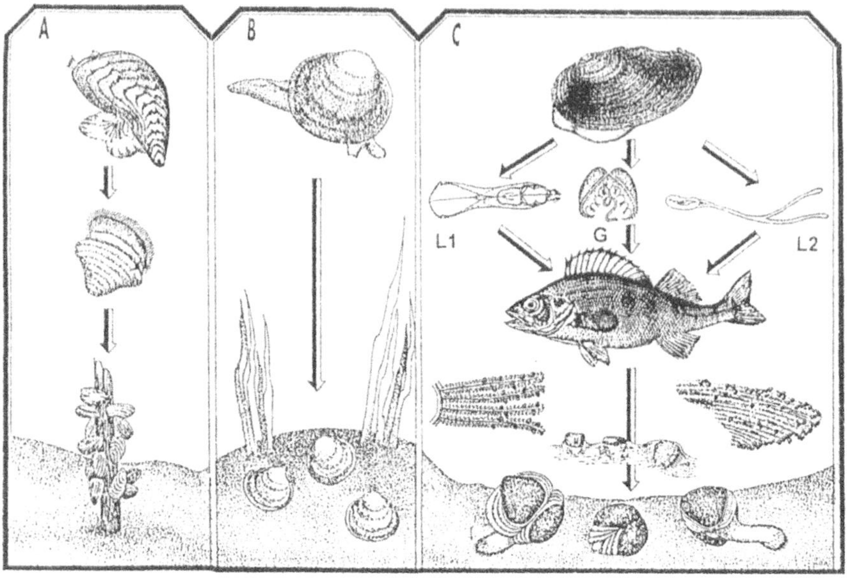

Fig. 6.1A–C. Developmental strategies in freshwater bivalves. **A** Planctic larvae (Pediveliger) undergo metamorphosis and settle at suitable sites (*Dreissena polymorpha*). **B** Eggs develop in brood pouches into young mussels which are released from the adult (Sphaeriidae). **C** Larvae (Glochidia or Lasidia) live as ectoparasites on teleost fish. *G* Glochidium; *L1* lasidium; *L2* lasidium after metamorphosis forming haustoria. (Drawings by Gregor Koterski)

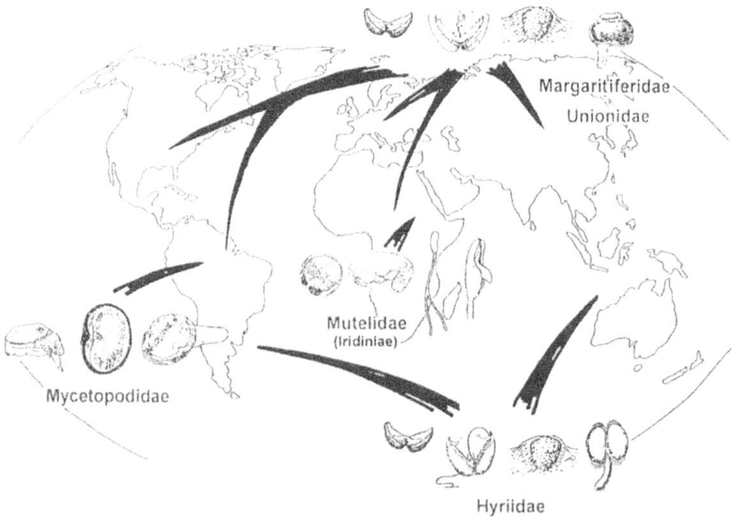

Fig. 6.2. Unionid larvae and their distribution. Species forming glochidia are found worldwide. Those forming lasidia are confined to South America and Africa. With the exception of *Mutela bourguignati* (Mutelidae = Iridinidae) forming haustoria all glochidia and South American lasidia (Mycetopodidae) develop within epithelial cysts formed by the host fish (Fig. 6.1c)

recognized by their swollen appearance. Increased oxygen consumption and reduced filtration rates in gravid females were shown by Tankersley and Dimock (1993a,b) and Tankersley (1996). Depending on the species, this state can last from a few weeks up to several months before the glochidia are released [short-term (tachytictic) and long-term (bradytictic) breeders] (Heard 1998). The release can occur in one lot within a few days or gradually over a longer period. Repetitive release and refilling during one spawning season were observed in several species of *Unio* (Engel 1990; Fleischauer-Rössing 1990; Hochwald and Bauer 1990; Hochwald 1997).

Two spawning periods per year have been found in *Cumberlandia monodonta* (Gordon and Smith 1990) and three are reported from Australian hyriids (Jones et al. 1986). In one of the Australian hyriids, *Velesunio angasi*, even all-year-round repetitive breeding was observed (Humphrey and Simpson 1985).

6.2.3 Numbers

The numbers of glochidia produced depend on both the size of the glochidia and the size of the female mussel. They range from a few thousand to several millions. Counts for some species are given in Table 6.1.

6.2.4 Morphology

The study of the glochidial morphology has been extended from light microscopy (Tucker 1927, 1928; Wood 1974) to SEM and TEM techniques. Giusti (1973), Giusti

Table 6.1. Number of glochidia found in a variety of unionoid species

Species	Number of larvae	Reference
Margaritifera margaritifera Linnaeus	3–4 million	Bauer (1989), Buddensiek (1991), Dettmer (1982), Young and Williams (1983a)
Unio tumidus Philipsson	ca. 200 000	Fleischauer-Rössing (1990), Maaß (1987), Wilms (1991)
Unio pictorum Linnaeus	ca. 200 000	Fleischauer-Rössing (1990), Maaß (1987), Wilms (1991)
Unio crassus Philipsson	9000–16 000	Bednarczuk (1986), Engel (1990), Engel and Wächtler (1990), Hochwald (1997)
Depending on the average size in diff. populations	81 000–100 000	
Anodonta cygnea Linnaeus	310 000–370 000	Claes (1987), Niemeyer (1992)
Anodonta woodiana	167–200 million	Wächtler (unpubl.)
Pseudanodonta complanata Roßmäßler	8000–16 000	Hüby (1988)
Pyganodon cataracta	ca. 1 million	Tankersley and Dimock (1993a,b)
Quadrula cylindrica strigilla	115 000	Yeager and Neves (1986b)

et al. (1975), Kinzelbach and Nagel (1986), Nezlin et al. (1994), Scharsack (1994), Shi-Anjing (1994), Pekkarinen and Englund (1995a,b), Pekkarinen and Valovirta (1996) and Araujo and Ramos (1998) give detailed comparative accounts mainly on European species. North American species with reference to those from other regions are treated in a detailed annotated literature survey by Watters (1994). M.C. Dreher-Mansur (unpubl.) summarizes information on glochidia from South American bivalves. The larval morphology of the Australian hyriid *Hyridella depressa* is given by Jupiter and Byrne (1997). The following description is mainly based on Scharsack (1994).

6.2.4.1 Shape and Size

Glochidia are triangular, spherical or hatch-shaped in outline with diameters from about 80 to 350 μm (Lefevre and Curtis 1910). The dimensions for a few species are given in Table 6.2.

Table 6.2. Glochidial sizes found in a variety of unionoid species

Species	Size in μm	Reference
Margaritifera margaritifera Linnaeus	60 × 80	Bauer (1989), Buddensiek (1991), Dettmer (1982), Young and Williams (1983a)
Unio tumidus Philipsson	210 × 200	Fleischauer-Rössing (1990), Maaß (1987)
Unio pictorum Linnaeus	210 × 200	Fleischauer-Rössing (1990), Maaß (1987), Wilms (1991)
Unio crassus Philipsson	220 × 195	Bednarczuk (1986), Engel (1990), Hochwald (1997)
Anodonta cygnea Linnaeus	310 × 310	Claes (1987), Niemeyer (1992)
Anodonta anatina Linnaeus	340 × 360	Niemeyer (1992)
Anodonta woodiana	390 × 400	Penzes (1992, pess. comm.), Wächtler (unpubl.)
Pseudanodonta complanata Roßmäßler	340 × 290	Hüby (1988), Kinzelbach and Nagel (1986)
Hyridella depressa	243 × 249	Jupiter and Byrne (1997)
[a]*Diplodon atratus* Sowerby 1839	260 × 230	Bonetto (1959, 1961)
[a]*Diplodon wagnerianus*	280 × 230	Bonetto (1959, 1961)
[a]*Diplodon hildae*	240 × 250	Bonetto (1959, 1961)
[a]*Prisodon (Triplodon) corrugatus*	310 × 270	Bonetto and Ezcurra (1965)
[a]*Castalia nehringi*	260 × 240	Ortman (1921a,b,c)

[a] Quoted from an unpublished comparative list of 55 South American Hyriidae by Dreher-Mansur (see also Dreher-Mansur 1998).

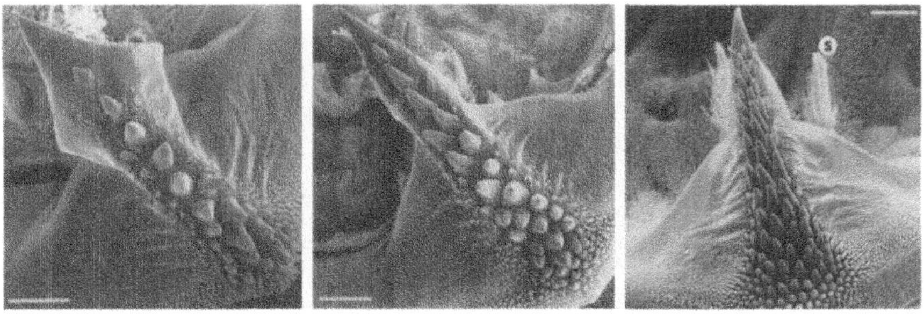

a b,c

Fig. 6.3a–c. Glochidial hooks with rows of teeth from **a** *Anodonta cygnea*, **b** *Anodonta anatina*, **c** *Pseudanodonta complanata*. *s* Sensory hairs (↔20 µm). (Scharsack 1994)

6.2.4.2 Shell

The glochidia are covered by two valves forming a straight hinge at their bases of up to several hundred microns in length. The calcium content of the shell is very low in most species (Kondo et al. 1997). Their apical edges are frequently armed with a hook which varies in size, shape and inclination. The outer surface of the hook carries rows of small teeth (Fig. 6.3). In *Margaritifera* the hooks are absent or very small (Panha 1995; Pekkarinen 1996; Araujo and Ramos 1998) and the edges are formed by a smooth elastic rim by which they adhere to the soft epithelia of the host fish gills. The hooks are also absent in *Diplodon iheringi* (Bonetto 1959, 1961; Dreher-Mansur 1998) and a variety of species in several regions (Lefevre and Curtis 1910; Prashad 1918a,b; Raulerson and Burbanck 1962). Hooked glochidia may attach to both gills and body surface, particularly to opercula and fins. Hookless glochidia normally attach to the gills only.

6.2.4.3 Adductor Muscle

Antagonistic to the elastic hinge, a well-developed adductor ensures the closure of the shells. It is capable of rapid contractions and relaxations. Ripe glochidia perform vivid snapping movements before attachment. These can be artificially induced by adding sodium chloride crystals or mucus from fish skin to a suspension of glochidia (Lefevre and Curtis 1910; Hoggarth and Gaunt 1988). In infestation studies this adductor reaction is a useful check for the fitness of the larvae. As soon as the glochidia become attached to the host fish, the adductor muscle stays contracted. It degenerates during metamorphosis and is replaced by the pair of adductors of the newly forming bivalve.

6.2.4.4 Sensory Hairs

Although detailed investigations of the sensory physiology are still missing, one can assume that chemical stimuli, possibly in combination with mechanical ones that lead to adductor contraction, are registered by a number of sensory hairs located in the glochidial softbody (Wood 1974). Size, shape and position of these structures may vary among species (Figs. 6.4, 6.6).

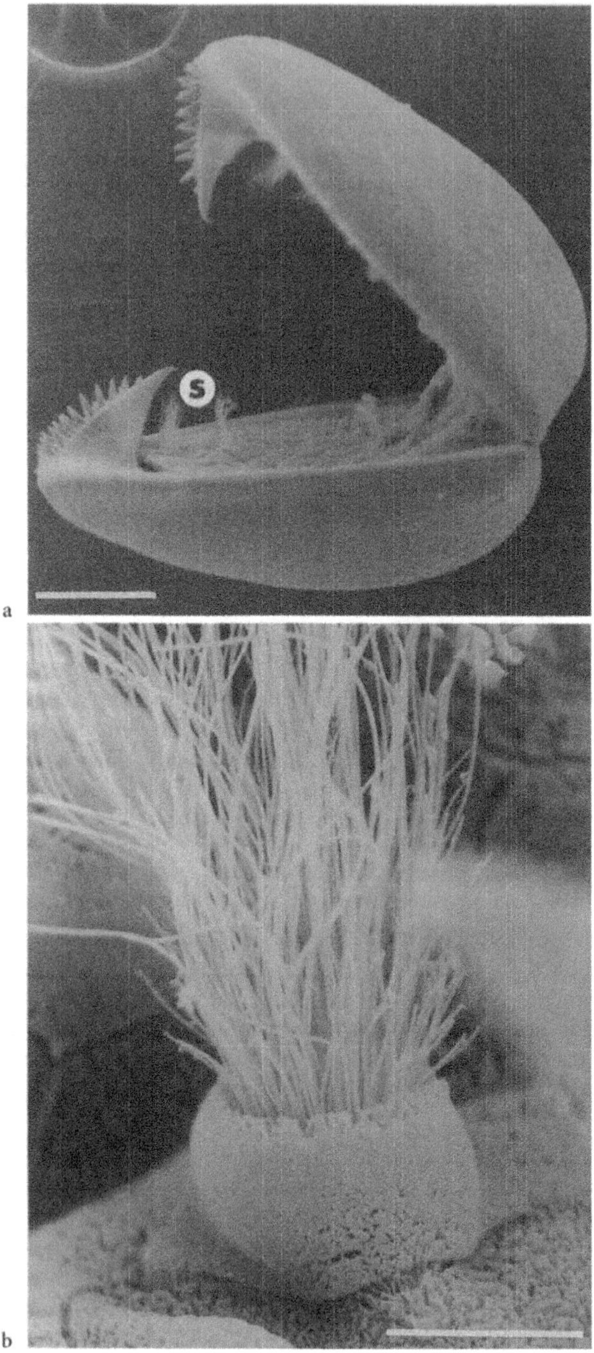

Fig. 6.4. **a** Sensory hairs (*s*) in *Anodonta anatina* (↔100 μm). **b** Bundle of sensory hairs enlarged (↔10 μm). (Scharsack 1994)

6.2.4.5 Ciliary Fields

In several sites of the inner surface of the glochidia, ciliary fields are located around pits or grooves in the mantle. In *Margaritifera*, long, densely packed cilia can be seen in the light microscope at both sides of the hinge region. They produce water currents which convey oxygen and chemical information (Scharsack 1994; Pekkarinen and Englund 1995b).

6.2.4.6 Larval Mantle

The inner surface of the glochidium is covered with an epithelium formed by large cells with apical microvilli likely to be involved in nutrient uptake from the fish tissue after cyst formation. (Schierholz 1889; Herbers 1914; Arey 1932a,b; Pekkarinen and Valovirta 1996). Scharsack (1994) observed the release of spherical particles from the apex of larval mantle cells. These particles are transported by lateral cilial bands and appear as storage granules in late parasitic stages (Fig. 6.5).

6.2.4.7 Larval Thread

While the structures described so far are found in all glochidia, the larval thread may be present or absent even in closely related species. In European species it is absent in those preferring more lotic conditions like *Margaritifera margaritifera*, *Unio crassus* and *Pseudanodonta complanata*, but it is present in *Anodonta cygnea*, *Anodonta anatina*, *Unio tumidus* and *Unio pictorum*.

In *Unio pictorum* the larval thread is kept in a longitudinal fold while it is contracted. Extended larval threads of many glochidia may form a floating web, pre-

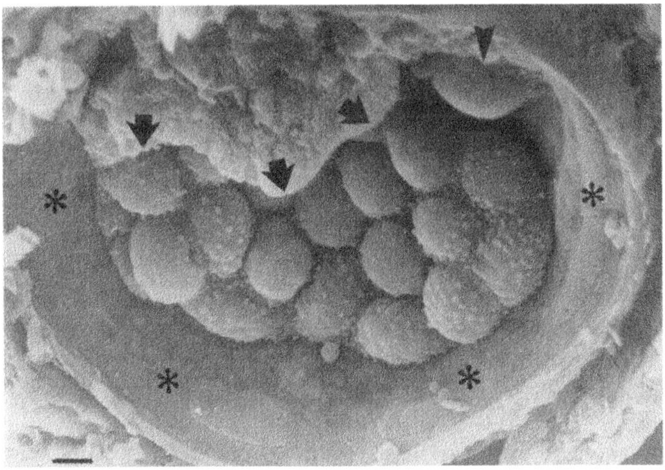

Fig. 6.5. Larval stage of *Margaritifera margaritifera* 3 months after cyst formation showing release of spherical particles from larval mantle cells (*middle*), definite mantle (*asterisks lower rim*) and growing gill papillae (*arrows top*) (↔20 μm). (Scharsack 1994)

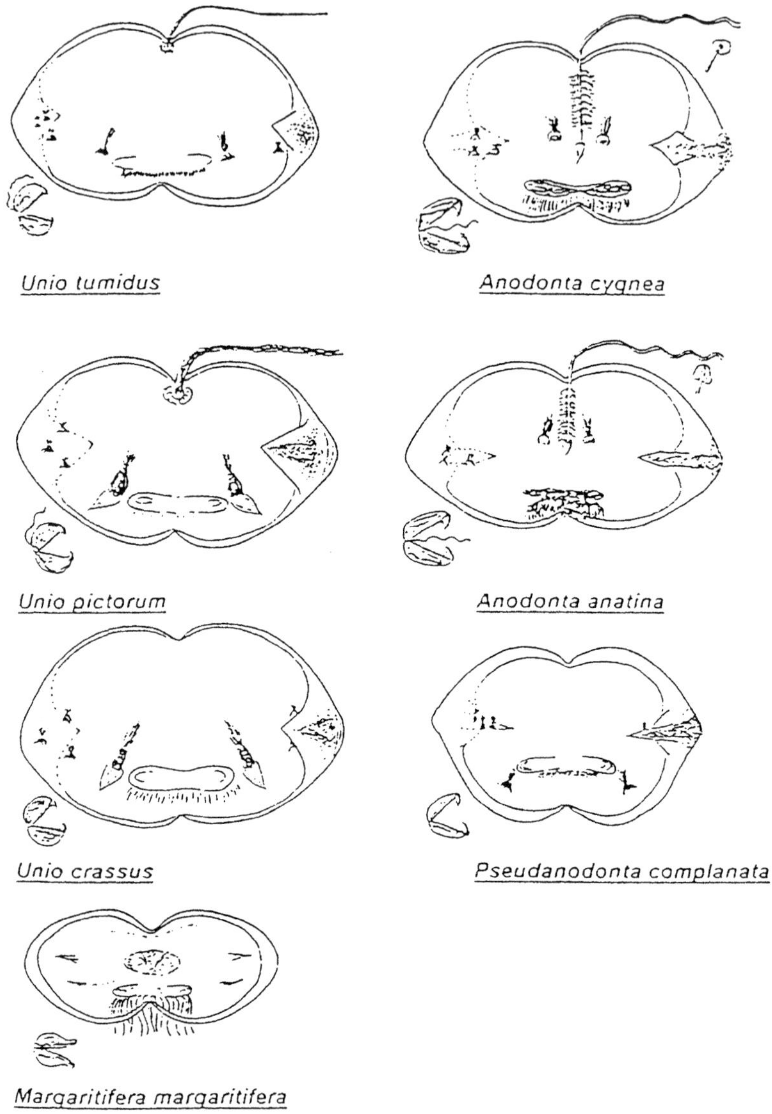

Fig. 6.6. Variability of internal structure in glochidia of several European species (hooks, sensory bristles, ciliary fields and larval thread). (Scharsack 1994)

venting the glochidia from sinking to the bottom (Wood 1974). The variability in form and position of the structures described is summarized in Fig. 6.6.

6.2.5 Interactions with the Host Fish

Glochidia have to come in contact with a host fish soon after being released from the marsupia. If kept in suspension under laboratory conditions for more than 3 days, the number of infective glochidia decreases to a small fraction (Sylvester et al. 1984). Survival times may range from 2 to 14 days (Mackie 1984). The chance to come in contact with a host fish may be raised by visually attracting the fish either by producing floating aggregates of glochidia (Matteson 1955; Yokley 1972) or by storing the glochidia in undulating well-visible structures which are actively taken up by the fish, as known from some Lampsilidae (van der Schalie 1970b; Kraemer 1970; Mackie 1984; Watters 1994; Roe 1998). In *Lampsilis perovalis* the glochidial aggregate remains fixed to the female by a transparent mucous strand and mimics in size, shape and movements a small fish (Haag et al. 1995). This kind of active host attraction by morphological changes may enable the bivalves to reduce the number of larvae necessary to complete their life cycles (Watters 1998). In some cases, release of glochidia is stimulated by the presence of fish (Zale and Neves 1982c). In the majority of species it is just the high number of glochidia which ensures an infestation (Table 6.1).

6.2.5.1 Suitable Host Fish

In regions with many species of both fish and bivalves, the host fish spectrum is known only in outline and is still under investigation (Nagel and Castagnolo 1993). For the 297 North American species the long list of hosts (Coker et al. 1921; Fuller 1974; Neves et al. 1985; Hoggarth 1992; Watters 1994) is being increasingly enlarged, but is far from being definite for most regions (Zale and Neves 1982a,b; Yeager and Neves 1986a; Yeager and Saylor 1995; Haag and Warren 1997; Howells 1997a,b; O'Dee 1998). Lampreys (Agnatha) as hosts are described by Wilson and Ronald (1967). Among bony fish (Osteichthyes) most hosts belong to the modern teleosts, but some species of the ancient orders Polypteriformes, Acipenseriformes, Lepisosteiformes and Amiiformes are also on the list (Altaba 1990; Watters 1994; Araujo and Ramos 1998). Amphibia as hosts were observed by Howard (1915, 1951), Seshaiya (1941) and recently by Watters (1997b) and Watters and O'Dee (1998), who successfully infested larval tiger salamanders (*Ambystoma tigrinum*) with glochidia of *Lampsilis cardium*.

From the comparative observations available, one can conclude that the larger glochidia with hooks are the more general and widely distributed type, whereas the smaller and hookless ones represent the more specialized type with restricted occurrence. *Anodonta anatina* is an example for a species with a wide host fish spectrum, whereas *Margaritifera margaritifera* can only develop on salmonids. In Central Europe it prefers the brown trout (*Salmo trutta*), whereas on the rainbow trout (*Oncorhynchus mykis*), which was introduced from North America, *Margaritifera* glochidia fail to develop soon after cyst formation. When brown trout of different origins were exposed to suspensions of *Margaritifera* glochidia, those from the same rivers as the pearl mussels were markedly better host fish (Dettmer 1982; Buddensiek 1995). On the other hand, *Margaritifera margaritifera* glochidia taken from a river with dense populations of both brown trout (*Salmo trutta*) and Atlantic salmon (*Salmo salar*) in Norway were found to develop successfully almost exclusively on salmon and were rejected by brown trout (K. Wächtler et al., unpubl.). Damage to fish by overinfection with glochidia appears to occur only under exceptional conditions, for instance in a West-Norwegian salmon farm, where the water inlet happened to be situated on top of a huge *Margaritifera* bank. Graded sensitivity to overinfection with glochidia of *Chamberlainia hainesiana* was observed by Panha (1993), who exposed fry of five possible host fish species from waters in Bangkok to concentrated glochidial suspensions. In Southern Europe the Asian *Anodonta woodiana* finds favourable conditions in some carp farms (Watters 1997a). As it can produce millions of relatively large hooked glochidia, it causes damage to the carp fry (Penzes 1992, pers. comm.).The mechanisms by which a fish may protect itself against glochidia are not understood in detail. Bauer and Vogel (1987) and Zotin and Ziuganov (1994) found lower infection rates for *Margaritifera* on older trout or after second infection. They could also show increasing immune reactivity. Hochwald (1997) found decreased infection rates of *Unio crassus* on *Scardinius erythrops* after previous infections by either *Unio crassus* or *Anodonta anatina*. On the other hand, the second infection on salmon in Norway was even more successful than the first (K. Wächtler et al., in prep.). Kirk and Layzer (1997) were able to induce successful development of glochidia to young mussels on several species of "non-host fish" after treating the fish with cortisol as an immunosuppressant.

Watters and O'Dee (1996) studied host-parasitic interactions in 15 fish species infected with glochidia of *Lampsilis fasciola* over several weeks. These authors come to the conclusion that the shedding of untransformed glochidia is due to an aquired immunity of these wild-caught fishes. Seshaiya (1969) was able to infest several species of murrels (*Ophiocephalus*) up to seven times with glochidia of *Lamellidens corrianus* before they were rejected at an early stage. Rejection from the wrong host normally occurs within hours or days after cyst formation (Scharsack 1994; see below).

6.2.5.2 Attachment

The host fish takes up the glochidia either by surface contact or by the water current drawn through the gills. Thus, the predominant site of attachment can be either exposed body parts like fins and opercula or the epithelium of the gills (Fig. 6.1c). As stated above, small glochidia almost exclusively attach to the gills (Fig. 6.7).

6.2.5.3 Cyst Formation

Soon after attachment even on some "non-hosts" the fish tissue forms one to several layers of cells around the larva. In the case of a successful development, this epithelial cyst provides both mechanical protection and nutrients from the host during the whole larval period (Arey 1932a,b) (Figs. 6.8, 6.9).

6.2.5.4 Growth

During the stay in the cyst, shell length may be unchanged (most unionids) or increase (*Margaritifera*). In both cases, a considerable increase in carbon content can be stated during the parasitic stage (see Sect. 6.5.2.1 and Table 6.3).

6.2.5.5 Metamorphosis

The development on the fish implies a complete reorganization of body structure from a parasitic larva to a young filter feeder. The main events during this metamorphosis are the replacement of the larval adductor by a pair of definitive ones, the

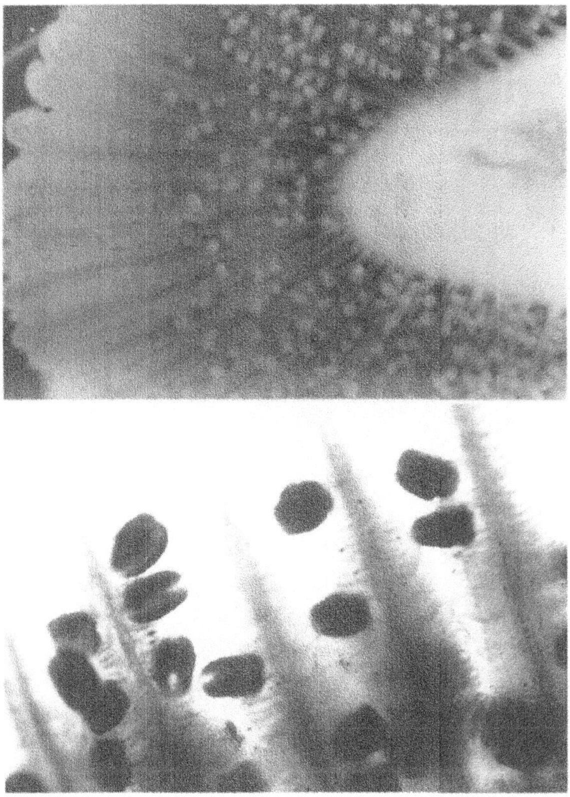

a

b

Fig. 6.7a,b. Glochidia of *Margaritifera margaritifera* encysted in gill filaments of *Salmo trutta* a few days before release. **a** Survey. **b** At higher magnification

formation of mantle tissue involved in larval food uptake and shell secretion, the formation of the digestive system with a pair of digestive glands, formation of the nervous system, the appearance of a large, ciliated foot and the initial stages of the paired gill apparatus.

Along with the differentiation of the organ systems the accumulation of nutritional reserves in form of spherical storage granules takes place (Fig. 6.5). Some steps of these processes have been studied in light and electronmicroscopy in *Margaritifera* by Scharsack (1994) and will be referred to below when the morphology of the juvenile mussel is described.

6.2.5.6 Length of Larval Period

To some degree influenced by the water temperature and depending on the glochidium size (Bauer 1994) the parasitic stage in unionids varies from 3 days (Seshaiya 1969) to 10 months. A period of about 20 days is found in many species. Populations of *Margaritifera margaritifera* with short and long developmental periods of 3 and

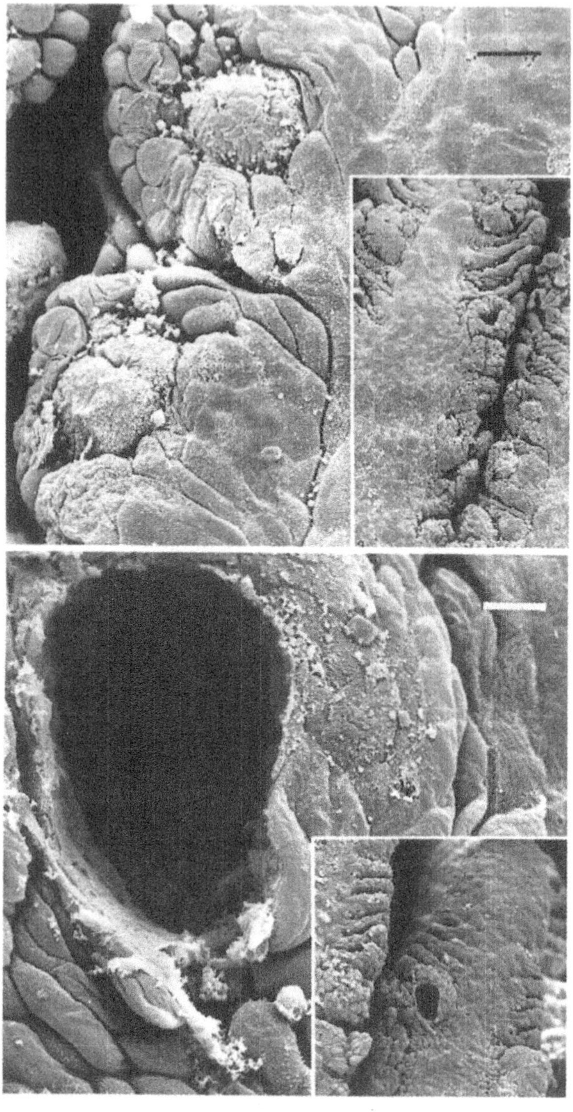

a

b

Fig. 6.8a,b. Epithelial cysts around glochidia of *Margaritifera margaritifera* on gill filaments of *Salmo trutta*. **a** Early stage 3 h after attachment. **b** Empty cyst after release of the glochidium (↔100 μm). *Insets* Adjacent gill filaments at lower magnification. (Scharsack 1994)

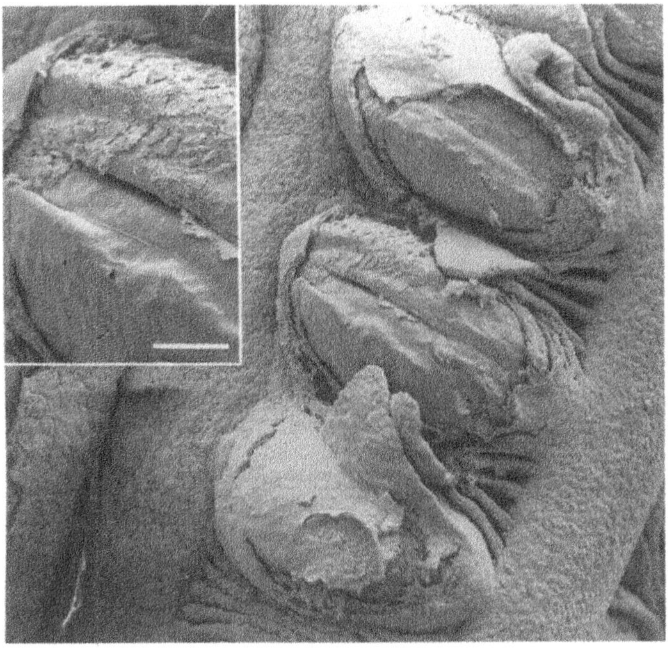

Fig. 6.9. Encysted juvenile *Margaritifera margaritifera* immediately before release about 10 months after infection (↔50 μm). (Scharsack 1994)

10 months, respectively, in the same geographical region have been described by Bauer (1987).

6.3 The Lasidium: a Larval Type with Two Developmental Options

6.3.1 Introductory Remarks on the Information Available

Lasidia are known only from Mycetopodidae and Mutelidae (Iridinidae) (Fig. 6.2).

Compared to the large number of detailed studies on glochidia over more than a century, the knowledge on lasidia is derived from a very small number of studies (Bonetto 1961; Parodiz and Bonetto 1963; Bonetto and Ezcurra 1965; Fryer 1959, 1961, 1970; Falk 1995; Herrmann 1995; Falk et al. 1996). As Kondo (1984) describes typical glochidia from a mutelid in Lake Tanganyika *Moncetia lavigeriana* which occurs together with unionids (Kondo 1990), it is not known how representative the descriptions below are for Mutelidae (Iridinidae) or even the Etherioidea species from Africa and South America, respectively.

In contrast to glochidia, which show a basic design over all geographic regions, the known lasidia from Mycetopodidae and Mutelidae (Iridinidae) pass early stages which resemble each other in the beginning, but differ markedly when development continues. In the Mycetopodidae a lasidium is found in the genera *Anodontites*, *Myce-*

topoda, *Monocondylaea* and *Leila* (Bonetto 1959, 1961; Parodiz and Bonetto 1963; Dreher-Mansur and Olazarri 1995). These authors provide many drawings with anatomical details particularly of *Anodontites trapezialis forbesianus*. Although a few specimens of this species were studied in this laboratory (Hanover), the documentation of the life cycle is still incomplete. For the African Mutelidae (Iridinidae) the situation is slightly better. Guided by detailed instructions from Dr. Fryer (Windermere) Falk and Herrmann have studied and documented the life cycle of *Mutela bourguignati* (Falk 1995; Herrmann 1995).

6.3.2 Early Observations

Among many South American species producing glochidia with structures as described above, von Ihering (1891) found in *Anodontites wynami* a larval type divergent in many respects, which he described as lasidium (Fig. 6.10). According to Parodiz and Bonetto (1963), it took more than half a century until the lasidium was rediscovered. A somewhat similar larva with a unique development (see below) was discovered by Fryer in *Mutela bourguignati* in Lake Victoria in Uganda (Fryer 1959, 1961, 1970; Fig. 6.10).

6.3.3 Formation, Storage and Release of Lasidia

Gravid Mycetopodidae or Mutelidae (Iridinidae) are, like Unionidae, recognized by swollen internal gills. The larva form from eggs in the marsupia and are kept there for a few weeks. In *Mutela bourguignati* released lasidia remain enclosed in the egg membrane for hours or even a day before infection.

6.3.4 Numbers

Herrmann (1995) studied the lasidial output of 57 gravid females of *Mutela bourguignati* (from Victoria Nile) over 17 days. These specimens were collected for controlled infections. His counts range from a few thousand to up to 388 800 lasidia per gravid female. He could determine neither a fixed spawning season nor the intervals between two spawning periods.

6.3.5 Morphology

6.3.5.1 Shape and Size

1. *Anodontites*
   All larva released from gravid females in the laboratory had the elongated shape of about 200 μm length and most of the structures depicted by Parodiz and Bonetto (1963; Fig. 6.11a).
2. *Mutela*
   The earliest stages released are about 150 μm in diameter covered by a transparent egg membrane through which a very long larval thread coiled up to 30 times around the body can be seen. A few hours later, the larva leaves the envelope and the larval thread, measuring up to several centimetres in length, is rolled out and ready to make contact with a host fish.

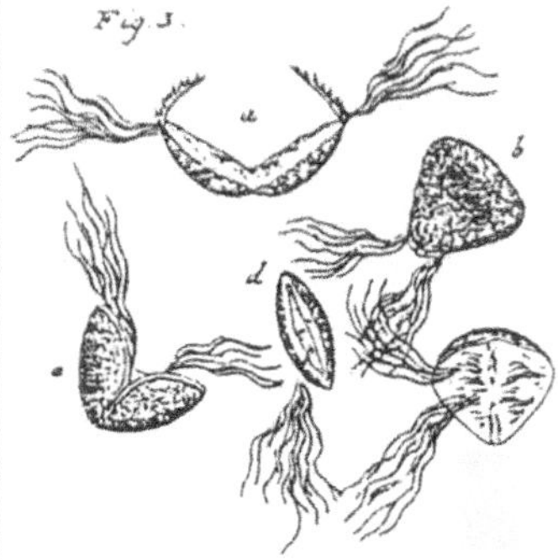

Men den mærkværdigſte Egyden er et Slags Hentæren, ſom mærkes hos mange af dem, iſær mod Høſten og Vinteren, og ſom hidrører af en utallig Mængde ſmaae Bivalver eller. Tveſkiel, der opholde ſig inden i Muslingens yderſte Par Geller, ſom af diſſe ere ſtærkt udſpendte, og have af de brune Skaller et Skier af ſamme Farve, formedelſt Gellerørenes Giennemſigtighed. Da diſſe, ſaavidt jeg veed, ikke tilforn ere beſkrevne, ſaa vil jeg forſøge at beſkrive dem, under Navnet Krogſkiæl, eller Glochidium: Animal — cirrhis longiſſimis inſtructum.

Teſta æqvilatera, æqvivalvis, intra marginem exteriorem hamata.

1. paraſiticum. Teſta triangulari, diaphana. Habitat intra Mytili anatini branchias exteriores.

Rathke 1797

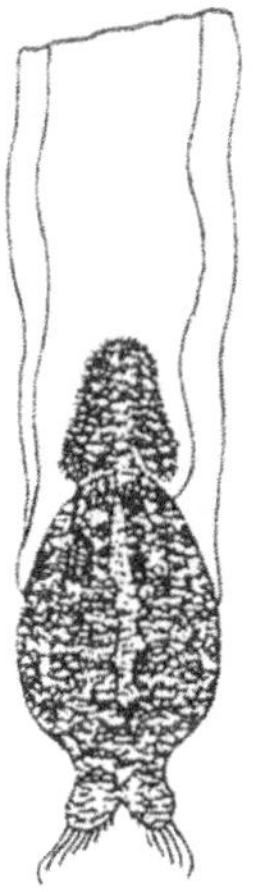

Reife Larve (*Lasidium*) von *Glabaris Wymanni* Lea.

Diese sonderbare Larve, die ich im Folgenden mit Rücksicht auf das borstentragende Hinterende Lasidium nennen werde, kenne ich also von zwei Arten südamerikanischer „Anodonten".

v. Ihering 1878

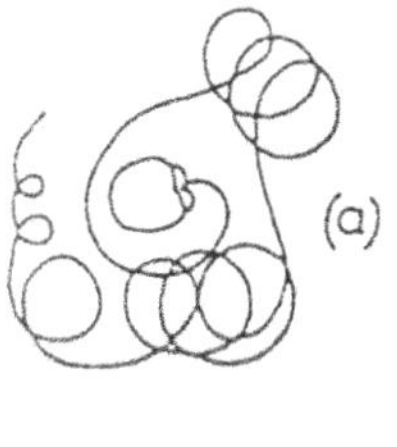

These larvae are in some ways similar to the typical molluscan veliger but differ in certain respects, most of the differences being correlated with the subsequent period of parasitism. Most remarkable of these larval features is the possession of an extremely long filamentous tentacle which resembles the tentacle of the glochidium of the Unionidae but is very much longer, being more than 70 times that of the larva itself.

Fryer 1959

Fig. 6.10. Original illustrations and descriptions of naiad larvae

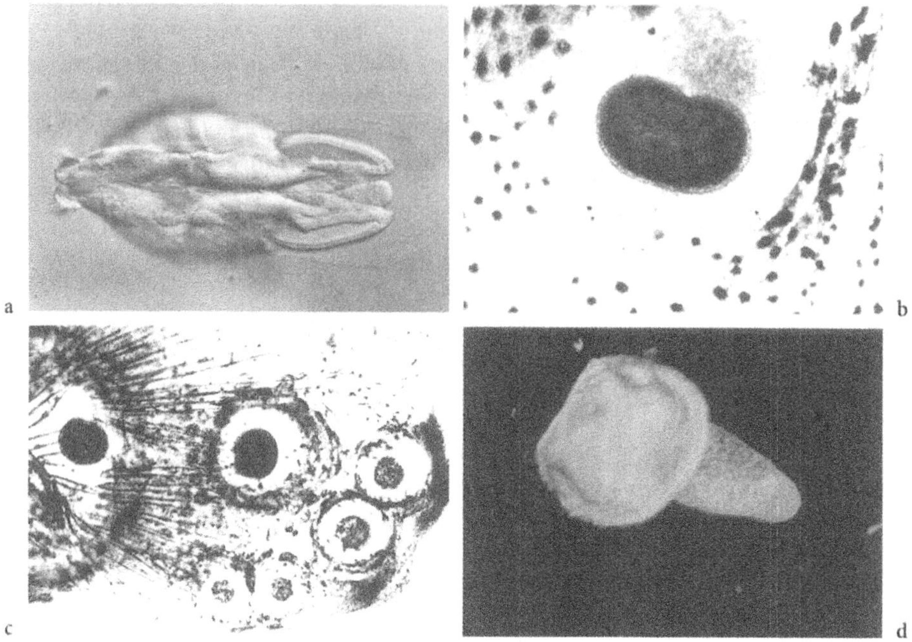

Fig. 6.11a–d. *Anodontites trapezialis*. **a** Lasidium posterior end with hooks to the left. **b** Epithelial cyst. **c** Lasidial cysts of *Anodontites* sp. on 2-day-old *Tilapia oriochromis niloticus* near Carnas (Costa Rica) photo: Morales. **d** Juvenile mussel of *Anodontites trapezialis*

6.3.5.2 Body Surface

The lasidial body is covered by a cuticula which in both species forms an undivided dorsal cup with a longitudinal fold. In *Mutela* it gradually changes from a spherical to a more elongated shape.

6.3.5.3 Adductor Muscle

As described by Fryer (1961), the lasidial adductor muscle is more complicated than in glochidia. It allows the elongated lasidium both to round up (shortening the longitudinal axis) and approach the free rims of each body side (shortening the lateral axis).

6.3.5.4 Adhesive Structures

Both larvae have in common an adhesive structure described as a pair of posterior lobes provided with several distal hooks (Fig. 6.11a). Anteriorly, the paired or unpaired lobuli also take part in attachment. In *Mutela* the hooks form two medial rows with a tuft of sensory hairs at each side (Fig. 6.12).

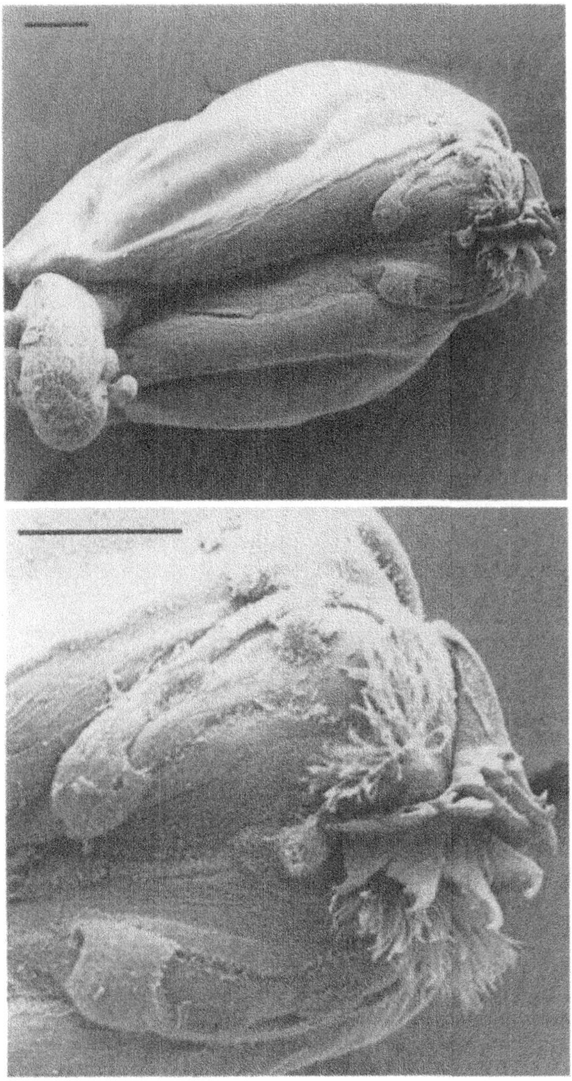

a

b

Fig. 6.12. **a** Lasidium of *Mutela bourguignati* seen from below with lobuli. Anterior lobes (*left*) with the paired posterior lobes carrying a pair of hooks medially and sensory hairs at both sides. These are the structures for the initial attachment on the host fish. **b** Posterior lobes and hooks enlarged (↔20 μm). (Falk 1995)

6.3.6 Interactions with the Host Fish

6.3.6.1 Suitable Host Fish

1. *Anodontites*
   As far as is known, the lasidia of the Mycepodida described can develop on a variety of fish (Bonetto and Ezcurra 1965).
2. *Mutela*
   As already described by Fryer (1959) *Mutela bourguignati* has a very narrow host spectrum; successful development is known only from *Barbus altianalis radcliffi*, and very likely from the catfish *Bagrus dogmae*. (Herrmann 1995). Infections of other fish species from the same river, like *Tilapia* sp. and *Mormyrus* sp., were either negative from the beginning or did not lead to young mussels (Herrmann 1995).

6.3.6.2 Attachment

Survival time of the lasidia is probably in the range of some days as in glochidia. Attachment is not limited to special body parts of the fish. Under laboratory conditions up to 1000 lasidia were found on one fish.

At this stage further development takes two absolutely divergent directions (Fig. 6.13, right).

6.3.6.3 Option One: Cyst Formation

In *Anodontites* the fish epithelium forms a cyst only hours after infection which is very similar to that around attached glochidia (Fig. 6.11b,c).

6.3.6.4 Option Two: Transformation into a Second Larval Stage

The lasidium of *Mutela* is the only known case in which infection does not lead to cyst formation. Instead, the attached larva undergoes a complete metamorphosis into a second larval stage fixed to the host by a pair of haustoria.

To the outstanding description of Fryer (1959, 1961) further details could be added to this unique mode of bivalve development by recent light and electronmicroscopical studies of Falk (1995), Herrmann (1995) and Falk et al. (1996), which will be dealt with here in some detail.

The host-parasite contact arouses both the fish and the larva. A few minutes after adding a lasidium suspension to the aquarium, *Barbus altianalis radcliffi* react by vivid swimming movements and operculum vibrations.

Like infective glochidia, the lasidia are immediately activated by the presence of the host or if NaCl crystals are added. They react by up and down movements of the anterior lobuli and by changing the larval shape from spherical to elongated. During this stretching, the cuticula rims of each side approach each other and the adhesive structures stand out at both ends. This gives the lasidium an appearance somewhat similar to a germing wheat grain (Fig. 6.12a).

The hooks on the posterior end ensure the initial attachment. Only a few hours later, the actual penetration of the host tissue occurs from the side of the anteriorly

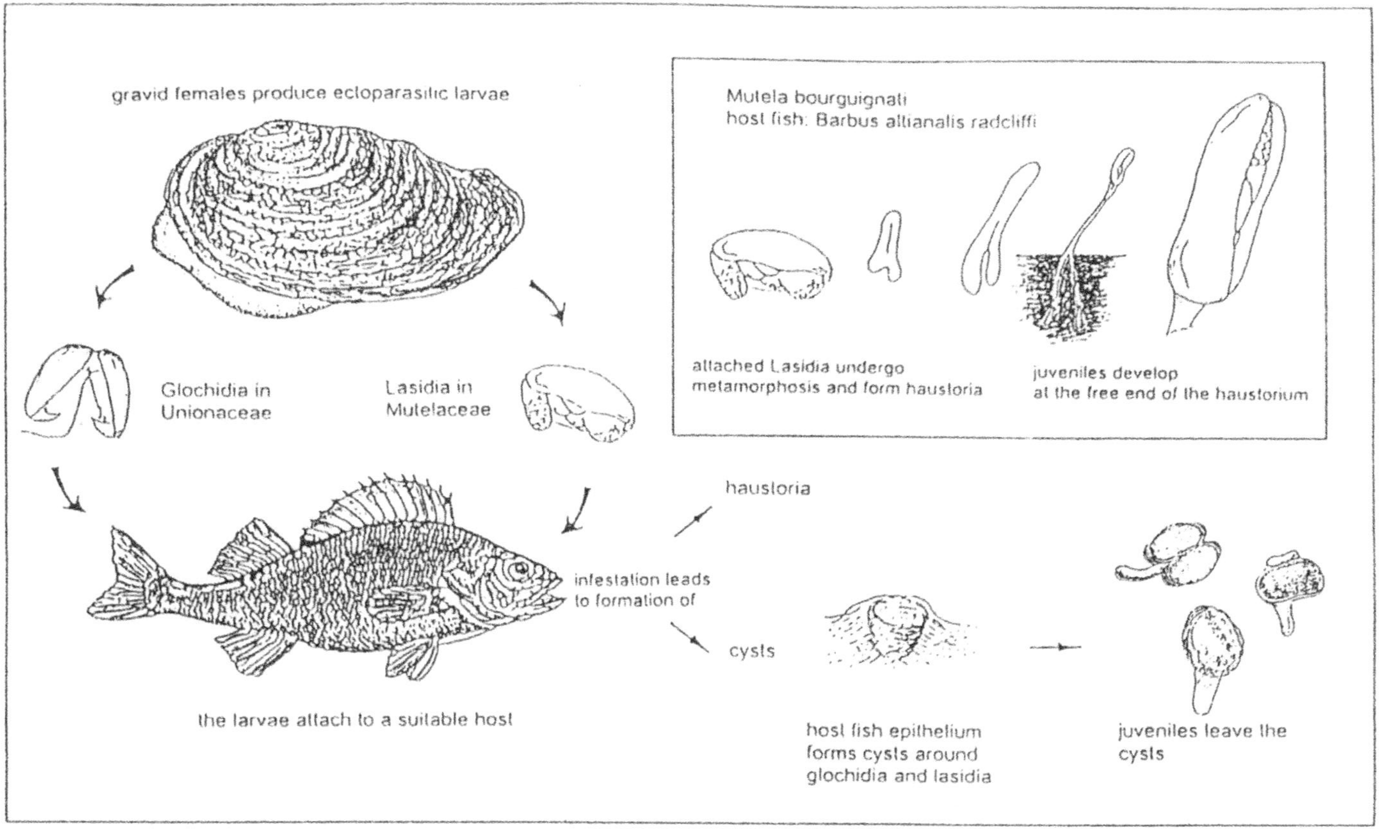

Fig. 6.13. Larval development of glochidia and lasidia with or without cyst formation. (Falk et al. 1996)

located lobuli where the fork-shaped haustoria form after the remaining larval organs, including the hooks, are covered by the cuticular shell, the rims of which come progressively closer to each other, forming a tube with a ventral suture. Within 2 days, only residues of lasidial structures including those from lobuli and hooks can be traced inside this tap-shaped structure protruding about 200 µm above the fish surface (Fig. 6.14a). From this stage on, differentiation and elongation take place in two directions, resulting proximally in a pair of haustoria and distally in a transparent stalk eventually forming the juvenile mussel at its tip (Figs. 6.14b, 6.15).

6.3.6.4.1 Haustorium Formation in *Mutela bourguignati*

Already at the second day after infection, a fork-shaped haustorium sends its ends 80 to 100 µm into the tissue. The whole structure is covered by a cuticula. Under this cuticula the surface, and particularly the tips of the haustoria, are covered with microvilli, which are likely to be in relation with uptake processes from the fish tissue. As can be seen in the light microscope, the host tissue reacts by characteristic pigment aggregations around the haustorial root. The paired ends of the haustoria form hollow epithelial tubes which unite outside the tissue to form the lumen of the stalk. Within 14 days the whole larva reaches a total length of up to 1.5 mm (Fig. 6.15).

6.3.6.4.2 Differentiation of the Young Mussel at the Distal End of a Stalk

Instead of growing in a cyst, in *Mutela* the young mussel forms outside the host tissue from a differentiation zone at the free end of the stalk. As seen in the light microscope, the differentiation zone is visible from the second day on. In the middle of the metamorphosed lasidium, traces of the hooks remain discernable in the young mussel for about 10 days after infection. After 2 weeks, the young mussel has almost all the typical bivalve structures and, although still fixed on the stalk, it already starts active filter feeding 1 week before it leaves the fish by breaking off the distal stalk (Fig. 6.15).

6.4 Direct Development in Naiads

Direct development from egg to young mussels omitting a parasitic stage has been reported for some American species, *Anodonta inbecilis*, *Strophitus undulatus*, *Obliquaria reflexa* and others (Parodiz and Bonetto 1963; Mackie 1984; Heard 1998) and recently for *Lasmigona subviridis* (Barfield and Watters 1999). *Grandidieria burtoni*, a unionid from Lake Tanganyika (Kondo 1990) is a case of direct development in Africa. In European unionoids this was obviously never observed. As at least in some cases it is assumed that the juvenile mussels carried for a while in the marsupia develop from glochidia, this ontogenetic mode is certainly a derived feature. It is interesting that the glochidia in their morphology resemble those of other species but they have obviously lost their infectiosity (Lefevre and Curtis 1911, 1912; Howard 1914a,b, 1915; Allen 1924). Barfield and Watters (1999) were able to collect juvenile mussels out of the marsupia of *Lasmigona subviridis*, thus confirming earlier observations on direct development. According to Lefevre and Curtis (1911, 1912) facultative parasitism may occur in *Strophitus undulatus* and *Obliquaria reflexa*. The

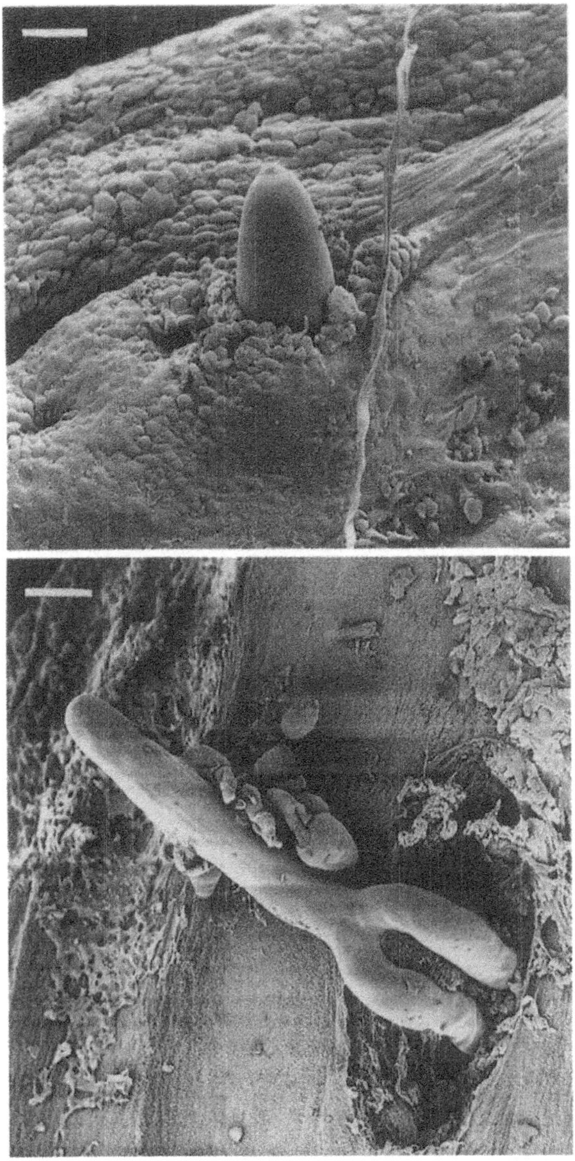

a

b

Fig. 6.14a,b. Lasidia of *Mutela bourguignati* soon after attachment on *Barbus altianalis radcliffii* forming a pair of haustoria in the fish skin. At the free end the juvenile mussel originates. **a** Stage first day (↔50 µm). **b** Fish skin partly removed sixth day (↔100 µm). (Falk 1995)

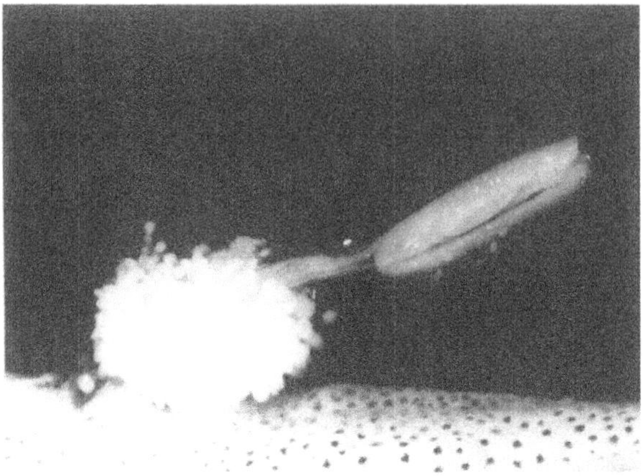

Fig. 6.15. In *Mutela bourguignati* the young mussel forms at the distal end of the haustorium (basis covered with colonial ciliata)

conditions under which either parasitic or direct development is chosen are not known.

6.5 From Ectoparasitism to Filter Feeding: the Biology of the Juvenile Mussel

6.5.1 Leaving the Host Fish

The release from the host starts by an opening up of the epithelial cyst in nearly all species with the interesting exception of *Mutela bourguignati*, which breaks off at the distal end of the haustorium.

6.5.2 Morphology

In spite of differences in the larval structures and the length of the parasitic period, the juvenile mussels dropped by the host are very similar in their basic structural design (Figs. 6.9, 6.11d).

6.5.2.1 Initial Size

In most species, the juvenile mussel is about the same size as the glochidium. In *Margaritifera margaritifera*, however, the juvenile mussel grows considerably from about 80 to 450 µm. In *Mutela bourguignati* a lasidium of 200 µm length changes into a juvenile mussel of 1.5 mm (Falk 1995).

Increase in carbon content can be confirmed in the course of the parasitic development even if no increase in size is noticeable. Some data are given in Table 6.3.

The relation between size of the glochidia and length of the parasitic period is discussed by Bauer 1994.

Table 6.3. Size and carbon content in glochidia and juveniles of four species of freshwater bivalves. Data from Kuhfuß (1992), Niemeyer (1992) and Hetzel (1993)

	Size in μm		Carbon content in μg	
	Gloch.	Juv.	Gloch.	Juv.
Margaritifera margaritifera	80	450	0.017	1.786
Anodonta anatina	340	340	0.630	1.650
Anodonta cygnea	340	340	0.530	1.430
Unio tumidus	220	220	0.221	0.387

6.5.2.2 Shells

The typical shell structure with an elastic hinge and an anterior and posterior adductor muscle ensuring closure of the two valves can already be seen while the juvenile is still in the cyst. Growth ring formation and shell growth depend very much on the species and the temperature preferred. Shell pores possibly improving the gas exchange are reported from some species. The glochidial shell can sit on top of the definite one giving for instance early free-living *Anodonta* stages their characteristic triangular shape (Fig. 6.16). Irregular protuberances on the umbo are found on the shells of young *Unio tumidus*, *Unio pictorum*, *Anodonta cygnea* and several other species.

6.5.2.3 Gills

In many, if not most, species benthic life starts before the gills are fully developed. In postlarval *Unio* two and in postlarval *Margaritifera* four pairs of gill papillae are found. There is vivid ciliary activity on the gill surface. In contrast to species spending the whole parasitic period in a cyst, gill formation in *Mutela bourguignati* is so far advanced that this species is already a filter feeder for the last third (about 1 week) of its parasitic life. It is, however, not known to which proportion *Mutela* feeds by filtration and to which proportion it takes up nutrients through the haustorium at this period.

6.5.2.4 Foot and Locomotion

Compared to adult bivalves, the juveniles are remarkably mobile. This is due to the elongated foot capable of relatively quick peristaltic movements (Figs. 6.11d, 6.16) (Wächtler 1986; Yeager et al. 1994).

6.5.2.5 Mantle

In juvenile mussels the interior of the shells is laid out with the ciliated epithelium of the mantle. This epithelium successively replaces the larval mantle which forms soon

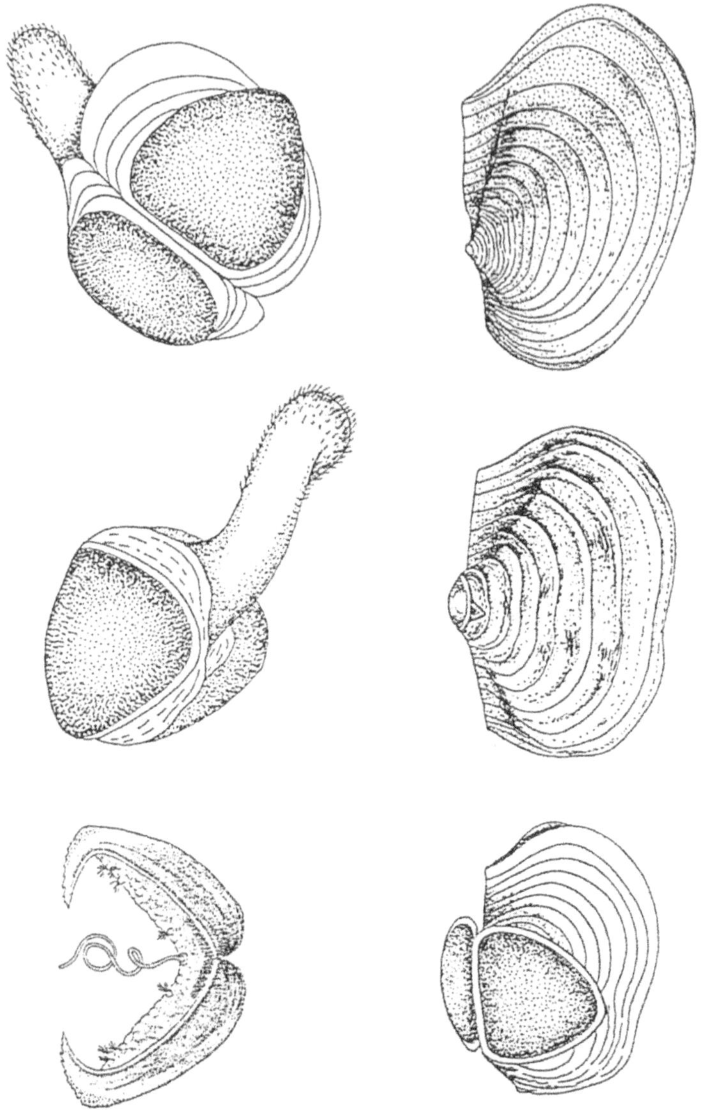

Fig. 6.16. *Anodonta cygnea:* glochidium and early postlarval stages. Juvenile mussels are released from the host fish with almost the same shell length as the glochidia. The valves of the glochidia (*dark*) persist for some time on top of the growing juvenile shell. (Redrawn from several authors)

after encystation, covering the interior with the huge spherical cells described above (Fig. 6.5).

6.5.2.6 Digestive System

The intestinal system consisting of a ciliated mouth opening surrounded by cirrae, a large digestive gland with diverticula and a stomach followed by an uncoiled intestinum, is complete by the end of the parasitic period. Particularly in the largely transparent juveniles of *Unio* and *Anodonta*, the vivid rotation of the crystalline stylus can be seen as soon as benthic life starts and the yellow or green content of the digestive gland indicates the onset of food uptake (Maaß 1987; Niemeyer 1992).

6.5.3 Ecology

As long as ecological conditions are favourable for both mussels and their hosts, unionoids may be among the most successful freshwater invertebrates with very high population densities. The risks of missing the host or being transported to the wrong place are compensated for by the production of a very large number of small larvae and the long life expectancy of the adult mussels, which is quite unusual among invertebrates (Ziuganov et al. 1998b). Reports on declining naiad populations worldwide from many regions (Young and William 1983b; Zettler 1996; Hochwald 1997) show, however, that this mode of development is less safe than one would have assumed some decades ago. That is why the further study of the ecological requirements of the juvenile mussel, which is obviously the most vulnerable developmental stage, deserves particular attention.

6.5.3.1 Habitat Selection

As in most lakes and rivers suitable benthos sites are not evenly distributed, there are considerable risks for the young mussels to be dropped from the host fish at an unfavourable place. Within a suitable site some adjustment is possible by active movements (Wächtler 1986). What habitats are actively chosen is difficult to study in the field and not known for many species. As survival under laboratory conditions is limited for many species, the influence of subthreshold factors is difficult to assess in long-term studies. What can be concluded from many observations is that, at least for many species, the habitat of choice at the early postparasitic stage is within or above the upper millimetres or centimetres of the sediment provided that it is well-aerated (Young and Williams 1983b; Neves and Widlack 1988; Buddensiek 1991; Buddensiek et al. 1993). It is just here that they can be exposed to sediment-bound contaminants from low but chronic inputs of toxicants hardly noticeable in the water (Salomons et al. 1987; Yeager et al. 1994).

6.5.3.2 Feeding

Feeding undergoes drastic changes during metamorphosis. On the host fish food uptake appears to occur via the dense microvillus felt of the larval mantle cells. In many species storage granules obviously produced by the larval mantle cells accumulate during the parasitic stage. These are used up during the early postparasitic life. This depletion may be the reason for a sudden breakdown of young mussels in laboratory cultures after a few weeks.

The active postparasitic food uptake occurs by water flow caused by cilia at several sites. The gill papillae certainly contribute to this particle stream but the structure mainly involved is the foot with a ciliary groove on its surface. If kept on sediment with small water-filled depressions, young *Margaritifera* can be observed to make probing movements with the tip of the foot on sand grains and the water surface, thus possibly taking up bacteria from the microfilm on both the water surface and the sediment. The role of the foot in early nutrient uptake was recently stressed by Yeager et al. (1993, 1994), who studied early benthic stages of *Villosa iris*. They distinguish between pedal-sweep and pedal-locomotory feeding behaviour, which the juveniles can prefer respectively according to the particle density of their surroundings. It is assumed that with increasing differentiation of the gills the adult mode of feeding is gradually adopted.

6.5.3.3 Growth

Increase in shell size in early postparasitic stages varies from species to species. It is, within limits, temperature-dependent (Gatenby et al. 1997). Some observations on the growth of juvenile *Unio pictorum* (over 12 months) and *Margaritifera margaritifera* (over 26 months) are given in Fig. 6.17.

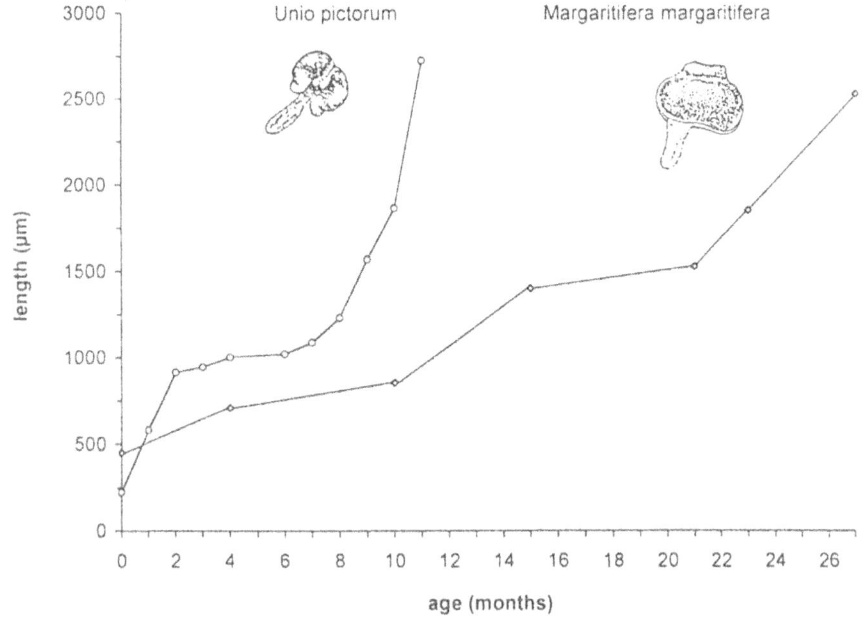

Fig. 6.17. Survival and growth of juvenile *Unio pictorum* and *Margaritifera margaritifera* caged after release from the host fish for 12 months and 26 months, respectively, in flowing pond water. Month *o* June $n = 460$ juveniles (T. Bauer 1992)

6.6 Comparative Remarks on Naiad Development

All possible modes of development known from marine bivalves can also occur in freshwater species: production of planktic larvae, viviparity and direct development. In naiads an additional mode evolved: larvae living as ectoparasites on fish and thus linking the life cycles of two species. This is a very efficient mode of dispersal by which these slowly moving benthos species are able to reach distant sites on the bottom of lakes and remote upstream habitats in brooks and rivers. It is certainly a derived trait which emerged late in bivalve phylogeny. If once attached to the host, the larvae are protected and well provided with resources. What makes the comparison of these ectoparasites so interesting is the fact that the young mussels, almost identical in shape and structure, can arise in two basically different ways either inside a cyst or at the end of a stalked haustorium. Compared to juvenile mussels from the cysts, those from the haustoria are three to five times larger and certainly more effective filter feeders when they start benthic life. This could mean an obvious advantage. However, considering its distribution, haustoria formation occurs in only one or possibly very few African species, whereas cyst formation is found worldwide both in species producing glochidia or lasidia. In most cases where bivalves are threatened with extinction, some parts of the life cycle can still be completed. The main reason for the breakdown of a population is often not reduced fertility, lack of hosts or increased adult mortality, but the incapability of the released juveniles to find adequate conditions in the upper millimetres or centimetres of the benthos near the water-sediment interface where they can take up detritus and components of the biofilm.

6.7 Summary

Naiads produce three kinds of larvae: glochidia, lasidia and (derived from lasidia) haustoria. These are produced in huge numbers. For further development, they have to be taken up by a host, which is in most cases a teleost fish. Once taken up by a host, they pass a stage as ectoparasites encysted (with only one known exception) by fish tissue. Larval morphology, the specificity of the host parasitic relations, length of the parasitic stage, cases of direct development, and the ecology of juveniles during the early postlarval period are reviewed. The exceptional development of the African mutelid *Mutela bourguignati*, where instead of encystation, haustorium formation takes place, is described in some detail and documented by scanning EM pictures.

Acknowledgements. I wish to thank Mrs. Hella Breitrück and Mrs. Martina Pohlmann for preparing the manuscript and the figures with patience and precision.

References

Allen E (1924) The existence of a short reproductive cycle in *Anodonta imbecillis*. Biol Bull 46:88–94

Altaba CR (1990) The last known population of the freshwater mussel *Margaritifera auricularia* (Bivalvia, Unionida): a conservation priority. Biol Conserv 52:271–286

Araujo R, Ramos MA (1998) Description of the glochidium of *Margaritifera auricularia* (Spengler 1793) (Bivalvia, Unionoidea). Philos Trans R Soc Lond B 353:1553–1559

Arey LB (1932a) The formation and structure of the glochidial cyst. Biol Bull 62:212–221
Arey LB (1932b) The nutrition of glochidia during metamorphosis. J Morphol 53:201–221
Barfield M, Watters GT (1999) Non-parasitic life cycle in the green floater, *Lasmigona subviridis* (Conrad, 1835). Trian Unionid Rep 16:22
Bauer G (1987) The parasitic stage of the freshwater pearl mussel. II. Suspectibility of brown trout. Arch Hydrobiol 76:403–412
Bauer G (1989) Die bionomische Strategie der Flußperlmuschel. Biol Unserer Zeit 19:69–75
Bauer G (1994) The adaptive value of offspring size among freshwater mussels (Bivalvia, Unionoidea). J Anim Ecol 63:933–944
Bauer G, Vogel C (1987) The parasitic stage of the freshwater pearl mussel. I. Host response to glochidiosis. Arch Hydrobiol 76:393–402
Bauer T (1992) Untersuchungen zur Hälterung einheimischer Unioniden unter besonderer Berücksichtigung der frühen postparasitären Phase von *Unio tumidus* Philipsson und *Unio pictorum* Linnaeus. Dipl Thesis, Universität Hannover
Bednarczuk J (1986) Untersuchungen zu Wirtsfischspektrum und Entwicklung der Bachmuschel *Unio crassus*. Thesis, Tierärztliche Hochschule Hannover
Bonetto A (1959) Contribución al conocimiento de las gloquidias del género *Diplodon*. Primer Congr. Sudamer. Zool La Plata 2:43–59
Bonetto A (1961) Nuevas notas sobre formas larvales de Nayades Sud y Centroamericanas. Physis 21:332–335
Bonetto A, Ezcurra I (1965) Estudio comparado de las formas larvales de Mutelidae Ortmann y su significacion sistematica y zoogeografica (Mollusca, Pelecypoda). An Segundo Congr Lat Am Zool 2:55–71
Buddensiek V (1991) Untersuchungen zu den Aufwuchsbedingungen der Flußperlmuschel *Margaritifera margaritifera* LINNAEUS (Bivalvia). Thesis, Universität Hannover
Buddensiek V (1995) The culture of juvenile freshwater pearl mussels *Margaritifera margaritifera* L. in cages: a contribution to conservation programmes and the knowledge of habitat requirements. Biol Conserv 74(1):33–40
Buddensiek V, Engel H, Fleischauer-Rössing S, Wächtler K (1993) Studies of the chemistry of interstitial water taken from defined horizons in the fine sediments of bivalve habitats in several northern German lowland waters. II: Microhabitats of *Margaritifera margaritifera* L., *Unio crassus* (Philipsson) and *Unio tumidus* Philipsson. Arch Hydrobiol 127:151–166
Carus CG (1832) Neue Untersuchungen über die Entwicklungsgeschichte unserer Flußmuschel. Verh K Leopoldinisch-Carolinischen Akad Naturforsch 16:1–87
Claes M (1987) Untersuchungen zur Entwicklungsbiologie der Teichmuschel *Anodonta cygnea*. Dipl Thesis, Universität Hannover
Coker RE, Shira AF, Clark HW, Howard AD (1921) Natural history and propagation of freshwater mussels. Bull US Bur Fish (Doc 893)37:75–181
Dettmer R (1982) Untersuchungen zur Ökologie der Flußperlmuschel (*Margaritifera margaritifera*) L. in der Lutter im Vergleich mit bayrischen und schottischen Vorkommen. Dipl Thesis, Universität Hannover
Dreher-Mansur MC (1998) Hyriidae (Unionoida) glochidia: a review of the discoveries and descriptions of unknown or little-known larvae. In: Bieler R, Mikkelsen PM (eds) World congress of malacology, Washington DC (Abstr) Unitas Malacologica 1998, p 207
Dreher-Mansur MC, Olazarri J (1995) Redescription, distribution and habitat preferences of *Anodontites ferrarisi* (Orbigny, 1835) revalidated (Bivalvia, Unionoida, Mycetopodidae). Iheringia Ser Zool 0(79):3–12
Engel H (1990) Untersuchungen zur Autökologie von *Unio crassus* (PHILIPSSON) in Norddeutschland. Thesis, Universität Hannover
Engel H, Wächtler K (1990) Folgen von Bachentkrautungsmaßnahmen auf einen Süßwassermuschelbestand am Beispiel eines kleinen Fließgewässers des Südlichen Drawehn (Lüchow-Dannenberg). Natur Landschaft 65:63–65
Falk L (1995) Das ektoparasitische Larvenstadium der afrikanischen Süßwassermuschel *Mutela bourguignati BOURGUIGNAT* 1885. Thesis, Tierärztliche Hochschule Hannover

Falk L, Hermann F, Wächtler K (1996) *Mutela bourguignati*: Eine afrikanische Süßwassermuschel mit abweichender Entwicklung. Verh Dtsch Zool Ges 89.1:326

Fleischauer-Rössing S (1990) Untersuchungen zur Autökologie von *Unio tumidus* Philipsson und *Unio pictorum* L. (Bivalvia) unter besonderer Berücksichtigung der frühen postparasitären Phase. Thesis, Universität Hannover

Fryer G (1959) Development in a mutelid lamellibranch. Nature (Lond) 183:1342–1343

Fryer G (1961) The developmental history of *Mutela bourguignati* (Ancey) *bourguignat* (Mollusca: Bivalvia). Philos Trans R Soc Edinb B 244:259–298

Fryer G (1970) Biological aspects of parasitism of freshwater fishes by crustaceans and molluscs. Symp Br Soc Parasitol 8:103–118

Fuller SL (1974) Clams and mussels (Mollusca: Bivalvia). In: Hart CW, Fuller LH (eds) Pollution ecology of freshwater invertebrates. Academic Press, New York, pp 215–273

Gatenby CM, Parker BC, Neves RJ (1997) Growth and survival of juvenile rainbow mussels, *Villosia iris* (Lea, 1829) (Bivalvia: Unionidae), reared on algal diets and sediment. Am Malacol Bull 14(1):57–66

Giusti F (1973) The minute shell structure of the glochidium of some species of the genera *Unio*, *Potamida* and *Anodonta* (Bivalvia, Unionacea). Malacologia 14:291–301

Giusti F, Castagnolo L, Moretti FL, Renzoni A (1975) The reproductive cycle and the glochidium of *Anodonta cygnea* L. from Lago Trasimeno (Central Italy). Monit Zool Ital 9:99–118

Gordon ME, Smith DG (1990) Autumnal reproduction in *Cumberlandia monodonta* (Unionidea: Margaritiferidae). Trans Am Microsc Soc 109:407–411

Haag WR, Warren ML Jr (1997) Host fishes and reproductive biology of six freshwater mussel species from the Mobile Basin, USA. J North Am Benthol Soc 16(3):576–585

Haag WR, Butler RS, Hartfield PD (1995) An extraordinary reproductive strategy in freshwater bivalves: prey mimicry to facilitate larval dispersal. Freshwater Biol 34(3):471–476

Heard WH (1998) Brooding patterns in freshwater mussels. Malacol Rev Bivalvia I Suppl 7:105–121

Herbers K (1914) Entwicklungsgeschichte von *Anodonta cellensis* Schröt. Z Wiss Zool 108:1–174

Herrmann F (1995) Untersuchungen zum Lebenszyklus der afrikanischen Süßwassermuschel *Mutela bourguignati* Bourguignat 1885. Thesis, Tierärztliche Hochschule Hannover

Hetzel U (1993) Reproduktionsbiologie – Aspekte der Viviparie bei Sphaeriidae mit dem Untersuchungsschwerpunkt *Musculium lacustre* (O. F. Müller 1774) (Bivalvia; Eulamellibranchiata). Thesis, Universität Hannover

Hochwald S (1997) Populationsökologie der Bachmuschel (*Unio crassus*). Bayreuther Forum Ökologie, Band 50, Bayreuth

Hochwald S, Bauer G (1990) Untersuchungen zur Populationsökologie und Fortpflanzungsbiologie der Bachmuschel *Unio crassus* Phil. 1788. Schriftenr Bayer Landesamtes Umweltschutz 97:31–49

Hoggarth MA (1992) An examination of the glochidia-host relationships reported in the literature for North American species of Unionacea (Mollusca: Bivalvia). Malacol Data Net 3(1–4):1–30

Hoggarth MA, Gaunt AS (1988) Mechanics of glochidial attachment (Mollusca: Bivalvia: Unionidae). J Morphol 198:71–81

Howard AD (1914a) Some cases of narrowly restricted parasitism among commercial species of fresh water mussels. Trans Am Fish Soc 44:41–44

Howard AD (1914b) A new record in rearing fresh water mussels. Trans Am Fish Soc 44:45–47

Howard AD (1915) Some exceptional cases of breeding among the Unionidae. Nautilus 29:4–11

Howard AD (1951) A river mussel parasitic on a salamander. Nat Hist Miscellanea 77:1–6

Howells RG (1997a) Status of freshwater mussels (Bivalvia: Unionidae) of the Big Ticket region of eastern Texas. Texas J Sci 49(3):21–34

Howells RG (1997b) New fish hosts for nine freshwater mussels (Bivalvia: Unionidae) in Texas. Texas J Sci 49(3):255–258

Hüby B (1988) Zur Entwicklungsbiologie der Fließgewässermuschel *Pseudanodonta complanata*. Thesis, Tierärztliche Hochschule Hannover

Humphrey CL, Simpson RD (1985) The biology and ecology of *Velesunio angasi* (Bivalvia: Hyriidae) in the Malaga Creek, Northern Territory. Report to the Office of Supervising Scientist for the Alligator Rivers Region, Kingston, Australia, Open File Record 38:476 pp, 2 appendices

Jones HA, Simpson RD, Humphrey CL (1986) The reproductive cycles and glochidia of freshwater mussels (Bivalvia: Hyriidae) of the Macleay River, northern New South Wales, Australia. Malacologia 27:185–202

Jupiter SD, Byrne M (1997) Light and scanning electron microscopy of the embryos and glochidia larvae of the Australian freshwater bivalve *Hyridella depressa* (Hyriidae). Invertebrate Reprod Dev 32(2):177–186

Kinzelbach R, Nagel KO (1986) Redescription of the glochidium of *Pseudanodonta complanata* (Bivalvia, Unionidae). Verh Naturwiss Ver Hamb NF 28:65

Kirk GS, Layzer BJ (1997) Induced metamorphosis of freshwater mussel glochidia on nonhost fish. Nautilus 110(3):102–106

Kondo T (1984) Hosts of the larvae of *Moncetia lavigeriana* (Bivalvia: Mutelidae) in Lake Tanganyika. Venus Jpn J Malacol 43:347–352

Kondo T (1990) Reproductive biology of a small bivalve *Grandidieria burtoni* in Lake Tanganyika. Venus Jpn J Malacol 49:120–125

Kondo T, Yoshihara M, Motomochi T, Yamaguchi H (1997) Calcification of glochidial shells of some Japanese unionid mussels. Venus Jpn J Malacol 56(2):169–175

Kraemer LR (1970) The mantle flap in three species of *Lampsilis* (Pelecypoda: Unionidae). Malacologia 10:225–282

Kuhfuß T (1992) Histologische Untersuchungen zur Wirt-Parasit Beziehung während der Entwicklung von Unionidenarten (Bivalvia). Dipl Thesis, Universität Hannover

Lefevre G, Curtis WC (1910) Reproduction and parasitism in the Unionidae. J Exp Zool 9(1):79–115

Lefevre G, Curtis WC (1911) Metamorphosis without parasitism in the Unionidae. Science 33:863–865

Lefevre G, Curtis WC (1912) Studies on the reproduction and artificial propagation of freshwater mussels. Bull US Bur Fish 30:105–201

Leydig F (1866) Mittheilung über den Parasitismus junger Unioniden an Fischen in Noll. Tübingen, Inaugural-Dissertation, Frankfurt-am-Main

Maaß S (1987) Untersuchungen zur Fortpflanzungsbiologie einheimischer Süßwassermuscheln der Gattung *Unio*. Thesis, Tierärztliche Hochschule Hannover

Mackie GL (1984) Bivalves. In: Wilbuer KM (Editor in Chief) The Mollusca, vol 7. Reproduction. Academic Press, London, pp 351–418

Matteson MR (1955) Studies on the natural history of the Unionidae. Am Midl Nat 53:126–145

Nagel KO, Castagnolo L (1993) Fish hosts for the glochidium of *Unio mancus*. Riv Idrobiol 30(2–3):339–346

Neves RJ, Widlack JC (1988) Occurrence of glochidia in stream drift and on fishes of the upper North Fork Holston River, Virginia. Am Midl Nat 119:111–120

Neves RJ, Zale AV, Weaver LR (1985) An evaluation of host fish suitability for glochidia of *Villosa vanuxemi* and *V. nebulosa* (Pelecypoda: Unionidae). Am Midl Nat 113(1):13

Nezlin LP, Cunjak RA, Zotin AA, Zivganov VV (1994) Glochidium morphology of the freshwater pearl mussel (*Margaritifera margaritifera*) and glochidiosis of Atlantic salmon (*Salmo salar*): a study by scanning electron microscopy. Can J Zool 72:15–21

Niemeyer B (1992) Vergleichende Untersuchungen zur bionomischen Strategie der Teichmuschelarten *Anodonta cygnea* L. und *Anodonta anatina* L. Thesis, Universität Hannover

O'Dee SH (1998) New or confirmed host identifications for 11 freshwater mussels. In: Bieler R, Mikkelsen PM (eds) World congress of malacology, Washington, DC (Abstr) Unitas Malacologica 1998, p 242

Ortmann AE (1921a) The anatomy of certain mussels from the upper Tennessee. Nautilus 34:81–91

Ortmann AE (1921b) Marsupium and Glochidium der südamerikanischen Muscheln aus der Unterfamilie der Hyriinae. Arch Molluskenkd 53:103–111

Ortmann AE (1921c) South American naiades; a contribution to the knowledge of the freshwater mussels of South America. Mem Carnegie Mus 8:451–671

Panha S (1993) Glochidiosis and juvenile production in a freshwater pearl mussel, *Chamberlainia hainesiana*. Invertebr Reprod Dev 24(3):157–160

Panha S (1995) Glochidium shell morphology of Thai amblemid mussels. Venus Jpn J Malacol / Kairuigaku-zasshi 54(3):225–236

Parodiz JJ, Bonetto AA (1963) Taxonomy and zoogeographic relationships of the South American najades (Pelecypoda: Unionacea and Mutelacea). Malacologia 1:179–213

Pekkarinen M (1996) Scanning electron microscopy, whole-mount histology, and histochemistry of two anodontine glochidia (Bivalvia: Unionidae). Can J Zool 74(11):1964–1973

Pekkarinen M, Englund VPM (1995a) Sizes of intramarsupial unionacean glochidia in Finland. Arch Hydrobiol 134(3):379–391

Pekkarinen M, Englund VPM (1995b) Description of unionacean glochidia in Finland, with a table aiding in their identification. Arch Hydrobiol 134(4):515–531

Pekkarinen M, Valovirta I (1996) Anatomy of the glochidia of the freshwater pearl mussel, *Margaritifera margaritifera* (L.). Arch Hydrobiol 137(3):411–423

Prashad B (1918a) Studies on the anatomy of Indian Mollusca. 1. The marsupium and glochidium of the genus *Physunio*. Rec Indian Mus 14:183–185

Prashad B (1918b) Studies on the anatomy of Indian Mollusca. 2. The marsupium and glochidium of some Unionidae and on the Indian species hitherto assigned to the genus *Nodularia*. Rec Indian Mus 15:143–148

Rathke J (1797) Om Dam-Muslingen. Naturhist Selskabets Skr Kjöbenhavn 4:139–179

Raulerson L, Burbanck WD (1962) The life cycle and ecology of *Elliptio hopetonensis* Lea. Assoc Southeast Biol Bull 9:39 (Abstr)

Reynolds JD, Debuse VJ, Aldridge DC (1997) Host specialisation in an unusual Symbiosis: European bitterlings spawning in freshwater mussels. Oikos 78(3):539–545

Roe KJ (1998) Molecular systematics of the "superconglutinate" producing mussels of the genus *Lampsilis* (Bivalvia: Unionidae). In: Bieler R, Mikkelsen PM (eds) World congress of malacology, Washington DC (Abstr) Unitas Malacologica 1998, p 280

Salomons W, De Rooij NM, Kerdijk H, Bril J (1987) Sediments as a source for contaminants? Hydrobiologia 149:13–30

Scharsack G (1994) Licht- und elektronenmikroskopische Untersuchungen an Larvalstadien einheimischer Unionacea (Bivalvia; Eulamellibranchiata). Thesis, Universität Hannover

Schierholz C (1889) Über Entwicklung der Unioniden. Denkschriften der Kaiserlichen Akademie der Wissenschaften, Wien. Math Naturwiss Classe 55:183–214

Seshaiya RV (1941) Tadpoles as hosts for the glochidia of the fresh-water mussel. Curr Sci Bangalore 10:535–536

Seshaiya RV (1969) Some observations on the life histories of South Indian freshwater mussels. Malacologia 9(1):286–287

Shi-Anjing (1994) Observation on reproductive cells and glochidium of *Anodonta woodiana* pacifica with scanning electron microscope. Sichuan Daxue Xuebao (Ziran Kexueban) 31(Spec Issue):170–174

Sylvester JR, Holland LE, Kammer TK (1984) Observations on burrowing rates and comments on host specificity in the endangered mussel *Lampsilis higginsi*. J Freshwater Ecol 2:555–559

Tankersley RA (1996) Multipurpose gills: effects of larval brooding on the feeding physiology of freshwater unionid mussels. Invertebr Biol 115(3):243–255

Tankersley RA, Dimock RV (1993a) Endoscopic visualization of the functional morphology of the ctenidia of the unionid mussel *Pyganodon cataracta*. Can J Zool 71:811–819

Tankersley RA, Dimock RV (1993b) The effect of larval brooding on the respiratory physiology of the freshwater unionid mussel *Pyganodon cataracta*. Am Midl Nat 130:146–163

Tucker ME (1927) Morphology of the glochidium and juvenile of the mussel *Anodonta imbecillis*. Trans Am Microsc Soc 46:286–291

Tucker ME (1928) Studies on the life cycles of two species of fresh-water mussels belonging to the genus *Anodonta*. Biol Bull 54:117–127

van der Schalie H (1970a) Hermaphroditism among North American freshwater mussels. Malacologia 10:93–112

van der Schalie H (1970b) Mussels in the Huron River above Ann Arbor in 1969. Sterkiana 39:17–22

von Ihering H (1891) *Anodonta* und *Glabaris*. Zool Anz 14:474–484

Wächtler K (1986) Zur Biologie der Flußperlmuschel *Margaritifera margaritifera* (L.). Naturwissenschaften 73:225–233

Watters GT (1994) An annotated bibliography of the reproduction and propagation of the Unionidea (primarily of North America). Ohio Biol Surv Misc Contrib 1:158

Watters GT (1997a) A synthesis and review of the expanding range of the Asian freshwater mussel *Anodonta woodiana* (Lea, 1834) (Bivalvia:Unionidae). Veliger 40(2):152–156

Watters GT (1997b) Glochidial metamorphosis of the freshwater mussel *Lampsilis cardium* (Bivalvia:Unionidae) on larval tiger salamanders, *Ambystoma tigrinum* ssp. (Amphibia: Ambystomidae). Can J Zool 75(3):505–508

Watters GT (1998) Why are freshwater mussels so big? Size, function, and phylogeny. In: Bieler R, Mikkelsen PM (eds) World congress of malacology, Washington DC (Abstr) Unitas Malacologica 1998, p 351

Watters GT, O'Dee SH (1996) Shedding of untransformed glochidia by fishes parasitzed by *Lampsilis fasciola* Rafinesque, 1820 (Mollusca: Bivalvia: Unionidae): evidence of acquired immunity in the field? J Freshwater Ecol 11(4):383–389

Watters GT, O'Dee SH (1998) Metamorphosis of freshwater mussel glochidia (Bivalvia: Unionidae) on amphibians and exotic fishes. Am Midl Nat 139(1):49–57

Wilms U (1991) Die Kieme als Brutraum. Histologische und histochemische Untersuchungen an den Süßwassermuscheln *Unio tumidus* Philipson und *Unio pictorum* Linnaeus. Dipl Thesis, Universität Hannover

Wilson KA, Ronald K (1967) Parasite fauna of the sea lamprey (*Petromyzon marinus* von Linné) in the Great Lakes region. Can J Zool 45:1083–1092

Wood EM (1974) Some mechanismus involved in host recognition and attachment of the glochidium larva of *Anodonta cygnea* (Mollusca: Bivalvia). J Zool Lond 173:15–30

Yeager BL, Neves RJ (1986a) *Quadrula cylindrica*. In: Hill DM (ed) Activity 3: Identification of fish hosts. Tennessee Valley Authority Cumberlandian Mollusk Conservation Program, Norris, Tennessee, pp 7–11, 21–35

Yeager BL, Neves RJ (1986b) Reproductive cycle and fish hosts of the rabbit's foot mussel, *Quadrula cylindrica strigillata* (Mollusca: Unionidae) in the upper Tennessee River drainage. Am Midl Nat 116:329–340

Yeager BL, Saylor CF (1995) Fish hosts for four species of freshwater mussels (Pelecypoda: Unionidae) in the upper Tennessee River drainage. Am Midl Nat 133(1):1–6

Yeager MM, Cherry DS, Neves RJ (1993) Interstitial feeding behavior of juvenile unionid mussels. Assoc Southeast Biol Bull 40:113 (Abstr)

Yeager MM, Cherry DS, Neves RJ (1994) Feeding and burrowing behaviors of juvenile rainbow mussels, *Villosa iris* (Bivalvia: Unionidae). J N Am Benthol Soc 13(2):217–222

Yokley P (1972) Life history of *Pleurobema cordatum* (Rafinesque, 1820) (Bivalvia: Unioniacea). Malacologia 11:351–364

Young MR, Williams J (1983a) Redistribution and local recolonization by the freshwater pearl mussel *Margaritifera margaritifera* (L.). J Conchol 31:225–234

Young MR, Williams J (1983b) The status and conservation of the freshwater pearl mussel *Margaritifera margaritifera* Linn. in Great Britain. Biol Conserv 25:35–52

Zale AV, Neves RJ (1982a) Reproductive biology of four freshwater mussel species (Mollusca: Unionidae) in Virginia. Freshwater Invertebr Biol 1:17–28

Zale AV, Neves RJ (1982b) Identification of a host fish for *Alasmidonta minor* (Mollusca: Unionidae). Am Midl Nat 107:386–388

Zale AV, Neves RJ (1982c) Fish hosts of four species of lampsiline mussels (Mollusca: Unionidae) in Big Moccasin Creek, Virginia. Can J Zool 60:2535–2542

Zettler M (1996) The aquatic malacofauna (Gastropoda et Bivalvia) in the catchment area of a North German Lowland River, the Warnow. Limnologica 26(3):327–337

Ziuganov VV, Kaliuzhin SM, Popkovitch EG, Neves RJ (1998a) Can pearl mussels (*Margaritifera margaritifera*) relocate to avoid inhospitable conditions? In: Bieler R, Mikkelsen PM (eds) World congress of malacology, Washington DC (Abstr) Unitas Malacologica 1998

Ziuganov VV, Beletsky VV, Neves RJ, Salan ESM, Lopez JCF, Portabales MAL, Gonzales RA (1998b) Extremes in longevity of the freshwater pearl mussel (*Margaritifera margaritifera*): a model system for long life. In: Bieler R, Mikkelsen PM (eds) World congress of malacology, Washington DC (Abstr) Unitas Malacologica 1998

Zotin AA, Ziuganov VV (1994) Immune basis of host-parasitic relationships between freshwater pearl mussels (Margaritiferidae) and salmonid fish. Izv Akad Nauk Ser Biol Moscow 0(4):701–708

7 Plasticity of Life-History Traits in *Unio crassus*

SUSANNE HOCHWALD

7.1 Introduction

The notably large variability of the European unionid *Unio crassus* was well documented by scientists of the last century. Instead of one species, possibly divided into several subspecies or races, they described more than 200 separate species (Roßmäßler and Kobelt 1835). Among populations of *Unio crassus*, differences in shape, length and weight of shells occur. Also the maximum life span and the growth pattern exhibit considerable plasticity. Sometimes this variability occurs even within the same river. As some of the features in question are life-history traits in the classical sense (Stearns 1992), this variability must have a direct bearing on the survival of mussel populations. This is gaining importance for conservation, since *Unio crassus* belongs, together with *Margaritifera margaritifera, M. auricularia* and *Pseudanodonta complanata*, to the most threatened naiad species in central Europe.

In this chapter the intraspecific plasticity and interrelations among life-history parameters of *Unio crassus* will be demonstrated. Its plasticity will be compared with other members of the Unionoidea, and special attention is given to the freshwater pearl mussel (*Margaritifera margaritifera*). Some underlying mechanisms and factors which cause the plasticity in *Unio crassus* will be identified, and consequences for the survival of the species will be shown.

7.2 Estimations of Body Growth Parameters

The *Unio crassus* populations of 34 brooks in Bavaria, Germany, were investigated. Individual ages were determined by counting the annual shell growth increments, which is quite reliable for the Bavarian stocks of *Unio crassus* (Hochwald 1988). Age- and length-frequency diagrams were generated for each population. The oldest and the longest specimen were taken to estimate the maximum age (a_{max}) and maximum length (l_{max}) in the population.

The body growth pattern was analyzed according to von Bertalanffy (1938) by measuring annual growth rings of 667 specimens along their longest axis and plotting the annual ring length at age t against the length at age t + 1 (Walford 1946). This Ford-Walford-Plot is a straight line, which inclines to the bisector. The point of contact with the bisector gives a theoretical end of growth, because at the bisector the time interval t equals the time interval t + 1. This is the accurate location, at which the growth process leads into the asymptote. From the regression equation (y = a + bx) two body growth parameters can be derived. The formula

$$(l_\infty) = a/1 - b$$

Ecological Studies, Vol. 145
G. Bauer and K. Wächtler (eds.) Ecology and Evolution of the Freshwater Mussels Unionoida

leads to a theoretical final length, the asymptotic length (l_∞).

The second growth parameter, the Bertalanffy body growth constant, is a measure of how fast an individual will reach the asymptotic length. This growth constant can be obtained by

$$k = -\ln b.$$

The Bertalanffy equation is commonly used to describe the growth process of mussels (Taylor 1959; Hughes 1970; Brousseau 1979; Haukioja and Hakala 1979; Bachelet 1980; Bayne and Worrall 1980; Griffiths 1981; MacDonald and Thompson 1985) and other poikilothermic animals (Frank 1975; Hughes and Roberts 1980; Roff 1984).

The parameters (l_∞) and k are average measures because they are derived from several individuals with different growth patterns, whereas the maximum length and maximum age are extreme values perceived from the longest and the oldest, individual of the population, provided these individuals have actually been found (which of course may not always be the case, mainly because of the burying and hiding habit of *Unio crassus*).

7.3 Interrelations Among Body Growth Parameters

7.3.1 Significance of the Body Growth Constant

According to Von Bertalanffy (1960), the body growth constant (k) not only describes the body growth process but can be interpreted as a coefficient of catabolism, because he found the growth constant to be strongly related to the excretion of nitrogen. His assumption was confirmed by Beverton and Holt (1959) in physiological experiments dealing with oxygen consumption of fish. For freshwater mussels Bauer et al. (1991) did some work to identify the meaning of this parameter: the body growth constants of several populations of three najad species were compared. Although there were overlaps, the frequency distributions of growth constants in all species could be separated from each other. The following order could be established: the lowest growth constants were present among the *Margaritifera margaritifera* populations, medium constants were recorded in *Unio crassus*, and the highest values were estimated for *Anodonta anatina*. For each species oxygen consumption of several individuals from two populations was measured under comparable conditions. A lower oxygen consumption in the freshwater pearl mussel compared to the two other species could be shown. Bauer (1992) therefore states that the metabolic rate of freshwater mussels is expressed by their body growth constant. He found this parameter to be a key factor for variation in life span and size of the freshwater pearl mussel. Now I shall aim to prove whether these results are also valid for *Unio crassus*.

7.3.2 Plasticity of Body Growth Parameters in *Unio crassus*

The largest degree of plasticity between the *Unio crassus* populations is evident in the body growth constant, varying between 0.05 and 0.31 (Fig. 7.1); also the plasticity in life span is considerable large. In populations with short-lived individuals the life span lasts only 8 years, whereas in long-lived stocks it may be almost three times longer, approaching 23 years. In comparison to the body growth constant and the maximum age, the length parameters l_{max} and l_∞ do not vary that much. In the population with

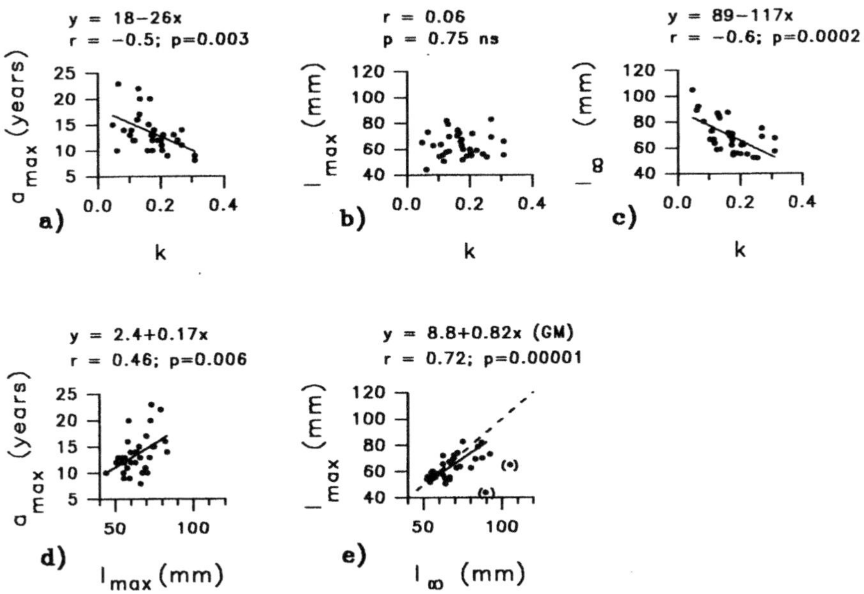

Fig. 7.1a–e. Relationships among the body growth parameters of 34 Bavarian populations of *Unio crassus*. Each *dot* refers to one population. The linear equation (y = a + bx), Pearsons correlation coefficient (*r*) and its level of significance (*p*) is given above the diagrams. a_{max} attainable maximum life span. l_{max} attainable maximum length. L_∞ asymptotic length. *k* Von Bertalanffy growth constant

the smallest growth pattern, the longest individual measured only 43 mm and in the largest-sized stock an individual with 83 mm was found.

The degree of plasticity of the investigated parameters in *Unio crassus* that was observed in the restricted area of Bavarian brooks (47°45′–50°17′N, 10°15′–13°25′W) equals the variation that Bauer (1992) found in the freshwater pearl mussel throughout its whole latitudinal distribution range reaching from Northern Spain up to the arctic circle (43°-66°33′N). According to former authors (Ehrmann 1933; Jäckel 1962), the latitudinal distribution of *Unio crassus* is suggested to be similar to that of the freshwater pearl mussel. If the whole distribution range is taken into account, it is quite likely that the plasticity of *Unio crassus* is much larger than presented in this study. Further north of the present study area, Timm (1994) found *Unio crassus* attaining a life span of up to 75 years. Timm and Mutvei (1993) recorded some shells with the extraordinarily large number of 90 growth increments. Age values obtained from shells of some Bavarian long-lived individuals were verified by H. Mutvei (pers. comm.). The longest individual that Timm (1994) found in Estonian rivers measured 97 mm.

7.3.3 Body Growth Constant Versus Life Span

The maximum age that can be attained in a *Unio crassus* population correlates negatively to the body growth constant. In populations with fast-growing individuals the life span is much shorter than in those with slow-growing specimens (Fig. 7.1a). This result corresponds to findings in the freshwater pearl mussel (Bauer 1992) and in *Anodonta anatina* (Franke 1993).

Bauer (1992) predicted the life span within several freshwater pearl mussel populations just by looking at the body growth constant of some of their shells. His underlying model was derived from separate populations. The study showed that the life of freshwater pearl mussels apparently terminates when the growth pattern has been completed.

Physiological investigations of marine bivalves confirm these findings. The anabolic-catabolic balance of metabolism changes with increasing age (size) of mussels, leading to a very small net energy gain (Joergensen 1976; Vahl 1981) and reduced tolerance to stress (Bayne and Newell 1983) in old individuals. This energetic bottleneck at old age, whose appearance in time is modulated by the growth curve, probably causes the termination of life span in mussels (Vahl 1981; Bauer 1992) and leads to the observed negative correlation between growth constant and age.

Life span shows a positive relationship to maximum size (Fig. 7.1d). This correlation was also found in the freshwater pearl mussel (Bauer 1992). For this species it might partially be due to a shortened growth curve caused by increased mortality rates at high levels of nitrate in the water. This could not be shown for *Unio crassus*, but my findings might be admittedly due to methodological problems with the estimation of correct mortality rates (Hochwald 1997). However, as the interrelations among the growth parameters are linked by metabolic rate, it is not unlikely that the relationship between life span and final size has an allometric component.

7.3.4 Body Growth Constant Versus Length Parameters

A negative correlation exists between the growth constant and the asymptotic length of *Unio crassus* (Fig. 7.1c), leading to a diminishing asymptotic size when growth rate is accelerated. This pattern is valid also for the freshwater pearl mussel (Bauer 1992) and it is possibly a rule among populations of one species or between species of a closely related group (Longhurst and Pauly 1987; Charnow and Berrigan 1991). It is not necessarily due to the fact that both parameters are obtained from the Ford-Walford-Plot (Beverton and Holt 1959). Between the asymptotic length (l_∞) and the maximum length (l_{max}) of *Unio crassus* populations, a significant positive relationship could be established (Fig. 7.1e). If x-axis measurement bias is obeyed by using the reduced major axis regression (Sokal and Rohlf 1995), neither the slope nor the axis of the regression line can be separated from the line $y = x$. Thus, the asymptotic length gives a fairly good prediction of the maximum length found actually outside. Both relations, the negative one between body growth constant and asymptotic length and the positive one between asymptotic length and maximum length, are present in *Unio crassus* (Fig. 7.1c,e) as well as in the freshwater pearl mussel (Bauer 1992). Surprisingly, the maximum length in *Unio crassus* shows no direct correlation to the body growth constant (Fig. 7.1b). This is a remarkable difference to the freshwater pearl mussel, where all the other growth parameters are strongly correlated with the body growth constant, implying overriding control of the metabolic rate about growth. In *Unio crassus* the body growth constant has a visible influence on life span, but no obvious effect on size.

7.4 Influence of Environmental Factors

Unio crassus is distributed in a broad range of different stream habitats, ranging from clear trout brooks, where it is sometimes associated with the freshwater pearl mussel, to trout brooks with higher productivity and content of calcium, where it lives alone or together with *Anodonta anatina* (Bauer et al. 1991) or further downstream to river sections with fish species like grayling (*Thymallus thymallus*) or barbel (*Barbus barbus*). In the downstream types of *Unio crassus* habitats also all of the other European unionids may occur (Wolff 1968). In this study, Bavarian trout and grayling brooks were investigated.

7.4.1 Water Temperature

The water temperature of 25 Bavarian brooks up to 200 km apart was compared. The effective temperature, which gives a fairly good determination of the mean temperature (Pallmann et al. 1940; Kundler 1954; Steubing 1965; Steubing and Fangmeyer 1992), was estimated by the sucrose inversion method. In each of the brooks, tubes filled with saccharose solution were exposed (Hochwald 1997). Dependent on exposition time and mean temperature within the brook, a part of the saccharose splits into glucose and fructose. This leads to a change in the rotation angle of polarized light after passage of the sugar solution. In each brook the exposition of sugar solutions was repeated three times during the main growth season in Bavaria. From these measurement periods, means and 95% confidence limits were calculated per brook. To check for methodological error, several controls were made in the laboratory (Hochwald 1997).

Comparision between the effective temperature and the body growth constant of the mussel populations within the brooks shows a significant positive impact of water temperature on the body growth rate (Fig. 7.2a). The higher the effective temperature, the faster the growth process strives for the asymptote. This temperature dependency of the body growth constant confirms the physiological importance of this parameter, as already mentioned.

With respect to the other body growth parameters of *Unio crassus*, no clear dependency on temperature can be detected (Fig. 7.2b–d), despite some dependencies on growth rate (Fig. 7.1). In all probability, the lesser base of the temperature data in comparison to the data presented in Fig. 7.1 is not the reason for these findings, as the intrinsic correlations within the growth pattern are not sensitive to data reduction (Hochwald 1997).

7.4.2 Physicochemical Parameters of Brook Water

To investigate further environmental impacts on the growth pattern of *Unio crassus*, measurements of physicochemical parameters of brook water were carried out. The intention was to identify parameters which give at least indirect information about the available amount of food or any other influence of water quality on the body growth pattern of the mussel populations.

Samples from 25 brooks were taken in four sampling excursions between May and August 1993. The following parameters of brook water were analyzed: turbidity, conductivity, oxygen content, pH, total organic carbon, biochemical oxygen demand,

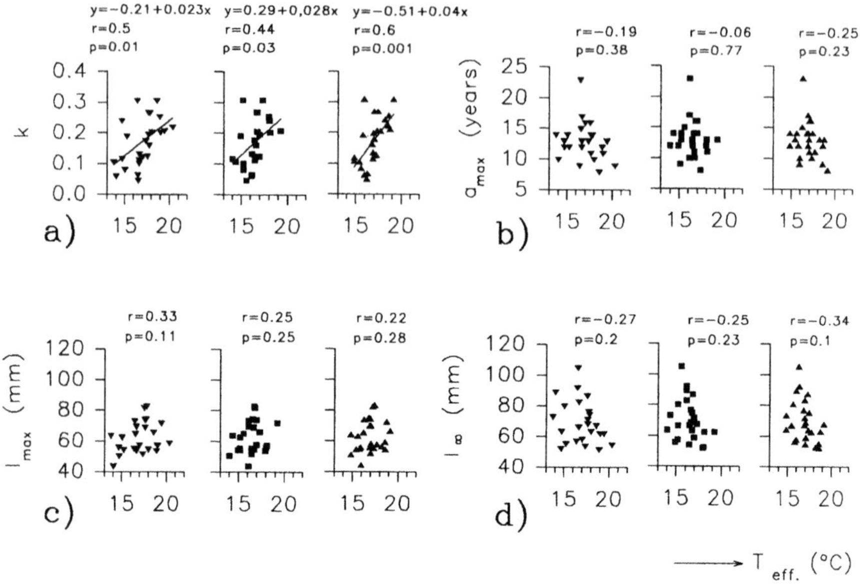

Fig. 7.2a–d. Dependency of body growth parameters from 25 populations of *Unio crassus* on brook water temperature ($T_{eff.}$). **a** k Von Bertalanffy growth constant. **b** a_{max} attainable life span. **c** l_{max} attainable maximum length. **d** l_∞ asymptotic length. *One datapoint* represents the mean of three simulaneous measurements in one population. Confidence limits of this values are too small to display in the diagrams. *Triangles down* Measurement interval of $T_{eff.}$ between May to mid-June; *squares* mid-June to mid-July; *triangles up* mid-July to mid-August

calcium, magnesium, chloride, sulphate, soluble phosphate, nitrate and ammonia. Means were calculated for each brook. Analysis of correlation coefficients showed that there were some intercorrelations within the physicochemical parameters, but an influence by them on the body growth parameters could barely be detected (Hochwald 1997). This was confirmed by a factor analysis, showing that the body growth parameters k, l_∞ and a_{max} lumped together with effective temperature on one factor. Only l_{max} showed another factor together with chloride, BOD_5 and also with effective temperature. Thus, the body growth parameters of *Unio crassus* are almost exclusively correlated to each other and to temperature. Other impacts could hardly be found.

7.5 The Impact of Growth Pattern on Reproduction

7.5.1 Allometric Constraints on Fertility

Gravid females can easily be identified by opening the mussels some millimetres and glancing through the gap between the valves. If they contain spawn, the outer demibranches of the females are thickened. To estimate fertility, gravid females were kept individually in aerated jars until glochidia were released. The gills were inspected again to ensure that no glochidia had remained within the marsupia, and the adult mussels were taken back to their original brooks. Twelve defined samples were taken out of the suspended glochidia, from which the mean and the 95% confidence interval were calculated. After log transformation of both variables, a significant relationship between fertility and body length of females could be established for two populations. This shows that there is an allometric constraint of body length on fertility, following a power function (Fig. 7.3). Bauer (1998) compares correlation coefficients between log fertility and log shell length in several Unionoidea. With a comparable data base he found a significant impact of body size on fertility in two out of nine populations of *Margaritifera margaritifera*. The correlation coefficient of all pooled females was $r^2 = 0.15, P < 0.001$ in pearl mussels. All *Unio crassus* data from Fig. 7.3 have an average correlation coefficient of $r^2 = 0.25, P < 0.001$. For all investigated species of the genus *Unio* Bauer calculated an overall correlation of $r^2 = 0.45$, $P < 0.001$ (Data from several authors given in Bauer 1994). Two *Anodonta* species (*A. anatina, A. grandis simpsoniana*) both had a correlation coefficient of $r^2 = 0.83, P < 0.001$ between log shell length and log fertility. Thus the comparision of the *Unio crassus* data to Bauer (1994, 1998) shows that among the investigated unionids the allometric constraint of body length on fertility is weakest in *Unio crassus*, but it seems to be slightly stronger than in *Margaritifera*.

There is some evidence that many species of the Unionoidea are able to uncouple the energetic demands for reproduction from other body functions and spend mainly surplus energy gained from increased food intake on this purpose (Haukioja and Hakala 1978; Bauer 1987; Jokela and Mutikainen 1995; Bauer 1998). One characteristic of this ability is a wide scatter around the power function of fertility on shell length, caused by several females that do not produce enough glochidia for a complete filling of their marsupia. Sometimes empty regions at the frontal edge of the marsupia or a lesser thickening in comparison to gills of other females can be recognized already during inspection of the outer demibranches of gravid *Unio crassus*.

7.5.2 Differences in Spawning Repetitions Among Individuals and Populations

The European *Unio* species have one tachytictic (short-time) breeding period in spring. A former study (Hochwald 1988) found females which completely paused reproduction for a year and, quite surprisingly, that *Unio crassus* is able to spawn at least twice within one reproductive period, which formerly was unknown for European naiads. For the American lampsiliid *Glebula rotundata*, multiple breeding was already known (Parker et al. 1984). Biannual gametogenesis, but no second brood, had been shown for the freshwater pearl mussel (Smith 1978) and Gordon and Smith (1990) recorded multiple spawnings for the related *Cumberlandia monodonta*. In Germany, Engel (1990) found one *Unio crassus* individuum spawning three times within one season. Nagel (1991) observed five individuals of *Unio crassus* spawning twice. Fleischauer-Rössing (1990) also found multiple breeding per season in the congenus species *Unio pictorum* and *Unio tumidus*. This confirms that the majority, if not even all, European *Unio* species have a multicycled tachytictic breeding behaviour.

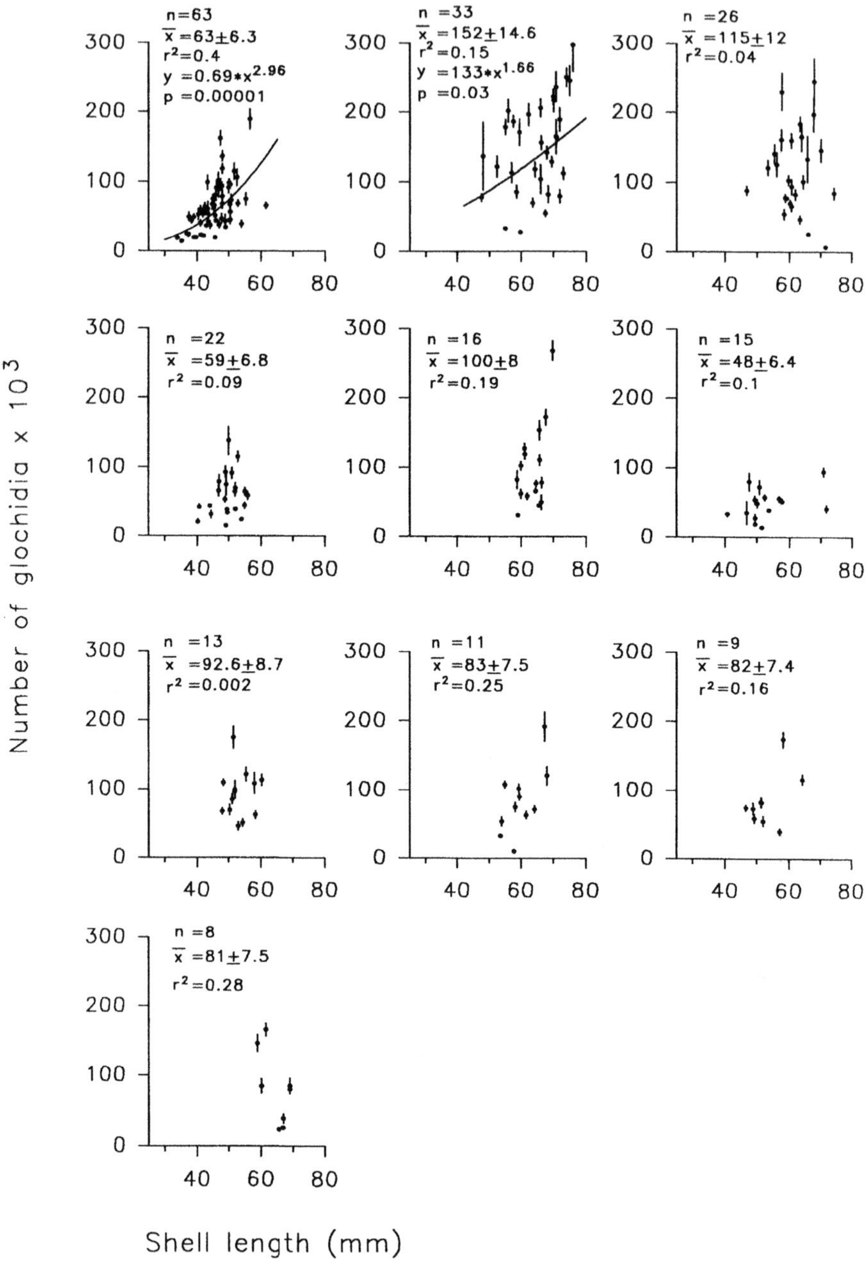

Fig. 7.3. Average fertility (number of glochidia in the marsupia) of females (±95% confidence limits) from ten populations of *Unio crassus*, plotted versus individual length. The number (*n*) of investigated females, the grand mean ($\bar{x}$) of a population (±95% confidence limits) and the explained variance (r^2) between log fertility and log length is given in the diagrams. A *regression line* shows a significant relationship between length and fertility due to the function $y = a^*x^b$

The following presentation will show results from a comparative study of the multiple breeding ability of *Unio crassus*. Individuals of *Unio crassus* in seven brooks were examinated weekly for gravidity. The colour of the thickened outer demibranches gives information about the ripeness of the spawn contained. Moderately thickened gills with dull surface and regular narrow stripes, the boundaries of the filaments, are filled with unripe eggs. With progressing ripeness the gills become still more turgid and their surface more glossy, irregular brown stripes occur and the regular fine pattern of filament boundaries diminishes. If a mussel found to be non-gravid between observations of gravidity, or changed from ripe to unripe this was assessed to be a new brood. The observations demonstrate that up to five spawnings may occur within one season (Fig. 7.4).

When populations were considered separately, no impact caused by the individual's age or length on the number of broods per season was found (Hochwald 1997). Consequently, all data were pooled, and age of individuals was standardized to the growth pattern according to Hughes and Roberts (1980). Again, no relationship between standardized age and mean number of spawnings could be detected in these individuals (Fig. 7.5). While an individual increases in age and length, it apparently does not change the number of broods per year. This supports the formerly mentioned view that an individual is free to spend as much energy towards reproduction as it has gained as surplus. However, old individuals in one population were observed to contain lower numbers of glochidia in their marsupia in the spawnings during the latter half of the reproductive period than in the former ones (Hochwald 1997).

A comparison of the mean number of spawnings between populations shows no relationship to the length parameters of the populations (Fig. 7.6a,b), but the number of spawnings diminishes with increasing maximum life span in a population (Fig. 7.6c). Between the growth constant and the mean number of spawnings in the population, no correlation is detectable (Fig. 7.6d). If data are divided into two groups – three populations with lower and four populations with higher body growth constants (Fig. 7.6d) – the pooled individuals in the "higher" group have significantly more spawnings than those of the "lower" group ($t = -8.28$, $df = 105$, $P < 0.001$). Due to methodological difficulties, the number of investigated populations is admittedly low and therefore it is difficult to confirm significance. So these findings must be interpreted solely as a trend.

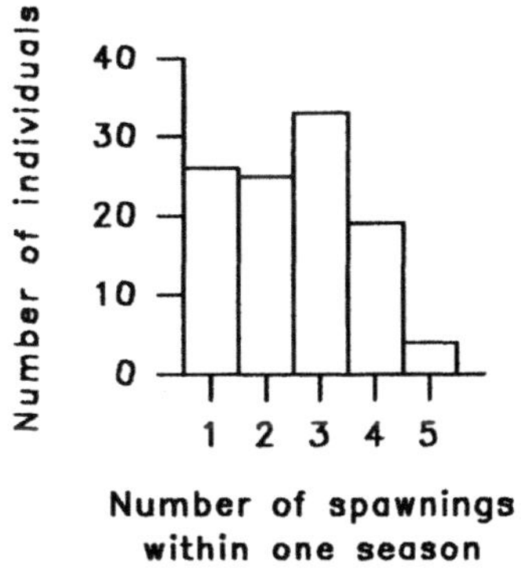

Fig. 7.4. Plasticity in number of spawnings within a brooding season. Pooled data for seven brooks

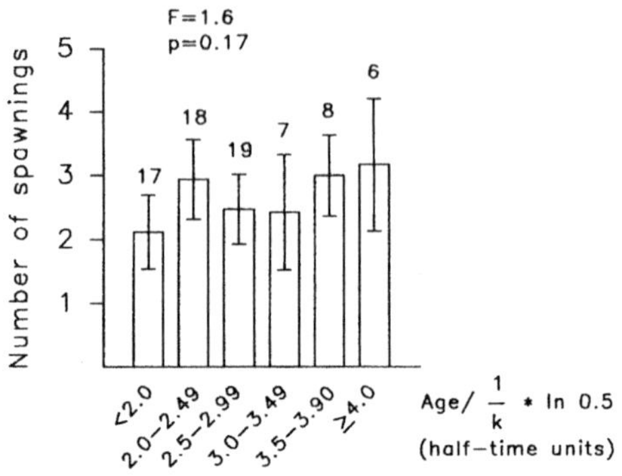

Fig. 7.5. Mean number (±95% confidence limits) of spawnings in different age classes. Data pooled. Age standardized for the growth pattern (Hughes and Roberts 1980). ONEWAY results and number of individuals investigated in each age class are given

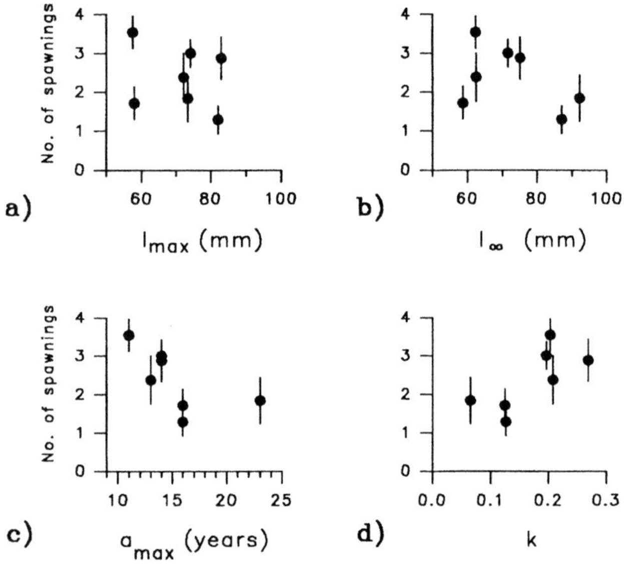

Fig. 7.6a–d. Relationship between the average number of spawnings (±95% con-fidence limits) and body growth parameters in seven populations. Spearman's cor- relation coefficient (r_s) and its level of signi-ficance (p) for **a** l_{max} (attainable maximum length): $r_s = -0.21, p = 0.64$. **b** l_∞ (asymptotic length): $r_s = -0.29, p = 0.53$. **c** a_{max} (attainable life span): $r_s = -0.76, p = 0.046$. **d** k (Von Bertalanffy growth rate): $r_s = 0.57, p = 0.18$

One should be cautious to explain this pattern with the "classical" tradeoff between reproductive effort and adult survival (Williams 1966). Because temperature affects metabolism and, by that, also the body growth constant and life span, it is rather likely that the number of spawning repetitions is also a trait that varies passively as a response to environmental factors like food and temperature. Differences in these factors between rivers lead to differences in the energetic condition of individuals in the populations. The whole energetic condition of mussels will be linked positively to reproduction and because of that to number of spawnings. This direct influence overrides a possibly existing effect of a variable allocation pattern between growth, survival and reproduction (Jokela and Mutikainen 1995) at different energy levels. A direct correlation between temperature and the number of spawning repetitions could not be shown, presumably because of the low number of investigated populations. Temperature dependency of the number of spawning repetitions of mussels is suggested by the work of Ropes and Stickney (1965), who presented a review showing that the marine *Mya arenaria* spawns once in the north and has a biannual spawning in the south of America's northwestern Atlantic coast.

7.6 Consequences of Life-History Variation for the Survival of Populations

Until the first half of this century, *Unio crassus* was the most common European naiad species (Israel 1913; Schnitter 1922). In the past four decades, a catastrophic decline of the species could be documented in Germany (Hochwald 1988; Falkner 1992; Jungbluth et al. 1995), Austria (Reischütz and Sackl 1991) and Poland (Piechocki and Dyduch-Falniowska 1993). The species is claimed to be rare also in The Netherlands (Wolff 1968), in Belgium (Van Goethem 1995), in Carinthia (Mildner and Troyer-Mildner 1992) in Switzerland (Vicentini 1998) and in the Gzech Republic (Beran 2000). Because *Unio crassus* lives completely buried in the substratum, juveniles can hardly be found. Their detection depends mainly on the kind of substratum and the size of the juveniles, which is dependent not only on their age but also on the body growth pattern within their population. Therefore, it was not possible to make a classification of populations just by their age structure. To determine the viability of the population the investigated stocks were divided into four categories (Fig. 7.7).

No pattern between these groups and the asymptotic length of the populations can be detected (Fig. 7.7c). A least significance difference procedure confirms that the maximum length is higher in group 2 than in extinct populations (Fig. 7.7a). Means of body growth constants show a decline from group 1 to group 3, and a slight rise in group 4 (Fig. 7.7d). Group 3 can be distinguished from groups 1 and 2, but not

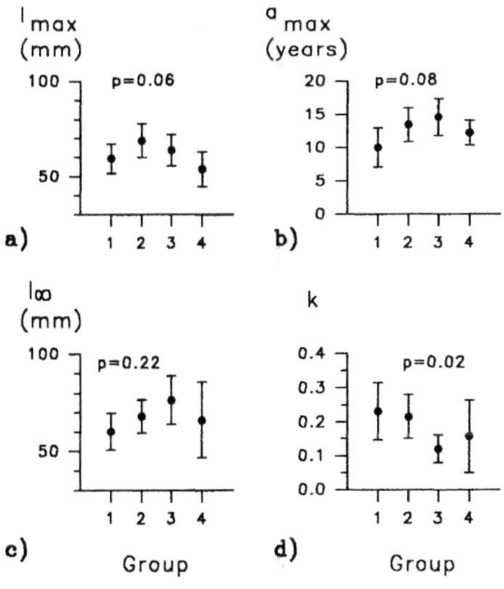

Fig. 7.7a–d. Differences in body growth parameters (mean ±95% confidence limits) between groups of stocks with different population status. *Group 1* four populations with more than 1000 individuals and occurrence of juveniles (<5 years) along a broad area within the population, *group 2* seven stocks with more than 50 animals and at least some juveniles in a restriced region; *group 3* nine completely overaged stocks with less than 50 individuals; *group 4* five extinct populations whose body growth parameters were measured by empty shells that have remained in the river. **a** Attainable maximum length. **b** Attainable life span. **c** Asymptotic length. **d** Von Bertalanffy growth rate. ONEWAY probabilities are given in each graph

from group 4. The life span shows a contrary pattern (Fig. 7.7b). It rises from group 1 to 3 and falls a little in 4. An LSD test tells group 1 from 4. Keeping in mind that the body-growth constant has a negative influence on the life span, this result shows that populations of slow-growing individuals with a long life span survive for a long time and can be detected commonly also if the population consists of only a handful of specimens that no longer have offspring. Stocks with fast-growing, short-living individuals do not survive that long, but suffer a hasty decline from viable populations (group 1) to complete extinction (group 4).

The complex reactions of the growth pattern and affiliated reproductive traits of *Unio crassus* show that the life-history strategy of *Unio crassus* contains traits with broad plasticity, which are tied together closely by typical tradeoffs and positive correlations. There apparently exists a link of metabolic rate and plasticity in some life-history traits, but the life history of *Unio crassus* is – other than the freshwater pearl mussel – not that much constrained by this underlying force. So the fitness of populations is not solely at the mercy of environmental temperature, but it may presumably be enhanced also by the productivity of its habitat that is much more variable than that of the freshwater pearl mussel. A higher food supply may affect predominantly the body length (Beverton and Holt 1959) and therefore reproduction. Nevertheless, the benefits of an advantageous chain of trait modulations at low temperature – having low metabolic rates that lead to a slow growth process with a long life span and therefore a large number of reproductive periods (years) and a fairly good resistance to temporarily high juvenile mortality – can, similarly to the freshwater pearl mussel (Bauer 1991), also be received by *Unio crassus* to a certain degree. In spite of its adaptability, the formerly common species *Unio crassus* has been brought to the verge of extinction by man. It is high time to take conservation measures, which hopefully will keep the species alive in Central Europe and – which should be possible because of its plastic life history – let it spread again in a broad variety of different waters.

7.7 Summary

Within several Bavarian populations of the European bivalve *Unio crassus*, investigations were made regarding life span and maximum attainable length gained from age- and length-frequency diagrams. For each population the asymptotic length and Bertalanffy body growth constant were calculated from Ford-Walford-Plots of annual growth increments. All parameters vary broadly among the populations. Amidst investigated environmental factors, solely mean brook temperature, measured by sucrose inversion method, was identified to have an accelerating influence on body growth rate. Life span and asymptotic length are shortened with elevated body growth constant. No correlation between body growth constant and maximum attainable length could be found, although maximum attainable length is linked positively to asymptotic length as well as to life span.

The constraint of body length on fertility seems to be moderate in *Unio crassus* in comparison to other *Unio*, *Anodonta* or *Margaritifera* species. Up to five spawnings within one reproductive period could be observed for *Unio crassus*. The number of spawnings was found to be independent of standardized age or size of the individual. However, a slight trend towards a declining average number of spawnings with rising life span is visible among populations. Because of the complicated building of relationships between life-history traits depending mainly on body growth rate as an expression of metabolic rate, it can be said that life-history variation in *Unio crassus* can be explained solely by a phenotypic reaction to environmental influences.

Unio crassus is threatened by extinction in most European countries. Populations that are able to withstand unfavourable environmental conditions are characterized by a long life span, a large maximum length and moderate to low body growth constants.

Acknowledgements. I would like to thank Prof. H. Zwölfer, Mrs. C. Schmidt, Mr. J. Kynast and Mr. O. Ansteeg. This study was partly supported financially by the Bavarian Ministry of Arts and Science (Project ID: Z 3/a-0734-192) and the Bavarian Board of Agriculture and Forestry (B/8-5/b/7 472).

References

Bachelet G (1980) Growth and recruitment of the tellinid bivalve *Macoma balthica* at the southern limit of its geographical distribution, the Gironde Estuary (SW France). Mar Biol 59:105–117

Bauer G (1987) Reproductive strategy of the freshwater pearl mussel *Margaritifera margaritifera*. J Anim Ecol 56:691–704

Bauer G (1991) Plasticity in life history traits of the freshwater pearl mussel. Consequences for the danger of extinction and for conservation measures. In: Seitz A, Loeschke V (eds) Species conservation: a population biological approach. Birkhäuser, Basel

Bauer G (1992) Variation in the life span and size of the freshwater pearl mussel. J Anim Ecol 61:425–436

Bauer G (1994) The adaptive value of offspring size among freshwater mussels (Bivalvia; Unionoidea). J Anim Ecol 63:933–944

Bauer G (1998) Allocation policy of female freshwater mussels. Oecologia 117:90–94

Bauer G, Hochwald S, Silkenat W (1991) Spatial distribution of freshwater mussels: the role of host fish and metabolic rate. Freshwater Biol 26:377–386

Bayne BL, Newell RC (1983) Physiological energetics of marine molluscs. In: Saleuddin ASM, Wilbur KM (eds) The Mollusca, vol 4. Physiology, part 1. Academic Press, New York

Bayne BL, Worrall CM (1980) Growth and production of mussels *Mytilus edulis* from two populations. Mar Ecol Prog Ser 3:317–328

Beran L (2000) Velevrub tupý (Unio crassus). Ochrana Přírody 55(7):208–209

Beverton RJH, Holt SJ (1959) A review of the lifespans and mortality rates of fish in nature and their relation to growth and other physiological characteristics. CIBA Found Colloq Ageing 5:142–180

Brousseau DJ (1979) Analysis of growth rate in *Mya arenaria* using the Von Bertalanffy growth equation. Mar Biol 51:221–227

Charnov EL, Berrigan D (1991) Evolution of life history parameters in animals with indeterminate growth, particularly fish. Evol Ecol 5:63–68

Ehrmann P (1933) Mollusca. In: Brohmer P, Ehrmann P, Ulmer G (eds) Die Tierwelt Mitteleuropas 2(1). Quelle and Meyer, Leipzig, pp 224–227

Engel H (1990) Untersuchungen zur Autökologie von *Unio crassus* (Philipsson) in Norddeutschland. PhD Thesis, Tierärztliche Hochschule, Hannover

Falkner G (1992) Rote Liste gefährdeter Schnecken und Muscheln (Mollusca) Bayerns. In: Rote Liste gefährdeter Tiere Bayerns. Schriften Bayer Landesamt Umweltschutz 111:47–56

Fleischauer-Rössing S (1990) Untersuchungen zur Autökologie von *Unio tumidus* Philipsson und *Unio pictorum* Linneus (Bivalvia) unter Berücksichtigung der frühen postparasitären Phase. PhD Thesis, Tierärztliche Hochschule, Hannover

Frank PW (1975) Latitudinal variation in the life history features of the black turban snail *Tegula funebralis* (Prosobranchia: Trochidae). Mar Biol 31:181–192

Franke G (1993) Zur Populationsökologie und Geschlechtsbiologie der Teichmuscheln *A. anatina* L. und *A. cygnea* L. Ms Thesis, University of Bayreuth

Gordon ME, Smith DG (1990) Autumnal reproduction in *Cumberlandia monodonta* (Unionoidea: Margaritiferidae). Trans Am Microsc Soc 109(4):407–411

Griffiths CL (1981) Population dynamics and growth of the bivalve *Choromytilus meridionalis* (Kr) at different tidal levels. Estuarine Coastal Shelf Sci 12:101–118

Haukioja E, Hakala T (1978) Life history evolution in *Anodonta piscinalis* (Mollusca Pelecypoda). Oecologia 35:253–266

Haukioja E, Hakala T (1979) Asymptotic equations in growth studies – an analysis based on *Anodonta piscinalis* (Mollusca: Unionidae). Ann Zool Fenn 15:60–65

Hochwald S (1988) Untersuchungen zur Populationsökologie und Fortpflanzungsbiologie der Bachmuschel *Unio crassus* Phil 1788. Ms Thesis, University of Bayreuth

Hochwald S (1997) Das Beziehungsgefüge innerhalb der Größenwachstums- und Fortpflanzungsparameter bayerischer Bachmuschelpopulationen (*Unio crassus* Phil 1788) und dessen Abhängigkeit von Umweltfaktoren. Bayreuther Forum Ökol 50:1–166

Hughes RN (1970) Population dynamics of the bivalve *Scrobicularia plana* (Da Costa) on an intertidal mud-flat in North Wales. J Anim Ecol 39:333–356

Hughes RN, Roberts DJ (1980) Reproductive effort of winkles (*Littorina* spp.) with contrasted methods of reproduction. Oecologia 47:130–136

Israel W (1913) Biologie der europäischen Süßwassermuscheln. KG Lutz, Stuttgart, pp 44–87

Jäckel SH (1962) Ergänzungen und Berichtigungen zum rezenten und quartären Vorkommen der mitteleuropäischen Mollusken. In: Brohmer P, Ehrmann P, Ulmer G (eds) Die Tierwelt Mitteleuropas 2(1). Quelle and Meyer, Leipzig, pp 206–207

Joergensen CB (1976) Growth efficiencies and factors controlling size in some mytilid bivalves especially *Mytilus edulis* L.: review and interpretation. Ophelia 15(2):175–192

Jokela J, Mutikainen P (1995) Phenotypic plasticity and priority rules for energy allocation in a freshwater clam: a field experiment. Oecologia 104:122–132

Jungbluth JH, Knorre D, Falkner G, Groh K, Schmidt G (1995) Rote Liste der Binnenmollusken [Schnecken (Gastropoda) und Muscheln (Bivalvia)] in Deutschland. Mitt Dtsch Malakozool Ges 56/57:1–17

Kundler P (1954) Zur Anwendung der Invertzuckermethode für standortkundliche Temperaturmessungen. Z Pflanzenernähr Düng Bodenkd 66:239–246

Longhurst AR, Pauly D (1987) Ecology of tropical oceans. Academic Press, New York

MacDonald BA, Thompson RJ (1985) Influence of temperature and food availability on the ecological energetics of the giant scallop *Placopecten magellanicus* I. Growth rates of shell and somatic tissue. Mar Ecol Prog Ser 25:279–294

Mildner P, Troyer-Mildner J (1992) Zum Bestand der gemeinen Flußmuschel *Unio crassus* Philipsson, 1788 (Mollusca: Bivalvia: Unionidae) in Kärnten. Carinthia II 182:101–112

Nagel KO (1991) Gefährdete Flußmuscheln in Hessen: Wachstum, Reproduktionsbiologie und Schutz der Bachmuschel (Bivalvia: Unionidae: *Unio crassus*). Z Angew Zool 78:205–218

Pallmann H, Eichenberger E, Hasler A (1940) Eine neue Methode der Temperaturmessung bei ökologischen oder bodenkundlichen Untersuchungen. Ber Schweiz Bot Ges 50:337–362

Parker RS, Hackney CT, Vidrine MF (1984) Ecology and reproductive strategy of a South Lousiana freshwater mussel *Glebula rotundata* (Lamarck) (Unionidae: Lampsilini). Freshwater Invertebr Biol 3(2):53–58

Piechocki A, Dyduch-Falniowska A (1993) Mieczaki (Mollusca). Wydawnictwo Naukowe PWN, Warsaw

Reischütz PL, Sackl P (1991) Zur historischen und aktuellen Verbreitung der gemeinen Flussmuschel *Unio crassus* Philipsson 1788 (Mollusca Bivalvia Unionidae) in Österreich. Linzer Biol Beitr 23:213–232

Roff DA (1984) The evolution of life history parameters in teleosts. Can J Fish Aquat Sci 41:989–1000
Ropes JW, Stickney AP (1965) Reproductive cycle of *Mya arenaria* in New England. Biol Bull 128:315–327
Roßmäßler EA, Kobelt W (1835) Ikonographie der Land- und Süßwassermollusken 15. Arnold, Dresden
Schnitter H (1922) Die Najaden der Schweiz. Rev Hydrobiol Suppl II:1–201
Smith DG (1978) Biannual gametogenesis in *Margartifera margaritifera* in northeastern North America. Bull Am Malacol Union 49–53
Sokal RR, Rohlf FJ (1995) Biometry. Freeman, New York
Stearns SC (1992) The evolution of life histories. Oxford University Press, Oxford, 249 pp
Steubing L (1965) Pflanzenökologisches Praktikum. Parey, Berlin
Steubing L, Fangmeier A (1992) Pflanzenökologisches Praktikum. Ulmer, Stuttgart
Taylor CC (1959) Temperature and growth: the Pacific razor clam. Int Counc Explor Sea 25:93–101
Timm H (1994) Big clams of the Estonian freshwaters: comparision of the age, shell length, and shell weight in different species and populations. Proc Estonian Acad Sci Biol 43(3):149–159
Timm H, Mutvei H (1993) Shell growth of the freshwater Unionid *Unio crassus* from Estonian rivers. Proc Estonian Acad Sci Biol 42(1):55–67
Vahl O (1981) Energy transformations by the iceland scallop, *Chlamys islandica* (O. F. Müller), from 70° N. I. The age specific energy budget. J Exp Mar Biol Ecol 53:281–296
van Goethem JL (1995) Report on the EIS mapping programme: Belgium 1990–92. Mitt Dtsch Malakozool Ges 56/57:41–42
Vicentini H (1998) Teich- und Flussmuscheln (Unionidae) in der Schweiz, mit besonderer Berücksichtigung derjenigen des Kantons Schaffhausen. Mitt Naturforsch Ges Schaffhausen 43:35–61
von Bertalanffy L (1938) A quantitative theory of organic growth (Inquiries of growth laws II). Hum Biol 10(2):181–213
von Bertalanffy L (1960) Principles and theory of growth. In: Nowinski WW (ed) Fundamental aspects of normal and malignant growth. Elsevier, Amsterdam, pp 137–259
Walford LA (1946) A new graphical method for describing the growth of animals. Biol Bull Woods Hole 90:141–147
Williams GC (1966) Natural selection, the costs of reproduction and a refinement of lack's principle. Am Nat 100:687–690
Wolff WJ (1968) The mollusca of the estuarine region of the rivers Rhine, Meuse and Scheldt in relation of the hydrogeography of the area. I. The Unionidae. Basteria 32(1–3):1–48

8 Life-History Data on the Virtually Unknown *Margaritifera auricularia*

RAFAEL ARAUJO and MARIA ANGELES RAMOS

8.1 Introduction

Published information on the declining and widely distributed freshwater pearl mussel *Margaritifera margaritifera* (L.) is currently flourishing in the scientific literature, being the centre of attention of most invertebrate conservation treatises (Young et al. this Vol.). The case of the other European species of the genus, the giant naiad *Margaritifera auricularia* (Spengler 1793; Fig. 8.1) is completely different. It was relatively abundant in large rivers in western Europe (Iberian Peninsula, France, Italy, England, Germany) and Morocco (Preece et al. 1983). Indeed, recent data on its former distribution in Spain (Araujo and Moreno 1999) demonstrate that it lived in several rivers other than the Ebro (northeast Spain), which is the only one known currently containing populations of this vanishing species. All available biological data on this nearly extinct species date from the past 3 years from one rediscovered population of about 2000 individuals in the *Canal Imperial de Aragón* (Imperial Canal), an old canal of the Ebro built in the 18th century, where the species has found one of its last refuges (Araujo and Ramos 1998a).

The only author to have studied live *M. auricularia* mussels previously was the German malacologist Fritz Haas, from a population in the main course of the Ebro (Haas 1916a,b, 1917). Thanks to Haas' papers, we know that the species was so abundant there that the nacre of the valves was collected to manufacture knife hilts. Haas' research produced a brief description of anatomy (Haas 1924), but no data on reproductive strategy, breeding season or larval morphology. He only noted that the mussels were not gravid between mid-July and early September and that they lived in a 70–120 m wide section of the Ebro River between stones and boulders, having a predilection for deep bottoms (5–7 m), and in the Imperial Canal (Haas 1916a,b, 1917). He also mentioned the species' decline.

In 1990, Altaba reported the presence of a few fresh empty valves of *M. auricularia* in an irrigation canal of the Ebro near the estuary in Tarragona (Catalonia, Spain). Since the rediscovery of the Imperial Canal population, there has been news of other *M. auricularia* specimens living in the main middle course of the Ebro in Zaragoza Province (Alvarez 1998; pens. obs.), in the lower course of the Ebro (Catalonia) and in one of the latter's irrigation canals in Catalonia (Altaba 1997).

8.2 The Population of *M. auricularia* in the Imperial Canal

In the Imperial Canal, a 200-year-old canal, 115 km long and with 3.5 m water depth, *M. auricularia* coexists with a very rich community of three unionid species: *Unio*

Ecological Studies, Vol. 145
G. Bauer and K. Wächtler (eds.) Ecology and
Evolution of the Freshwater Mussels Unionoida

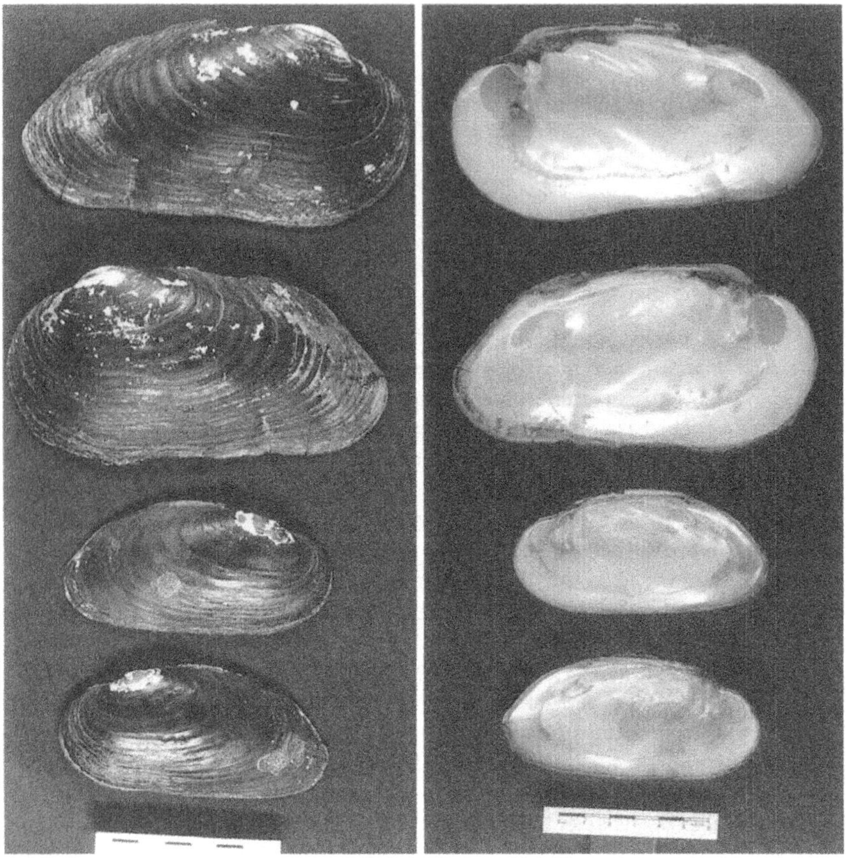

Fig. 8.1. Spanish specimens of *M. auricularia* (*above*) and *M. margaritifera*

elongatulus C. Pfeiffer, *Potoucida littoralis* (Lamarck) and *Anodonta cygnea* (Linnaeus), which live in different densities along much of the channel. The estimated number of *M. auricularia* individuals is by now around 2000, showing a patchy distribution as is common among unionoids. The largest subpopulation has a density of four mussels m^{-2}.

M. auricularia is a very sedentary species, living partially buried in clay-sand and gravel bottoms, sometimes in an almost vertical position with the hind portion exposed (especially in the reproductive season) and sometimes horizontal, showing only the umbonal and ligament area.

According to the reported data on physicochemical features of the *M. margaritifera* habitat, this species prefers soft waters poor in lime, with the exception of the Irish ecophenotype, *M. margaritifera durrovensis* (Chesney et al. 1993). On the contrary, *M. auricularia* is a hardwater species, calcium levels in the Ebro River being around $150\,mg\,l^{-1}$. In the Imperial Canal water velocity is $0.6\,m\,s^{-1}$ and temperature ranges between 7.4 (March) and 22.9 °C (August) (Araujo and Ramos 2000).

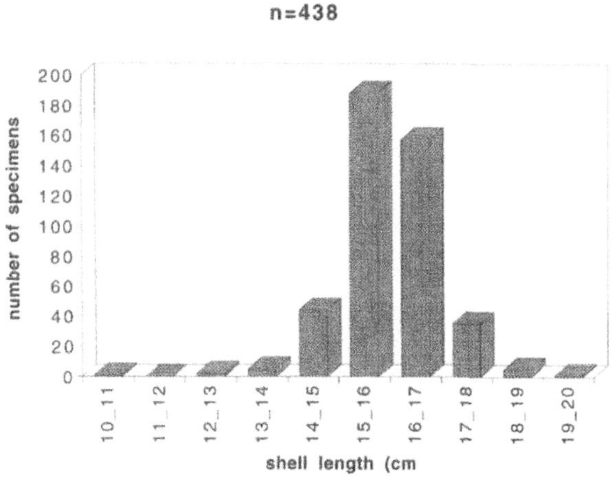

Fig. 8.2. Size structure of the *M. auricularia* population in the Imperial Canal

Table 8.1. Summarized data on reproduction of the *M. auricularia* population in the Imperial Canal

Oogenesis	Autumn-winter
Spermatogenesis	Autumn-winter
Release of spermatozoa	Winter
Eggs in marsupium	December to February
Release of glochidia	February-March
Number of released	?
Size of glochidia	127–144 μm
Host fish	Sturgeons
Place of attachment	Gill filaments
Release of juveniles	April
Size of juveniles	190 μm

All known *M. auricularia* specimens living in the channel are large (10–20 cm), suggesting that there has been no juvenile recruitment of the population for many years (Fig. 8.2) (Araujo and Ramos 2000). There may be two main reasons for the absence of juveniles: (1) the absence of the specific host fish, and (2) ecological factors involved in the juvenile life stages (sediment type, water quality, predators etc.). Although finding juvenile stages of unionoids is widely known to be a hard task, small and medium-sized specimens of the other three naiad species were found in the Canal.

8.3 Reproductive Behaviour

M. auricularia, like all known species of the genus *Margaritifera* (Heard 1970; Smith 1979; Bauer 1994), is a short-term brooder (= tachytictic), incubating the brood in the four gills.

Oogenesis probably takes place during autumn so that the female gonad is full of ripe oocytes at the beginning of winter (Table 8.1). The same occurs in the male gonad, as specimens in December present a great amount of spermatocytes, sperm morulae and mature spermatozoa. Release of spermatozoa was detected in the aquarium at the beginning of March (Fig. 8.3).

Preliminary data on the population sex ratio show that most (if not all) of the mussels in the population are hermaphrodites (Grande, Araujo and Ramos, in press; Fig. 8.4). The female gonad, which is larger than the male gonad, has ripe gametes at

Fig. 8.3. Release of *M. auricularia* spermatozoa

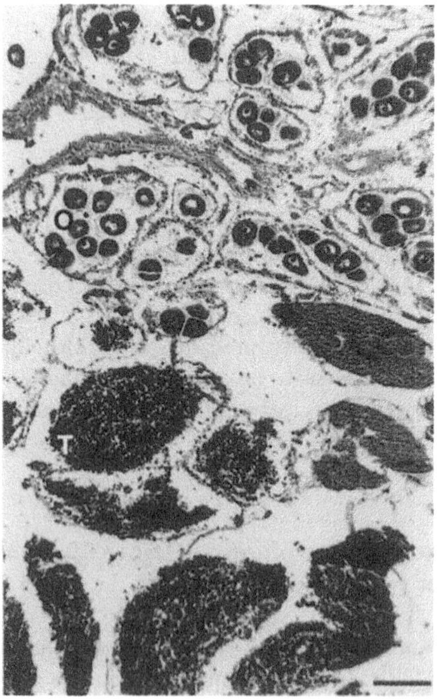

Fig. 8.4. Section of hermaphroditic gonad of *M. auricularia*. *O* ovary; *T* testicle. *Bar* 150 µm

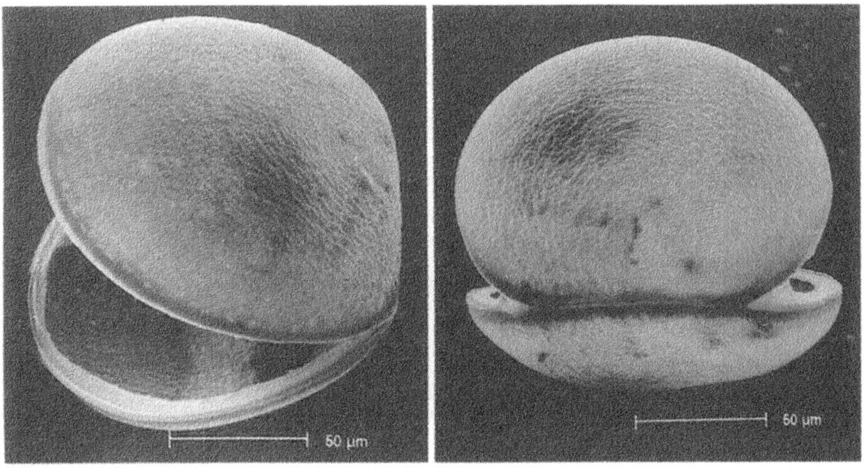

Fig. 8.5. *M. auricularia* glochidium

the end of winter, while the gills of the same individuals were full of developing embryos. Thanks to Haas (1917), we know that *M. auricularia* is not gravid between mid-July and the beginning of September, the only months he could work with the species. Fertilization of the ova probably occurs at the beginning of the year. Mature glochidia, which in other naiads develop in 2 weeks (at 20–24 °C) from the fertilized egg stage (Bruenderman and Neves 1993), are released into the water only once a year during February and March, with a peak in mid-March (Araujo et al. 2000; Araujo and Ramos 2000). In comparison, *M. margaritifera* glochidia are released during a period of between 1–4 weeks in a year (Young and Williams 1984a), although two annual reproductive periods have been documented in North American populations (Smith 1978). Observations in the wild and in aquaria showed that *M. auricularia* releases the glochidia from the exhalant opening with pumping movements. The glochidium of *M. auricularia* (Fig. 8.5) was recently described by Araujo and Ramos (1998b) as the biggest of all known species in the family (127–144 mm in length). It presents very minute teeth in the ventral border covered by a rim of periostracum. Nevertheless, it is the hookless type of glochidium specific to gill filament attachment in the host fish and therefore probably incapable of attaching itself to fish fins.

8.4 Two Species Declining Together

Electrofishing is probably the best way to assess the incidence of the glochidia of different naiad species on fishes living in the same basin, although it has been proposed that natural glochidiosis in fish gills in rivers is normally low (Bruenderman and Neves 1993).

Using this technique, we realized that none of the autochtonous or exotic fish species occurring with *M. auricularia* presented symptoms of infection with its glochidia during the glochidial release season although they are sensitive to the glochidia of other naiads (Araujo et al. 2000). This, and the large size of all the mussels

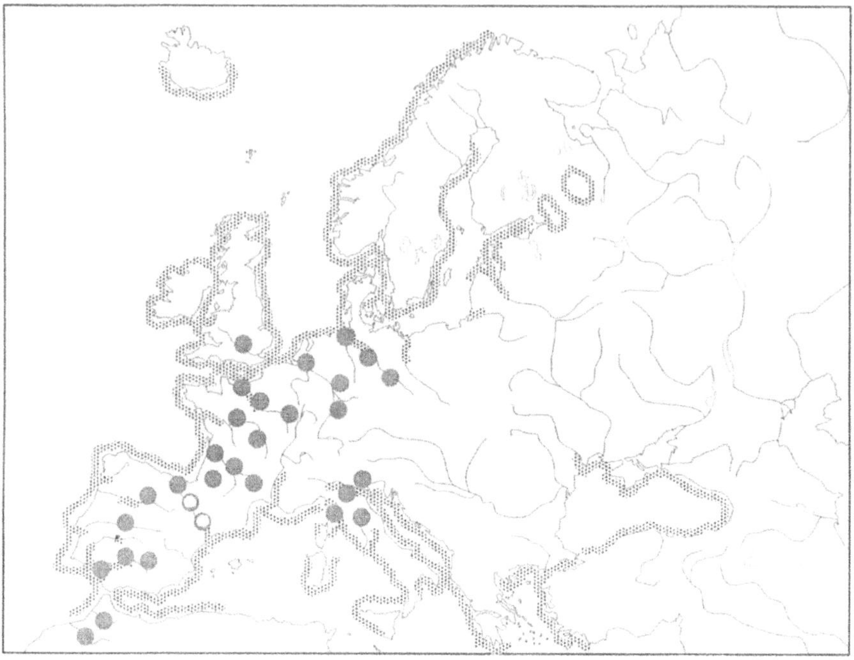

Fig. 8.6. Historical distribution of *Acipenser sturio* (*dotted areas*) (after Lelek 1987; Elvira et al. 1991) and *M. auricularia* (*large dots*) (Preece et al. 1983; Araujo and Moreno 1999). *White dots* indicate currently known localities of living populations of *M. auricularia*

in the population, suggest that the actual absence of the specific host could be one of the main reason for the lack of juveniles.

The sturgeon *Acipenser sturio* (Linnaeus) occurred naturally in Spain and all over the former distribution range of *M. auricularia* (Fig. 8.6). Indeed, fossil records of both species have been found together in Pleistocene deposits (Preece 1988). There is some information about exploitation of *A. sturio* in Spain until the mid-1960s in the Ebro and Guadalquivir Rivers (Elvira et al. 1991), but catches declined following the construction of reservoirs. With the data currently available, it is feasible that the absence of *A. sturio* in Spain, and especially in the Ebro basin for many years, might be one of the reasons for the decline in *M. auricularia*.

Congeneric and allopatric species of fishes reported or suspected as suitable glochidial hosts of endangered freshwater mussels have been proposed as a useful tool to restore their populations (Bruenderman and Neves 1993). Aquarium experiments showed that glochidia of *M. auricularia* can pass through the metamorphosis in the gills of exotic sturgeons (Fig. 8.7a) in 30 days at 23–24°C (690 degree days) (Araujo and Ramos 2000). The glochidia encysted in all the gill filaments and over their entire length are very frequent in the tips (Fig. 8.7b). During metamorphosis they change shape but do not increase in length as much as has been reported for *M. margaritifera* larvae (Young and Williams 1984b; Bauer 1987; Ziuganov et al. 1994).

After this process, the released live juveniles are 190 μm long. They have a finely ciliated foot and an almost spherical shape produced by the organogenetic process,

Fig. 8.7. a Section of a metamorphosed glochidium of *M. auricularia* in the gill filament of *Acipenser baeri*. **b** Encysted *M. auricularia* glochidia in *Acipenser baeri* gill filaments. **c** Recently released *M. auricularia* juvenile. *Bar* 100 μm

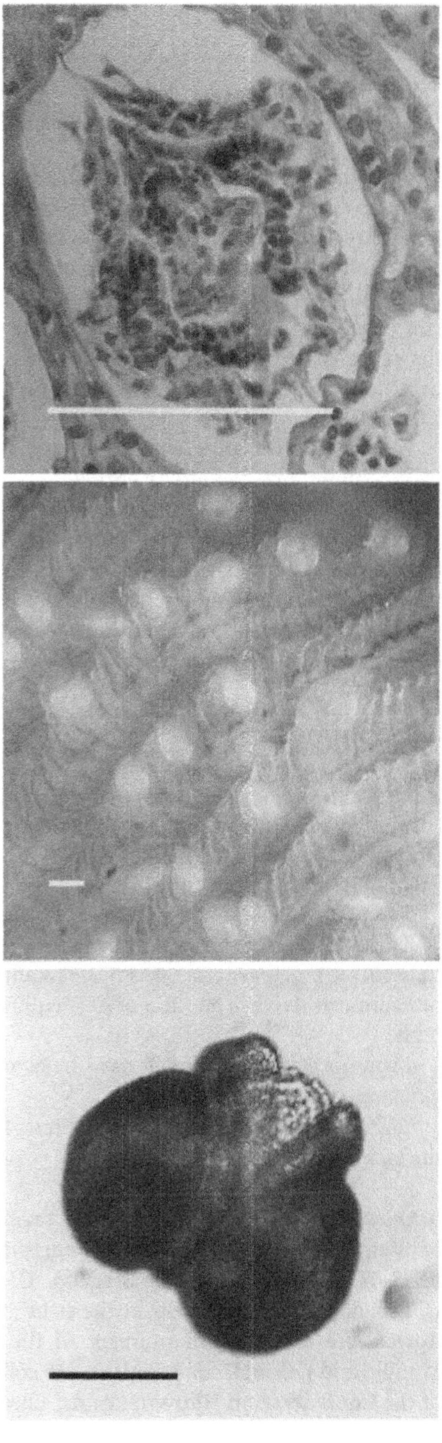

and an extra, very thin, border of shell material all around the old glochidial shell (Fig. 8.7c).

8.5 Conservation of an Endangered Species

M. auricularia is protected by European legislation under the Bern Convention and Habitats Directive and also by Spanish national legislation. It is certainly one of the most threatened invertebrates in the world, already thought to be extinct. The rediscovery of a living fertile population obliges us to put all our efforts into maintaining it for future generations. The dramatic decline of *M. auricularia* in Europe has run parallel to the decline of some other European and North American mussels, the reasons for which have been reviewed by Bogan (1993): habitat alteration or destruction, decline or extinction of host fishes, commercial exploitation and alien species.

In all reports dealing with freshwater fauna conservation, and particularly with naiad populations, changes in agricultural and socio-economic practices are urgently postulated (Bogan 1998; Chesney and Oliver 1998; Gibson 1998; Killeen et al. 1998; Valovirta 1998; Costello et al. 1998). New legislation, environmental standards and policies must be updated in order to stop the continuous destruction of freshwater habitats, which are a non-renewable resource. *M. auricularia*, the large naiad and a symbol of the pristine stage of European rivers, can be another spur to change our current ideas about the meaning of progress.

8.6 Summary

All available biological data on the vanishing *M. auricularia* are from a recently rediscovered Spanish population of about 2000 specimens living in the Canal Imperial de Aragón, a 200-year-old channel of the Ebro River. *M. auricularia* lives in water with calcium levels being around 150 mg l^{-1} and temperature ranging between 7.4 (March) and 22.9 °C (August).

This species is a short-term brooder (= tachytictic), incubating the brood in the four gills. Gametogenesis takes place in autumn-winter and glochidia release occurs in February-March. The ratio of hermaphrodite specimens in the population is very high.

Metamorphosed juveniles have been obtained from induced infestations with glochidia on exotic sturgeons.

The actual absence of the specific host fish could be one of the main reasons for the lack of juvenile recruitment in the population.

Acknowledgements. We would like to thank the MIMAM-CSIC Project Demografía, hábitat y ciclo vital de *Margaritifera auricularia* and the Fauna Ibérica Project (DGES PB95-0235) for support, D. Bragado, C. Grande, J. M. Remón, S. Jiménez and D. Erpenbeck for their help in the field and in the laboratory, the Department of Agriculture and the Environment of the Regional Government of Aragón (S. P. Munilla and J. Guiral) for permission to collect, and A. Núñez and J. A. Martínez Orús of the Confederación Hidrográfica del Ebro for their help.

References

Altaba CR (1990) The last known population of the freshwater mussel *Margaritifera auricularia* (Bivalvia, Unionoida): a conservation priority. Biol Conserv 52:271–286

Altaba CR (1997) Al l'mit de l'extinció: *Margaritifera auricularia* (Bivalvia: Unionoida). Butll Inst Cat Hist Nat 65:137–148

Alvarez RM (1998) Salvar la almeja de rio: *Margaritifera auricularia*, una especie en peligro de extinción. Trébede, Mensual Aragonés Anál Opin Cult 21:7–14

Araujo R, Moreno R (1999) Former Iberian distribution of *Margaritifera auricularia* (Spengler) (Bivalvia: Margaritiferidae). Iberus 17(1):127–136

Araujo R, Ramos MA (1998a) *Margaritifera auricularia* (Unionoidea, Margaritiferidae), the giant freshwater pearl mussel rediscovered in Spain. Graellsia 54:129–130

Araujo R, Ramos MA (1998b) Description of the glochidium of *Margaritifera auricularia* (Spengler, 1793) (Bivalvia, Unionoidea). Philos Trans R Soc Lond B 353:1553–1559

Araujo R, Ramos MA (2000) Status and conservation of the relict giant European freshwater pearl mussel *Margaritifera auricularia* (Spengler, 1793) (Bivalvia: Unionoidea). Biol conserv 96(2):233–239

Araujo R, Bragado D, Ramos MA (2000) Occurrence of glochidia of *Margaritifera auricularia* (Spengler, 1793) and other mussel species (Bivalvia: Unionoida) in drift and on fishes in an ancient channel of the Ebro River, Spain. Arch Hydrobiol 158(1):157–160

Bauer G (1987) The parasitic stage of the freshwater pearl mussel (*Margaritifera margaritifera* L.) II. Susceptibility of brown trout. Arch Hydrobiol Suppl 76 4:403–412

Bauer G (1994) The adaptive value of offspring size among freshwater mussels (Bivalvia: Unionoidea). J Anim Ecol 63:933–944

Bogan AE (1993) Freshwater bivalve extinctions (Mollusca: Unionoida): a search for causes. Am Zool 33:599–609

Bogan AE (1998) Freshwater molluscan conservation in North America: problems and practices. In: Killeen I, Seddon MB, Holmes AM (eds) J Conchol Spec Publ 2:223–230

Bruenderman SA, Neves RJ (1993) Life history of the endangered fine-rayed pigtoe *Fusconaia cuneolus* (Bivalvia: Unionidae) in the Clinch River, Virginia. Am Malacol Bull 10(1):83–91

Chesney HCG, Oliver PG, Davis GM (1993) *Margaritifera durrovensis* Phillips, 1928: taxonomic status, ecology and conservation. J Conchol 34:267–299

Chesney HCG, Oliver PG (1998) Conservation issues for Margaritiferidae in the British Isles and western Europe. In: Killeen I, Seddon MB, Holmes AM (eds) J Conchol Spec Publ 2:231–242

Costello MJ, Moorkens EA, Larkin M, Kurz I, Dowse J (1998) Management priorities for the river Nore (Ireland) to conserve the pearl mussel, *Margaritifera durrovensis* Philips, 1928. In: Killeen I, Seddon MB, Holmes AM (eds) J Conchol Spec Publ 2:257–264

Elvira B, Almodóvar A, Lobón-Cerviá J (1991) Recorded distribution of sturgeon (*Acipenser sturio* L., 1758) on the Iberian Peninsula and actual status in Spanish waters. Arch Hydrobiol 121(2):253–258

Gibson M (1998) Issues involved in protecting important populations of *Margaritifera margaritifera*. In: Killeen I, Seddon MB, Holmes AM (eds) J Conchol Spec Publ 2:243–244

Grande C, Araujo R, Ramos MA The gonads of *Margaritifera auricularia* (Spengler 1793) and *Margaritifera margaritifera* (L. 1758) (Bivalvia: Unionoidea) J Molluscan Stud (in press)

Haas F (1916a) Sobre una concha fluvial interesante (*Margaritana auricularia*, Spengler) y su existencia en España. Bol Soc Arag Cienc Nat XV(2):33–45

Haas F (1916b) Spanischer Brief III. Nachr Dtsch Malakol Ges 48(32–44)

Haas F (1917) Estudios sobre las Náyades del Ebro. Bol Soc Arag Cienc Nat XVI:71–82

Haas F (1924) Anatomische Untersuchungen an europäischen Najaden. I. Arch Molluskenkd 56:66–82

Heard WH (1970) Hermaphroditism in *Margaritifera falcata* (Gould) (Pelecypoda: Margaritiferidae). Nautilus 83(3):113–114

Killeen IJ, Oliver PG, Fowles AP (1998) The loss of a freshwater pearl mussel (*Margaritifera margaritifera*) population in NW Wales. In: Killeen I, Seddon MB, Holmes AM (eds) J Conchol Spec Publ 2:245–250

Lelek A (1987) Threatened fishes of Europe. Eur Committee for the Conservation of Nature and Natural Resources, Council of Europe (ed). Aula-Verlag, Wiesbaden

Preece RC (1988) A second British interglaciar record of *Margaritifera auricularia*. J Conchol 33(1):50–51

Preece RC, Burleigh R, Kerney MP, Jarzembowski EA (1983) Radiocarbon age determination of fossil *Margaritifera auricularia* (Spengler) from the River Thames in West London. J Archaeol Sci 10:249–257

Smith DG (1978) Biannual gametogenesis in *Margaritifera margaritifera* (L.) in northeastern North America. Bull Am Malacol Un 1978:49–53

Smith DG (1979) Marsupial anatomy of the demibranch of *Margaritifera margaritifera* (Lin.) in northeastern North America (Pelecypoda: Unionacea). J Molluscan Stud 45:39–44

Valovirta I (1998) Conservation methods for populations of *Margaritifera margaritifera* (L.) in Finland. In: Killeen I, Seddon MB, Holmes AM (eds) J Conchol Spec Publ 2:251–256

Young M, Williams J (1984a) The reproductive biology of the freshwater pearl mussel *Margaritifera margaritifera* (Linn.) in Scotland I. Field Studies. Arch Hydrobiol 99(4):405–422

Young M, Williams J (1984b) The reproductive biology of the freshwater pearl mussel *Margaritifera margaritifera* (Linn.) in Scotland II. Laboratory Studies. Arch Hydrobiol 100(1): 29–43

Ziuganov V, Zotin A, Nezlin L, Tretiakov V (1994) The freshwater pearl mussels and their relationships with salmonid fish. VNIRO, Moscow

Part III
Populations and Communities

9 Factors Affecting Naiad Occurrence and Abundance

GERHARD BAUER

9.1 Introduction

This topic will be treated on two levels: the level of populations and the level of communities. Within their areas of distribution, species' population densities generally are determined by immigration, emigration, age-specific mortality rates and age-specific fertility. There is certainly migration of glochidia carrying fish, which will depend on the fish species and the duration of the parasitic stage. However, apart from catastrophic events like spates washing adults downstream (see Sect. 9.2.1.4), there are still no data available about the extent to which migration processes influence population dynamics among naiads.

There is increasing knowledge about age-specific mortality and fertility. For further considerations (Chap. 13, this Vol.) one should bear in mind that these age-specific patterns are not only important for determining population density, rather they are the key factors shaping the evolution of life histories (Stearns 1992). Theory predicts that the reproductive value, i.e. the probability of current and future reproduction, should be optimized in each stage or age class, as far as permitted by constraints. A number of generalizing models on the evolution of life histories are built on age-specific mortality patterns; primarily its intensity and variability (e.g. Pianka 1972; Southwood et al. 1974; Stearns 1976; Greenslade 1983); but the method can be turned around and life-history theory can be used to predict mortality rates (Myers and Doyle 1983).

In this respect, the extremely long-lived naiad species (Chap. 5, this Vol.) are of particular interest. The possibility of attaining a long life, and thus a high number of reproductive periods, should create a strong selective pressure for extended longevity. In these cases we would expect very low mortality rates compared to species which are a priori short-lived. On the other hand, due to the complicated life cycle, comprising a parasitic larval stage produced in very high numbers (Bauer 1994; Chap. 7, this Vol.), one would expect high mortality occurring somewhere before maturity.

On the community level, a number of habitat characteristics will become important, determining whether a naiad species can occur in a particular environment or not. Furthermore, mortality factors may act differently on different species, and, finally, interspecific competition might be a factor structuring naiad communities.

Ecological Studies, Vol. 145
G. Bauer and K. Wächtler (eds.) Ecology and
Evolution of the Freshwater Mussels Unionoida

9.2 The Population Level

9.2.1 Stage- and Age-Specific Mortality

9.2.1.1 Larval Stage

The tremendous mortality rates of glochidia are described in Jansen et al. (this Vol.). This mortality is mainly due to the inefficient mode of host infection. Glochidia are not able to swim actively, they are very short-lived and are not capable of selecting a proper host. As far as we know (Falk 1995; Herrmann 1995; Chap. 6, this Vol.), this also holds true for the haustorium and lasidium of the Etherioidea. Thus, the life cycle of probably all naiads is marked by this very high larval mortality.

Surprisingly little has occurred during naiad evolution to increase larval survival. There is no case where an actively swimming and host-selecting larva was developed, as occurs in many other parasitic animals. The comparatively few adaptations to improve survival of larvae, as described in Chapters 5, 11 and 13 (this Vol.), concern the adult. Unfortunately, little is known about their efficiency. Jansen and Hanson (1991) and Jansen et al. (this Vol.), estimated glochidial survival (i.e. the rate of successfully attaching glochidia) of the freshwater pearl mussel (*Margaritifera margaritifera*) and *Anodonta grandis simpsoniana* at comparable host densities [4.5 $10\,m^{-2}$ (Bauer 1988), or 5.3 $10\,m^{-2}$, respectively]. Survival of the *Anodonta* glochidia was ca. 70 per million and thus 1 order of magnitude higher than among freshwater pearl mussels. This may be due to the ability of *Anodonta* to perceive the presence of hosts, as described in Chapter 13. On the other hand, freshwater pearl mussels have evolved no mechanism for improved host infection.

9.2.1.2 Parasitic Stage

While attached to a host, a main problem for glochidia is the host's immune response. Mortality rates reported for this stage are between nearly zero (Bauer 1987b) and 80 to nearly 100% (Young and Williams 1984a; Teutsch 1997; Chap. 11, this Vol.). They depend on the age of the glochidium, the host species, the previous infections of the host and the infection intensity (review in Chap. 11, this Vol.).

9.2.1.3 Juvenile Stage

When dropping off the host, most species apparently migrate into the substrate (Yeager et al. 1994; Chap. 6, this Vol.), before they come to the surface at a size of 1 or 2 cm. Neves and Widlak (1987) report for some North American species mortality rates of >40% per year. Young and Williams (1984b) calculated mortality rates of 95% per year for juvenile freshwater pearl mussels in Scotland. This mortality is believed to result from the mussels falling into unfavourable habitats and it is certainly also due to predation.

9.2.1.4 Adult Stage

Although there are very few data on adult mortality, it seems that under natural conditions this stage is least likely to be affected by mortality. In Central Europe there

Table 9.1. Loss estimates of marked freshwater pearl mussels from October 1994 till October 1995. The mussels were controlled every October until 1997, to make sure that living individuals not found in 1995 (but later) are considered. The difference between both size classes is not significant (Fisher's test)

	Size class		
	60–100 mm	>100–130 mm	Total
No. marked in 1994	76	76	152
No. lost in 1995	4	3	7
Loss rate	5.3%	3.9%	4.6%

was no predator for naiads (Stone et al. 1982) and since the Ice Age, when *Dreissena polymorpha* died out here, there has been also no encrusting competitor (Chap. 10, this Vol.). Accordingly, mortality rates between 5 and 12% per year are reported for adult *A. anatina* in England (Negus 1966), and for the same species in Finland values between 75 and 25% for young (5 years and younger) and between 50 and 0% for older mussels (Jokela and Mutikainen 1995). This species attains a maximum life span of ca. 10 years, approximately the same as *Unio crassus*, where Hochwald (1997) found values between 24 and 38% per year.

For comparison, mortality was estimated in a freshwater pearl musssel population where the maximum life span is around 110 years (pers. observ.). In October 1994, pearl mussels were collected at three sites in a small river (ca. 1 m wide and 10 to 30 cm deep) in North Bavaria. The mussels were marked individually and then returned to their original site. The three sites were searched carefully every following October until 1997. In some cases, living mussels disappeared for 1 or even 2 years, but then were found again, apparently having been deeply buried. Therefore it is not possible to calculate loss rates from 1996 to 1997. However, rates from 1994 to 1995 should give reliable estimates, because animals not found in 1995 but later are considered. Table 9.1 shows that the loss rates are around 4%, and that they are not increased in the highest size (= age) class. Actually, these values are not exact mortality rates because they also contain losses due to downstream drift, which can be considerable: at one site where in 1995 only 56 marked mussels occurred, 1 year later 50 unmarked mussels appeared, most of them having been washed downstream. Thus, the annual mortality rates of freshwater pearl mussels will be in the range of very few percent.

Hardly any predator is reported for hyriids in Australia and New Zealand, whereas in America the muskrat is a natural predator of naiads, but its effect on naiad populations generally is considered to be very low, particularly in large lakes and rivers (Chap. 10, this Vol.). Very few diseases are known from naiads (Pauley 1968); however, heavy infection by trematodes may lead to castration (Zale and Neves 1982; Jokela et al. 1993; Chap. 2, this Vol.).

9.2.1.5 Survivorship Curve

Evidently, stage-specific mortality may vary considerably depending on species and conditions. Unfortunately, little is known about the mechanisms underlying this vari-

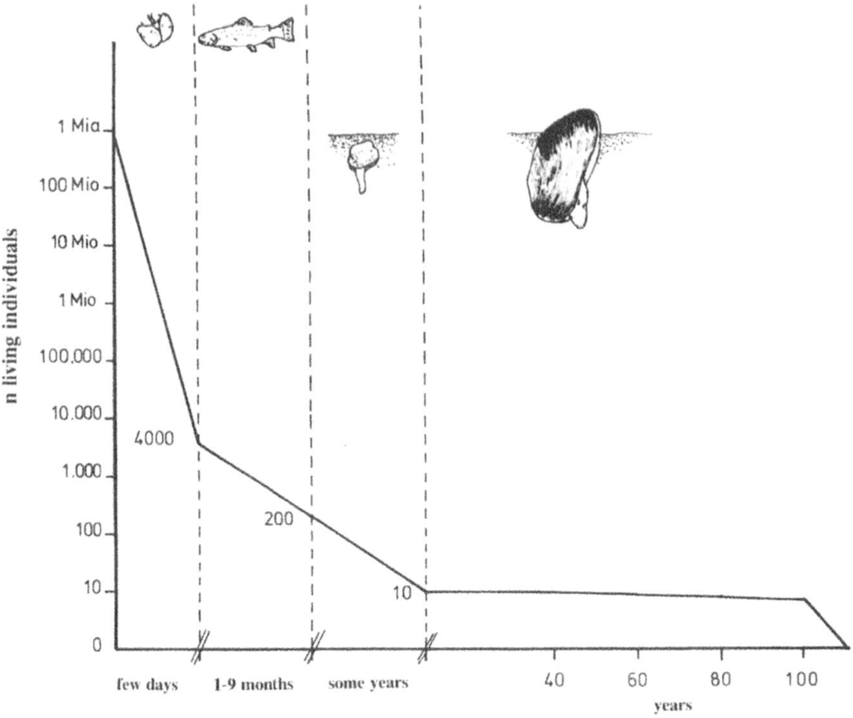

Fig. 9.1. Approximate survivorship curve of the freshwater pearl mussel

ability. Haukioja and Hakala (1978) suggest variable survival of juvenile *A. anatina* due to variable resource availability. Vannote and Minshall (1982) conclude that the long-term population dynamics of *Margaritifera falcata* is mainly caused by periodic floods, perhaps approaching a 50–100-year event, which cause high mussel mortality by bed scour. Glochidial survival until the juvenile stage will depend on complicated interactions between the population dynamics of mussels and hosts, both affected by a large number of extrinsic and intrinsic factors. Furthermore, plastic responses of life span, size and fertility (Chaps. 5, 7, this Vol.) contribute to complicating the situation. In spite of this, some general conclusions can be drawn, as demonstrated in Fig. 9.1, for the freshwater pearl mussel:

1. Free glochidia suffer the highest mortality rates due to the inefficient mode of host infection.
2. Mortality during the parasitic and juvenile stage is still considerable.
3. Mortality is lowest during the adult stage.

Finally, it must be added that the mortality values given here refer to healthy populations, i.e. populations not under severe threats. They should therefore approximately represent the conditions under which the respective species evolved. An exception are the adult mortalities given for the European *Anodonta* and *Unio* (Sect.

9.2.1.4). In some of these cases, predation by the introduced muskrat might have been an additional mortality factor. However, as will be shown later (Chap. 16, this Vol.), also mortality in all other stages may be increased by human activities.

9.2.2 Age-Specific Fertility

In general, fertility is high, but varies between taxa (Chaps. 6, 8, 18, this Vol.). These differences are due to the different glochidial size (Bauer 1994), However, they are also attributed to different marsupial space (Chap. 2, this Vol.). Fertility may increase with increasing food supply (Chaps. 2, 18, this Vol.), which may be the reason for the high population densities of euryecious species frequently observed in eutrophic waters (Chaps. 17, 18, this Vol.).

With respect to age specifity, it is important that fertility usually increases with size. Thus, since naiads grow asymptotically (Chaps. 18, 21, this Vol.), fertility increases with age (Bauer 1998). For example Hochwald (this Vol.), demonstrates that in most *Unio crassus* populations there is a significant relationship between female size (and thus age), and the number of glochidia in the marsupia. Furthermore, in this species no effect of female age on the number of spawnings per reproductive season was found. There is only scant evidence of a decline in fertility of very big (= very old) individuals (Downing et al. 1993). Since mortality rates in adults are generally low (Sect. 9.2.1.4), even in the highest age classes there will be sufficient individuals to contribute substantially to the reproduction of a population.

9.3 The Community Level

9.3.1 Habitat Characteristics

A number of macro- and microhabitat factors have been shown to influence mussel assemblages. Watters (1992) found that the number of fish species may be predicted by the area of the drainage basin, and he found a linear correlation between number of fish species in a river and the number of naiad species. This indicates that the higher diversity of host fish may be a factor for the species-rich naiad communities usually found in large rivers.

Hornbach (this Vol.), investigating a species-rich mussel fauna in the St. Croix River (Minnesota), attributes the major influence on naiad assemblages to a dam providing a fish migration barrier. Within areas above or below this barrier, stream gradient seems to play a secondary role in explaining the noted variance. According to Strayer (1993) stream size, hydrologic variability, calcium concentration, physiographic province and the presence/absence of a tide were important predictors for the presence or absence of naiad species.

On a small spatial scale, water velocity, depth, substrate type and trophic level have been shown to have some influence on naiad assemblages, but no generalizations are possible (for examples see Chaps. 12, 13, this Vol.).

9.3.2 Predation

Nothing is known about species-specific effects of predation in the juvenile stage. In the adult stage, predation is of minor importance in many areas (Chaps. 2, 20, this Vol.; Sect. 2.1.4). However, in North America, (and in this century also in Europe), the muskrat preys naturally on naiads. As Meike and Hanson (this Vol.) show, this predator may be highly selective due to species-specific shell thickness and differences in energetic value (meat weight). The effect of the foraging behaviour of this predator must be expected to be strongest on mussel assemblages in small rivers, since predation is most effective there.

9.3.3 Interspecific Competition

As generally in ecology, competition is frequently supposed to be important for the structure of a community (e.g. Davis et al. 1981), but its real effect has rarely been demonstrated. Among naiads, competition for food seems to be highly unlikely (Hanson et al. 1988). However, there may be interspecific competition for hosts. Hochwald (1997) shows that mortality of *Unio crassus* glochidia on rudd (*Scardinius erythrophthalmus*) is increased if the fish had been infected before with glochidia from *Anodonta anatina*. Accordingly Terberger (1999) found that glochidia of *A. cygnea* survived at lower rates if rudd had been previously infected with glochidia from *A. anatina*. (In both cases the young mussels from the first infection had already left the fish before the second infection was conducted). Apparently, antibodies produced by a fish against a particular naiad species are also active against other species. The strength of this factor may depend on the fish species, on the mussel species and on the infection intensities. Its importance in the real world is so far unknown.

Competition with one species may have different effects on different taxa. The encrusting competitor zebra mussel (*Dreissena polymorpha*) causes stress and symptoms of starvation in infested naiads. Baker and Hornbach (1997) showed that *Actinonais ligamentina* (Lampsilinae) was more sensitive to infestation than *Amblema plicata* (Ambleminae). They attribute this to the higher metabolic rate of *A. ligamentina* and suggest that zebra mussel introduction could drastically alter naiad communities in North America.

9.4 Summary

Mortality rates of free naiad larvae are very high because of the inefficient mode of host infection. Out of one million, less than 100, or even less than 10, depending on the mussel species, attach successfully to a host. Also during the parasitic and juvenile stages, considerable mortality may occur, in the former due to the host's immune response, in the latter mainly due to failure in finding a suitable habitat. Mortality is lowest during the adult stage, approaching values of less than 5% $year^{-1}$ in very long-lived species. Fertility ususally increases with size (age).

Factors affecting species' communities include macro- and microhabitat characteristics, species-specific predation rates by muskrats and interspecific competition.

References

Baker SM, Hornbach DJ (1997) Acute physiological effects of zebra mussel (*Dreissena polymorpha*) infestation on two unionid mussels, *Actinonaias ligamentina* and *Amblema plicata*. Can J Fish Aquat Sci 54:512–519

Bauer G (1987a) The parasitic stage of the freshwater pearl mussel.II. Susceptibility of brown trout. Arch Hydrobiol Suppl 76/4:403–412

Bauer G (1987b) The parasitic stage of the freshwater pearl mussel.III. Host relationships. Arch Hydrobiol Suppl 76/4:413–423

Bauer G (1988) Threats to the freshwater pearl mussel in central Europe. Biol Cons 45:239–253

Bauer G (1994) The adaptive value of offspring size among freshwater mussels (Bivalvia; Unionoidea). J Anim Ecol 63/4:933–944

Bauer G (1998) Allocation policy of female freshwater pearl mussels. Oecologia 117:90–94

Davis GM, Heard WH, Fuller SLH, Hesterman C (1981) Molecular genetics and speciation in *Elliptio* and its relationships to other taxa of North American Unionidae (Bivalvia). Biol J Linn Soc 15:131–150

Downing J, Rochon Y, Pérusse M, Harvey H (1993) Spatial aggregation, body size, and reproductive success in the freshwater mussel *Elliptio complanata*. J N Am Benthol Soc 12(2):-148–156

Falk L (1995) Das ektoparasitische Larvenstadium der afrikanischen Süßwassermuschel *Mutela bourguignati* Bourguignat 1885. PhD Thesis, Tierärztliche Hochschule Hannover, 128 pp

Greenslade PJM (1983) Adversity selection and the habitat templet. Am Nat 122:352–365

Hanson JM, MacKay WC, Prepas EE (1988) The effects of water depth and density on the growth of a unionid clam. Freshwater Biol 19:345–355

Haukioja E, Hakala T (1978) Life history evolution in *Anodonta piscinalis*. Oecologia 35:253–266

Herrmann F (1995) Untersuchungen zum Lebenszyklus der afrikanischen Süßwassermuschel *Mutela bourguignati* Bourguignat 1885. PhD Thesis, Tierärztliche Hochschule Hannover, 96 pp

Hochwald S (1997) Das Beziehungsgefüge innerhalb der Größenwachstums- und Fortpflanzungsparameter bayerischer Bachmuschelpopulationen (*Unio crassus* Phil.1788) und dessen Abhängigkeit von Umweltfaktoren. PhD Thesis University Bayreuth, 172 pp

Jansen WA, Hanson JM (1990) Estimates off the number of glochidia produced by clams (*Anodonta grandis simpsoniana*), attaching to yellow perch (*Perca flavescens*), and surviving to various ages in Narrow Lake, Alberta. Can J Zool 69:973–977

Jokela J, Mutikainen P (1995) Phenotypic plasticity and priority rules for energy allocation in a freshwater clam: a field experiment. Oecologia 104:122–132

Jokela J, Uotila L, Taskinen J (1993) Effect of the castrating trematode parasite *Rhipidocotyle fennica* on energy allocation of fresh-water clam *Anodonta piscinalis*. Funct Ecol 7:332–228

Myers RA, Doyle RW (1983) Predicting natural mortality rates and reproduction-mortality trade offs from fish life history data. Can J Fish Aquat Sci 40:612–620

Negus C (1966) A quantitative study of growth and production of unionid mussels in the River Thames at Reading. J Anim Ecol 35:513–532

Neves RJ, Widlak JC (1987) Habitat ecology of juvenile freshwater mussels (Bivalvia: Unionidae) in a headwater stream in Virginia. Am Malacol Bull 5(1):1–7

Pauley GB (1968) A disease of the freshwater mussel, *Margaritifera margaritifera*. J Invertebr Pathol 12:321–328

Pianka ER (1972) r- and K-selection or b- and d-selection? Am Nat 106:453–464

Southwood TRE, May R, Hassel MP, Conway GR (1974) Ecological strategies and population parameters. Am Nat 108:791–804

Stearns SC (1976) Life history tactics. Quart Rev Biol 51:3–47

Stearns SC (1992) The evolution of life histories. Oxford University Press, Oxford

Stone NM, Earll R, Hodgson A, Mather JG, Parker J, Woodward FR (1982) The distributions of three sympatric mussel species (Bivalvia: Unionidae) in Budworth Mere, Cheshire. J Molluscan Stud 48:266–274

Strayer DL (1993) Macrohabitats of freshwater mussels in streams of the northern Atlantic slope. J N Am Benthol Soc 12:236–246

Terberger J (1999) Wirtsbeziehung der Teichmuschel *Anodonta*. Ms Thesis, Univerisity Freiburg

Teutsch S (1997) Populationsbiologische Untersuchung an *Anodonta*: Fortpflanzungsstrategien und Charakterisierung von Teilpopulationen innerhalb eines Stillgewässers. MS Thesis, University Freiburg, 82 pp

Vannote RL, Minshall GW (1982) Fluvial processes and local lithology controlling abundance, structure, and composition of mussel beds. Proc Natl Acad Sci USA 79:4103–4107

Watters GT (1992) Unionids, fishes, and the species-area curve. J Biogeogr 19:481–490

Yeager MM, Cherry DS, Neves RJ (1994) Feeding and burrowing behaviors of juvenile rainbow mussels, *Villosa iris*. JN Am Benthol Soc 13:217–222

Young M, Williams J (1984a) The reproductive biology of the freshwater pearl mussel *Margaritifera margaritifera* (Linn.) in Scotland I. Field studies. Arch Hydroiol 99,4:405–422

Young M, Williams J (1984b) The reproductive biology of the freshwater pearl mussel *Margaritifera margaritifera* (Linn.) in Scotland II. Laboratory studies. Arch Hydroiol 100,1:29–43

Zale AV, Neves RJ (1982) Reproductive biology of four freshwater mussel species (Mollusca: Unionidae) in Virginia. Freshwater Invertebr Biol 1(1):17–28

10 Effect of Muskrat Predation on Naiads

ELKE ZAHNER-MEIKE and J. MARK HANSON

10.1 Introduction

The freshwater mussel fauna in Europe is very different from that in North America. There are only ten species of freshwater mussels (Unionacea: Margaritiferidae and Unionidae) in Central Europe (Badino et al. 1991; Bogan 1993) whereas the North American fauna is very diverse, with about 290 recognized species and a high degree of endemism (Williams et al. 1993; Turgeon et al. 1998; Tyrrell and Hornbach 1998). While both freshwater mussel faunas are especially threatened by human activities, they face differing threats from introduced species due to marked differences in their evolutionary history. European naiads evolved in the absence of a large mammalian predator but, in some areas, evolved in the presence of an encrusting competitor (i.e., the zebra mussel, *Dreissena polymorpha*). Conversely, the North American naiads evolved in the presence of a large mammalian predator (i.e. the muskrat, *Ondatra zibethicus*) but in the absence of the encrusting competitors.

The muskrat is one of the few animals, other than man, to prey extensively upon freshwater mussels. During the last century, muskrats spread rapidly throughout Europe, subjecting the freshwater mussel populations, already severely stressed by human activities, to a new source of mortality. This study examines and compares the effects of muskrat predation on freshwater mussel communities in Europe and North America. We report the results of new studies in New Brunswick, Canada, and southwestern Germany and evaluate these results in the context of the different history of exposure to muskrat predation for freshwater mussel populations in North America and Europe.

10.1.1 The Study Species

The muskrat, *Ondatra zibethicus*, is a large rodent native to North America, that inhabits a wide range of wetland habitats: from fens and marshes to large lakes and rivers (Johnson 1925; Marcström 1964; Proulx and Gilbert 1983; Clough 1987). The early literature describes the muskrat either as a strict vegetarian (Bursik 1914; Butler 1940; Takos 1947) or as a species that occasionally eats animal food (Prell 1924; Johnson 1925; Enders 1932; Ching and Chih-Tang 1965; Van Troostwijk 1976; Willner et al. 1980). Many of these early studies focused on marsh-dwelling populations, and freshwater mussels typically are not found in marshes (Convey et al. 1989; Lacki et al. 1990). Muskrats living in small lakes and along rivers and streams, however, often include animal materials, especially freshwater mussels, in their diets (Headlee 1906;

Ecological Studies, Vol. 145
G. Bauer and K. Wächtler (eds.) Ecology and
Evolution of the Freshwater Mussels Unionoida

Johnson 1925; Enders 1932; Van Cleave 1940; Fuller 1974; Willner et al. 1980; Convey et al. 1989; Jokela and Mutikainen 1995).

The muskrat was first introduced into Europe in 1905 when five individuals were released near Prague (Ulbrich 1930; Hoffmann 1952; Danell 1978, 1996). These muskrats reproduced rapidly and their descendants spread along the waterways of the continent. The first invaders of Germany were recorded from 1914 to 1915 (Hoffmann 1952). The rapid and widespread colonization was due to the high dispersion ability, tremendous adaptability and a high reproductive rate of the species (Hoffmann 1952; Frank and Härle 1967; Van Troostwijk 1976). A female usually has two litters per year (Erickson 1963; Becker 1973; Akkermann 1975; Proulx and Gilbert 1983) with an average of six to seven young per litter (Hoffmann 1952; Marcström 1964; Becker 1973; Akkermann 1975; Van Troostwijk 1976). Moreover, the potential natural enemies of the muskrat in central Europe have been largely exterminated by human activities (Hoffmann 1952; Hochwald 1990). Consequently, the muskrat is now firmly established as a prominent feature of the European wetland fauna.

In New Brunswick, Canada, we evaluated the effects of muskrat predation on freshwater mussel populations in two habitats. The freshwater mussel community of a small river consisted only of *Margaritifera margaritifera*. This provided the opportunity to examine the potential impact of muskrat predation on the severely threatened *M. margaritifera* populations in Europe. In central Europe the abundance of freshwater pearl mussels (= eastern pearlshells in North America, Turgeon et al. 1998) has decreased by over 90% since the early 1900s (Bauer 1988). The reasons for this dramatic decline are complex and locally different. In the past few decades, the main problem has been the eutrophication, which alters the sediment characteristics and leads to increased mortality of newly settled juveniles (Bauer 1991; Bogan 1993). In North America the habitat of *M. margaritifera* is still in relatively good shape so that the threats to continued existence are less severe than in central Europe. The second habitat studied in New Brunswick was lacustrine (a small pond and two small reservoirs). The species *Pyganodon cataracta* dominated the lacustrine mussel community. *Pyganodon cataracta* is a widespread species and its conservation status is considered to be stable (Williams et al. 1993). This species and *M. margaritifera* represent extremes in terms of foraging costs for muskrats due to differences in shell thickness and costs of transport in running versus still waters. These differences were evaluated in terms of sizes of animal eaten and meat rewards (for animals of the same size) and can serve as a model to explain some of the patterns in species selection that are seen in areas where freshwater mussels species of widely differing shell thickness cooccur.

In southwestern Germany the effects of muskrat predation on the freshwater mussel community were studied in a single habitat type. Unlike the New Brunswick locations, there were three freshwater mussel species present in the small streams in Germany. These three species (*Anodonta anatina*, *Unio crassus* and *Unio pictorum*) face many of the same threats as *M. margaritifera* in southern and central Europe. Overall, their numbers are in decline but, until recently, their conservation status was not considered to be as threatened as that of *M. margaritifera*. This might change as the abundance of the muskrat and its influence on aquatic habitats in Europe increases.

10.2 Effect of Muskrat Predation on a Population of *Margaritifera margaritifera* and *Pyganodon cataracta* in New Brunswick, Canada

10.2.1 Methods of the Canadian Studies

10.2.1.1 Predation on *Margaritifera margaritifera*

Live *Margaritifera margaritifera* and empty shells discarded by muskrats were collected from two separate sites in the Little River (a tributary of the Petitcodiac, New Brunswick, Canada). Site 1 (45° 52.00 W; 64° 59.23 W) was 10 to 18 m wide, up to 60 cm deep (summer low water), and 285 m of river was surveyed. Site two (45° 57.37 N; 64° 57.50 W) was about 13 km downstream, 15 to 20 m wide, up to 70 cm deep, and mussels were collected from a 200-m stretch of river. The bottom at both sites consisted mostly of cobble and rock with sand and gravel near shore and around the bases of rocks and boulders. The water was clear and moderate in velocity. Trees or thick brush grew almost to the water's edge for most of the study areas. Emergent vegetation was sparse and, when it existed, occupied a narrow (<1 m) band along shore. There were occasional patches of water buttercup (*Ranunculus aquatilis*), the only submersed vegetation, at both sites. Old abandoned fields occurred along a small part of the shoreline at both sites. Thus, both sites represented marginal habitats for muskrat.

Site 1 was an area frequented by muskrats in 1997 and 1998 whereas no evidence of muskrat activity was detected at site 2 in either year. Therefore, site 2 was used as a no-predation control for site 1. At site 1, we collected all mussel shells from animals consumed by muskrats for the full 285-m stretch of river on 30 July 1997, 20 August 1997, 19 June 1998 (no mussel predation by muskrats detected) and 30 July 1998. The spring freshet removed shells of any mussels consumed after 20 August 1997. The shells of mussels eaten by muskrats were easily distinguished from those killed by stranding because muskrats broke one valve (occasionally both) in a characteristic pattern (Triplet 1983; Convey et al. 1989; Hanson et al. 1989) and frequently only consumed the viscera, leaving the adductor muscles intact (J.M. Hanson, pers. observ.). A shell was counted as an individual eaten by muskrat only if the hinge was present. Shells were recovered on shore immediately adjacent to the water, in the water near shore, and beside large rocks in the river (usually in discrete piles). For site 1, five 10 × 10-m quadrats were arbitrarily marked in the areas adjacent to middens and the bottom was carefully searched visually and by digging in sand and gravel around rocks and boulders. When encountered, locations with deposits of sand and fine gravel were sampled with a 30-cm-wide push net (to a depth of 5–6 cm). The sediments were sieved through the 6-mm mesh netting, which retained all mussels about 10 mm long and larger. Sieving of sediments is almost a necessity when attempting to collect mussels <25 mm long (Hanson et al. 1988; Amyot and Downing 1991; Richardson and Yokely 1996; Obermeyer et al. 1997). We used natural landmarks (outcrops and ledges of bedrock), to delimit the quadrat sizes in site 2. As a result, the area searched on 22 August 1997 consisted of a single 15 × 30-m and two 10 × 15-m quadrats. We returned to site 2 on 24 June 1998 and again found no evidence of muskrat predation.

10.2.1.2 Predation on *Pyganodon cataracta*

Muskrat middens containing *Pyganodon cataracta* shells were discovered in three locations in the Petitcodiac River system: (1) on the shore of an impounded section of a small tributary, Turtle Creek (46° 02.68 N; 64° 52.54 W), (2) around a small (about 600 m^2) artificial pond (46° 04.76 N; 65° 04.35 W) on the North River (a large tributary of the Petitcodiac River), and (3) at the edge of a small (about 5 ha) reservoir (45° 45.92 N; 65° 02.08 W) in the headwaters of the Little River (also a large tributary of the Petitcodiac River). It was possible to quantitatively collect live specimens to compare sizes eaten with those in the natural habitat only at the Turtle Creek site. No evidence of muskrat predation on freshwater mussels was detected in the main headpond of the Petitcodiac River, although muskrats were frequently observed in the marsh areas of the headpond.

The Turtle Creek site was a flooded section of the Petitcodiac Headpond, an impoundment of a formerly tidal section of the Petitcodiac River. The impoundment is periodically drained, which strongly affects the structure of the freshwater mussel community. The headpond was drained and some saltwater was allowed into the impoundment in spring 1988, which appeared to eliminate all freshwater mussels. Further, the headpond was again drained for over 30 days in 1990 and 1991. These disturbances have resulted in a freshwater mussel community in which *Pyganodon cataracta* is overwhelmingly dominant and few animals > age-6-years (in 1997) were present (J.M. Hanson, unpubl. data). Most recently, the impoundment was partially drained on 31 May 1998 and we discovered an active muskrat midden on 2 June 1998 while we sampled the part of the mussel community stranded by the drawdown. We collected all mussels in a large quadrat (263 m^2) that encompassed the midden and represented water 0 to 5.0 m deep when the impoundment was full. Three other quadrats were sampled (combined area 401 m^2) and mean density was calculated. The site was subsequently visited on 23 July 1998 (the impoundment was refilled on 18–19 June 1998) but no evidence of muskrat predation on mussels was found, in part because nearly all mussels formerly present at this site had been killed by the drawdown, the macrophyte community had been eliminated, and cattle had been allowed access to the shoreline (which eliminated most of the emergent vegetation). Thus, the site represented marginal muskrat habitat at the end of this study.

The pond and small reservoir samples were used to determine the sizes of *Pyganodon cataracta* that could be eaten when the full range of sizes of this species was available to the muskrats. A second purpose was to obtain samples of large animals for meat-weight and shell–weight relationships to compare to those of *Margaritifera margaritifera*. In the case of the artificial pond, we collected all shells from the entire perimeter of the pond on 23 June and 23 July 1998. The pond itself had a steep shoreline, clear water and sand bottom covered with organic debris. The small reservoir was sampled on 30 July 1990. We collected shells from one large midden and several small middens near the dam. The water was very dark with dissolved humic materials. Although qualitative sampling detected numerous living *Pyganodon* in the shallows (<70 cm deep) of the reservoir, we were unable to quantitatively sample the mussel population due to time limitations. Consequently, we used only the length distribution to estimate the maximum sizes of *Pyganodon cataracta* that could be eaten when the full range of sizes of this species was available to muskrats and to contrast these sizes with those of *M. margaritifera*, which attains a similar adult size. A narrow

10-m-wide band of emergent vegetation surrounded the reservoir and most of the water surface was covered by the yellow pond lily (*Nuphar variegatum*). Both sites represented relatively good muskrat habitats when compared to Turtle Creek in July 1998 and to the two river sites where we sampled the *Margaritifera* population.

10.2.1.3 Laboratory and Statistical Analyses

All shells or intact animals were returned to the laboratory, where total length (to the nearest mm) was measured with calipers. Total wet weight (water drained from shell to 0.01 g), meat weight (wet weight to 0.01 g), and shell weight (blotted dry, to 0.01 g) were recorded for a random subsample of both species.

Regression equations were determined between the three morphological variables and total length using the least-squares method. Variables were transformed as needed to comply with the requirements of linear regression. The length-frequency distributions of mussels recovered on muskrat middens were compared to the length-frequency distributions of mussels collected from the natural habitat by means of contingency tables.

10.2.2 Results of the Studies in Canada

10.2.2.1 Species and Size Selection

Species selection by muskrats was not a factor in the New Brunswick studies. *Margaritifera margaritifera* was the only species present at the two stream sites studied. Species other than *Pyganodon cataracta* were extremely rare in the three lacustrine sites. *Pyganodon* was the only species collected from middens at the two reservoir locations. Muskrat predation clearly was size-selective for both mussel species; however, the pattern differed substantially.

The size of *M. margaritifera*, eaten by muskrats differed significantly ($\chi^2 = 205.02$, df = 16, $P \ll 0.0001$) from the size left in the environment (Fig. 10.1). Of particular interest was the fact that almost all individuals <75 mm long were eliminated from the predation site but were common at the control site. Moreover, *Margaritifera* >110 to 115 mm long appeared to be immune to muskrat predation, i.e., this appeared to be the size threshold above which muskrats could not handle the mussels. When compared to the predation site, the distributions of living animals also differed

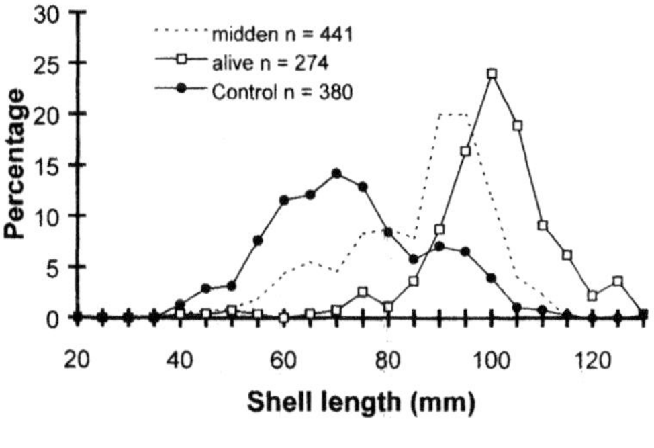

Fig. 10.1. Length frequency distributions for *Margaritifera margaritifera* collected from muskrat middens on the Little River (*midden*), from the Little River adjacent to the muskrat middens (*alive*), and from a site devoid of muskrat predation (*Control*)

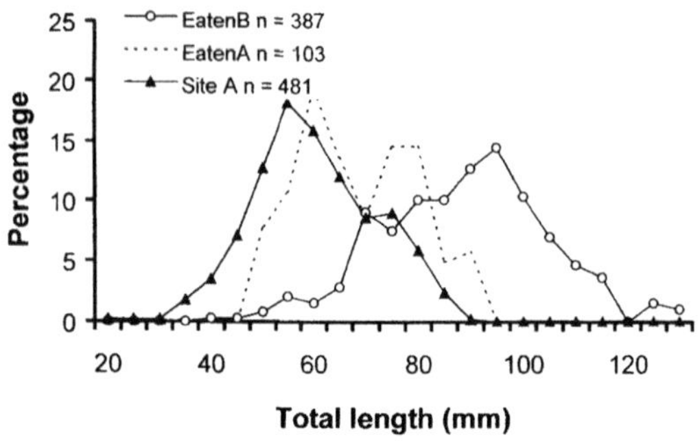

Fig. 10.2. Length-frequency distributions for *Pyganodon cataracta* collected from muskrat middens on the shores of a small pond and reservoir (*EatenB*), from muskrat middens on the shore of Turtle Creek (*EatenA*) and from Turtle Creek itself (site A)

significantly ($\chi^2 = 155.85$, df = 16, $P \ll 0.0001$) from that of the control site, much of the population at the control site consisted of animals <80 mm long and animals >100 mm long were scarce. This scarcity of large animals was most likely a sampling artefact of the three quadrats sampled because we collected individuals as large as 125 mm long in areas adjacent to the quadrats (J. M. Hanson, unpubl. data); however, these samples were not collected quantitatively and could not be included in the size distribution for the control site.

The size distributions of *Pyganodon* eaten by muskrats at Turtle Creek differed significantly ($\chi^2 = 53.11$, df = 13, $P \ll 0.0001$) from that of animals in the environment (Fig. 10.2). In contrast with the case for *Margaritifera*, the muskrats selectively removed the largest animals when preying upon *Pyganodon*. Animals larger than 90 mm were not present at the Turtle Creek site but samples collected from the artificial pond and reservoir showed that muskrats could eat *Pyganodon* as large as 130 mm (solid line with open circles, Fig. 10.2).

Muskrat predation and overgrowth by the macroalgae *Chara* (Hanson et al. 1988) resulted in almost complete elimination of freshwater mussels from the small pond site. We collected numerous *Pyganodon* shells from middens around the pond on 23 June 1998 and about 30 large, living, animals from the pond. We collected a large number of empty shells but could find no living animals on 23 July 1998. In a subsequent visit to the pond (10 May 1999), five living *Pyganodon* and several dozen intact, empty, shells were found in the water and the empty shell of one newly consumed animal was found on shore. Thus the mussel population in the pond has been reduced to a very low density.

10.2.2.2 Morphology and Energetic Value

The difference in the pattern of size-selective predation by muskrat on the two mussel species was consistent with the difference in morphology between the species (Table 10.1). For *Margaritifera*, the animal changes shape, becoming heavier with increased length. Moreover, the proportion of total weight represented by the shell increased with increasing length while that of the meat decreased. The pattern was different for *Pyganodon*. The slope of the total weight-length regression did not differ from 3 but meat weight represented proportionately more of the total weight as length increased and the proportion represented by shell weight actually decreased. For all sizes, *Pyganodon* yielded more meat yet had a lower transport cost (total weight) than *Mar-*

Table 10.1. Variation in total weight, shell weight and meat weight for selected sizes (TL = total length in mm) of *Margaritifera margaritifera* (Ma) and *Pyganodon cataracta* (Py) from the Petitcodiac River watershed (predicted from regression equations). All weights are in g, wet weight

TL	Total weight		Shell weight		Meat weight	
	Ma	Py	Ma	Py	Ma	Py
50	6.68	6.27	3.72	1.95	2.98	4.21
60	12.23	10.78	7.12	3.24	5.06	7.44
70	20.39	17.05	12.33	4.97	7.93	12.06
80	31.76	25.36	19.83	7.21	11.69	18.32
90	46.94	36.00	30.17	10.00	16.47	26.46
100	66.58	49.24	43.90	13.41	22.37	36.85
110	91.33	65.38	61.65	17.48	29.52	49.66
120	121.89	84.69	84.04	22.26	38.02	65.21

garitifera. The total weight of a small (50 mm) *Margaritifera* was 1.07 times that of the same-sized *Pyganodon* and by 110 mm the weight of a *Margaritifera* was 1.40 times that of the same-sized *Pyganodon*. In terms of meat, the pattern reversed: the meat of a *Pyganodon* 50 mm long was 1.41 times as heavy as that of a *Margaritifera* and the difference increased to 1.68 times for animals 110 mm long. These values would represent a large difference between species in the value (meat weight) and handling cost (total weight) of mussels of the same size if the two clam species occurred in the same habitat. Based on the size of animal eaten by muskrat, *Margaritifera* > about 110 mm long were too heavy for muskrat to handle. This corresponds to a total weight of about 90 g, of which only 30 g was meat. In contrast, a 110-mm-long *Pyganodon* weighed 65 g, of which 50 g was meat. Clearly, it would be more profitable for muskrats to preferentially prey on the thin-shelled species such as *Pyganodon* instead of the thick-shelled species *Margaritifera* where the two occur in the same mussel beds. Although these two species did not occur together in the locations sampled in this study, thin- and thick-shelled species commonly cooccur elsewhere (e.g. Neves and Odom 1989; Watters 1994, 1995; Tyrrell and Hornbach 1998). We suggest that many of the observed differences in both species and sizes of mussels eaten between in situ communities and muskrat middens could be explained if the species-specific total weight and meat weight to length relationship were known for the species inhabiting the mixed-species mussel beds. To the best of our knowledge, similar between-species comparisons of total weight and meat weights at length have not been attempted elsewhere.

10.3 Effect of Muskrat Predation on the Populations of Unionid Mussels in Germany

10.3.1 Methods of the German Studies

Live mussels and empty shells were collected from two small brooks, Fischgießen and Rinnbach, in the Ortenau region of southwestern Germany. The brooks were normally between 1 and 2 m wide, about 40 cm deep (summer normal water), and had a sandy substrate. Samples were collected from four sites: three 60-m-long sites (S I, S II, S III), each separated by about 200 m, in the Fischgießen and a single 200-m-long site in the Rinnbach. Three freshwater mussel species, *Anodonta anatina*, *Unio crassus* and *Unio pictorum*, occurred in all collection sites. Collection of mussels and empty shells was done visually by touch while searching the substrate through a box fitted with a transparent bottom. Each intact valve, where the second valve was missing, was assumed to represent one clam consumed by the muskrats. Evidence that the mussel was killed by a muskrat included: the location of the shell on a midden in the shallow water or at the water's edge; scratches and tooth marks on the shell; and the remnants of soft body parts in recently killed mussels (mostly remnants of the adductor muscles). Only thin-shelled mussel species, such as *Anodonta anatina*, had the shell broken by the muskrat (mostly only one valve). We feel these data represented a minimum estimate of numbers of mussels eaten by muskrats because muskrats occasionally break both valves and some shells were probably missed due to being buried in sediments, washed away by the current or were hidden in shoreline vegetation.

The freshwater mussel population in the habitat and muskrat middens in the Fischgießen were sampled at irregular intervals between August 1995 and October 1997 and those in the Rinnbach between September 1996 and July 1997. Sampling of the habitat was not possible at times due to high water conditions, poor visibility, ice cover, or when the brooks became overgrown with vegetation during summer.

On each collection date, we identified each specimen to species and recorded the numbers detected for each site. Live mussels were returned to their original location. Individuals of *Unio pictorum* were treated differently as part of a more detailed study of the biology of this species. The total length of each live individual was measured with callipers to the nearest mm and each specimen received a unique identification number written on synthetic lacquer (Cowisit synthetic lacquer of the CWS company) applied to one valve. This number was necessary to decide in the repeated samplings if an individual mussel had already been registered. After this treatment the animal was returned to its original capture location.

The shells collected from muskrat middens were taken back to the laboratory where the numbers of each species and total length of each *Unio pictorum* were recorded.

A series of contingency tables were used to detect differences in the species composition, and in length frequencies of *U. pictorum*, from samples collected from the natural habitat compared with samples collected from muskrat middens.

10.3.2 Results of the Studies in Germany

10.3.2.1 Numbers of Mussels Eaten by Muskrats

Muskrat feeding activity was detected in all four collection sites. During the study period, muskrats consumed at least 824 mussels in the Fischgießen site I (137 *Anodonta anatina*, 300 *Unio crassus*, 387 *Unio pictorum*), 737 mussels in the Fischgießen site II (123 *A. anatina*, 319 *U. crassus*, 295 *U. pictorum*), 162 mussels in the Fischgießen site III (122 *A. anatina*, 25 *U. crassus*, 15 *U. pictorum*) and 724 mussels (5 *A. anatina*, 502 *U. crassus*, 217 *U. pictorum*) in the Rinnbach site.

Muskrats killed 10 to 60% of the *Unio pictorum* individuals initially present and marked during this study. Of the 460 *Unio pictorum* in the Fischgießen site I, 173 in the Fischgießen site II, 10 in the Fischgießen site III, and 173 in the Rinnbach site that we marked, 277 (60%) in the Fischgießen site I, 77 (45%) in the Fischgießen site II, 1 (10%) in the Fischgießen site III, and 49 (28%) in the Rinnbach site were later recovered from muskrat middens.

10.3.2.2 Species-Selective Predation

The initial species composition of the mussel community differed between the four collection sites (Fig. 10.3): Fischgießen site I was dominated by *Unio pictorum;* the dominant species in Fischgießen site II were *Unio pictorum* and *Unio crassus; Anodonta anatina* dominated Fischgießen site III; and *Unio crassus* dominated the Rinnbach site. In addition, *Anodonta anatina* was noticeably rare in the Rinnbach site. We are not in a position to say if these species compositions are the result of prior muskrat predation because we have no data for these sites from the time before our examination.

Muskrats showed clear species-specific preferences at some times but not at others (Fig. 10.3). In all cases where there was species selection, the species eaten in higher proportions was *U. crassus* while that eaten in lower proportions was *U. pictorum.* In the Fischgießen site I, there was no significant difference between the species composition between living mussels and shells ($\chi^2 = 0.827$, df = 2, $P = 0.661$) during the first 7 months but the species composition then diverged ($\chi^2 = 6.360$, df = 2, $P = 0.042$), muskrats ate *Unio crassus* individuals in higher proportion relative to abundance in habitat by the end of the study period (Fig. 10.3). A different pattern was observed in the Fischgießen site II. Here the species composition differed ($\chi^2 = 9.192$, df = 2, $P = 0.010$) between the living mussels and the shells collected from middens in the first part of the study but the difference disappeared ($\chi^2 = 0.306$, df = 2, $P = 0.858$) by the end of the study (Fig. 10.3). In the Fischgießen site III the species composition did not differ between the living mussels and shells collected from middens for either period (August 1995 to March 1996: $\chi^2 = 1.203$, df = 2, $P = 0.548$; August 1995 to October 1997: $\chi^2 = 0.741$, df = 2, $P = 0.691$) (Fig. 10.3). Similarly to site I, muskrats in the Rinnbach showed a clear preference for *U. crassus* ($\chi^2 = 26.274$, df = 2, $P < 0.0001$) at the end of sampling but not at the beginning of the study ($\chi^2 = 0.710$, df = 2, $P = 0.701$) (Fig. 10.3).

10.3.2.3 Size-Selective Predation on *Unio pictorum*

When comparing the sizes of mussels present in the habitat with those eaten by muskrats (Fig. 10.4), the size distribution for the natural *Unio pictorum* population was taken from the beginning of the examination period because the muskrat's influence was lowest at this time and to minimize the effects of seasonal growth. For this same reason, the shell data for the comparisons were taken from the first midden collection (August or September) until the end of the following February because growth was minimal during this period. There was no significant difference between the length distribution in the natural clam population and the mussels killed by

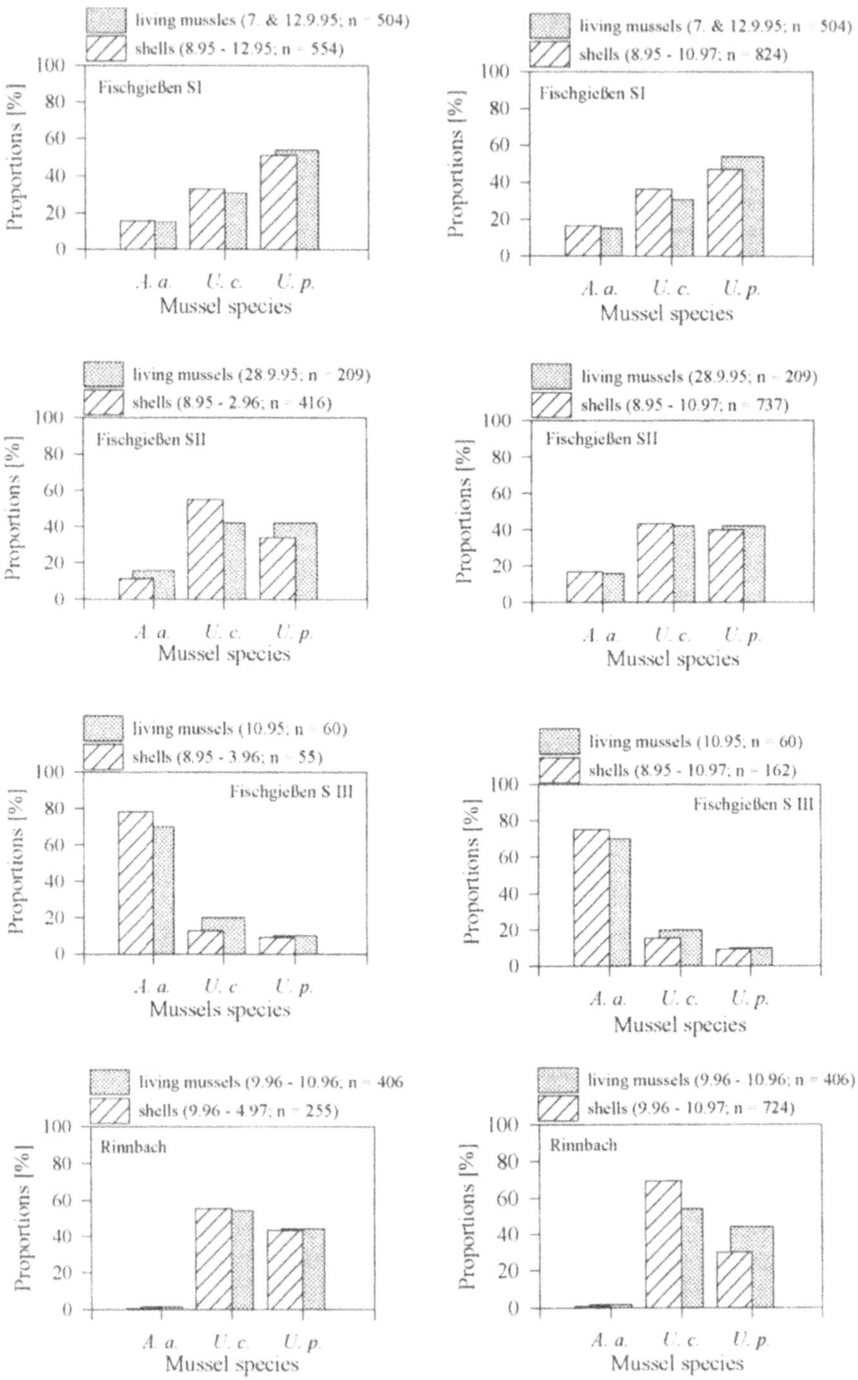

Fig. 10.3. Species distribution (*Anodonta anatina*, *Unio crassus* and *Unio pictorum*) of the living mussels in comparison with the collected shells on the different collection sites (*Fischgießen S I, S II, S III* and *Rinnbach*) and in different periods of time

Fig. 10.4. Length distribution of the living *Unio pictorum* in comparison with the length distribution of shells'

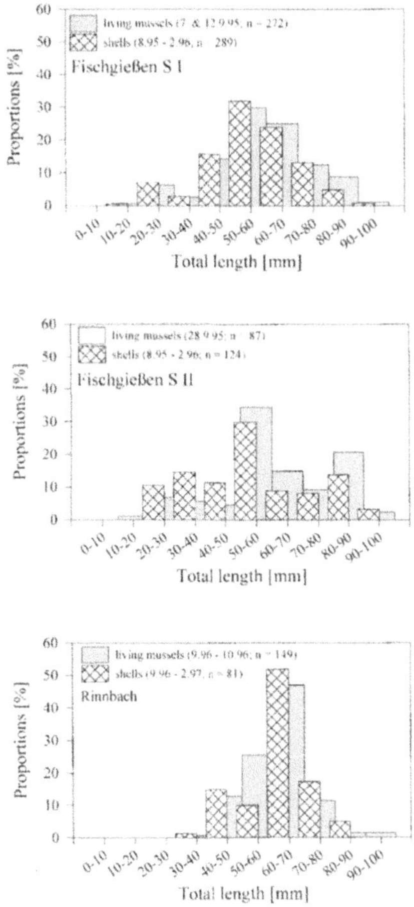

muskrats at both Fischgießen sites (Fischgießen site I: $\chi^2 = 4.399$, df = 6; $P = 0.623$; Fischgießen site II: $\chi^2 = 9.657$, df = 6; $P = 0.140$) (Fig. 10.4). On the other hand, the length distribution of the living mussels in the Rinnbach was different from that of shells collected from muskrat middens ($\chi^2 = 8.860$, df = 3, $P = 0.030$), the muskrats consumed proportionately more of the larger individuals (Fig. 10.4).

10.3.2.4 Seasonal Pattern of Muskrat Predation

The seasonal pattern of muskrat predation appeared to differ between collection sites. In the three Fischgießen sites, muskrats mostly consumed mussels during the colder months (Fig. 10.5, above). In contrast, mussel consumption at the Rinnbach site seemed to be limited mainly to summer months (Fig. 10.5, below).

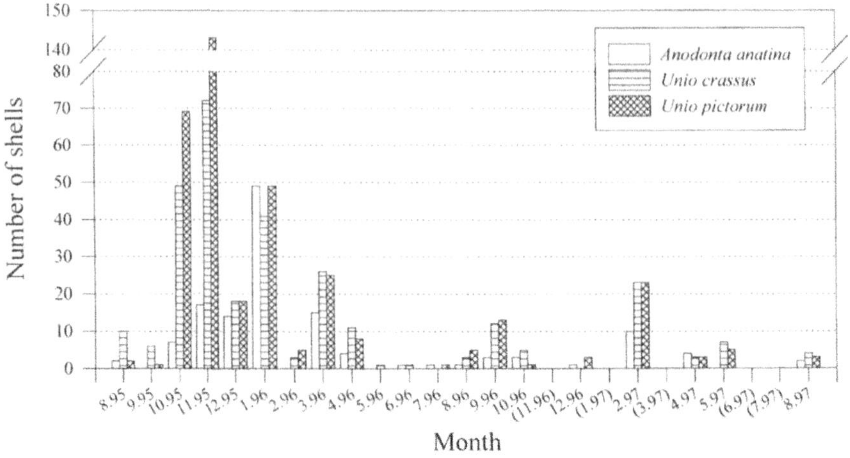

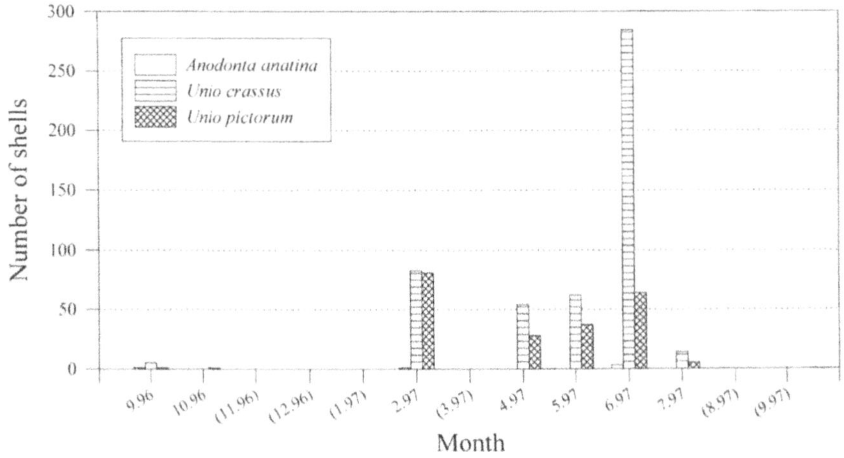

Fig. 10.5. Shells collected in the Fischgießen site I (*above*) and in the Rinnbach (*below*) (in the months *in brackets* a collection was impossible)

10.4 Discussion

10.4.1 Muskrat Predation Rates and Persistence of Endangered Mussel Populations

Muskrat predation rates on freshwater mussels and the impact of this predation on the persistence of populations are highly variable and site-dependent. The numbers of mussel shells recovered from individual middens varys from a few dozen to hundreds or thousands per year (Van Cleave 1940; Brander 1955; Hanson et al. 1989; Watters 1994; Jokela and Mutikainen 1995; Tyrrell and Hornbach 1998; this study). The frequency of collection is, of course, an important determinant of the number of shells collected from a midden. The impact of this feeding on species abundance, or even continued existence of a species, needs to be placed in context of the size of the natural population.

In our New Brunswick study, the impact of muskrat consumption of a few hundred *Pyganodon cataracta* varied between the near elimination of the entire population in a small pond (aided by overgrowth by the macroalgae *Chara*) (Sect. 10.2.2.1) and an amount representing an insignificant proportion of a population consisting of millions of animals in an impoundment (i.e. the Petitcodiac Headpond, J. M. Hanson, unpubl. data). Similarly, Hanson et al. (1989) estimated the muskrats in a small lake ate 37 000 mussels (*Pyganodon grandis simpsoniana*) in 1 year. In this case, the predation represented 3% of the population and was limited to one of three basins in the lake. The authors concluded that this level of predation would not lead to extermination of the population because the biomass consumed was less than the annual production of the surviving population. In contrast, the consumption of a few hundred individuals in the small brooks in Germany represented 10 to 60% of the population for species whose conservation status is uncertain (Sect. 10.3.2.1). This level of consumption, if widespread, could represent a serious threat to the continued existence of these species in central and southern Europe. Muskrat predation is not a threat to the continued existence of most North American species (Haag and Warren 1998) except where one of the few remaining locales of an endangered species coincides with areas of active muskrat feeding on mussels (e.g. Neves and Odom 1989; Bruenderman and Neves 1993; Hogarth et al. 1995).

One common finding of both current and published work is that freshwater mussels living in small streams are most susceptible to the effects of muskrat predation. In large waterbodies, muskrats can largely eliminate mussels from areas near shore (Headlee 1906; Van Cleave 1940; Jokela and Mutikainen 1995; Tyrrell and Hornbach 1998), but those distant from the feeding areas are untouched either because they are too far from shore or in water too deep for muskrats to forage effectively. Consequently, there is continuous recruitment to the near shore area. In contrast, there are no refuges from muskrat predation for mussels living in small streams except, perhaps, for individuals buried deeply in the sediments. Although it is unlikely that muskrats can remove every individual of a species from a stretch of habitat, they may reduce local populations to a density below which successful reproduction is unlikely. Dioecious freshwater mussels need to be within some relatively small distance of each other to fertilize the eggs successfully (McMahon 1991; Graf 1997). Intense predation could result in the few remaining animals being too far apart to assure fertilization success. This is an important consideration because many freshwater mussel species in Central Europe are restricted to a small number of animals in isolated populations that are dominated by a few old age classes (Bauer 1988). In these cases, the loss of even a few individuals to muskrat predation could mean that reproduction becomes unlikely (Hochwald 1990) and the population will disappear once these few old individuals die.

The pattern of muskrat predation on *Margaritifera* in New Brunswick (Little River) suggests that muskrat predation may pose a serious threat to the remaining populations of this genus in Central and Southern Europe. Muskrats ate 500 to 600 *Margaritifera* in a short stretch of the Little River. The impact of this predation was small because the river supports a large *Margaritifera* population (it probably exceeds 500 000 animals; J.M. Hanson, unpublished data). The impact of muskrat predation in the area near middens, however, was clear; there was almost complete elimination of individuals <80 mm long (see Sect. 10.4.2). Many populations of *Margaritifera* spp. in Europe are reduced to a few hundred individuals living in a relatively short section of a river or stream (e.g. Bauer 1988; Altaba 1990; Young 1991; Morkens and Costello 1994). Consumption of several hundred individuals by muskrats could eliminate the entire remaining adult populations. Moreover, muskrats consumed many small individuals in the New Brunswick river, which means they could also jeopardize efforts to stock juveniles into currently vacant habitat (e.g. Bauer 1988; Altaba 1990; Buddensiek 1995). While it is not clear that muskrat predation has resulted in recruitment failure of any of the endangered *Margaritfera* populations in Europe, they clearly represent a threat previously unaccounted for.

10.4.2 Size- and Species-Selective Predation

Muskrats almost always show some size selection when preying upon freshwater mussels and, if more than one species is present, they usually show species selection because size and species selection are interrelated processes. The sizes and species of mussels eaten by muskrats depend very much on the characteristics of the naiad species and associated handling costs specific to the location (Convey et al. 1989; Hanson et al. 1989; Neves and Odom 1989; Watters 1994; Jokela and Mutikainen 1995); important characteristics are shell thickness and weight, strength of adductor muscles and burrowing behaviour and anchoring ability of the naiad species. As with any predator, a muskrat will attempt to maximize its energy and critical nutrient gains while minimizing costs, including handling costs and the threat of predation on the muskrat itself (Jokela and Mutikainen 1995; Tyrrell and Hornbach 1998). The basic data needed to make this type of evaluation are, however, seldom available. As information on species-specific values and foraging costs become available, it will become possible to understand and even predict the predation patterns by muskrats on mixed-species mussel beds.

Relative shell thickness is an obvious factor determining both the sizes eaten and species selectivity of muskrats foraging in mixed-species mussel beds. This study showed there can be large difference in energetic value (meat weights) available to muskrats between two mussel species that attain the same maximum size but differ substantially in shell thickness (Table 10.1). Muskrat ate the largest sizes available of the thin-shelled *Pyganodon cataracta* (up to as 130 mm long). Similarly, muskrat in Narrow Lake, Alberta (Canada), also ate the largest available individuals of *Pyganodon grandis simpsoniana* (again a thin-shelled species) (Convey et al. 1989; Hanson et al. 1989). In fact, muskrats can eat thin-shelled mussels to a length of about 150 mm for *Anodonta cygnea* (Libois 1988). In contrast, muskrats could not handle thick-shelled *Margaritifera* larger than about 100 mm long in the Little River (Sect. 10.2.2.1). This difference in maximum size eaten is consistent with numerous reports that thin-shelled species are consumed preferentially over thick-shelled species. Moreover, it is the largest-sized individuals of thin-shelled species that are taken first while medium-sized individuals of thick-shelled species are consumed, the largest individuals of thick-shelled species being too heavy to carry and open, even when abundant (Van Cleave 1940; Neves and Odom 1989; Watters 1994, 1995). An important effect is that the strength of adductor muscles increases with increasing mussel size (Tyrrell and Hornbach 1998). Shell weight is not, however, the only factor in determining size and species selection by muskrats; habitat variables can be very important. For example, muskrats in a stream in Finland did not eat individuals of the thin-shelled *Anodonta piscinalis* longer than about 90 mm (Jokela and Mutikainen 1995). The difference in foraging and transport costs between running and still waters probably explains this lower size threshold.

Species selection by muskrat was not a factor in the New Brunswick study sites because there was either only one mussel species present or the community was overwhelmingly dominated by one species (see Sect. 10.2.2.1). In fact, the interaction between size and species shell characteristics has rarely been studied directly in North America (one exception is Tyrrell and Hornbach 1998). The few studies that report species selection tend to report only qualitative measures of the sizes of animals that were consumed (e.g. Headlee 1906; Van Cleave 1940; Neves and Odom 1989; Watters 1994); nevertheless, a clear pattern emerges. Large, heavy-shelled species are consumed in lower proportions than found in the nearby mussel beds and the large individuals of thin-shelled species are quickly eliminated. For intermediate sizes, the species composition does not seem to matter, the muskrats seem to eat them in proportion to their availability.

The pattern of species selection by muskrats at our sites in Germany seems to resemble the second case (Sect. 10.3.2.2). There were three mussel species present in both German brooks, which is considered unusual in these small waters because the three mussel species appear to differ in their habitat requirements. In these study sites, the muskrats showed a clear preference for *Unio crassus* over *Anodonta anatina*, which was unexpected because Akkermann (1972) found that muskrats typically selected the thin-shelled *Anodonta* over *Unio*. Our results indicated that the muskrat's species selection was principally based on relative abundance of the mussel species. It is possible that morphological characteristics (e.g. shell thickness and size, strength of adductor muscles) of the three naiad species were too similar to be a strong factor for a clear species preference. Furthermore, mussels greater than 100 mm long (which seems to be a rough threshold for handling thick-shelled mussel species) were not found at any site and this, along with the very small size of the brooks and simple substrate, most likely resulted in there being little energetic difference between the species.

Muskrats rarely eat freshwater mussels less than 30 to 40 mm long, regardless of species (Hanson et al. 1989; Neves and Odom 1989; Jokela and Mutikainen 1995; Tyrrell and Hornbach 1998; this study). Muskrats locate mussel prey primarily by touch, at least when feeding at night (J.M. Hanson, pers. observ.). The small mussels either are buried too deep in the sediment for the muskrat to detect or simply represent too little return for the effort of carrying the shell ashore. Failure to detect a lower size limit when studying muskrat predation is usually due to a failure to adequately sample the smaller mussels present in the habitat. Visual (and tactile) searches by humans are quite inefficient for collecting mussels < about 35 mm long, which results in the abundance of mussels <30 to 40 mm long being severely underestimated (Hanson et al. 1988; Amyot and Downing 1991; Hove and Neves 1994; Richardson and Yokely 1996). Consequently, the apparent consumption of small mussels will be overestimated.

Muskrats appeared to use two different techniques to open mussel shells. Muskrats bit a piece out of one valve of thin-shelled taxa such as *Anodonta* and *Pyganodon*

(Headlee 1906; Prell 1924; Johnson 1925; Akkermann 1972; Convey et al. 1989; Hanson et al. 1989; Sect. 10.2) but are not reported to break the shell in other species (Apgar 1887; Brander 1955; Neves and Odom 1989; Hogarth et al. 1995; Sect. 10.3). At first glance, differences in shell thickness between species is the obvious reason for the difference in shell-opening behaviour until one considers predation on *Margaritifera*. All of the *Margaritifera* shells collected from the New Brunswick were broken in the same manner as described for thin-shelled species. One possibility is that *Margaritifera* shells are weak for their thickness, but this is not consistent with their habitat. *Margaritifera* often lives in habitats subject to significant freshets that can displace these animals and most likely are responsible for the erosion on the umbo that is typical of this species. Therefore, it is unlikely that the shell of *Margaritifera* is weak compared to other species with similarly thick shells. Since direct comparisons of relative shell strength have not been done, however, this question remains unanswered.

Studies on single-species communities have yielded important insights into the consequences of size-selective predation by muskrats on freshwater mussels (Convey et al. 1989; Hanson et al. 1989; Jokela and Mutikainen 1995; this study). In the absence of confounding species effects, there has been clear documentation of the effects of size-selective predation on topics ranging from quantitative measurement of changes in growth rate (Hanson et al. 1989; Jokela and Mutikainen 1995) to modelling of the evolution of life-history traits. The data of Hanson et al. (1989) and Jansen and Hanson (1991) were combined to examine the development of indeterminate growth (Heino and Kaitala 1996) and to test predictions of life-history models on age at first maturity (Heino and Kaitala 1997). For species that do not have a maximum size that is above the muskrat-predation threshold, phenotypic selection for slow growth (by removal of fastest-growing individuals in an age group for a cohort over many years) is unidirectional and can eliminate all of the large animals from a population over time by reducing the average growth of a cohort to zero or even negative values. This effectively truncates the size distribution (Hanson et al. 1989). In a study conducted by Jokela and Mutikainen (1995) for *Anodonta piscinalis* in Finland, the fastest-growing individuals were removed at young ages while the slowest-growing individuals were preferentially removed at old ages. In this case, the populations had individuals too large for muskrats to handle, just as found for *Margaritifera* (Sect. 10.2). Thus, the effect of muskrat predation on mussel growth can be neutral when the mussels exceed the predator's upper size limit and if the predation rate on intermediate-sized individuals is not excessive (Hanson et al. 1989). The influence of a size-selective predation by muskrats on freshwater mussels represents the same process evaluated in studies of the effect of fishing on the life-history traits of commercial and recreationally important fishes (Hanson and Chouinard 1992; Krohn and Kerr 1997; Kristiansen and Svasand 1998). One of the critical questions is whether the pattern of size-selective mortality has led to genetic selection for slow growth (Handford et al. 1977; Favro et al. 1979; Beacham 1983). Although the predation rates are high enough for this to happen (Heino and Kaitala 1997), no author has concluded that muskrat predation has resulted in genetic selection for slow growth in freshwater mussels. The lack of a genetic effect was attributed to there being a large pool of individuals of normal growth not being exploited by the muskrats (Hanson et al. 1989; Jokela and Mutikainen 1995).

10.4.3 Seasonal Pattern of Muskrat Predation

Most studies of muskrat predation on freshwater mussel report seasonal differences in predation rates but this is location-specific. The data of the German studies (Sect. 10.3.2.4) illustrate this point well. Muskrats fed upon mussels primarily during the colder months in the Fischgießen sites but mussel consumption was highest during the summer in the Rinnbach site. There was no evidence of muskrat predation on *Margaritifera* during spring and early summer in New Brunswick but it clearly occurred during mid- to late summer. A review of published studies yields a similar inconsistency. Some authors (e.g. Butler 1885; Brander 1955; Marcström 1964) report mussel consumption by muskrats only during winter and early spring and attribute it to a scarcity of suitable plant food. In contrast, Hanson et al. (1989) observed that the lowest predation rates occur during spring-early summer, the highest in autumn, with no evidence of mussel consumption during winter. Yet other authors found active shell middens during the whole year, even in the summer (Johnson 1925; Akkermann 1972; Neves and Odom 1989).

As with any predator, there is the question of identifying which factors are most important in determining when, if at all, muskrats prey on various food types, including freshwater mussels. Muskrat food selection is thought to depend (simultaneously) on: diversity, relative abundance and seasonal value of the various food types; movements of the animals (including distance of the food from the feeding platform); necessary energetic intake (especially high during periods of reproductive activity or low environmental temperature); individual preferences and experience; and predator avoidance (Butler 1940; Takos 1947; Proulx and Gilbert 1983; Clough 1987; Lacki et al. 1990; Hjälten 1991; Campbell and MacArthur 1994, 1996; Clark 1994). In our German studies, we observed that the number of mussels killed by muskrat were not only different between two brooks but also in several sections within one brook (see Sect. 10.3.2.1): the predation rates were highest in the section with the greatest mussel density. In our small, 1- to 2-m-wide brooks, transport and searching costs were very low and reduced further by the simple substrate. Mussels living on sandy sediment most likely are easier to find and dig out than those living in gravel and cobble habitats. There is an interrelation between the availability and relative nutrient or energy quality of all available food resources; the election of the main resource is a result of the weighing up between costs and profits, which may vary between individuals (Butler 1940; Marcström 1964; Ching and Chih-Tang 1965; Convey et al. 1989; this study). In New Brunswick, the predation on *Margaritifera* was most likely the result of seasonal dispersal of the juvenile muskrats. There did not appear to be sufficient emergent or submergent vegetation in the river to support muskrats year-round and yet, in July, some animals clearly had moved into this area and consumed large numbers of mussels. It is highly unlikely that muskrats could overwinter in this area due to a lack of other food types but suitable habitat occurs several kilometres upstream. Dispersal migrations of several kilometres are not unusual for muskrats (Becker 1973; Van Troostwijk 1976; Marinelli and Messier 1993).

10.4.4 Consequences of the Muskrat Invasion of Europe

The opinions about the importance of the European invasion of muskrats for wetland habitats and naiad populations are very different. The impact of muskrats on aquatic systems of central and southern Europe appears to be strong and negative while their impact on habitats in northern Europe and North America is largely neutral to positive.

One reason muskrats are not considered a serious threat to freshwater habitats in North America and northern Europe is that they are exploited for their fur. Furthermore, populations generally do not get out of control because there are numerous natural predators including (but not limited to): mink (*Mustela vison*), bob cat (*Felis rufus*), red fox (*Vulpes vulpes*), coyote (*Canis latrans*), large owls, eagles, large hawks, and American alligator (*Alligator mississippiensis*) (Johnson 1925; Valentine et al. 1972; Todd et al. 1982; Jones and Thereberge 1983; Cash et al. 1985; Litvaitis et al. 1986; Arnold and Fritzell 1987; Clough 1987; Proulx et al. 1987; Lacki et al. 1990; Ewins and Andress 1995; Soper and Payne 1997; J.M. Hanson, pers. observ.).

In southern and central Europe, muskrat densities are high and they continue to spread in distribution. There are few predators on muskrats in this region, they are not trapped for their fur, and they were introduced into an empty niche (Ulbrich 1930; Hoffman 1952; Van Troostwijk 1976). Consequently, muskrats have become serious pests by damaging structures used in water supplies, fisheries, agriculture and the canal traffic network. Moreover, the muskrats disturb the ecology of wetlands and produce drastic changes in vegetation (Lynch et al. 1947; Ching and Chih-Tang 1965; Clark 1994). Finally, it is in southern and central Europe that freshwater mussel species are the most threatened with extirpation or extinction. Hence, the overall impact of the muskrat introduction on wetlands in central Europe is strongly negative, and managers must try to control muskrat abundance to protect valuable aquatic resources and infrastructure (Hoffmann 1952; Marcström 1964; Ching and Chih-Tang 1965; Van Troostwijk 1976). This is especially important for endangered freshwater mussels because, after humans, the muskrat is the greatest threat to endangered freshwater mussel populations of Europe (Brander 1955). So the muskrat introduction to Europe is a classic example of what can happen when animals are transferred to regions where they are not indigenous.

10.5 Summary

The effects of muskrat predation on naiads in two different habitats in New Brunswick, Canada, and in two brooks in southwestern Germany were examined and compared. We investigated the muskrat's predation rates, the species and size selection by muskrat, the morphology and energetic value of the mussels, and the seasonal pattern of muskrat predation. In our Canadian study sites we evaluated these effects on two different freshwater mussel species (*Margaritifera margaritifera*, *Pyganodon cataracta*), in our German study sites on three species (*Anodonta anatina*, *Unio crassus* and *Unio pictorum*).

The impact of muskrat consumption of a few hundred *Pyganodon cataracta* in New Brunswick varied between the near elimination of the entire population in a small pond and an amount representing an insignificant proportion of a population consisting of millions of animals in an impoundment. In the German collection sites muskrats killed 10 to 60% of the *Unio pictorum* individuals.

In our Canadian study species selection by muskrat was not a factor. Muskrat predation was here clearly size-selective but the pattern differed substantially between *M. maragaritifera* and *P. cataracta*: *Margaritifera* >119 to 115 mm long appeared to be immune to muskrat predation and almost all individuals <75 mm were eliminated, whereas the muskrat selectively removed the largest animals (>90 mm) of *Pyganodon*. Muskrats in the German study showed clear species-specific preferences at some times but not at others. In all cases where there was species selection, the species eaten in higher proportions was *U. crassus*. The data in one brook showed no size-selective predation by muskrat, whereas in the other brook the muskrats consumed proportionately more of the larger individuals.

The seasonal pattern of muskrat predation in our German brooks appeared to differ between collection sites: in one brook muskrats mostly consumed mussels during the colder months, in contrast, mussel consumption in the second brook seemed to be limited mainly to summer months.

The difference in the pattern of size-selective predation by muskrat on the two mussel species in Canada was consistent with the difference in morphology between the species: for all sizes, *Pyganodon* yielded more meat yet had a lower transport cost (total weight) than *Margaritifera*. These values would represent a large difference between species in the value (meat weight) and handling cost (total weight) of mussels of the same size.

The impact of muskrats on aquatic systems of central and southern Europe appears to be strong and negative while their impact on habitats in northern Europe and North America is largely neutral to positive. Muskrat populations are expanding rapidly in Europe and they represent an additional stress on already highly stressed and endangered mussel populations. In contrast, the muskrat is only a threat to endangered North American species when areas of high muskrat activities coincide with last remnant populations of these species.

References

Akkermann R (1972) Süßwassermuscheln als tierische Zukost des Bisam, *Ondatra zibethicus*. Bonn Zool Beitr 1(23):61–65

Akkermann R (1975) Untersuchungen zur Ökologie und Populationsdynamik des Bisams (*Ondatra zibethicus* L.) an einem nordwestdeutschen Verlandungssee – III. Verhalten und Populationsdynamik. Z Angew Zool 62:281–338

Altaba CR (1990) The last known population of the freshwater mussel *Margaritifera auricularia* (Bivalvia, Unionida): a conservation priority. Biol Conserv 52:271–286

Amyot JP, Downing JD (1991) Endobenthic and epibenthic distribution of the unionid mollusc *Elliptio complanata*. J N Am Benthol Soc 10:280–285

Apgar AC (1887) The muskrat and the *Unio*. J Trenton Nat Hist Soc 1:58–59

Arnold TW, Fritzell EK (1987) Food habits of prairie mink during the waterfowl breeding season. Can J Zool 65:2322–2324

Badino G, Celebrano G, Nagel KO (1991) *Unio elongatulus* and *Unio pictorum* (Bivalvia: Unionidae): molecular genetics and relationship of Italian and central European populations. Boll Mus Reg Sci Nat Torino 9(2):261–274

Bauer G (1988) Threats to the freshwater pearl mussel *Margaritifera margaritifera* L. in central Europe. Biol Conserv 45:239–253

Bauer G (1991) Plasticity in life history traits of the freshwater pearl mussel, consequences for the danger of extinction and for conservation measures. In: Seitz A, Loeschke V (eds) Species conservation: a population biological approach. Birkhäuser, Basel, pp 103–120
Beacham TD (1983) Variability in median size and age at sexual maturity of Atlantic cod, *Gadus morhua*, on the Scotian shelf in the northwest Atlantic Ocean. Fish B 81:303–321
Becker K (1973) Populationsstudien an Bisamratten (*Ondatra zibethicus* L.) – IV. Fortpflanzungsbiologie der Weibchen. Z Angew Zool 60:343–363
Bogan AE (1993) Freshwater bivalve extinctions (Mollusca: Unionidae): a search for causes. Am Zool 33:599–609
Brander T (1955) Über die Bisamratte, *Ondatra z. zibethica* (L.), als Vernichter von Najaden. Arch Hydrobiol 50:92–103
Bruenderman SA, Neves RJ (1993) Life history of the endangered fine-rayed pigtoe *Fusconaia cuneolus* (Bivalvia: Unionidae) in the Clinch River, Virginia. Am Malacol B 10:83–91
Buddensiek V (1995) The culture of juvenile freshwater pearl mussels *Margaritifera margaritifera* L. in cages: a contribution to conservation programmes and the knowledge of habitat requirements. Biol Conserv 74:33–40
Bursik A (1914) Ist die Bisamratte schädlich? Österr Fischerei Z 17:234
Butler AW (1885) Observations on the muskrat. Am Nat 8:1044–1055
Butler L (1940) A quantitative study of muskrat food. Can Field Nat 54:37–40
Campbell KL, MacArthur RA (1994) Digestibility and assimilation of natural forages by muskrat. J Wildl Manage 54:633–641
Campbell KL, MacArthur RA (1996) Seasonal changes in gut mass, forage digestibility, and nutrient selection of wild muskrat (*Ondatra zibethicus*). Physiol Zool 69(5):1215–1231
Cash KJ, Austin-Smith PJ, Banks D, Harris D, Smith PC (1985) Food remains from bald eagle nests sites on Cape Breton Island, Nova Scotia. J Wildl Manage 49:223–225
Ching C, Chih-Tang Y (1965) Food and food basis of the muskrat, *Ondatra zibethicus* Linnaeus. Acta Zool Sin 17(1):352–363
Clark WR (1994) Habitat selection by muskrats in experimental marshes undergoing succession. Can J Zool 72:675–680
Clough GC (1987) Ecology of island muskrats, *Ondatra zibethicus*, adapted to upland habitat. Can Field Nat 101:63–69
Convey LE, Hanson JM, Mackay WC (1989) Size-selective predation on unionid clams by muskrats. J Wildl Manage 53:654–657
Danell K (1978) Population dynamics of the muskrat in a shallow Swedish lake. J Anim Ecol 47:697–709
Danell K (1996) Introductions of aquatic rodents: lessons of the muskrat *Ondatra zibethicus* invasion. Wildl Biol 2:213–220
Enders RK (1932) Food of the muskrat in summer. Ohio J Sci 32:21–30
Erickson HR (1963) Reproduction, growth, and movements of muskrats inhabiting small water areas in New York State. N Y Fish Game J 10:90–117
Ewins PJ, Andress RA (1995) The diet of bald eagles, *Haliaeetus leucocephalus*, wintering in the lower Great Lakes basin. Can Field Nat 109:418–425
Favro LD, Kuo PK, McDonald JF (1979) Population-genetic study of the effects of selective fishing on the growth rate of trout. J Fish Res Board Can 36:552–561
Frank F, Härle A (1967) Derzeitiger Stand und voraussichtliche Entwicklung des Bisambefalls (*Ondatra zibethicus*) in der Bundesrepublik Deutschland (nach den Meldungen der Pflanzenschutzämter der Bundesländer). Nachricht Dtsch Pflanzenschutz (Braunschweig) 19(8):123–125
Fuller SLH (1974) Clams and mussels (Mollusca: Bivalvia). In: Hart CW Jr, Fuller SLH (eds) Pollution ecology of freshwater invertebrates. Academic Press, New York, pp 216–273
Graf DL (1997) Sympatric speciation of freshwater mussels (Bivalvia: Unionoidea): a model. Am Malacol B 14:35–40
Haag WR, Warren ML Jr (1998) Role of ecological factors and reproductive strategies in structuring freshwater mussel communities. Can J Fish 55:297–306

Handford P, Bell G, Reimchen T (1977) A gillnet fishery considered as an experiment in artificial selection. J Fish Res Board Can 34:954–961

Hanson JM, Chouinard GA (1992) Evidence that size-selective mortality affects growth of Atlantic cod (*Gadus morhua* L.) in the southern Gulf of St Lawrence. J Fish Biol 41:31–41

Hanson JM, Mackay WC, Prepas EE (1988) Population size, growth, and production of a unionid clam, *Anodonta grandis simpsoniana*, in a small, deep Boreal Forest lake in central Alberta. Can J Zool 66:247–253

Hanson JM, Mackay WC, Prepas EE (1989) Effect of size-selective predation by muskrats (*Ondatra zebithicus*) on a population of unionid clams (*Anodonta grandis simpsoniana*). J Anim Ecol 58:15–28

Headlee TJ (1906) Ecological notes on the mussels of Winona, Pike, and Center Lakes of Koscuisko County, Indiana. Biol B 11:305–318

Heino M, Kaitala V (1996) Optimal resource allocation between growth and reproduction in clams: why does indeterminate growth exist? Funct Ecol 10:245–251

Heino M, Kaitala V (1997) Should ecological factors affect the evolution of age at maturity in freshwater clams? Evol Ecol 11:67–81

Hjälten J (1991) Muskrat (*Ondatra zibethicus*) territoriality and the impact of territorial choice on reproduction and predation risk. Ann Zool Fenn 28:15–21

Hochwald S (1990) Bestandsgefährdung seltener Muschelarten durch den Bisam (*Ondatra zibethica*). Schriftenr Bayer Landesamtes Umweltschutz 97:113–114

Hoffmann M (1952) Die Bisamratte. Die Neue Brehm-Bücherei. Akademische Verlagsgesellschaft Geest & Portig, Leipzig

Hogarth MA, Rice DL, Lee DM (1995) Discovery of the federally endangered freshwater mussel, *Epioblasma obliquata obliquata* (Rafinseque, 1820) (Unionidae), in Ohio. Ohio J Sci 95: 298–299

Hove MC, Neves RJ (1994) Life history of the endangered James spinymussel *Pleurobema collina* (Conrad, 1837) (Mollusca: Unionidae). Am Malacol B 11:29–40

Jansen WA, Hanson JM (1991) Estimates of the number of glochidia produced by clams (*Anodonta grandis simpsoniana* Lea), attaching to yellow perch (*Perca flavescens*), and surviving to various ages in Narrow Lake, Alberta (Canada). Can J Zool 69:973–977

Johnson Ch E (1925) The muskrat in New York: its natural history and economics. Syracuse Univ Bull 24(25), Roosevelt Wild Life Bull 3(2):205–320

Jokela J, Mutikainen P (1995) Effect of size-dependent muskrat (*Ondatra zibethica*) predation on the spatial distribution of a freshwater clam, *Anodonta piscinalis* Nilsson (Unionidae, Bivalvia). Can J Zool 73:1085–1094

Jones DM, Thereberge JB (1983) Variation in red fox, *Vulpes vulpes*, summer diet in northwest British Columbia and southwest Yukon. Can Field Nat 97:311–314

Kristiansen TS, Svasand T (1998) Effect of size-selective mortality on growth of coastal cod illustrated by tagging data and individual-based growth and mortality model. J Fish Biol 53:688–705

Krohn MM, Kerr SR (1997) Declining weight-at-age in northern cod and the potential importance of the early years and size-selective fishing mortality. NAFO Sci Counc Stud 29:43–50

Lacki MJ, Peneston WT, Adams KB, Vogt FD, Hooper JC (1990) Summer foraging patterns and diet selection of muskrats inhabiting a wet fen. Can J Zool 68:1163–1167

Libois RM (1988) Densité et biomasse des Unionidae (Mollusca) de l'étang de Virelles (Hainaut, Belgique). Ann Soc R Belg 118:61–68

Litvaitis JA, Sherburne JA, Bissonette JA (1986) Bobcat habitat use and home range size in relation to prey density. J Wildl Manage 50:110–117

Lynch JJ, O'Neil T, Lay DW (1947) Management significance of damage by guse and muskrat. J Wildl Manage 11(1):50–76

Marcström V (1964) The muskrat *Ondatra zibethicus* L. in northern Sweden. Viltrevy 2(6):329–407

Marinelli L, Messier F (1993) Space use and the social system of muskrats. Can J Zool 71:869–875

McMahon RF (1991) Mollusca: Bivalvia. In: Thorpe JH, Covich AP (eds) Ecology and classification of North American freshwater invertebrates. Academic Press, New York, pp 315–399

Morkens EA, Costello MJ (1994) Imminent extinction of the Nore freshwater pearl mussel *Margaritifera durrovensis* Phillips: a species unique to Ireland. Aquat Conserv: Mar Freshwater Ecosyst 4:363–365

Neves RJ, Odom MC (1989) Muskrat predation on endangered freshwater mussels in Virginia. J Wildl Manage 54:934–941

Obermeyer BK, Edds DR, Prophet CW, Miller EJ (1997) Freshwater mussels (Bivalvia: Unionidae) in the Verdigris, Neosho, and Spring River basins of Kansas and Missouri, with emphasis on species of concern. Am Malacol B 14:41–55

Prell H (1924) Die Bisamratte als wirtschaftlicher Feind der Perlfischerei. Naturwissenschaften 12:733–736

Proulx G, Gilbert FF (1983) The ecology of the muskrat, *Ondatra zibethicus*, at Luther Marsh, Ontario. Can Field Nat 97:377–390

Proulx G, McDonnell JA, Gilbert FF (1987) The effects of water level fluctuations on muskrat, *Ondatra zibethicus*, predation by mink, *Mustela vison*. Can Field Nat 101:89–92

Richardson TD, Yokely P Jr (1996) A note on sampling technique and evidence of recruitment in freshwater mussels (Unionidae). Arch Hydrobiol 137:135–140

Soper LR, Payne NF (1997) Relationship of introduced mink, an island race of muskrat, and marginal habitat. Ann Zool Fenn 34:251–258

Takos MJ (1947) A semi-quantitative study of muskrat food habits. J Wildl Manage 11:331–339

Todd CS, Young LS, Owen RB, Gramlich FJ (1982) Food habits of the bald eagle in Maine. J Wildl Manage 46:636–645

Triplet P (1983) Régime carné du rat musqué *Ondatra zibethica* dans le Marquenterre (Somme). Mammalia 47:129–130

Turgeon DD, Bogan AE, Coan EV, Emerson WK, Lyons WG, Mikkelsen PM, Quinn JF, Roper CFE, Rosenberg G, Roth B, Sweeney MJ, Scheltema AH, Thompson F, Vecchione M, Williams JD (1998) Common and scientific names of aquatic invertebrates from the United States and Canada: molluscs, 2nd edn. American Fisheries Society Spec Publ, Bethesda, Maryland

Tyrrell M, Hornbach DJ (1998) Selective predation by muskrats on freshwater mussels in two Minnesota rivers. J N Am Benthol Soc 17(3):301–310

Ulbrich J (1930) Die Bisamratte, Lebensweise, wirtschaftliche Bedeutung, Gang ihrer Ausbreitung in Europa und Bekämpfung. Dissertation der Sächsischen Technischen Hochschule, Dresden

Valentine JM, Walthers JR, McCartney KM, Ivy LM (1972) Alligator diets on the Sabine National Wildlife Refuge, Louisiana. J Wildl Manage 36:809–815

Van Cleave HJ (1940) Ten years of observation on a fresh-water mussel population. Ecology 21:363–370

Van Troostwijk WJD (1976) The musk-rat (*Ondatra zibethicus*) in the Netherlands, its ecological aspects and their consequences for man. Rijksinstitut voor Naturbeheer Verhandling, Arnheim, Leersum

Watters TG (1994) Sampling freshwater mussels populations: the bias of muskrat middens. Walkerana 7:63–69

Watters TG (1995) Small dams as barriers to freshwater mussels (Bivalvia: Unionidae) and their hosts. Biol Conserv 75:79–85

Williams JD, Warren ML Jr, Cummings KS, Harris JL, Neves RJ (1993) Conservation status of freshwater mussels of the United States and Canada. Fisheries 18(9):6–22

Willner GR, Feldhamer GA, Zucker EE, Chapman JA (1980) *Ondatra zibethicus*. Mamm Spec 141:1–8

Young MR (1991) Conserving the freshwater pearl mussel (*Margaritifera margaritifera* L.) in the British Isles and Continental Europe. Aquat Conserv Mar Freshwater Ecosyst 1:73–77

11 Glochidial Mortality in Freshwater Mussels

WOLFGANG JANSEN, GERHARD BAUER, and ELKE ZAHNER-MEIKE

11.1 Introduction

A crucial factor in the reproduction of parasites is the rate of successful host infection. In most parasitic species subtle mechanisms for the effective transmission of infectious stages have evolved. These include vector transmission, highly mobile propagules that are able to locate and infect new hosts actively, attraction of hosts, extremely resistant propagules that may spend an extended dormant period "waiting" for a suitable host, or parasite transmission by direct contact between host individuals (Anderson 1981; Begon et al. 1988). The precise mode of host infection must be considered an important evolutionary constraint (Calow 1983). Except for the African and South American families Mutelidae and Mycepodidae, which produce parasitic larvae with a specific morphology (Wächtler et al., Chap. 6, this Vol.; Hoeh et al., Chap. 14, this Vol.), in the Unionoidea, infection is achieved via a unique larval stage, the glochidium. Glochidial anatomy is well adapted for efficient attachment to epithelia of aquatic vertebrates (Lefevre and Curtis 1910; Arey 1924a). Three morphological types of glochidia can be distinguished based on the number and shape of hooks on their ventral apex. The most common (e.g. genera *Margaritifera, Quadrula, Elliptio*) form of glochidia is hookless, referring to the absence of a spined hook projecting inward from the ventral edge of each valve present in hooked glochidia (e.g. genera *Anodonta, Unio, Lasmigona*). A third form of glochidia (axe-head) with distinctly flared valve margins carrying double hooks is known from the North American genus *Proptera* (Lefevre and Curtis 1910). Intriguing mechanisms of host attraction have been reported from some freshwater mussels (also referred to as clams or naiades) to increase the chances of glochidial infection (Kat 1984; Haag et al. 1995). However, little is known about the fate of the sometimes millions of glochidia produced by a single parent mussel, of which very few will develop into free-living clams. Mainly because of our limited knowledge of the host species of most naiades, a lack of studies that simultaneously and quantitatively sample host fish and gravid mussels, difficulties in the species identification of glochidia or juvenile clams and methodological problems of quantitatively collecting mussels soon after metamorphosis (Hanson et al. 1988a; Nalepa and Gauvin 1988), mortality estimates during these critical life history stages are largely lacking.

In this chapter we summarize data on larval and early juvenile mortality of glochidia-producing Unionoidea. Our approach is chronological, in that we address the life history and associated mortality factors for consecutive developmental stages including their transitional time periods. The first stage in Unionoidea life history we considered is the appearance of mature glochidia. Embryos and early larvae, numbers

Ecological Studies, Vol. 145
G. Bauer and K. Wächtler (eds.) Ecology and Evolution of the Freshwater Mussels Unionoida

of which may be affected by the nutritional status or parasite load of the parent mussel (Taskinen and Valtonen 1995), were not addressed.

11.2 Factors Contributing to the Mortality of Unionoidea Early Life Stages

11.2.1 Larval Development and Release of Glochidia

Many species among the Unionoidea, particularly within the large subfamily Anodontinae, are long-term or bradytictic breeders (Heard 1975) that retain their glochidia for many months and often throughout the winter until spring. Thus, mussels from northern latitudes may carry mature glochidia for periods of 6–10 months (Arey 1921; Heard 1975; Paterson and Cameron 1985; Jansen and Hanson 1991; Hruška 1992; Niemeyer 1992). Over such a long period, adult naiades will experience a considerable proportion of their annual mortality, in particular, because predation risk may be increased for gravid clams. At least in some naiad species, most individuals remain completely buried in the substrate except during periods of glochidia release (Okland and Andersen 1985; Amyot and Downing 1991; Downing et al. 1992) when they expose shell and mantle well above the sediment (Zale and Neves 1982a; Haag et al. 1999) and perform conspicuous movements (Kraemer 1970). These behavioral adaptations, designed to promote larval survival (see Sect. 2.3, this Vol.), most likely render naiades more vulnerable to visually orientated predators during this time of the reproductive cycle. Muskrats, for example, are a major predator on freshwater mussels (Zahner-Meike and Hanson, this Vol.), and may consume 7.5% of the annual glochidia production (Hanson et al. 1989). Tachytictic, or short-term breeders (Heard 1975) carry glochidia over only a relatively short period during spring and summer, and may have multiple gravid periods (Harms 1909; Lefevre and Curtis 1910; Allen 1924; Neves and Widlak 1988; Engel 1990; Hochwald 1988, 1997; Fleischauer-Rössing 1990) that in *Unio crassus* can last as little as 19 days (Lang 1998). In these cases, glochidial mortality as a result of adult predation is probably less severe than in bradytictic species.

Once the parent mussel releases the glochidia, the availability of a suitable host becomes critical. A number of fascinating adaptations have been evolved among some Unionoidea to increase the likelihood of host detection and glochidial attachment and, thus, to reduce larval mortality. Tens of thousands to many million glochidia are released into the water by a parent mussel (McMahon 1991; Bauer 1987a). In some species glochidial release occurs over a short time period and is well synchronized among members of a population at a time when host abundance and/or behavior increases the probability of successful infection (Bauer 1979; Trdan and Hoeh 1982; Zale and Neves 1982a; also see Sect. 2.2, this Vol.). In a less predictable habitat where host detection becomes more critical, naiades may use a sit-and-wait tactic in host finding, and among the members of a population glochidial release may extend over several month (Jansen and Hanson 1991; Jokela et al. 1991).

Except for species that produce conglutinates (see below), mature glochidia are expelled individually (Matteson 1948; Bauer 1979; Neves and Widlak 1988; Engel 1990; Jokela and Palokangas 1993) and/or in loosely bound mucous masses that dissolve more or less rapidly to expose the larvae (Lefevre and Curtis 1910; Arey 1921; Matteson 1948; Bauer 1979; Jokela and Palokangas 1993; Haag and Warren 1997). This also applies to glochidia that possess a so-called attachment thread, i.e. genera *Anodonta* and *Unio* (Forel 1866; Schierholz 1888; Lefevre and Curtis 1910; Arey 1921; Wächtler et al., Chap. 6, this Vol.). A number of mechanisms are apparently involved in this function (Wood 1974a,b): for individual glochidia suspended in the water column, the thread reduces sinking time by more than 30%. Furthermore, and similar to some threadless glochidia (Yokely 1972; Mattesson 1948), glochidia with a thread often form entangled masses that are transitionally suspended in a mucilaginous net some distance off the bottom substrate (e.g. Watters and O'Dee 1997a). Floating glochidia have a better chance of host attachment (Wood 1974b; also see Sect. 2.2, this Vol.), particularly if their fish hosts are attracted to suspended particles, as has been suggested by Yokely (1972) and Jokela and Palokangas (1993). Also, if one larva of a glochidial mass makes host contact, other larvae will also be pulled towards the fish, thus facilitating their attachment (Forel 1866; Herbers 1913; Wood 1974b). Despite these obvious benefits, most mussel species do not possess an attachment thread (Lefevre and Curtis 1910; Wood 1974a), and no clear pattern exists relating its presence or absence to glochidia size or the presence/absence of hooks. Thus, the question remains, what selective forces make the thread an effective attachment aid in some species and not in others?

In many species within the unionid subfamilies Lampsilinae and Ambleminae, glochidia are bound together in discrete packages by a mucoid or gelatinous matrix, primarily consisting of infertile eggs (Barnhart 1997). Upon expulsion, these so-called conglutinates (if they contain the glochidia from a single marsupial demibranch or a part thereof) or superconglutinates (if they contain all glochidia from the entire gill; Haag et al. 1995; Hartfield and Hartfield 1996) may dissolve rapidly to release the glochidia (Yeager and Neves 1986) or, more commonly, remain intact for several days. Uniquely for these taxa, shape and coloration of both types of conglutinates often mimic fish food items such as leeches, flatworms, insect larvae (Chamberlain 1934; Neves and Widlak 1988; Hartfield and Hartfield 1996; Haag and Warren 1997), or even fish eggs (Haag and Warren 1997) or forage fish (Haag et al. 1995; O'Brian and Brim Box 1999). The hookless glochidia of mussels that release conglutinates are typically gill parasites. Upon consumption by a fish, glochidia are released from the conglutinate and come into direct contact with suitable attachment sites on the gills or within the buccal cavity, while some larvae may also be swallowed and killed (Chamberlain 1934). Field observations by Neves and Widlack (1988) that report greater prevalence and intensity of glochidial infestations of species that produce conglutinates compared to those that do not, support the idea that this type of mimicry is a successful strategy to increase glochidial survival during the initial parasitic phase. Clams that produce (super)conglutinates or invest in other adaptations that promote the chances of glochidial infection (see below) generally have a relatively low fecundity (Kat 1984). Thus, this strategy also seems to reduce the energy costs of reproduction.

In addition to the production of conglutinates, several species of Lampsilinae have evolved a further mechanism of host attraction that exploits predator-prey interactions. In these mussels, the pigmented muscular mantle margins of females resemble small fish or invertebrates and periodically undulate, flap or pulsate to lure potential fish hosts to the gravid clam (Kraemer 1970; Kraemer and Swanson 1985; Williams et al. 1993). Fish attacking these lures come into direct contact with a dense stream of glochidia expulsed by the parent mussel (Kat 1984). An additional and supplementary interpretation of the bellows-like mantle flap movements is that they keep glochidia suspended in the water after their release (Kraemer 1970). Although light intensity can stimulate mantle flapping (Kraemer 1970), and it has been suggested that the distinct mantle eyespots of some species possess a photosensory function that aids in the detection of potential hosts (Clarke 1981), there is no histological evidence for such a capacity (Kraemer and Swanson 1985). The primary mode of host detection by adult mussels probably is via chemical cues (e.g. Jokela and Palokangas 1993), which, at least in some species, can be enhanced by other stimuli. In laboratory experiments, for example, Teutsch (1997) and Schneidt (1998) exposed gravid *Anodonta cygnea* and *Anodonta anatina* either to known host fish (*Gasterosteus aculeatus* and *Leuciscus cephalus*, respectively), to water from an aquarium containing fishes or to tap water. In both species, the continued exposure to live fish resulted in the highest and most sustained glochidia release, whereas mussels in the tap-water treatment showed no response (the small numbers of glochidia found were believed due to a stress-related release within the first 36 h after transfer), and the results from the fish-water treatment differed between species and experiments. These results suggest that the release of glochidia functionally responded to the concentration of chemical substances from the fish or from its metabolic products.

Whereas the production of conglutinates or prey mimicry is primarily found in mussels from lotic habitats, host contact in clams from lentic environments is often facilitated by the synchronization of their reproductive cycle with yearly cyclic events in the biology of their hosts. These mussels rely on behavioral interactions related to fish migration, reproduction, territoriality or foraging in order to promote glochidial attachment and, thus, to decrease larval mortality (Kat 1984; Jansen 1991; McMahon 1991; Jokela et al. 1991). Irrespective of how first contact is facilitated, the initial attachment to an host is simply mechanical, with the glochidium closing its valves on any available surface (Lefevre and Curtis 1910).

11.2.2 Free-Living Glochidia

Except for glochidia conglutinates that regularly remain intact and attach to instream structures like macrophytes to continue their mimicry for some time after they have broken off from the parent mussel (Haag et al. 1995), larvae unable to fasten to a host upon the initial release will eventually sink onto the bottom substrate. Since valve snapping does not provide sufficient lift, glochidia cannot actively swim (Paling 1968) and remain immobile on the sediment (Lefevre and Curtis 1910) with limited chances of host attachment. The probability of resuspension from the sediment is very low, at least for the very small glochidia of *M. margaritifera* (Wenz 1990). In their stream environment, mussels of the genus *Margaritifera* often coexist with only one or a few potential salmonid hosts (Awakura 1968; Bauer 1979, 1987a; Young and Williams 1984a; Cunjak and McGladdery 1991; Ziuganov et al. 1990). Many salmonids consume mostly invertebrate drift and may select a feeding station that provides maximum drift (Chapman 1966). For these reasons, glochidia that stay adrift in the water column for some distance after their release will increase their chances of host attachment.

That glochidia of *M. margaritifera* actually remain in the drift for substantial distances was recently demonstrated from our own experiments. In two separate trials, seven 12-cm-long plastic tubes (7-cm diameter, closed at one end with 0.03-mm-mesh gauze) and 33 caged brown trout (*Salmo trutta*, 17–20 cm total length, 4–5 fish/cage, seven cages) were placed onto the sediment of two small streams at distances between 2 and 272 m downstream of the point of glochidia release. Close to 3.23 and 2.24 million glochidia, in the tube and the fish experiment, respectively, suspended in 10 l of water were introduced into the stream. Approximately 45 min later all tubes and fish were removed from the creek to be inspected for glochidia in the laboratory. A total of 354 and 130 glochidia were recovered from the tubes and the trout, respectively. In both experiments, the number of glochidia decreased at a similar rate with increasing distance downstream and a few larvae were found inside the tubes or attached to the fish 200 m and more from the point of release (Fig. 11.1). Maximum drift distances of glochidia in a typical pearl mussel stream with an average current speed of $0.4\,m{\cdot}s^{-1}$ are most likely limited to a few hundred metres. This is suggested by the data of Wenz (1990), who found that juvenile trout caught within a mussel bed were all infested with glochidia, whereas fish captured 500 m below the bed were free of glochidia.

Overall, glochidial drift apparently increases the chances of encountering the ventilating current of pelagic fish hosts, and thus is adaptive in reducing mortality in the preparasitic stage. Substantial numbers of glochidia may occur in the drift (Clark and Stein 1921; Zale and Neves 1982a,b; Young and Williams (1984a), and maximum drift densities of 8 and 117 individuals$\cdot m^{-3}$ of water have been reported for North American lampsiline and amblemine species (Neves and Widlak 1988). For individual species, densities may range between 0.01 and 1.3 (*Fusconaia cuneolus*; Bruenderman and Neves 1993), 1.5–15.9 (*Pleurobema collina*; Hove and Neves 1994), 1–56 (*Elliptio dilatata*; Jirka and Neves 1992) and 3-64 glochidia$\cdot m^{-3}$ (*Actinonaias ligamentina*; Jirka and Neves 1992). Importantly, Neves and Widlak (1988) found that the number of drifting glochidia was positively correlated with water velocities, thus increasing the chances of interaction with their drift-feeding host fishes. In contrast to their lotic relatives, entering the drift may be disadvantageous for mussels living in lakes. Lewis (1985), for example, speculated that reproductive plasticity in *Pyganodon grandis* from a lake experiencing spring flushing was related to the avoidance of extensive glochidial drift.

Glochidia are incapable of active locomotion (Lefevre and Curtis 1910), and once in the sediment these larvae depend on a potential host to make contact or to be resuspended into the water column. Since glochidia lack storage tissues, they will soon die if unable to attach to an host. Glochidia of *Margaritifera falcata*, for example, survived for 11 days in containers submersed in their natural habitat at a temperature of 11 °C (Murphy 1942). In laboratory tests on five European mussel species (*M. margaritifera*, *A. anatina*, *A. cygnea*, *U. crassus*, *U. pictorum*) we found that individual glochidia may survive (indicated by valve closure or contraction of the adductor muscle after addition of NaCl-solution) for 10 (*M. margaritifera*) to 17 (*U. pictorum*) days (Fig. 11.2). These maximum values were recorded in water of 5 °C, and more typical ambient conditions of 10 and 16 °C resulted in considerably earlier mortality, particularly in *A. cygnea* (Fig. 11.2). Furthermore, at the latter temperatures, median survival time of glochidia ranged between 2.5 days for *A. cygnea* and 5 days for *U. pictorum* (Fig. 11.2). At least for glochidia of *U. pictorum* at concentrations between

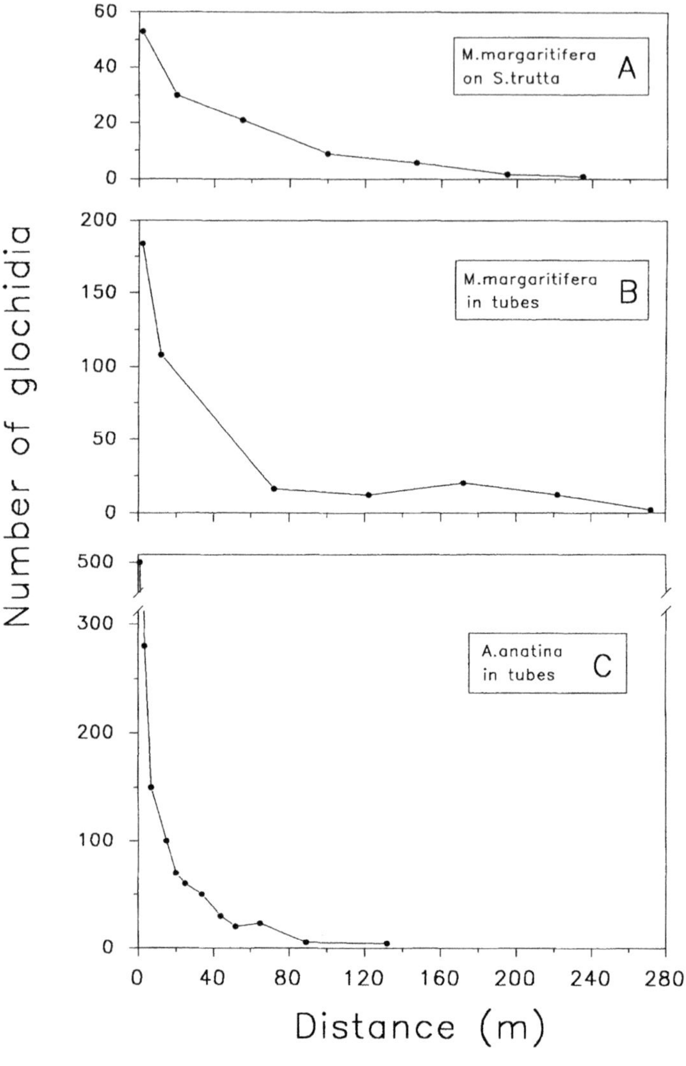

Fig. 11.1A–C. Drift of glochidia from *Margaritifera margaritifera* and *Anodonta anatina*. Glochidia released in numbers of 2240000 (**A**), 3230000 (**B**), and 60000 (**C**) were recovered at each downstream distance either from four caged brown trout (*Salmo trutta*, total length 17–20 cm; **A**) or from individual plastic tubes (7 cm diameter, 0.03 cm screen at downstream end; **B** and **C**). (Data for *A. anatina* from Trübsbach 1998)

30000 and 180000 larvae·l^{-1}, these survival rates and the time course of mortality were independent of glochidial densities. The above results are similar to those obtained in comparable experiments on *U. crassus* (Hochwald 1988; Engel 1990) and *A. cygnea* (Teutsch 1997), two species that also exhibited a strong temperature effect on survival times (Table 11.1). Metabolic rate of aquatic invertebrates is closely linked to ambient

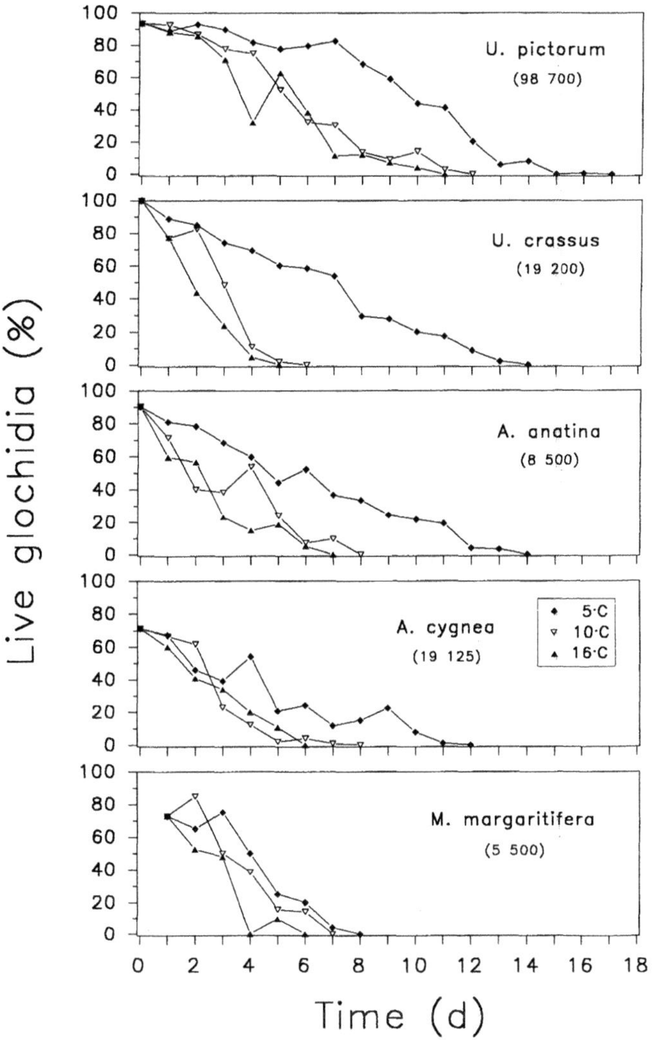

Fig. 11.2. Survival of glochidia of *Unio pictorum*, *Unio crassus*, *Anodonta anatina*, *Anodonta cygnea* and *Margaritifera margaritifera* at temperatures of 5, 10 and 16 °C in 500-ml beakers. Each *symbol* gives the percentage of live glochidia as a mean of six samples of 0.8 ml water; initial densities (number of glochidia·l^{-1}) are indicated in *brackets*

temperature (Ivleva and Popenkina 1969). As suggested from the shape of the curves in Fig. 11.2, free-living glochidia that do not feed and that have very limited energy storage show a negative relationship between survival time and temperature.

However, the reaction to a salt solution may not be an ecologically meaningful measure of glochidial viability. Using glochidia from the same parent mussel as in the survival experiment, Teutsch (1997) demonstrated that only a small percentage of

Table 11.1. Summary of three measures of vitality for glochidia of *Anodonta cygnea* at different water temperatures. Presented is the number of days to which 50% of the glochidia survived and 50% of the glochidia responded to the presence of a host fish by closing their valves (VC). For survival, the maximum number of days (Max) is also given. Attachment refers to the number of glochidia encapsulated on a fish 3 days postrelease. Initial densities are approximately 30 000 glochidia·l^{-1}. (Data from Teutsch 1997)

Temperature (°C)	Survival		Reaction to host presence	
	50% (days)	Max (days)	50% VC (days)	Attachment (%)
5	9.5	11	8.0	18
10	6.5	9	6.3	9
18	5.0	7	4.9	8

these larvae could successfully attach to a known host fish 3 days after their release (Table 11.1); this percentage decreased further with higher temperature. Similarly, Tedla and Fernando (1969) found that the percentage of viable glochidia of *Lampsilis radiata* rapidly decreased with increasing temperature, and less than 1% survived after 36h at 20 °C. In the congeneric *L. higginsi*, none of the glochidia held at 25 °C survived after 72h (Sylvester et al. 1984). In contrast to these findings, Botnariuc and Tudorancea (1967) found attachment of glochidia of *A. anatina* 5–6 days after their release. However, these authors gave no indications of the success of attachment, and old glochidia (i.e. 24-h postrelease) that initially attach to an host are known to drop off within the next day (Hochwald 1988). The frequency of valve snapping, which is characteristic for freshly released glochidia of *M. margaritifera*, starts to decline a few hours postrelease, and after 24h glochidia appear "lifeless" (Young and Williams 1984b). For glochidia of *M. falcata*, Meyers and Milleman (1977) report a normal life span under laboratory conditions of only a few hours, and Tompa (1979) found that glochidia of *Lasmigona compressa* live only 1–2 days after their release. This corresponds well with observations that in commercial mussel propagation good artificial infections are obtained within approximately 12h after removal of glochidia from the marsupium (Arey 1921). The above data suggest that survival of infectious glochidia under natural conditions is probably limited to 1 or 2 days. From an evolutionary point of view, the short glochidial life span seems to be part of a life history that produces individual offspring as energetically cheaply as possible, irrespective of the large differences in glochidial size found among different species (Bauer 1994). Since the chances of attaching to a host, and thus of reaching the parasitic stage, are highest during a short time immediately after glochidial release, there is no selective advantage in prolonging the duration of this particular life history stage.

Apart from intrinsic factors, larval mussels may not survive to become parasitic because of predation. Very little is known about potential predators of glochidia, particularly invertebrates. Perhaps feeding on glochidia is associated with a high cost/benefit ratio because of long handling times due to their large physical dimensions relative to their small weight (e.g. 0.066mg in *A. anatina*, Negus 1966) and their low energy density. Furthermore, although glochidia vary in size and shape (Lefevre and Curtis 1910; Arey 1921; Coker et al. 1921), most are very small at the time of release, measuring between 50 and 400 μm (McMahon 1991; Bauer 1994).

Combined with their immobility and mostly inconspicuous colour, individual glochidia, unlike conglutinates, are probably not actively preyed upon by relatively large, mainly visually orientated predators like fish (but see Harms 1907a). This hypothesis is supported by the almost complete absence of glochidia in freshwater fish diets despite the multitude of available studies. This includes our own diet analyses of more than 500 yellow perch caught during a time of heavy infestations from relatively large glochidia of *P. grandis* (Jansen 1991). Perhaps glochidia are distasteful to many fish species, as has been suggested by Latter (1891). To our knowledge, the only two published reference to glochidia predation by fish are by Dartnall and Walkey (1979) and Hove and Neves (1994). Whereas the latter authors report the consumption of sloughed glochidia of *Pleurobema collina* by *Catostomus commesoni* and *Hypentelium nigricans* after artificial infestations, Dartnall and Walkey (1979) found that sticklebacks (*G. aculeatus*) actively prey upon glochidia of *A. cygnea* in experimental situations and that "their stomach contents may contain appreciable numbers of glochidia under natural feeding regimes".

11.2.3 Glochidial Attachment and Encapsulation

The majority of known hosts of freshwater mussels are fish. In North America, for example, where the Unionidea reach their greatest diversity (Williams et al. 1993), only very few species are known to successfully infect amphibians (Howard 1915; Watters 1997; Barnhart et al. 1998; Waters and O'Dee 1998). This equally applies to European species (Faussek 1901) and mussels from the Southern Hemisphere (Seshaiya 1941, 1969; Walker 1981). Glochidia may also attach to invertebrate hosts like glossiphonid leeches (Seshaiya 1941) or decapod crustaceans (Panha 1990), but are unlikely to complete metamorphosis (Watters and O'Dee 1998) except where development is very rapid (Walker 1981). Overall, little research has been devoted to establish non-piscine vertebrate hosts, and almost no data are available on natural infections. The following aspects of parasite-host relationships involve exclusively fish hosts.

Although chemical cues are important for host attachment, such sensitization by mucous, blood or fish tissue is not a prerequisite for successful attachment (Wächtler et al., Chap. 6, this Vol.). Glochidia will initially fasten to any available surface, including inanimate materials (Arey 1921; Wood 1974b). For more permanent attachment, however, certain properties of the attachment surface seem to be of importance (Wood 1974b). This should, however, not be interpreted in favour of some type of host recognition. Under natural conditions glochidia frequently attach to unsuitable hosts (e.g. Awakura 1968; Meyers and Millemann 1977; Trdan and Hoeh 1982; Neves et al. 1985; Maaß 1987; also see Sect. 11.2.4).

Once the valves have closed on host tissue, they do not open again and glochidia are thus unable to switch attachment sites (Lefevre and Curtis 1910; Paling 1968). The valves are closed by a single adductor muscle which may contract for several hours (Wood 1974a), during which time glochidia must successfully encapsulate in order to start the parasitic phase (Pflugfelder 1951). Glochidia that just clasp onto the outermost cell layers may remain attached to the host for up to 14 days without encapsulating (Pflugfelder 1951), and eventually will drop off without showing any signs of further development (Arey 1932a; Pflugfelder 1951). Nezlin et al. (1994) report that between 2 and 5% of glochidia of *M. margaritifera* do not elicit a reaction in Atlantic salmon gill epithilium cells, do not encapsule, and disappear within 2 days. Complete encapsulation, in which the glochidium becomes entirely enclosed by cells of the host (for details see Wächtler et al., Chap. 6, this Vol.), is generally faster in the relatively small hookless glochidia (2–16h: Harms 1909; Lefevre and Curtis 1910; Arey 1921; Tedla and Fernando 1969; Meyers et al. 1980; Nezlin et al. 1994) than in the large hooked larvae (10–36h: Faussek 1901; Harms 1909; Lefevre and Curtis 1910; Pflugfelder 1951; Hüby 1988). Apart from glochidia type, temperature, nutritional status of a host (Pflugfelder 1951) and the location of attachment (Arey 1932a) have been known to affect the time to encapsulation.

For a given species, the initial distribution of glochidia on a host is largely determined by the probability of contact with a particular body part/area, which in the case of gills is related to areas of maximum water flow (Paling 1968), and the differential attachment success on these locations (Young and Williams 1984a). The first factor relates strongly to the morphology and the behaviour of the host, while the second factor largely depends on the interaction between glochidia type/size and fish size, and, thus, the thickness of its fins and gills. The few available data from natural infestations show that on percids (Fig. 11.3) and cyprinids glochidia are preferentially found on the pectoral and caudal fins, indicating that the active role in locomotion combined with the foraging mode of the respective species increases the fins susceptibility to glochidia attachment (Dudgeon and Morton 1984; Hüby 1988; Jansen 1991).

Generally, hooked glochidia are able to attach to a broader spectrum of hosts (Bauer 1994) and host body parts than hookless glochidia. Apart from the hooks, which are covered with spines and serve as effective attachment aids (Herbers 1913; Arey 1924a), this capacity is most likely related to the generally larger size of hooked glochidia (Bauer 1994). Although hookless glochidia have occasionally been found on fish fins and scales (Harms 1908; Matteson 1948; Meyers and Millemann 1977), they

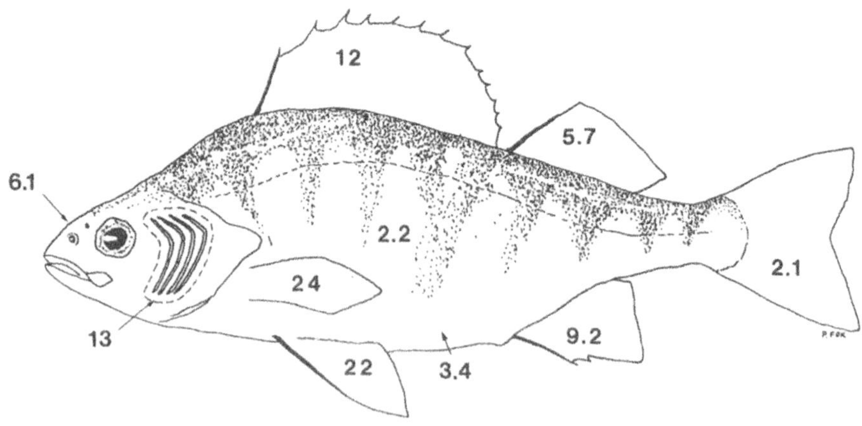

Fig. 11.3. Percentage distribution of glochidia of *Pyganodon grandis* on fins and distinct body regions of *Perca flavescens* during the time of maximum intensity of infestation. *Numbers* represent means for 22 fish infested by 1080 glochidia. (Data from Jansen 1991)

may not survive on these anatomical regions (Matteson 1948). Hookless glochidia, whose shell shape and adductor muscle placement provide a large area of sweep (Hoggarth and Gaunt 1988), most frequently attach to fish gills (Lefevre and Curtis 1910; Arey 1924a; Murphy 1942; Matteson 1948; Meyers and Millemann 1977; Young and Williams 1984a). There, and on other delicate epithelia, the attachment of hookless glochidia is facilitated by minute (0.5–1.5 μm) teeth located on the valve margins (Arey 1924a; Araujo and Ramos 1998; O'Brien and Brim Box 1999). According to Arey (1924a), these teeth, which are also found on some axe-head and hooked glochidia, are simply smaller versions of the hook spines. Hooked glochidia, particularly of the genus *Unio*, may also primarily attach to the gills of their hosts (*U. tumidus* and *U. pictorum*: Maaß 1987; Fleischauer-Rössing 1990; *U. crassus*: Engel 1990; *A. anatina* and *A. cygnea*: Schneidt 1998). However, the relative proportion of glochidia attaching to the gills and to other body parts may be host-specific (Maaß 1987; Hüby 1988; see below) and may also vary with host size (Harms 1908). More commonly, glochidia with hooks are found on the fins and other soft, scaleless epithelia including the eyes and the nasal openings (Pflugfelder 1951; Giusti et al. 1975; Dartnall and Walkey 1979; Claes 1987; Hüby 1988; Jansen 1991; Fig. 11.3), but have also been found on scales or the opercula (Herbers 1913; Campbell 1974; Dudgeon and Morton 1984; Claes 1987; Jansen 1991). In these and other areas where early dislodgement is likely to occur, attachment of anodontine glochidia (and perhaps hooked glochidia in general) is further facilitated by their great gripping strength (Hoggarth and Gaunt 1988).

Overcrowding of glochidia on a host, which often occurs after prolonged exposure of fishes under laboratory conditions, commonly results in increased larval mortality (Harms 1909; Lefevre and Curtis 1910; Maaß 1987). However, under natural conditions, at 4–65 glochidia (Trdan 1981; Berrie and Boize 1985; Jansen 1991), the mean intensity of infection is usually low. Only in the Margaritiferidae (Awakura 1968; Young and Williams 1984a; Moog et al. 1998) and a few Ambleminae and Lampsilinae (Surber 1913) is the mean intensity of infection known to reach several hundred glochidia. One possible reason for the relatively low natural intensities of infection is that in mussel species with common hosts, competition for attachement sites may be reduced by spatial (i.e. encapsulating on different anatomical areas) and temporal (i.e. parasitizing at different times of the year) differences in infection patterns (Stern and Felder 1978; Trdan and Hoeh 1982). Therefore, glochidial mortality due to overcrowding is usually of little relevance in the wild. Nevertheless, as will be addressed in the following section, glochidia may still suffer considerable mortality during their parasitic phase.

11.2.4 Fish Hosts and Parasitic Phase

Considering the long history of scientific interest in the Unionidea (Leeuwenhoek 1694; Mery 1710; Rathke 1797, all cited in Herbers 1913), including their reproductive biology and ecology (e.g. Carus 1832; Forel 1866; Leydig 1866; Braun 1878; Schierholz 1888), and the both historic and renewed commercial importance of many species (Lefevre and Curtis 1910; Coker 1919; Coker et al. 1921; Wächtler 1986; Williams et al. 1993), relatively little is known about their hosts. This particularly applies to the almost 300 North American species (Williams et al. 1993) for some of which no host information is available. Even for the less than 15 species endemic to Europe (see Nagel et al. 1998) many questions remain about the natural host spectrum. This situation is particularly regrettable because studies on both local and regional scales have shown that the distribution and diversity of freshwater mussels may strongly depend on the migration range and species richness of their fish hosts (Smith 1985; Theler 1987; Watters 1992, 1996a; Lee et al. 1998), and thus of the probability of glochidial survival. Known fish hosts of primarily Nearctic freshwater mussels have been reviewed by Fuller (1974), Gordon and Layzer (1989), Hoggarth (1992) and Watters (1994), and more recent data are provided in the addenda of Watters (1994) and in Michaelson and Neves (1995), Haag and Warren (1997) and Haag et al. (1999). Host relationships of mussels from other biogeographical regions have not been comprehensively reviewed. Some summary information exists for hosts of European naiades (Nagel 1985; Bauer 1987b; Hochwald 1997).

Although glochidia apparently have no capacity of active host selection (see Sect. 11.2.3), the fact that not all native fish species within the reach of a mussel population harbour glochidia (Trdan 1981; Berry and Boize 1985; Hove and Neves 1994), and that the prevalence and intensity of infestation may differ substantially between host species (Karna and Millemann 1978; Trdan 1981; Jokela et al. 1991) indicates a certain host specificity. The degree of this specificity differs between mussel species, and distinctions have been made between host generalists, that use a taxonomically wide array of fish species, and host specialists, for which host use may follow familial or even generic lines (Howard and Anson 1923; Arey 1932b; Zale and Neves 1982a; Yeager and Neves 1986; Haag and Warren 1997; O'Brien and Brim Box 1999).

In addition to the (mechanical) loss of poorly attached, non-encapsulated glochidia (see Sect. 11.2.3), there have been numerous reports of partial or complete glochidial sloughing by the host (Harms 1908; Lefevre and Curtis 1910; Yokley 1972; Zale and Neves 1982a; Fleischauer-Rössing 1990). This process, described in detail by Arey (1932b), results in considerable glochidial mortality and may occur at different times during larval development. Often, rejection occurs during (Meyers et al. 1980; Young and Williams 1984a; Yeager and Neves 1986; Claes 1987; Hüby 1988; Hove and Neves 1994; Haag and Warren 1997) or shortly after (Surber 1913; Maaß 1987; Hove and Neves 1994; Haag and Warren 1997) capsule formation. For different hosts and mussel species, time to complete loss of glochidia may range between 1 and 8 weeks (e.g. Matteson 1948; Neves et al. 1985; Hochwald 1988, 1997; Hove and Neves 1994; Haag and Warren 1997) but also 14 weeks have been reported (Young and Williams 1984a). In these cases, developing glochidia may die as pre-metamorphosed juveniles (Kitchel 1985, cited in Hove and Neves 1994) while inside the capsule, with the empty valves eventually dropping off the host (Maaß 1987).

Glochidial rejection is not limited to non-host fish. Many confirmed fish hosts show a gradual decline in the intensity of glochidial infestation over the course of the parasitic period (Meyers et al. 1980; Young and Williams 1984a,b; Bauer and Vogel 1987; Hüby 1988; Engel 1990; Niemeyer 1992; Lechner 1997; Moog et al. 1998), and the occurrence of pre-metamorphosed juveniles has also been described (Hove and Neves 1994). Furthermore, considerable differences in the intensity of glochidial infestation have been documented between individuals (Young and Williams 1984a; Bauer and Vogel 1987; Haag and Warren 1997), populations or races (Wächtler et al. 1987; Engel and Wächtler 1989; Hochwald 1997) and size classes or ontogenetic stages

(Tedla and Fernando 1969; Young and Williams 1984b; Bauer 1987b,c; Jansen 1991). This suggests that immunological reactions are involved in the parasite-host relationship (Lefevre and Curtis 1910). Meyers et al. (1980) and Bauer and Vogel (1987) found a serum response indicative of specific glochidial antibodies 10–12 and 7 weeks, respectively, postinfection with glochidia of *M. margaritifera*. Bauer and Vogel (1987) speculated that the immune response has two components: a general tissue response soon after infection that results in sloughing of tissue (also see Fustish and Millemann 1978), and a humoral response that is responsible for subsequent mortalities. The tissue response is believed to be the basis of natural immunity, whereas the humoral component is the source of acquired immunity (Arey 1932b; Watters and O'Dee 1996). The underlying serum reaction often seems to be developed unequally, resulting in individual differences in host suitability (Bauer and Vogel 1987). These differences may be related to the specific history of parasite contact and, thus, the presence or magnitude of an acquired immunity. The existence of aquired immunity is supported by observations that previously infected fish show increased glochidial mortality in subsequent infections (Arey 1924b, 1932b; Seshaiya 1969; Claes 1987; Bauer and Vogel 1987; Fleischauer-Rössing 1990; Hochwald 1997; see Fig. 11.4). However, such an immune reaction does not always occur (Young et al. 1987; Hochwald 1997) and if it does, may require multiple previous infections (Reuling 1919; Howard and Anson 1923; Arey 1924b, 1932b; Seshaiya 1969; Bauer 1987c). Furthermore, there are conflicting results with regard to the degree of specificity of the antibodies. For example, Reuling (1919) and Hochwald (1997) found that host fish previously infected with glochidia were less suitable as hosts for other mussel species in subsequent infections, whereas Watters (1996b) and Watters and O'Dee (1997b)

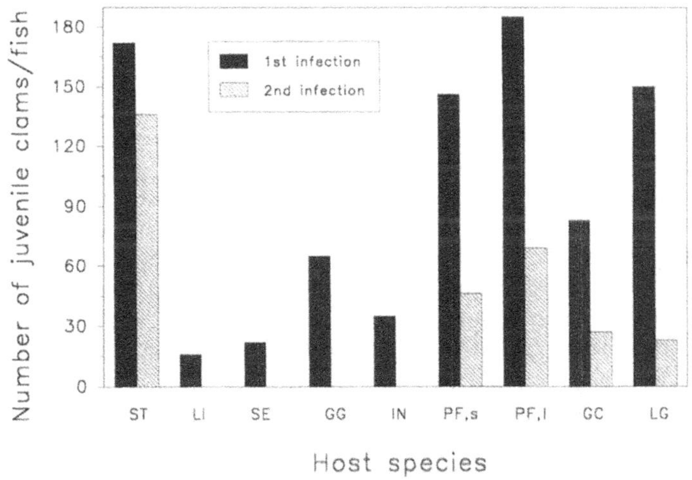

Fig. 11.4. Number of juvenile mussels of *Anodonta cygnea* successfully metamorphosed on eight host species after a first (*solid bar*) and a second (*stippled bar*) infection 5 weeks later; *ST Salmo trutta*; *GG Gobio gobio*; *LI Leuciscus idus*; *SE Scardinius erythrophthalmus*; *GC Gymnocephalus cernua*; *PF Perca fluviatilis*, small (*s*) fish of 10–13 cm total length and large (*l*) fish of 15–20 cm total length; *LG Lepomis gibbosus*. For both infections, fish were exposed to 50 000 glochidia in aerated 10-l aquaria for 30 min. (Data from Claes 1987)

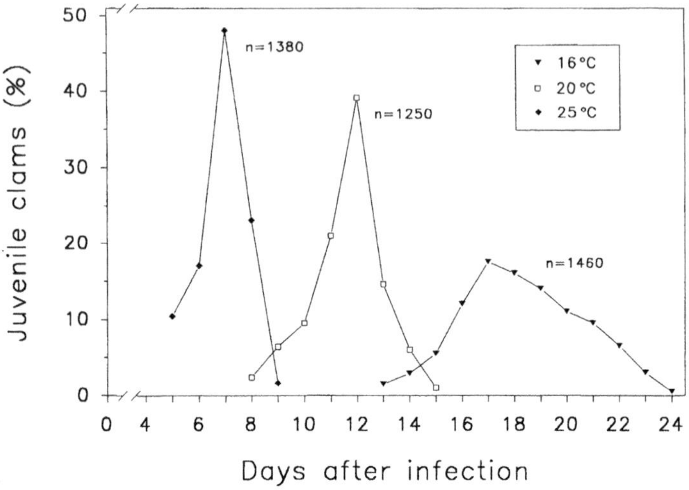

Fig. 11.5. Percentage of metamorphosed mussels collected on each day after laboratory infections of ten *Perca fluviatilis* with glochidia of *Anodonta cygnea* at water temperatures of 16, 20 and 25 °C. The total number of mussels (*n*) obtained during each experiment is indicated. (Data from Claes 1987)

demonstrated a highly specific immune reaction of *M. salmoides* towards glochidia of *Lampsilis cardium*.

The outcome of glochidial infections may also be determined by the initial rate of infestation. Within normal limits, glochidial mortality on individual fish is often negatively related to infection rate (Bauer 1987b; Hochwald 1997), suggesting that the immune system can be overwhelmed. Temperature may also affect the survival of glochidia on the host, since both glochidial development (Harms 1907b; Lefevre and Curtis 1910; Claes 1987, also see Fig. 11.5) and host immune response (Avtalion et al. 1973; Hruška 1992) are temperature-dependent. Hochwald (1997) speculated that possible differences in the temperature sensitivity of these two processes may result in more or less suitable temperatures for successful glochidial metamorphosis. Similar species-specific differences may also explain frequent observations that the period of glochidial development may differ between hosts (Hochwald and Bauer 1990; Yeager and Saylor 1995; Haag and Warren 1997).

Apart from factors that modulate immunological reactions directly, other, more indirect effects like host stress, general health or nutritional status have been suggested to affect both the length of the glochidial phase (Harms 1907b; Herbers 1913) and glochidial mortality (Bednarczuk 1986; Fleischauer-Rössing 1990; Lechner 1997). Finally, some caution must be exercised with respect to the relevance of laboratory experimental results concerning the immunology of the parasite-host relationship and its consequences for the suitability of fish as a natural host (Hochwald 1997). Only recently, Watters and O'Dee (1996) obtained first evidence for the natural occurrence of acquired immunity, and our overall knowledge of the immunological aspects of glochidia-host interactions is still very limited.

In addition to immune reactions, other mechanisms can affect the length of and the mortality rate during the parasitic phase. Some mussel species, particularly within the Margaritiferidae (Murphy 1942; Negus 1966; Young and Williams 1984a; Bauer and Vogel 1987) but also among the genera *Lampsilis*, *Epioblasma* and *Quadrula* (Howard and Anson 1923; Heath et al. 1998), grow considerably during the parasitic phase. As a consequence, delicate gill epithelia may rupture and, as Bauer (1979) has shown at high infestation rates of *M. margaritifera*, the larvae will be lost while still inside the capsule. Even glochidia that do not appreciably grow during their parasitic phase may affect host bioenergetics. Soon after encapsulation, glochidia begin their anatomical transformation into the adult body plan (Harms 1909). During this time, glochidia are known to increase their carbon content (Niemeyer 1992; Wächtler et al., Chap. 6, this Vol.), and, thus, their energy density. Therefore, largely depending on the mussel species and the duration of encapsulation, variable amounts of energy and nutrients must be obtained from the host to complete development. Some of these resources come from the digestion of larval organs (Harms 1909; Pflugfelder 1951). However, most nutrients are obtained from bitten host tissue and liquids and cells infiltrating the capsule (Faussek 1895, 1901; Harms 1909; Herbers 1913; Blystad 1924; Arey 1932c). Neves et al. (1985) suggested that early rejection of glochidia may occur because of deficient nutrient supplies. Little is known about the exact nutritional requirements of glochidia and their impact on the host fish, but at high infection rates, hosts have been reported to die as a result of glochidial growth (Meyers and Millemann 1977). The possibility that glochidia can suffer from malnutrition is supported by the fact that anatomical position affects successful glochidial development, suggesting that locations differ in their nutritional qualities (Herbers 1913). For example, Matteson (1948), Engel and Wächtler (1989), Fleischauer-Rössing (1990), Hochwald (1997) and Lechner (1997) found that mortality of glochidia on (highly vascularized) gills is lower than on fins. Finally, at least for bradytictic species, the length of the breeding period has been shown to affect the time period to detachment, and, indirectly, the mortality during the parasitic phase (Corwin 1920; Tedla and Fernando 1969; Zale and Neves 1982a). This suggests that the older glochidia released by the parent mussel in the spring develop faster and have better survival rates (Zale and Neves 1982a).

Regardless of the exact cause of parasite rejection and death, the few available quantitative data show that even on known host fish with no or little previous exposure, glochidial mortality during the parasitic stage may differ between mussel species and can be very high (Table 11.2). These data partially contrast with the previously held notion that in the wild, losses of glochidia on suitable hosts are relatively low once encapsulation has occurred (e.g. Howard and Anson 1923).

11.2.5 Glochidial Detachment and Early Postparasitic Stage

Although it is generally assumed that naiades soon detach from their host after metamorphosis is complete, juveniles may remain encapsulated for many days (Tedla and Fernando 1969). A number of reports claim that young mussels actively rupture the capsule wall by movements of their foot (Harms 1909; Lefevre and Curtis 1910; Herbers 1913; Arey 1932a; Karna and Millemann 1978). However, Pflugfelder (1951)

Table 11.2. Mortality estimates of glochidia during their parasitic phase. Except for the field data by Young and Williams (1984b), all estimates are from laboratory experiments on endemic host fish with no or little previous exposure to clams. Initial numbers of glochidia were taken on day 1 postinfection

Clam species	Host	Mortality (%)	Source
	a) Individual fish		
Margaritifera margaritifera	*Salmo trutta*	24–78	Bauer (1987b)
	Salmo trutta	53–100	Young and Williams (1984a)
	Salmo salar	26–95	Young and Williams (1984a)
Anodonta cygnea	*Perca fluviatilis*	78	Teutsch (1997)[a]
	b) Means for groups of fish		
Margaritifera margaritifera	*Salmo trutta*	92.5[b]	Young and Williams (1984a)
	Salmo salar	63.8[b]	Young and Williams (1984a)
	Salmo trutta/salar	87[c]–95	Young and Williams (1984b)
Anodonta anatina	*Gasterosteus aculeatus*	83.6	Teutsch (1997)[a]
Anodonta cygnea	*Gasterosteus aculeatus*	88.1	Teusch (1997)[a]
	Phoxinus phoxinus	86.3	Teutsch (1997)[a]
	Stizostedion lucioperca	97.4	Teutsch (1997)[a]
Pseudoanodonta complanata	*Gasterosteus aculeatus*	11.4	Hüby (1988)[a,d]
	Perca fluviatilis	44.0	Hüby (1988)[a,d]
	Pungitus pungitus	5.4	Hüby (1988)[a,d]
	Salmo trutta	19.5	Hüby (1988)[a,d]
	Stizostedion lucioperca	30.7	Hüby (1988)[a,d]
Unio crassus	*Leuciscus cephalus*	26[e]	Hochwald and Bauer (1990)
	Phoxinus phoxinus	0–23[f,g]	Hochwald and Bauer (1990)
	Cottus gobio	36[h]	Hochwald and Bauer (1990)

[a] Mortalities were calculated from infection rates 2–3 days before the appearance of the first juveniles; all data read from graph.
[b] To day 113 postinfection.
[c] Prevalence of infection was ignored in calculation.
[d] Means of three fish held at 12 °C.
[e] To day 21 of 33-day-long parasitic period.
[f] To day 21 of 26-day-long parasitic period.
[g] Two groups of fish.
[h] To day 21 of 32-day-long parasitic period.

doubts such an active role and suggests that detachment is achieved by histolytical processes within the inner layer of the epithelium and the gradual loss of outer layers during encapsulation. Perhaps both the thinning of the capsule wall and movement of the mussel combine to liberate the juvenile mussel (Waller and Mitchell 1989). Since the host may detach glochidia by rubbing body and fins against suitable objects (Wesenberg-Lund 1939; Pflugfelder 1951), active detachment by the clam would reduce mortality from mechanical damage. A more active role of the mussel during its release may also reduce mortality in a different way. It has been suggested that mussels are able to time their detachment according to ambient temperature, thereby increasing the likelihood of encountering favourable habitat conditions (Hruška 1992), which are vital for the early survival of free-living clams (Wächtler et al., Chap. 6, this Vol.).

11.3 Quantitative Estimates of Glochidial Mortality

As we have identified in Sections 11.2.1–11.2.5, there are many potential causes of glochidial mortality during the relatively short period before mussels start their benthic life. From the few studies that have attempted to quantify mortality rates in glochidia, it is obvious that survival during this life history stage is extremely low. Estimates of overall mortality indicate that out of every billion glochidia produced, only 10 (for *M. margaritifera*) to 18 000 (for *P. grandis*) survive to age 1 or age 2 years, respectively (Table 11.3). The more than 3 orders of magnitude difference between these mortality rates is in itself remarkable. Compared to *M. margaritifera* (Bauer 1998), *P. grandis* grows fast, is relatively short-lived, and has few reproductive periods as a bradytictic breeder (Hanson et al. 1988a,b; Jansen and Hanson 1991). *P. grandis* represents a life history strategy of priority allocation into reproduction, as typically found in anodontid clams (Bauer 1998). This strategy strongly differs from that of *M. margaritifera* and probably the other species of the family Margaritiferidae, which is characterized by a priority allocation into somatic tissue (Bauer 1998). Thus, the above mortality estimates from two species that represent opposites within the range of Unionidea life histories most likely span much of the interspecific variation in mortality of glochidia-producing freshwater mussels. Unfortunately, quantitative estimates of mortality rates for glochidia released as conglutinates or of glochidia from mussel species with morphological and/or behavioral features to attract potential hosts (see Sect. 11.2.1) are lacking. If these are successful adaptations for more efficient host contact, glochidial survival in these species is expected to be relatively high.

Among the different stages in the life of glochidia and juvenile clams identified in Table 11.3, mortality is most severe during the few hours between parental release and encapsulation on a host. For some species (i.e. *M. margaritifera*) establishment of freshly released mussels is considered to be poor (Young and Williams 1984b). Interestingly, for a drift distance of 100 m, which is close to the median value observed in field experiments (see Sect. 11.2.2), our calculations of preparasitic glochidial mortality in *M. margaritifera* (Appendix 11.1) result in almost identical rates compared to the estimates from field measurements by Young and Williams (1984b; Table 11.3). Once mussels have settled in the sediment, mortality until maturity is very low (Young and Williams 1984b), estimated at approximately 10% (Table 11.3). Mortality is only slightly higher after reproduction has started (Negus 1966; Bauer 1983, see below). If these survival rates equally apply to a wider range of species, the preparasitic stage certainly is the most critical in the life cycle of the Unionidea.

The very low survival of their offspring has been interpreted as evidence for the relative inefficiency of reproduction in freshwater clams (Neves and Widlak 1988). However, as evidenced by some of the extremely dense mussel beds reported from the time before massive human exploitation (Smith 1899; Israel 1913; Coker 1919), reproductive rate did not limit (local) population size. Perhaps, phylogenetically constrained by a free-living larval stage with a low survival rate, the

Table 11.3. Summary of literature data on mortality estimates of glochidia (G) for the time period between their release by the parent clam to the settlement of juveniles in the sediment. Mortality rates for juvenile clams until maturity (age 5 and 6 years) are given for comparison. All estimates are from field collections except for the laboratory experiments of studies 2–6 for the parasitic stage

Parameter measured	Clam species	Mortality (%)	Reference
	a) Preparasitic stage		
G recovered from stream drift on fish	*M. margaritifera*	99.9998[a]	Current
G attached to host / G produced	*M. margaritifera*	99.9996	Young and Williams (1984a)
G attached to host / G produced	*Pyganodon grandis*	99.9931	Jansen and Hanson (1991)
G attached to host / G produced	*Unio crassus*	99.9987	Hochwald (1988)
	b) Parasitic stage		
G attached in fall / G attached in spring	*M. margaritifera*	95.2	Young and Williams (1984a)
G attached on day n + x / G attached on day n	*M. margaritifera*	63.8–92.5	Young and Williams (1984b)
G attached on day n + x / G attached on day n	*Anodonta anatina*	83.6	Teutsch (1997)
G attached on day n + x / G attached on day n	*Anodonta cygnea*	86.3–97.4	Teutsch (1997)
G attached on day n + x / G attached on day n	*P. complanata*	5.4–44.0	Hüby (1988)
G attached on day n + x / G attached on day n	*Unio crassus*	0–35.8	Hochwald and Bauer (1990)
	c) Postparasitic stage		
YOY clams in sediment / G at detachment	*M. margaritifera*	95.0	Young and Williams (1984b)
Clams age 2 / G attached to host	*Pyganodon grandis*	73.4	Jansen and Hanson (1991)
Clams age 5 / clams age 2	*Pyganodon grandis*	9–18	Jansen and Hanson (1991)
Clams age 6 / clams age 2	*Fusconaia ebena*	<10[b]	Payne and Miller (1989)
	d) Overall		
YOY clams in sediment / G produced	*M. margaritifera*	99.999999	Young and Williams (1984a)
Clams age 2 in sediment / G produced	*Pygonodon grandis*	99.9982	Jansen and Hanson (1991)

YOY = young-of-year; M. = *Margaritifera*; P. = *Pseudoanodonta*.
[a] Assuming a drift distance of 100 m (see text).
[b] Rough estimate based on the data in Fig. 1 of Payne and Miller (1978), authors state that mortality "is negligible".

Unionidea have evolved as organisms that are generally long-lived (Comfort 1957), highly fecund (Bauer 1987a; Hanson et al. 1989; Hochwald 1997), and, once mature, have an age-specific survivorship of 80% and more (Negus 1966; Bauer 1983). In most freshwater mussels these life-history traits result in a very high life-time fecundity, that apparently compensates for the inefficiency of host infection and associated high larval mortality. Freshwater mussels often comprise a major fraction of the total biomass of invertebrates (Fisher and Tevesz 1976; James 1985; Hanson et al. 1988b). This successful group of animals, which for most of their life are mainly sedentary, has adapted its reproductive strategy to ensure population survival, while maintaining a larval stage that provides opportunities for faster and more distant dispersal.

11.4 Summary

This chapter summarizes known and potential causes of glochidial mortality during a number of distinct stages in the life history of these larvae of freshwater clams (Unionoidea). Between the time mature glochidia are inside the marsupia of the parent mussel, are released into the water to spent a brief free-living existance before they attach and encapsule onto a fish host, to their parasitic phase and eventual detachment from the host, glochidia die mainly because of predation, starvation, mechanical failure to attach to a host and immunological host reactions. Of the many thousands to millions of glochidia expelled by individual females of different mussel species, only very few reach the free-living juvenile stage. The few available quantitative mortality rates for the entire glochidial stage range between 99.9982 and 99.999999%. At least for species that lack special adaptations for host attraction which are supposed to promote glochidial survival, by far most of the overall mortality occurs during the preparasitic stage, making it the most critical period in the life history of glochidia-producing Unionidea.

Acknowledgements. We thank Dave Strayer for advice on the nomenclature of North American species and Thomas Watters for help with the literature. Cheryl Podemski provided helpful comments on the manuscript which are greatly appreciated. We are thankful to Pearl Fok for drawing Fig. 11.3.

References

Allen E (1924) The existence of a short reproductive cycle in *Anodonta imbecillis.* Biol Bull 46:88–94

Amyot J-P, Downing JA (1991) Endo- and epibenthic distribution of the unionid mollusc *Elliptio complanata.* J N Am Benthol Soc 10:280–285

Anderson RN (1981) Population ecology of infectious disease agents. In: May RM (ed) Theoretical ecology. Blackwell, Oxford, pp 318–355

Araujo R, Ramos MA (1998) Description of the glochidium of *Margaritifera auricularia* (Spengler 1793) (Bivalvia, Unionidea). Philos Trans R Soc Lond B 353:1553–1559

Arey LB (1921) An experimmental study on glochidia and the factors underlying encystment. J Exp Zool 33:463–499

Arey LB (1924a) Glochidial cuticulae, teeth, and mechanics of encystment. J Morphol 39:323–335

Arey LB (1924b) Observations on an acquired immunity to a metazoan parasite. J Exp Zool 38:377–381
Arey LB (1932a) The formation and structure of the glochidial cyst. Biol Bull Mar Biol Lab Woods Hole 62:212–221
Arey LB (1932b) A microscopical study of glochidial immunity. J Morphol 53:367–379
Arey LB (1932c) The nutrition of glochidia during the metamorphosis: a microscopical study of the sources and manner of utilization of nutritive substances. J Morphol 53:201–221
Avtalion RR, Wojdani A, Malik Z, Rshahrabani, Duczyminer M (1973) Influence of environmental temperature on the immune response of fish. Current Top Microbiol Immunol 61:1–35
Awakura T (1968) The ecology of parasitic glochidia of the fresh-water pearl mussle, *Margaritifera laevis* (Haas). Sci Rep Hokkaido Fish Hatchery 23:1–21
Barnhart MC (1997) Conglutinates and fish hosts of the western fanshell, *Cyprogenia aberti*. Triannu Unionid Rep 12:2
Barnhart C, Riusech F, Baird M (1998) Host of salamander mussel, *Simpsonaias ambigua*, and snuffbox, *Epioblasma triquetra*, from the Meramee River system, Mo. Triannu Unionid Rep 16:34
Bauer G (1979) Untersuchungen zur Fortpflanzungsbiologie der Flußperlmuschel (*Margaritana margaritifera*) im Fichtelgebirge. Arch Hydrobiol 85:152–165
Bauer G (1983) Age structure, age specific mortality rates and population trend in the freshwater pearl mussel (*Margaritifera margaritifera*) in North Bavaria. Arch Hydrobiol 98: 523–532
Bauer G (1987a) Reproductive strategy of the freshwater pearl mussel *Margaritifera margaritifera*. J Anim Ecol 56:691–704
Bauer G (1987b) The parasitic stage of the freshwater pearl mussel (*Margaritifera margaritifera* L.) III. Host relationships. Arch Hydrobiol Suppl 76:413–423
Bauer G (1987c) The parasitic stage of the freshwater pearl mussel (*Margaritifera margaritifera* L.) II. Susceptibility of brown trout. Arch Hydrobiol Suppl 76:403–412
Bauer G (1991) Plasticity in life history traits of the freshwater pearl mussel – cnsequences for the danger of extinction and for conservation measures. In: Seitz A, Löschke V (eds) species conservation: a population-biological approach. Birkhäuser, Basel, pp 103–120
Bauer G (1994) The adaptive value of offspring size among freshwater mussels (Bivalvia; Unionoidea). J Anim Ecol 63:933–944
Bauer G (1998) Allocation policy of female freshwater pearl mussels. Oecologia 117:90–94
Bauer G, Vogel C (1987) The parasitic stage of the freshwater pearl mussel (*Margaritifera margaritifera* I. Host response to glochidiosis. Arch Hydrobiol Suppl 76:393–402
Bauer G, Hochwald S, Silkenat W (1991) Spatial distribution of freshwater mussels: the role of host fish and metabolic rate. Freshwater Biol 26:377–386
Bednarczuk J (1986) Untersuchungen zu Wirtsfischspektrum und Entwicklung der Bachmuschel Unio crassus. Dissertation am Institut für Zoologie, Tierärztliche Hochschule Hannover
Begon M, Harper JL, Townsend CR (1988) Ecology. Blackwell, Oxford
Berrie AD, Boize BJ (1985) The fish hosts of *Unio* glochidia in the River Thames. Verh Int Ver Limnol 22:2712–2716
Blystad CN (1924) Significance of larval mantle of fresh-water mussels during parasitism, with notes on a new mantle condition exhibited by Lampsilis luteola. Bull US Bur Fish 39:203–219
Botnariuc N, Tudorancea C (1967) Beiträge zur Populationsdynamik bei *Anodonta piscinalis* im Flachsee Jijila (Überschwemmungsgebiet der Donau). Arch Hydrobiol Suppl 30:400–419
Braun M (1878) Postembryonale Entwicklung von Anodonta. Zool Anz 1:7–10
Bruenderman SA, Neves RJ (1993) Life history of the endangered fine-rayed pigote *Fusconaia cuneolus* (Bivalvia: Unionidae) in the Clich River, Virginia. Am Malacol Bull 10:83–91
Calow P (1983) Life-cycle patterns and evolution. In: Russell-Hunter WD (ed) The Mollusca, vol 6. Ecology. Academic Press, New York, pp 649–678

Campbell AD (1974) The parasites of fish in Loch Leven. Proc R Soc Edinb 74:347–364
Carus CG (1832) Neue Untersuchungen über die Entwickelungsgeschichte unserer Flussmushel. Verh K Leopoldinisch-Carolinischen Akad Naturforsch 16:1–87
Castognolo L (1977) Ciclo biologico riproduttivo di *Anodonta cygnea* L. e *Unio elongatulus* Pfeiff. nel Po di Tolle. Riv Idrobiol 16:3–13
Chamberlain TK (1934) The glochidial conglutinates of the Arkansas Fanshell, *Cyprogenia aberti* (Conrad). Biol Bull 66:55–61
Chapman DW (1966) Food and space as regulators of salmonid populations in streams. Am Nat 100:345–357
Claes M (1987) Untersuchungen zur Entwicklungsbiologie der Teichmuschel *Anodonta cygnea*. Diplomarbeit, Institut für Zoologie, Tierärztliche Hochschule Hannover
Clark HW, Stein S (1921) Glochidia in surface towings. Nautilus 35:16–20
Clarke A (1981) The freshwater mollusks of Canada. National Museum of Natural Sciences, National Museums of Canada, Ottawa
Coker RE (1919) Fresh-water mussels and mussel industries of the United States. Bull US Bur Fish 36:11–89
Coker RE, Shira AH, Clark HW, Howard AD (1921) Natural history and propagation of fresh-water mussels. Bull US Bur Fish 37:75–181
Comfort A (1957) The duration of life in molluscs. Proc Malacol Soc Lond 32:219–241
Corwin RS (1920) Raising freshwater mussels in enclosures. Trans Am Fish Soc 49:81–84
Cunjak RA, McGladdery SE (1991) The parasite-host relationship of glochidia (Mollusca: Margaritiferidae) on the gills of young-of-the-year Atlantic salmon (*Salmo salar*). Can J Zool 69:353–358
Dartnall H, Walkey M (1979) The distribution of glochidia of the Swan mussel, *Anodonta cygnea* (Mollusca) on the three-spined stickleback *Gasterosteus aculeatus* (Pisces). J Zool Lond 189:31–37
Downing WL, Shotsell J, Downing JA (1992) Non-annual ecternal annuli in the freshwater mussels *Anodonta grandis grandis* and *Lampsilis radiata siliquoidea*. Freshwater Biol 28:309–317
Dudgeon D, Morton B (1983) The population dynamics and sexual strategy of *Anodonta woodiana* (Bivalvia: Unionacea) in Plover Cove Reservoir, Hong Kong. J Zool Lond 201:161–183
Dudgeon D, Morton B (1984) Site selection and attachment duration of *Anodonta woodiana* (Bivalvia: Unionacea) glochidia on fish hosts. J Zool Lond 204:355–362
Engel H (1990) Untersuchungen zur Autökologie von *Unio crassus* (Philipsson) in Norddeutschland. Dissertation, Fachbereich Biologie, Universität Hannover
Engel H, Wächtler K (1989) Some peculiarities in developmental biology of two forms of the freshwater bivalve *Unio crassus* in northern Germany. Arch Hydrobiol 115:441–450
Faussek V (1895) Ueber den Parasitismus der *Anodonta*-Larven in der Fischhaut. Biol Zentralbl 15:115–125
Faussek V (1901) Ueber den Parasitismus der *Anodonta*-Larven. Ver Int Congr Zool Berlin 1901, pp 761–766
Fisher JB, Tevesz MJ (1976) Distrubiton and population density of *Elliptio complanata* (Mollusca) in Lake Pocotopaug, Connecticut. Veliger 18:332–338
Fleischauer-Rössing S (1990) Untersuchungen zur Autökologie von *Unio tumidus* Phillipsson und *Unio pictorum* Linnaeus unter besonderer Berücksichtigung der frühen postparasitären Phase. Dissertation, Fachbereich Biologie, Universität Hannover
Forel FA (1866) Beiträge zur Entwicklungsgeschichte der Najaden. Inaugural-Abhandlung, Medizinischen Fakultät, Universität Würzburg
Fuller S (1974) Clams and mussels (Mollusca: Bivalvia). In: Hart CW Jr, Fuller SLH (ed) Pollution ecology of freshwater invertebrates. Academic Press, New York, pp 228–237
Fustish CA, Millemann RE (1978) Glochidiosis of salmonid fishes. II. Comparison of tissue response of coho and chinook salmon to experimental infection with *Margeritifera margeritifera* (L.) (Pelecypoda: Margaritanidae). J Parasitol 64:155–157

Giusti F, Castagnolo L, Farina L, Renzoni A (1975) The reproductive cycle and the glochidium of *Anodonta cygnea* L. from Lago Trasimeno (central Italy). Monit Zool Ital (NS) 9:99–118

Gordon ME, Layzer JB (1989) Mussels (Bivalvia: Unionidea) of the Cumberland River: review of life histories and ecological relationships. Biological Report 89(15), US Department of the Interior Fish and Wildlife Service, Washington, DC

Haag WR, Warren ML Jr (1997) Host fishes and reproductive biology of 6 freshwater mussel species from the Mobile basin, USA. J N Am Benthol Soc 16:576–585

Haag WR, Butler RS, Hartfield PD (1995) An extraordinary reproductive strategy in freshwater bivalves: prey mimicry to facilitate larval dispersal. Freshwater Biol 34:471–476

Haag WR, Warren ML Jr, Shillingsford M (1999) Host fishes and host-attracting behavior of *Lamsilis altilis* and *Villosa vibex* (Bivalvia: Unionidae). Am Midl Nat 141:149–157

Hanson J, Mackay W, Prepas E (1988a) Population size, growth, and production of a unionid clam, *Anodonta grandis simpsoniana*, in a small, deep Boreal Forest lake in central Alberta. Can J Zool 66:247–253

Hanson J, MacKay W, Prepas E (1988b) The effects of water depth and density on the growth of a unionid clam. Freshwater Biol 19:345–355

Hanson J, MacKay W, Prepas E (1989) Effect of size-selective predation by muskrats (*Ondatra zebithicus*) on a population of unionid clams (*Anodonta grandis simpsoniana*). J Anim Ecol 58:15–28

Harms W (1907a) Zur Biologie und Entwicklungsgeschichte der Flußperlmuschel (*Margaritana margaritifera* Dupuy). Zool Anz 31:814–824

Harms W (1907b) Über die postembryonale Entwicklung von *Anodonta piscinalis*. Zool Anz 31:801–814

Harms W (1908) Die postembryonale Entwicklung von *Unio pictorum* und *Unio tumidus*. Zool Anz 32:693–703

Harms W (1909) Postembryonale Entwicklungsgeschichte der Unioniden. Zool Jahrb Anat 28:325–386

Hartfield P, Hartfield E (1996) Observations on the conglutinates of *Ptychobranchus greeni* (Conrad, 1834) (Mollusca: Bivalvia: Unionidea). Am Midl Nat 135:370–375

Heard W (1975) Sexuality and other aspects of reproduction in *Anodonta* (Pelecypoda: Unionidae). Malacologia 15:81–103

Heath D, Hove M, Benjamin R, Endris M, Kenyon R, Kurth J (1998) *Quadrula fragosa* exhibit unusual reproductive behaviors. Triannu Unionid Rep 16:32

Herbers K (1913) Entwicklungsgeschichte von *Anodonta cellensis* Schröt. Z Wiss Zool 108:1–174

Hochwald S (1988) Untersuchungen zur Populationsökologie und Fortpflanzungsbiologie der Bachmuschel *Unio crassus* Phil. 1788. Diplomarbeit, Lehrstuhl Tierökologie I, Universität Bayreuth

Hochwald S (1997) Das Beziehungsgefüge innerhalb der Größenwachstums- und Fortpflanzungsparameter bayerischer Bachmuschelpopulationen (*Unio crassus* Phil. 1788) und dessen Abhängigkeit von Umweltparametern. Dissertation, Lehrstuhl für Tierökologie I, Universität Bayreuth; Bayreuther Forum Ökologie, Band 50, 166 S

Hochwald S, Bauer G (1990) Untersuchungen zur Populationsökologie und Fortpflanzungsbiologie der Bachmuschel *Unio crassus* (Phil.) 1788. Schriftenr Bayer Landesamt Umweltsch 97:31–49

Hoggarth MA (1992) An examination of the glochidia-host relationships reported in the literature for North American species of Unionacea Mollusca: Bivalvia). Malacol Data Net 3:1–30

Hoggarth MA, Gaunt AS (1988) Mechanics of glochidial attachment (Mollusca: Bivalvia). J Morphol 198:71–81

Hove MC, Neves RJ (1994) Life history of the endangered James spinymussel *Pleurobema collina* (Conrad 1837) (Mollusca: Unionidae). Am Malacol Bull 11:29–40

Howard AD (1915) Some exceptional cases of breeding among the Unionidae. Nautilus 29:4-11
Howard AD, Anson BJ (1923) Phases in the parasitism of the Unionidae. J Parasitol 9:70-84
Hruška J (1992) The freshwater pearl mussel in South Bohemia: evaluation of the effect of temperature on reproduction, growth and age structure. Arch Hydrobiol 126:181-191
Hüby B (1988) Zur Entwicklungsbiologie der Fließgewässermuschel *Pseudoanodonta complanata*. Dissertation, Institut für Zoologie, Tierärztliche Hochschule Hannover
Israel W (1913) Biologie europäischer Süsswassermuscheln. KG Lutz, Stuttgart
Ivleva I, Popenkina M (1969) On the temperature dependence of metabolism in poikilothermic animals. (From: Biologiya Morya Fiziologicheskie osnovy ekologii vodnykh zhivotnykh 15:29-51, 1968), Fish Res Board Can Transl Ser 1224
Jackson JW (1925) The distribution of *Margaritifera margaritifera* in the British Isles. J Conchol 17:195-211, 270-278
James MR (1985) Distribution, biomass and production of the freshwater mussel *Hyridella menziesi* (Gray) in Lake Taupo, New Zealand. Freshwater Biol 15:307-314
Jansen WA (1991) Seasonal prevalence, intensity of infestation, and distribution of glochidia of *Anodonta grandis simpsoniana* Lea on yellow perch, *Perca flavescens*. Can J Zool 69:964-972
Jansen WA, Hanson JM (1991) Estimates of the number of glochidia produced by clams (*Anodonta grandis simpsoniana* Lea), attaching to yellow perch (*Perca flavescens*), and surviving to various ages in Narrow Lake, Alberta. Can J Zool 69:973-977
Jirka KJ, Neaves RJ (1992) Reproductive biology of four species of freshwater mussels (Mollusca: Unionidae) in the New River, Virginia and West Virginia. J Freshwater Ecol 7:35-44
Jokela J, Palokangas P (1993) Reproductive tactics in *Anodonta* clams: parental host recognition. Anim Behav 46:618-620
Jokela J, Valtonen ET, Lappalainen M (1991) Development of glochidia of *Anodonta piscinalis* and their infection of fish in a small lake in northern Finland. Arch Hydrobiol 120:345-355
Karna DW, Millemann RE (1978) Glochidiosis of salmonid fishes. III. Comparative susceptibility to natural infection with *Margaritifera margaritifera* (L.) (Pelecypoda: Margaritanidae) and associated histopathology. J Parasitol 64:528-537
Kat PW (1984) Parasitism and the Unionacea (Bivalvia). Biol Rev 59:189-207
Kraemer LR (1970) The mantle flap in three species of *Lampsilis* (Pelecypoda: Unionidae). Malacologia 10:225-282
Kraemer LR, Swanson CM (1985) Functional morphology of "eyespots" of mantle flaps of *Lampsilis* (Bivalvia: Unionacea): evidence for their role as effectors, and basis for hypothesis regarding pigment distribution in bivalve mantle tissue. Malacologia 26:241-251
Lang K (1998) Die Bachmuschel *Unio crassus* (Philipsson, 1788) im Niederbach: Bestandssituation, Fortpflanzungsbiologie und Wanderverhalten. Diplomarbeit, Lehrstuhl für Ökologie, Morphologie und Evolutionsbiologie, Universität Freiburg
Latter OH (1891) Notes on Anodonta and Unio. Proc Zool Soc Lond 1891:52-59
Lechner S (1997) Populationsökologische Untersuchungen der Bachmuschel *Unio crassus* (Philipsson 1788) im Einzugsgebiet der hessischen Kinzig. Diplomarbeit, Lehrstuhl für Ökologie, Morphologie und Evolutionsbiologie, Universität Freiburg
Lee HL, DeAngelis D, Koh HL (1998) Modeling spatial distribution of the unionid mussels and the core-satellite hypothesis. Water Sci Technol 38:73-79
Lefevre G, Curtis W (1910) Reproduction and parasitism in the unionidae. J Exp Zool 9:79-115
Lewis J (1985) Breeding cycle of the freshwater mussel *Anodonta grandis* Say. Can J Zool 63:2477-2478
Leydig F (1866) Mittheilung über den Parasitismus junger Unioniden an Fischen. In: Noll FC (ed) Der Main in seinem unterem Laufe. Inaugural-Dissertation, Universität Tübingen, Frankfurt am Main, Joh. Christ. Hermann'sche Verlagsbuchhandlung
Maaß S (1987) Untersuchungen zur Fortpflanzungsbiologie einheimischer Süßwassermuscheln der Gattung Unio. Dissertation, Institut für Zoologie, Tierärztliche Hochschule Hannover

Matteson MR (1948) Life history of *Elliptio complanatus* (Dillwyn, 1817). Am Midl Nat 40:690–723

McMahon RF (1991) Mollusca: Bivalvia. In: Thorp JH, Covich AP (eds) Ecology and classification of North American freshwater invertebrates. Academic Press, New York, pp 315–399

Meyers TR, Millemann RE (1977) Glochidiosis of salmonid fishes. I. Comparative susceptibility to experimental infection with *Margaritifera margaritifera* (L.) (Pelecypoda: Margaritanidae). J Parasitol 63:728–733

Meyers TR, Millemann RE, Fustish CA (1980) Glochidiosis of salmonid fishes. IV. Humoral and tissue responses of coho and chinook salmon to experimental infection with *Margaritifera margaritifera* (L.) (Pelecypoda: Margaritanidae). J Parasitol 66:274–281

Michaelson DL, Neves RJ (1995) Life history and habitat of the endangered dwarf wedgemussel *Alasmidonta heterodon* (Bivalvia: Unionidae). J North Am Benthol Soc 14: 324–340

Moog O, Ofenböck T, Nesemann H, Stundner C (1998) The freshwater pearl mussel *Margaritifera margaritifera* (L.) in Austria: fundamental conservation measures for an endangered species. Verh Int Ver Limnol 26:2438–2443

Murphy G (1942) Relationship of the fresh-water mussel to trout in the Truckee River. Cal Fish Game 28:89–102

Nagel K-O (1985) Glochidien und Fortpflanzungsbiologie von Najaden des Rheins (Bivalvia-Unionidae-Anodontinae). Mainzer Naturwissenschaftl Archiv, Beiheft 5:163–174

Nagel K-O, Badino G, Celebrano G (1998) Systematics of European najades (Bivalvia: Margaritiferidae and Unionidae): a review and some new aspects. Malacol Rev Suppl 7:83–104

Nalepa TF, Gauvin JM (1988) Distribution, abundance, and biomass of freshwater mussels (Bivalvia: Unionidae) in lake St. Clair. J Great Lakes Res 14:411–419

Negus CL (1966) A quantitative study of growth and production of unionid mussels in the river Thames at Reading. J Anim Ecol 35:513–531

Neves RJ, Widlak JC (1988) Occurrence of glochidia in stream drift and on fishes of the Upper North Fork Holston River, Virginia. Am Midl Nat 119:111–120

Neves RJ, Weaver LR, Zale AV (1985) An evaluation of host fish suitability for glochidia of *Villosa vanuxemi* and *V. nebulosa* (Pelecypoda: Unionidae). Am Midl Nat 113:13–19

Nezlin LP, Cunjak RA, Zotin AA, Ziuganov VV (1994) Glochidium morphology of the freshwater pearl mussel (*Margaritifera margaritifera*) and glochidiosis of Atlantic salmon (*Salmo salar*): a study by scanning electron micoscropy. Can J Zool 72:15–21

Niemeyer B (1992) Vergleichende Untersuchungen zur biomischen Strategie der Teichmuschelarten *Anodonta cygnea* L. und *Anodonta anatina* L. Dissertation, Fachbereich Biologie, Universität Hannover

O'Brien CA, Brim Box J (1999) Reproductive biology and juvenile recruitment of the shinyrayed pocketbook, *Lampsilis subangulata* (Bivalvia: Unionidae) in the Gulf costal plain. Am Midl Nat 142:129–140

Okland J, Andersen A (1985) The first records of the freshwater mussel *Pseudoanodonta complanata* (Bivalvia: Unionidae) in Norway with remarks on other large mussels in freshwater. Fauna 28:95–100

Paling JE (1968) A method of estimation the relative volume of water flowing over the different gills of a freshwater fish. J Exp Biol 48:533–544

Panha S (1990) The site survey and the study on reproductive cycles of freshwater pearl mussels in the central part of Thailand. Venus 49:240–257

Paterson CG, Cameron IF (1985) Comparative energetics of two populations of the unionid, *Anodonta cataracta* (Say). Freshwater Invertebr Biol 4:79–90

Payne BS, Miller AC (1989) Growth and survival of recent recruits to a population of *Fusconaia ebena* (Bivalvia: Unionidae) in the lower Ohio River. Am Midl Nat 121:99–104

Pflugfelder O (1951) Über das Parasit-Wirtsverhältnis von Anodonta-Glochidien und Spiegelkarpfen (*Cyprinus carpio* L.). Z Parasitenkd 15:119–133

Reuling FH (1919) Aquired immunity to an animal parasite. J Infect Dis 24:337–346
Schierholz C (1888) Über Entwicklung der Unioniden. Denkschr K Akad Wiss Wien Math-Naturwiss Kl 55:183–214
Schneidt A (1998) Sympatrische und allopatrische Anodonta-Vorkommen in Südbaden: Bestandssituation, Fortpflanzungsstrategie und Konkurrenz. Diplomarbeit, Lehrstuhl für Ökologie, Morphologie und Evolutionsbiologie, Universität Freiburg
Seshaiya RV (1941) Tadpoles as hosts for the glochidia of the freshwater mussel. Curr Sci 10:535–536
Seshaiya RV (1969) Some observations on the life-histories of South Indian freshwater mussels. Malacologia 9:286–287
Smith DG (1985) Recent range expansion of the freshwater mussel *Anodonta implicata* and its relationship to clupeid fish restoration in the Connecticut River system. Freshwater Invertebr Biol 4:105–108
Smith HM (1899) The mussel fishery and pearl-button industry of the Mississippi River. Bull US Fish Comm 18:289–314
Stern EM, Felder L (1978) Identification of host fishes for four species of freshwater mussels (Bivalvia: Unionidae). Am Midl Nat 100:233–236
Surber T (1913) Notes on the natural hosts of fresh-water mussels. Bull US Bur Fish 32:110–116
Sylvester JR, Holland LE, Kammer TK (1984) Observations on burrowing rates and comments on host specificity in the endangered mussel *Lampsilis higginsi*. J Freshwater Ecol 2:555–559
Taskinen J, Valtonen ET (1995) Age-, size-, and sex-specific infection of *Anodonta piscinalis* (Bivalvia: Unionidae) with *Rhipidocotyle fennica* (Digenea: Bucephalidae) and its influence on host reproduction. Can J Zool 73:887–897
Tedla S, Fernando C (1969) Observations on the glochidia of *Lampsilis radiata* (Gmelin) infesting yellow perch, *Perca flavescens* (Mitchell) in the Bay of Quinte, Lake Ontario. Can J Zool 47:705–712
Teutsch S (1997) Populationsbiologische Untersuchungen an *Anodonta*: Fortpflanzungsstrategien und Charakterisierung von Teilpopulationen innerhalb eines Stillgewässers. Diplomarbeit, Lehrstuhl für Ökologie, Morphologie und Evolutionsbiologie, Universität Freiburg
Theler JL (1987) Prehistoric freshwater mussel assemblages of the Mississippi River in southwestern Wisconsin. Nautilus 101:143–150
Tompa AS (1979) Life-cycle completion of the freshwater clam *Lasmigona compressa* (Bivalvia: Unionidae) in an experimental host, *Lebistes reticulatus*. Veliger 22:188–190
Trdan RJ (1981) Reproductive biology of *Lampsilis radiata siliquoidea* (Pelecypoda: Unionidae). Am Midl Nat 106:243–248
Trdan RJ, Hoeh WR (1982) Eurytopic host use by two congeneric species of freshwater mussel (Pelecypoda: Unionidae: Anodonta). Am Midl Nat 108:381–388
Trübsbach J (1998) Fortpflanzungsbiologie und Mortalität der gemeinen Teichmuschel, *Anodonta anatina*. Diplomarbeit, Lehrstuhl für Ökologie, Morphologie und Evolutionsbiologie, Universität Freiburg
Walker KF (1981) The ecology of freshwater mussels in the River Murray. Aust Water Res Tech Pap 63:1–110
Waller DL, Mitchell LG (1989) Gill tissue reactions in walleye *Stizostedion vitreum vitreum* and common carp *Cyprinus carpio* to glochidia of the freshwater mussel *Lampsilis radiata siliquoidea*. Dis Aquat Org 6:81–87
Watters GT (1992) Unionids and the species-area curve. J Biogeogr 19:481–490
Watters GT (1994) An annotated bibliography of the reproduction and propagation of the Unionoidea (primarily of North America). Ohio Biological Survey, Misc Cont No 1, Columbus, Ohio, vi + 158 pp
Watters GT (1996a) Small dams as barriers to frehwater mussels (Bivalvia, Unionoida) and their hosts. Biol Conserv 75:79–85
Watters GT (1996b) And immunity for all. Triannu Unionid Rep 10:14

Watters GT (1997) Glochidial metamorphosis of the freshwater mussel *Lampsilis cardium* (Bivalvia: Unionidae) on larval tiger salamanders, *Ambystoma tigrinum* ssp. (Amphibia: Ambystomidae). Can J Zool 75:505–508

Watters GT, O'Dee SH (1996) Shedding of untransformed glochidia by fishes parasitized by *Lampsilis fasciola* Rafinesque, 1820 (Mollusca: Bivalvia: Unionidae): evidence of acquired immunity in the field ? J Freshwater Ecol 11:383–389

Watters GT, O'Dee SH (1997a) *Utterbackia imbecillis* goes fishing with nets. Triannu Unionid Rep 13:40

Watters GT, O'Dee SH (1997b) No aquired immunity against a congener. Triannu Unionid Rep 13:39

Watters GT, O'Dee SH (1998) Metamorphosis of freshwater mussel glochidia (Bivalvia: Unionidae) on amphibians and exotic fishes. Am Midl Nat 139:49–57

Wenz G (1990) Ökologische Untersuchungen an Perlbächen. Diplomarbeit, Institut für Zoology, Universität Bayreuth

Wesenberg-Lund C (1939) Biologie der Süsswassertiere. Julius Spinger, Wien

Williams JD, Warren ML Jr, Cummings KS, Harris JL, Neves RJ (1993) Conservation status of freshwater mussels of the United States and Canada. Fisheries 18:6–22

Wood EM (1974a) Development and morphology of the glochidium larva of *Anodonta cygnea* (Mollusca: Bivalvia). J Zool 173:1–13

Wood EM (1974b) Some mechanisms involved in host recognition and attachment of the glochidium larva of *Anodonta cygnea* (Mollusca: Bivalvia). J Zool 173:15–30

Wächtler K (1986) Zur Biologie der Flußperlmuschel *Margaritifera margaritifera* (L.). Naturwissenschaften 73:225–233

Wächtler K, Dettmer R, Buddensiek V (1987) Zur Situation der Flußperlmuschel (*Margaritifera margaritifera* (L.)) in Niedersachsen: Schwierigkeiten, eine bedrohte Tierart zu erhalten. Ber Naturhist Ges Hannover 129:209–224

Yeager BL, Neves RJ (1986) Reproductive cycle and fish hosts of the rabbit's foot mussle, *Quadrula cylindrica strigillata* (Mollusca: Unionidae) in the upper Tennessee River drainage. Am Midl Nat 116:329–340

Yeager BL, Saylor CF (1995) Fish hosts for four species of freshwater mussels (Pelecypoda: Unionidae) in the upper Tennessee River drainage. Am Midl Nat 133:1–6

Yokely P (1972) Life history of *Pleurobema cordatum* (Rafinesque 1820) (Bivalvia: Unionacea). Malacologia 11:351–336

Young M, Williams J (1984a) The reproductive biology of the freshwater pearl mussel *Margaritifera margaritifera* (Linn.) in Scotland I. Field studies. Arch Hydrobiol 99:405–422

Young M, Williams J (1984b) The reproductive biology of the freshwater pearl mussel *Margaritifera margaritifera* (Linn.) in Scotland II. Laboratory studies. Arch Hydrobiol 100:29–43

Young M, Purser GJ, Al-Mousawi B (1987) Infection and successful reinfection of brown trout [*Salmo trutta* (L.)] with glochidia of *Margaritifera margaritifera* (L.). Am Malacol Bull 5:125–128

Zale A, Neves R (1982a) Fish hosts of four species of lampsiline mussels (Mollusca: Unionidae) in Big Moccasin Creek, Virginia. Can J Zool 60:2535–2542

Zale AV, Neves RJ (1982b) Reproductive biology of four freshwater mussel species (Mollusca: Unionidae) in Virginia. Freshwater Invertebr Biol 1:17–28

Ziuganov VV, Nezlin LP, Zotin AA, Rozanov AS (1990) Host-parasite relationships between glochidia of *Margaritifera margaritifera* (Margaritiferidae, Bivalvia) and species of fish from the European North of the U.S.S.R. Parazitologija 24:315–321

Appendix 11.1. Calculation of glochidial mortality of *Margaritifera margaritifera* from literature data for the time period between parental release and successful encapsulation on a fish host (*Salmo trutta*)

1. Relationship between the (log) density of parasitising glochidia per 10 m^2 creek area (Y) and the (log) number of adult clams (X): $\log Y = \log X - 0.6$ ($n = 13$ creeks, $r^2 = 0.94$, $P < 0.001$; Bauer 1991)
2. $\log Y / \log X = -0.6$
3. $Y/X = 0.25$
4. $Y = 0.25 \times X$

Assuming that in a hypothetical creek (1 m wide) 32% of all clams are gravid with an average of 4 million glochidia (Bauer 1987a) and that glochidia drift over distance D, the corresponding mortality rates can be calculated as:

5. total glochidial production (N): $N = 4 \times 10^6 \times 0.32 \times X$
6. number of parasitizing glochidia (n): $n = 0.025 \times X \times D$
7. mortality rate $n/N = 0.025 \times X \times D / 4 \times 10^6 \times 0.32X$
   $= 0.000000019 \times D$

For drift distances of 10, 50, 100 and 200 m (see Sect. 2.2), mortality rates (%) are:

10 m: 99.99998
50 m: 99.99991
100 m: 99.99981
200 m: 99.99962

12 Macrohabitat Factors Influencing the Distribution of Naiads in the St. Croix River, Minnesota and Wisconsin, USA

DANIEL J. HORNBACH

12.1 Introduction

Freshwater mussels (superfamily Unioniacea) are widely distributed throughout North America and are important components of the biodiversity of streams and large river systems. There are currently 297 species and subspecies of unionid mussels living in North America but, of those, 44 species are federally endangered and another 69 species may be endangered (Fish and Wildlife Service 1991; Bogan 1993; Neves 1993; Williams et al. 1993). Approximately 19 species of freshwater mussels in North America are presumed extinct. Unionid mussels are ecologically significant because of their diversity and, as filter-feeders, they are important in processing seston and as a trophic link to higher predators (Vannote et al. 1980). Mussels are one of the few organisms that can process the fine particulate organic matter that makes up a large portion of the energy basis in riverine systems (Wallace et al. 1977). Unionids are also considered important biological monitors (McMahon 1991).

Despite the importance of mussels in aquatic systems and the threat of continued loss of biodiversity within this group, little is known about the factors that regulate their distribution and abundance. Raffaelli et al. (1992) suggested that a primary need for understanding and prediction in aquatic systems is the elucidation of the mechanisms that generate biologically significant patterns at different spatiotemporal scales. Hildrew and Giller (1992) point out that population and community structure and diversity gradients span a large range of spatial and temporal scales.

At the largest spatial scales, broad zoogeographical factors such as glaciation and geographic isolation are fairly well established in describing major differences in mussel community composition in North America (Baker 1928; van der Schalie 1945; Johnson 1970, 1980; Graf 1997). For example, Watters (1992, 1993) showed that unionid species diversity can be positively correlated with drainage area. Since the slope of this relationship is the same for fish and mussels, he concludes that the diversity of mussels for a given river depends on the diversity of fishes.

At a smaller spatial scale, several studies have examined the role of various microhabitat characteristics on the distribution and abundance of unionids. These factors include water velocity (Fuller 1974; Horne and McIntosh 1979; Salmon and Green 1983; Way et al. 1989; Holland-Bartels 1990), depth (Brönmark and Malmquist 1982; Stern 1983; Doolittle 1988; Strayer 1983; Johnson 1995), and substrate type (Harman 1972; Brönmark and Malmquist 1982; Vannote and Minshall 1982; Salmon and Green 1983; Strayer and Ralley 1991). Green (1971) suggested that trophic factors, rather than physical or chemical factors, are most important in accounting for among-species variance in bivalves (unionids and fingernail clams). Supporting this hypothesis, Bauer et al. (1991) found *Margaritifera margaritifera* in areas of low food

Ecological Studies, Vol. 145
G. Bauer and K. Wächtler (eds.) Ecology and Evolution of the Freshwater Mussels Unionoida

availability compared to two species of unionids found in areas of higher food availability. They explained this difference in distribution by *M. margaritifera* having lower metabolic rates than the two species of unionids. Strayer and Ralley (1993), however, indicated that the predictive power of discriminant models based on microhabitat variables is so low that they question the adequacy of a traditional microhabitat approach to unionid ecology.

While many studies have examined zoogeographic factors and microhabitat factors, fewer studies have explicitly examined intermediate levels of scale between the large geographic range and the microhabitat range. One example of studies conducted at the macrohabitat level is that of Vaughn (1993). She showed that at the scale of an individual river, species distributions are bimodal with a group of abundant "core" species and a group of rare species, whereas at the scale of a major river drainage the distributions are unimodal, with increased numbers of rare species and fewer core species. Strayer (1993), examining unionid populations in streams of the North Atlantic slope, showed that stream size, stream gradient, hydrologic variability, calcium concentration, physiographic province and the presence or absence of a tide had some predictive power in describing the species composition of mussel assemblages. However, only stream size and tidal influence were effective predictors of mussel distributions. DiMaio and Corkum (1995) examined the macrohabitat characteristic of hydrological variability and found that specific mussel assemblages were related to different flow regimes. Morris and Corkum (1996) found distinct relationships between unionid assemblage structure and the nature of the riparian zone surrounding the river. Haag and Warren (1998) emphasized the importance of reproductive strategies of mussels, the relationship between mussels and their fish hosts, and stochastic processes that may influence these relationships, in structuring mussel communities.

This chapter examines the distribution of mussels in the St. Croix River, Minnesota and Wisconsin, USA. This river harbors dense and diverse assemblages of unionid mussels. While studies have been conducted by various researchers to examine the distribution of mussels in the St. Croix, few have been published. I examined this large set of unpublished data to examine whether particular mussel species assemblages are associated with macrohabitat phenomenon such as location of tributaries and river gradient.

12.2 Materials and Methods

12.2.1 Study Location

The St. Croix River is a small sixth-order stream approximately 276 km long with a 20000-km^2 drainage area (Fago and Hatch 1993; Fig. 12.1). This river forms much of the boundary of the states of Minnesota and Wisconsin and empties into the Mississippi River south of Minneapolis, Minnesota. Historically, the river supported 41 species of mussels, two of which are on the US federally endangered species list (*Quadrula fragosa* and *Lampsilis higginsi*). A number are also on the state endangered species list for either Minnesota or Wisconsin. All of the species are representative of the upper Mississippi River mussel species assemblage (Fuller 1980). In 1968–1972 the St. Croix River was designated a National Wild and Scenic River by the US Congress.

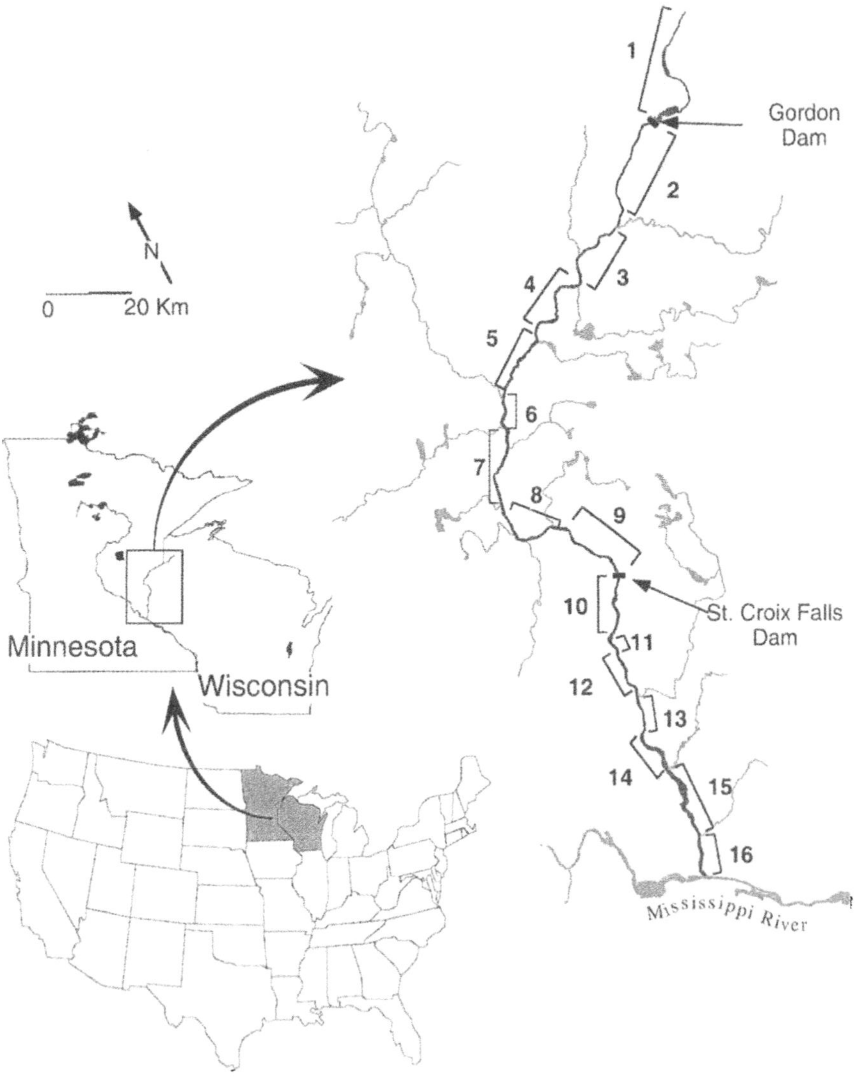

Fig. 12.1. Map of the study area, St. Croix River, MN and WI. River sectors are based on changes in river gradient (Fig. 12.2) and the location of major tributaries

The physiography of the St. Croix watershed allows it to be partitioned into several discrete units. The hierarchy of these units makes it attractive for examining spatial scale phenomena. To examine the distribution of macroinvertebrates, fish, and changes in water quality along the length of the river, James Perry and his students at the University of Minnesota (Holmberg et al. 1997) divided the St. Croix River into 16 major river segments based on stream gradient (Fig. 12.2) and the location of major tributaries (Fig. 12.1). This a priori categorization of the river was used to examine the changes in mussel assemblages along the length of the river at a fairly large scale of organization.

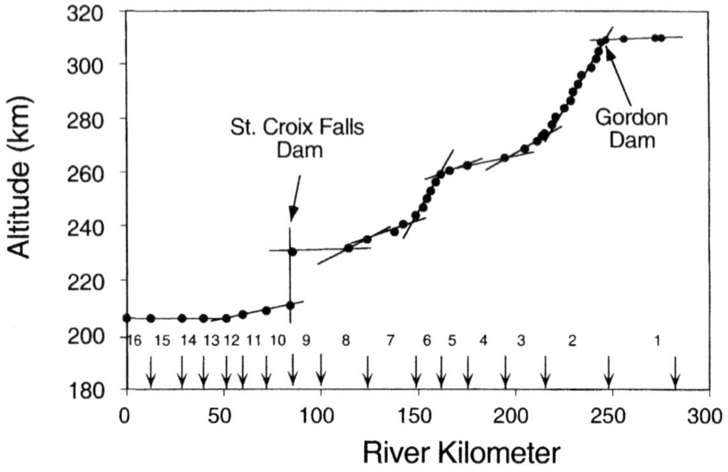

Fig. 12.2. Relationship between river kilometer and elevation of the St. Croix River, MN and WI. *Lines* show major changes in river gradient. *Numbers at the bottom* indicate the river sectors used in this chapter

12.2.2 Data Sources

There have been many studies of the mussels of the St. Croix Drainage. In the mainstem of the river studies have been conducted by Fuller (1980), Busacker (1988), Doolittle (1988), Heath (1989), Havlik (1993), Dunn (1994), Hove (1995), Doolittle and Heath (1997) and Hornbach (1991, 1992, 1995a,b, 1998), Hornbach and Deneka (1996), and Hornbach et al. (1994, 1995a,b, 1996, 1997). These studies had different goals and employed a variety of sampling techniques. Most of these reports were supplied to various state and federal government agencies and the raw data can be obtained from the Minnesota Department of Natural Resources, St. Paul, MN, the National Park Service, St. Croix National Riverway, St. Croix Falls, WI and the Wisconsin Department of Natural Resources, Madison WI. In the dataset are included only studies in which all mussel species in a given area were collected. If the studies were designed to collect only particular species, for example in searches for federally endangered mussels, they were not included in the dataset.

12.2.3 Data Analysis

Species names follow Turgeon et al. (1988) and subfamilies follow Cummings and Mayer (1992). Various statistics were calculated concerning the mussel assemblages in each of the river sectors including: total number of mussels collected, species richness (number of species), species diversity (Shannon-Wiener diversity index [$H' = -\sum_{i=1}^{s} p_i \log(p_i)$, where p_i is the proportion of species i and s is the total number of species)] and the evenness index $\left(E = \frac{H'}{\log(s)}\right)$ (Zar 1984). To further examine changes in mussel assemblages among the river sectors, a correspondence analysis was conducted (SAS Institute 1994), a technique that allows for the examination of patterns of counts in contingency tables. In this analysis, the contingency table constructed was of the number of individuals of each species (rows) found within each river sector (columns). The canonical coefficients found through correspondence analysis can be used to perform a cluster analysis allowing for an examination of the relationships among either species or river sectors (Thioulouse 1989, 1990). This analysis used a hierarchical cluster analysis calculating the average distance among clusters (SAS Institute 1994).

12.3 Results

The datasets used in this analysis included 46 140 mussels representing 40 species of mussels (Table 12.1). Two species of mussels that have been found in the St. Croix River (*Venusticoncha ellipsiformes* and *Arcidens confragosis* – Bell Museum of Natural History, University of Minnesota, St. Paul, MN) were not included in any of the studies used to construct the dataset for this chapter (Table 12.1). Because the number of species found in a sector is largely explained by the number of individuals collected (Fig. 12.3) much of the variability in the number of species collected (Table 12.1) is most likely due to differences in sampling effort. Consequently, considerable care must be taken when examining the variation among mussel assemblages in various parts of the river. Seven species of mussels had very broad distributions, being found in 15 or 16 of the river sectors, while nine species had very restricted distributions, being found in only one or two (Table 12.1).

Mussel species diversity varied among the river sectors (Fig. 12.4), with the lowest diversity in sectors 5, 6, and 14. Part of this variability could be related to the difference in sampling effort among river sectors (Table 12.1) which produced the difference in the number of species found among sectors (Fig. 12.3); but the diversity index

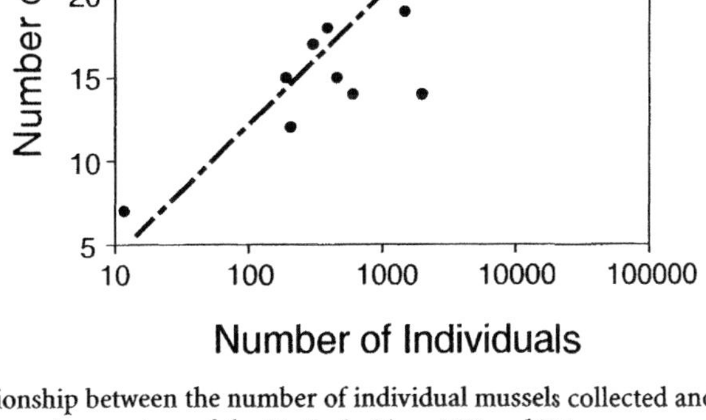

Fig. 12.3. Relationship between the number of individual mussels collected and the number of species found in various sectors of the St. Croix River, MN and WI

Table 12.1. Number of mussels found in various sectors of the St. Croix River, MN and WI

Species	River sector																Total no. of mussels	Total no. of locations
	1	2	3	4	5	6	7	8	9	10	11	12	13	14	15	16		
Actinonaias ligamentina	521	186	226	671	155	329	120	1803	2	355	7	4	0	2	31	42	4 454	15
Alasmidonta marginata	13	19	3	30	1	5	1	51	0	45	0	0	1	0	5	1	175	12
Amblema plicata plicata	3	71	7	109	1	2	10	34	0	142	21	9	152	3223	6 492	2654	12 930	15
Anodontoides ferussacianus	22	0	0	0	0	0	0	0	0	0	0	0	0	0	0	0	22	1
Cumberlandia monodonta	0	0	0	0	0	0	20	41	0	6	1	0	0	0	0	0	68	4
Cyclonaias tuberculata	1	7	8	64	3	5	12	208	0	162	0	0	0	0	1	10	481	11
Ellipsaria lineolata	0	0	0	0	0	0	0	0	0	106	0	0	0	0	29	13	148	3
Elliptio crassidens crassidens	0	0	0	0	0	0	0	0	0	0	0	0	0	0	11	72	83	2
Elliptio dilatata	474	43	40	196	21	32	76	1462	3	279	27	8	0	19	416	2652	5 748	15
Epioblasma triquetra	0	0	2	0	0	0	0	0	0	238	9	0	0	0	0	0	249	3
Fusconaia ebena	0	0	0	0	0	0	0	0	0	0	0	0	0	0	0	27	27	1
Fusconaia flava	205	28	22	103	3	7	7	108	0	1 481	96	118	214	554	1 273	662	4 881	15
Lampsilis cardium	226	8	6	26	2	7	6	125	0	92	9	3	7	19	106	46	688	15
Lampsilis higginsi	0	0	0	0	0	0	0	0	0	6	0	0	0	0	62	53	121	3
Lampsilis siliquoidea	245	137	5	59	9	10	3	186	1	30	3	0	0	53	92	191	1 024	14
Lasmigona complanata	0	0	0	0	0	0	0	1	0	16	1	0	46	8	18	4	94	7
Lasmigona compressa	0	0	0	0	0	0	0	2	0	3	0	0	0	0	0	0	5	2
Lasmigona costata	17	52	10	50	3	26	2	174	1	14	0	0	0	0	3	0	352	11
Leptodea fragilis	0	0	1	4	0	0	1	36	0	143	12	3	11	32	49	28	320	11
Ligumia recta	31	9	38	30	2	12	5	66	0	31	0	0	1	0	8	53	286	12
Megalonaias nervosa	0	0	0	0	0	0	0	0	0	0	0	0	0	0	13	68	81	2
Obliquaria reflexa	0	0	0	0	0	0	0	10	0	123	5	4	162	158	896	780	2 138	8
Obovaria olivaria	0	0	3	18	0	1	19	104	0	162	8	0	3	1	3	0	322	10
Plethobascus cyphyus	0	0	0	0	0	0	0	0	0	0	0	0	0	0	0	3	3	1
Pleurobema sintoxia	0	9	3	20	0	0	4	83	0	214	6	3	5	17	222	577	1 163	12
Potamilus alatus	0	0	0	4	0	2	1	24	0	138	20	12	19	34	124	66	444	11
Potamilus ohioensis	0	0	0	0	0	0	0	0	0	5	1	1	12	4	1	0	24	6
Pyganodon grandis	55	10	0	1	0	1	0	4	0	6	0	1	19	133	104	153	487	11
Pyganodon imbecillis	0	0	0	0	0	0	0	2	0	1	0	0	2	38	10	3	56	6
Pyganodon suborbiculata	0	0	0	0	0	0	0	0	0	0	0	0	0	0	0	15	15	1
Quadrula fragosa	0	0	0	0	0	0	0	0	0	33	0	0	0	0	0	0	33	1
Quadrula metanevra	0	0	0	0	0	0	0	0	0	405	12	1	12	0	64	47	541	6
Quadrula pustulosa	1	1	7	55	1	1	9	289	1	546	24	10	227	135	317	194	1 818	16
Quadrula quadrula	0	0	0	0	0	0	0	0	0	4	0	0	0	8	72	9	93	4
Simpsonaias ambigua	0	0	0	4	0	0	0	0	0	0	0	0	0	0	0	0	4	1
Strophitus undulatus	158	5	1	7	1	8	3	66	1	8	3	0	5	3	41	29	339	15
Toxolasma parvus	0	0	2	0	0	0	0	4	0	14	0	0	4	1	13	24	62	7
Tritigonia verrucosa	0	0	0	0	0	0	0	0	0	167	11	5	6	0	17	2	208	6
Truncilla donaciformis	0	0	1	0	0	0	0	0	0	393	1	0	15	7	28	1	446	7
Truncilla truncata	0	0	0	3	0	0	0	199	3	5 105	16	5	53	15	219	819	5 707	10
Total no. of mussels	1972	585	385	1454	202	448	299	5082	12	10 473	293	187	976	4464	10 740	8568	46 140	
Total no. of species	14	14	18	19	12	15	17	24	7	33	21	15	21	21	31	31		

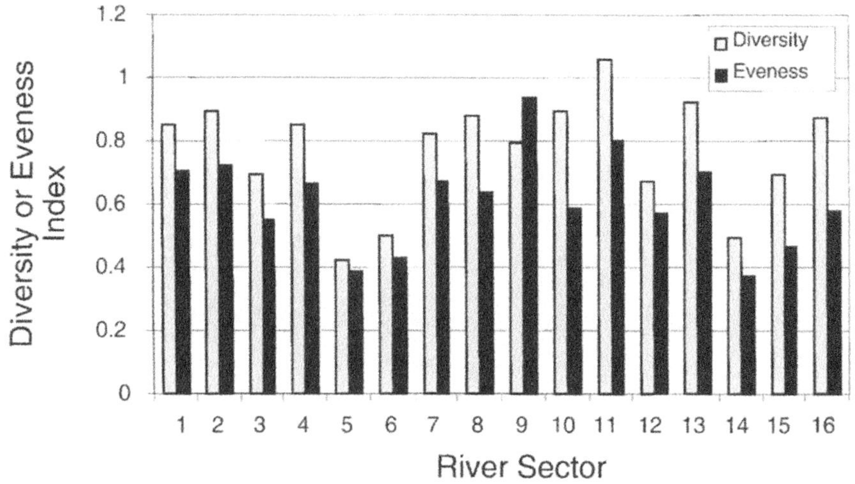

Fig. 12.4. Variation in species diversity (Shannon-Wiener index) and evenness in various sectors of the St. Croix River, MN and WI

is a function not only of the number of species, but also of the evenness of the distribution of these species. Since evenness also varied among sectors (Fig. 12.4), factors other than sampling effort may be responsible for the differences in species diversity among river sectors. There was no significant correlation between Shannon-Wiener diversity index and the number of mussels collected (Spearman rank correlation = 0.25, $P = 0.35$).

Figure 12.5 shows the results of a correspondence analysis with rows and columns in a crosstabs table corresponding to species and river sectors, respectively. In correspondence analysis the inertia value reflects the variation accounted for in the canonical dimensions. In this analysis the first three inertia values are 0.58, 0.51, and 0.16, accounting for 85% of the total inertia. Figure 12.5 is based on values derived from the first two canonical dimensions. Species or river sectors that have similar profiles in a crosstabs table are represented by being close on the graph. The distance between species points on the one hand and river sector points on the other has no meaning, but the direction from the origin has meaning. For example, both river sector 10 and the species *Quadrula fragosa* and *Truncilla truncata* are found in the upper left quadrant of Fig. 12.5. All of the 33 federally endangered *Q. fragosa* and 89% of the 5707 *T. truncata* in the dataset are from sector 10 (Table 12.1).

The cluster analyses, based on the first three canonical scores are shown in Fig. 12.6 (clustering of the river sectors) and Fig. 12.7 (clustering of species). Figure 6 shows that the river displayed two major clusters based on the species of mussels found. One cluster is found above the 18-m-high dam at St. Croix Falls, WI. This dam is at a site where historically there was a 10-km stretch of steep rapids (1.81 $m\,km^{-1}$ gradient: Fago and Hatch 1993). The cluster above the dam is broken into two subclusters, sectors 2–6 and sector 1 with sectors 7–9. Sectors 2–6 are found above the Kettle River rapids, that has an average gradient of 1.05 $m\,km^{-1}$ (Fago and Hatch 1993) and in areas of relatively high stream gradients (Fig. 12.2). Sectors 1 and 7–9 are of lower gradient and include areas of the river that are part of the reser-

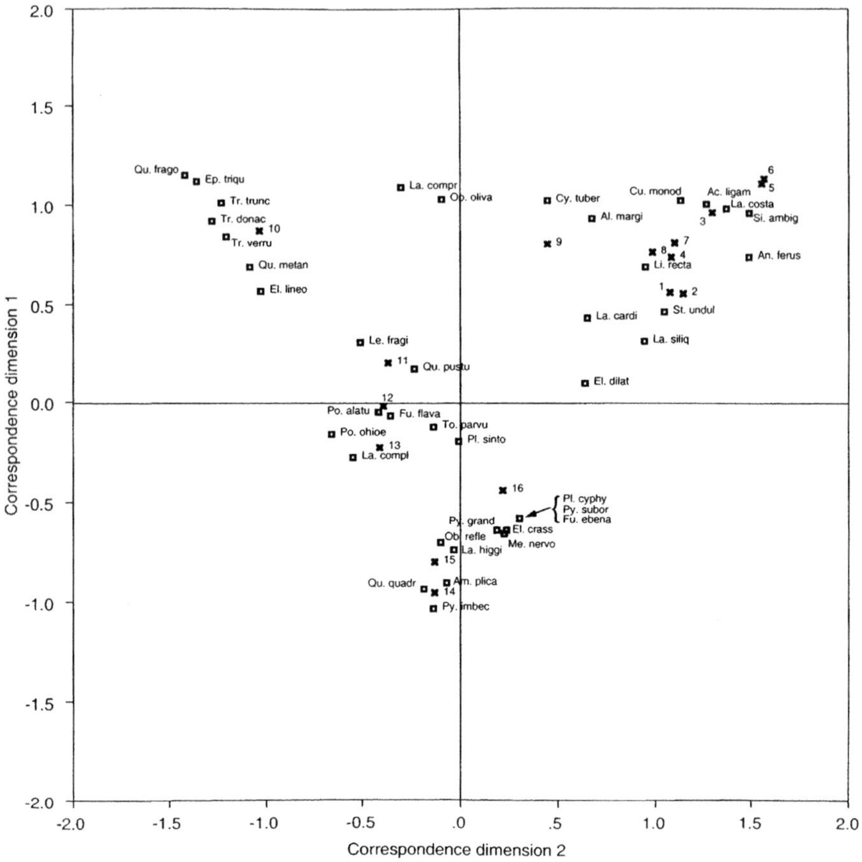

Fig. 12.5. Correspondence analysis of the relationship among the species of mussels found and the river sectors of the St. Croix River, MN and WI. *Abbreviations* are based on the first two letters of the genus and first five letters of the species name from Table 12.1. *Squares* species values and *x* represent sector values

voir behind the hydroelectric dam at St. Croix Falls or the Gordon dam (Fig. 12.1). In the cluster below the St. Croix Falls dam there are also two subclusters (Fig. 12.6). These include sectors 10–13 that have a moderate gradient and are riverine (Fig. 12.2) and sectors 14–16 that include the more lacustrine, Lake St. Croix (Fago and Hatch 1993).

There are two major clusters of species within the river (Fig. 12.7). The upper cluster (A) includes two subclusters. The upper subcluster (A1) includes species that have >60% of the individuals (average was 79.2%) in sectors 1–9, above the dam at St. Croix Falls. In this subcluster one species belongs to the subfamily Ambleminae, four are in the subfamily Anodontinae, one in the subfamily Cumberlandinae, and the rest in the subfamily Lampsilinae. There are three groupings within this subcluster. One of these (A1a) includes two species (*A. ligamentina* and *L. costata*) that are found in all sectors but have their highest relative abundances in sector 8. The other two are associated with either high stream gradient (A1b – sectors 2–6) or low stream gradi-

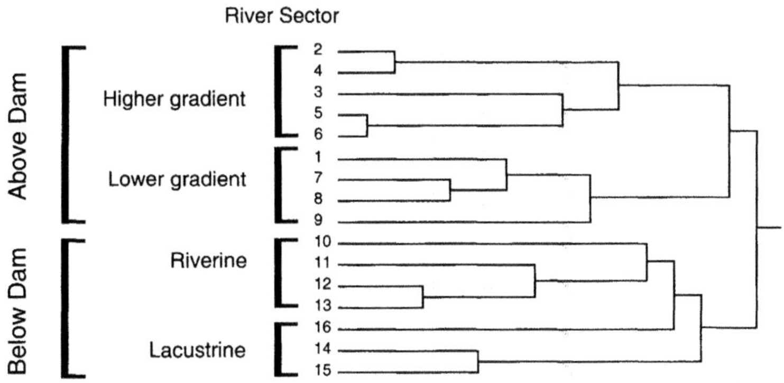

Fig. 12.6. Cluster analysis of river sectors of the St. Croix River, MN and WI, based on a correspondence analysis

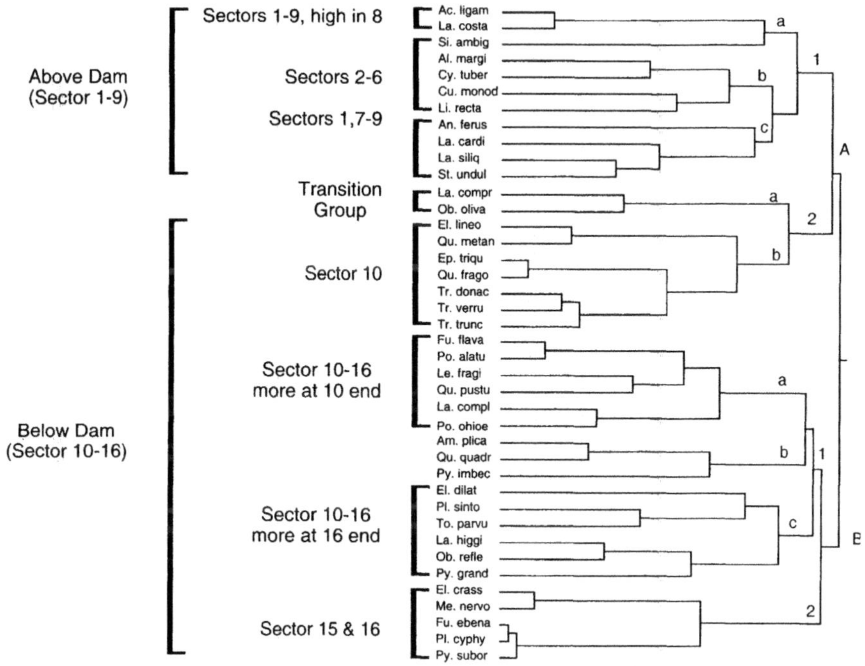

Fig. 12.7. Cluster analysis of mussel species in the St. Croix River, MN and WI, based on a correspondence analysis. *Abbreviations* are based on the first two letters of the genus and first five letters of the species name from Table 12.1. *Letters on the branch points* are referred to in the text

ent (A1c – sectors 1 and 7–9). For four species, 90% or more of the individuals collected were taken from above the dam (Table 12.1 – *A. ferussacianus*, *A. ligamentina*, *C. monodonta*, *L. costata* and *S. ambigua*). The lower subcluster (A2) contains two groups, one (A2a) with two species, *L. compressa* and *O. olivaria*, that are found in similar relative abundances above and below the dam (40 and 56% found

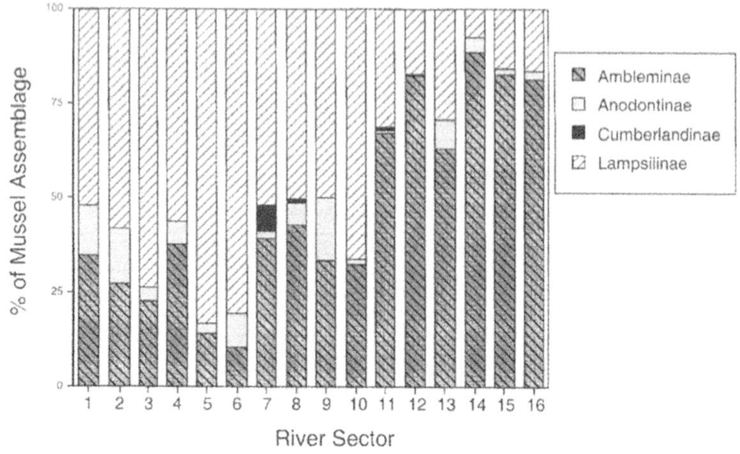

Fig. 12.8. Variation in mussel subfamilies in various sectors of the St. Croix River, MN and WI

above the dam, respectively). However, below the dam >90% of the individuals from these two species were found in sector 10. All of the other species linked in the subcluster with *L. compressa* and *O. olivaria* (A2b) had at least 96% of the individuals found in sector 10. Consequently, the lower subcluster of the upper cluster in Fig. 12.7 links the species found predominately above the dam with species found in sector 10. The second major cluster (B) includes species that had >60% of their individuals (average was 93.5%) from below the dam at St. Croix Falls (Table 12.1). There are two major subclusters within this cluster. The lower subcluster (B2) includes five species (*E. crassidens crassidens, M. nervosa, F. ebena, P. cyphyus*, and *P. suborbiculata*) which are quite rare and found only in the lower reaches of the river (all five species together account for about 0.5% of all mussels collected). The other subcluster (B1) below the dam is further subdivided into groups representing a gradation of species found from sector 10 downstream to sector 16 (B1a – B1c).

Changes observed in the mussel assemblage from the headwaters to the mouth of the St. Croix River (Table 12.1; Fig. 12.7) are also reflected in changes in the relative proportions of mussels found in the various subfamilies of unionoids (Cummings and Mayer 1992). There was a dominance of lampsilines in the upper reaches of the river and a switch to dominance by amblemines in the lower reaches of the river (Fig. 12.8).

12.4 Discussion

Within the St. Croix River there are clearly defined clusters of unionid species at the km-level scale of examination used in this study (Figs. 12.5, 12.6, and 12.7). These clusters are found associated with river sectors that Holmberg et al. (1997) established using an a priori classification based on stream gradient and location of major tributaries. Many studies use microhabitat characters to explain the distribution of mussels in river systems (e.g., Harman 1972; Horne and McIntosh 1979; Brönmark and Malmquist 1982; Vannote and Minshall 1982; Salmon and Green 1983; Strayer

1983; Way et al. 1989: Holland-Bartels 1990; Strayer and Ralley 1991; Johnson 1995). However, Strayer and Ralley (1993) suggested that while they could account for the presence and absence of some species based on microhabitat characters, the predictive power was low. From this, they suggested that habitat factors working at the spatial scale of hundreds to thousands of meters, such as that used in this study, might be important.

In the St. Croix River the major influence on the distribution of mussels appears to be the presence of the dam/historic falls at St. Croix Falls (Figs. 12.6 and 12.7). The river above the dam is narrower and shallower and has a much higher gradient than the river below the dam (Fig. 12.2). Strayer (1983) found that macrohabitat characteristics such as stream size and surface geology were good predictors of mussel assemblages. Despite the high predictive power, he found that these macrohabitat characters accounted for little of the variability in species richness. Similarly, in this study, mussel assemblage diversity and evenness varied considerably (Fig. 12.4). Clearly, little of this variability can be explained by river sector and consequently macrohabitat characters used to define river sector apparently have little influence on diversity and evenness.

Strayer (1993) found that five macrohabitat characters (stream size, hydrologic variability, calcium concentration, physiographic province, and the presence/absence of a tide) were important predictors of the presence or absence of various species of mussels in Atlantic slope streams of the US. In the St. Croix River, it is unlikely that stream size (based on drainage area) accounts for differences in the mussel assemblages above and below the dam. Of the 20000-km^2 drainage for the entire watershed, 81% is found above the dam (Fago and Hatch 1993). DiMaio and Corkum (1995) supported Strayer's (1993) study and showed that differences in mussel assemblages are related to the hydrologic variability of a stream. In the St. Croix River, the annual discharge is fairly stable over periods of 20–90 years (Fago and Hatch 1993). The catchment area in the upper reaches of the river has highly permeable stratified and gravel soils reducing the impact of spate events. Further, during periods of low precipitation, groundwater provides a counterbalance-stabilizing flow. Consequently, it seems unlikely that hydrologic variability is responsible for the differences in mussel assemblages above and below the dam. Morris and Corkum (1996), like Strayer (1993), found that physiographic province, as reflected through grassy or forested riparian zones, explained differences in mussel species assemblages among streams. Again, since the St. Croix River has a forested riparian zone throughout its length (Fago and Hatch 1993), it is unlikely that variation in the riparian zone accounts for differences in mussel assemblages. Of the five macrohabitat characters that Strayer implicated as major factors accounting for the variations in mussel assemblages, only calcium content differs above and below the dam in the St. Croix River. Although most of the bedrock is overlain with glacial till and outwash, there is an increase in carbonate content from upstream to downstream, resulting in an increased calcium content and pH in downstream regions (Graczyk 1986).

Strayer (1993) also investigated stream gradient as a macrohabitat factor that might be important in explaining mussel assemblage structure and found it to be a poor predictor of mussel assemblage structure. In the St. Croix River, however, the differences in the gradient above and below the dam could help explain the differences in the assemblages. It may be that the gradient at the dam, and historically at the falls, is so great that it has influenced the distribution of fish hosts between these major sectors of the river. Thirty-eight of the 40 species of mussels are found below the dam (Table 12.1). There were two species in the dataset not recorded below the dam (*Anodontoides ferussaciancus* and *Simpsonaias ambigua*). *Simpsonaias ambigua* has been found below the dam (Bell Museum of Natural History, University of Minnesota, St. Paul, MN) and *A. ferussaciancus* is found in other tributaries of the upper Mississippi River (Cummings and Mayer 1992) and thus their absence below the dam is probably not significant. There were 11 species not found above the dam, and it is possible that restricted fish dispersal could be responsible for their absence.

Other studies focusing on macrohabitat factors that influence mussel assemblages have generally been conducted in areas with relatively low mussel density and assemblage diversity. The St. Croix River has a very diverse mussel assemblage with relatively high mussel density. Many locations harbor assemblages of 20–30 species (Table 12.1) with numerous locations having densities >20 mussels m^{-2} and some locations having densities of up to 200 mussels m^{-2} (Hornbach and Deneka 1996; Hornbach et al. 1996; unpubl. data). Because of the high population density and assemblage richness it is possible that other factors not found to be significant in other studies could be important in the St. Croix. Competition for fish hosts, food, and habitat could influence species distributions at high density but few studies have been conducted on the role of these factors in mussel abundance and distribution (Bauer et al. 1991).

In this study some species had wide distributions, being found in most river sectors while others had narrow distributions (Table 12.1). Possible explanations for this difference include chance, differences in microhabitat requirements, and fish host distribution. Some species in the St. Croix with restricted distributions include the endangered species *Lampsilis higginsi* and *Quadrula fragosa*. Interestingly, species from the same genera, *L. cardium* and *Q. pustulosa*, are widely distributed (Table 12.1). Another restricted species, *Fusconaia ebena*, is rare in the upper Mississippi drainage presumably because its fish host, the skipjack herring (Watters 1994b), is restricted from its normal migration to the Gulf of Mexico by the navigation dams on the upper Mississippi (Fago and Hatch 1993). However, other species with restricted distributions in the St. Croix, e.g., *M. nervosa* and *Q. quadrula* (Table 12.1), have wide distributions and abundant populations in the upper Mississippi River (Hornbach et al. 1992) and thus their distribution in the St. Croix most likely has to do with limitations of suitable habitat.

The mussel assemblages above and below St. Croix Falls exhibit a shift from amblemines below the dam to lampsilines above the dam (Fig. 12.8). This shift does not appear to be due to an inherent difference in the number of widely or narrowly distributed species within subfamilies. On average all subfamilies (except the Cumberlandinae with only one species) had species found in seven to nine river sectors and there was no significant difference among subfamilies in the number of locations in which they were found (Anova, $F = 0.55$, 3.38 df, $p = 0.65$). Other aspects of the biology of lampsilines and amblemines must be examined in order to explain the shift in the dominant mussel subfamily with stream sector. Lampsilines and amblemines may differ in many respects including reproduction, fish host specialization, and shell sculpture although there is great variability in these factors within the subfamilies as well.

Lampsilines tend to have unsculptured shells, while most amblemines are sculptured. Watters (1994a) suggested that there is a tradeoff in these two sculpture types. Shell sculpturing helps in anchoring but it costs in terms of burrowing. Watters (1994a) also noted that nonsculptured forms are found more often in headwaters, and this is consistent with the findings in this study with more lampsilines upstream (Fig. 12.8). Strayer and Ralley (1993) indicated that a lack of sandy patches may limit mussel distribution in rocky substrates. The substrate upstream of the dam is, on average, coarser and more variable than that downstream (D.J. Hornbach, unpubl. data). Thus, if substrates are limiting, reburying may be important and responsible for an increased number of nonsculptured lampsilines in areas upstream with high gradients (Fig. 12.2).

Bauer et al. (1991) suggested that the lower metabolic rate of *Margaritifera margaritifera* may allow it to survive in areas of low productivity compared to other species of unionids. While such physiological information is not available for many mussel species, Baker and Hornbach (1997) compared the seasonal rates of metabolism and feeding in *Amblema plicata plicata* (an amblemine) and *Actinonaias ligamentina* (a lampsiline). *Actinonaias ligamentina* had higher weight-specific grazing rates and higher weight-specific metabolic rates than *A. p. plicata*. *Amblema p. plicata* had a higher condition index than *A. ligamentina*. *Amblema p. plicata* is found in areas of the St. Croix River subjected to higher levels of suspended solids (Fago and Hatch 1993), which would result in poor food quality. The slower metabolic and feeding rates may allow *A. p. plicata* to thrive in these areas. If this is true for other amblemines, it may help to explain their success in the lower reaches of the St. Croix River.

Kondo (1989, 1993) said that summer brooders like amblemines tend to be host-specific, whereas winter brooders (long-term) tend to be less host-specific. Neves et al. (1985) also suggested that amblemines, found predominately on cyprinid fish, tend to exhibit a great degree of host-fish specificity (Neves and Widlak 1988). Fago and Hatch (1993) stated that there are 18 species of fish found above but not below the dam at St. Croix Falls. Of these, 11 are cyprinids. From this, one might hypothesize a greater number of amblemines above the dam but this was not so (Fig. 12.8). However, differences in fish density or fish infestation rates might explain the variation in subfamily distribution that does not correlate with fish host presence. For example, most amblemines release conglutinates (packets of glochidia). Lampsilines, on the other hand, release clouds of free glochidia when struck by a host, but may release conglutinates if no host if present (G.T. Watters, pers. comm.). This disparity may result in differential fish infestation rates. Thus, differences in fish host specificity, fish host density, and reproductive habits may influence the distribution of the subfamilies of mussels in the St. Croix. Haag and Warren (1998) found that patterns of mussel community variation were correlated with fish host distribution. They indicated that host specialists which do not have modified mantle structures to attract fish (i.e., many amblemines) are often restricted to large streams while host generalists (i.e., many anodontines) and host specialists with displaying structures (i.e., many lampsilines) are often found in headwater areas. Their observations are consistent with the findings in this study.

Lee and DeAngelis (1997) developed a model for mussel invasion into new habitats. They show that if the age of maturity of a species is delayed, then the velocity of the wavefront of invasion is lower. Lampsilines generally have faster growth rates than amblemines, and thus one may hypothesize that they also have a younger age of maturity than amblemines. If this is true, they may have had the ability to more rapidly invade the upstream areas of the St. Croix River after glaciation. Barriers to migration that arose later, such as the St. Croix Falls, may then have arisen and prevented further migration of slower-migrating species. This is consistent with the study of Graf (1997), who showed that the mussel fauna above the St. Anthony Falls on the Mississippi River includes more lampsilines than amblemines. If the St. Anthony Falls and St. Croix Falls act similarly as reinvasion barriers, then the model of Lee and DeAngelis may provide a plausible explanation for the difference in assemblage structure. However, although amblemines do not dominate assemblages upstream of the dam, many species are represented in this portion of the river. This argues for microhabitat or macrohabitat factors rather than differential invasion as the reasons for variation in species assemblages along the length of the river.

This study reinforces the ideas of Strayer (1993), DiMaio and Corkum (1995), and Morris and Corkum (1996) that macrohabitat factors may be important in explaining variation in mussel assemblages. In the St. Croix River, the presence of a migration barrier appears to be a major factor in explaining mussel assemblage variability. The differential dominance of subfamilies in these areas requires further study, but may be related to differences in reproductive strategies of various species of mussels, as suggested by Haag and Warren (1998). Within areas above or below this barrier, stream gradient seems to play a secondary role in explaining the noted variance. Therefore, the effects of stream gradient on assemblage structure also requires additional examination.

12.5 Summary

Many factors influence the distribution of naiads in rivers. These factors act at a variety of scales from microhabitat to macrohabitat. Over the past 20 years, the mussel resources in the St. Croix River, a Wild and Scenic River in the US, have been extensively studied. Some of these studies were focused on conservation efforts, while others were related to basic research or mussel relocation associated with construction projects. The results of these unpublished projects are used to examine if macrohabitat factors, including stream gradient and the presence of major tributaries, influence mussel assemblages in the St. Croix River. Much of the upper portion of the river is a relatively shallow, fast-flowing stream, while the lower portion of the river, near the confluence with the Mississippi River, is a deep lake-like area. There is a region of geologic discontinuity between the upper and lower regions of the river that historically contained a rapids and for the past 100 years has been occupied by a hydroelectric dam.

The studies reviewed examined over 46000 individuals representing 40 species of mussels. Ten of these are listed as either federal or state (Wisconsin and Minnesota) endangered or threatened species. Some stretches of the St. Croix River contained over 30 species of mussels. Based on the relative abundance of the species in the various sectors of the river a cluster analysis was conducted to test whether the community structure of mussels varies along the length of the river. Two major clusters were detected, one above the St. Croix Falls Dam and one below the dam. The dam may provide a barrier to fish migration and may explain the difference in community structure. The cluster below the dam was further divided into two larger clusters, one for the riverine portion and one for the lacustrine portion of the river (i.e., Lake St. Croix).

Mussels of the subfamily Ambleminae were more prevalent in the lower reaches of the river while members of the Lampsilinae were more prevalent in the upstream reaches. There are differences in reproductive habits and fish hosts between these two subfamilies, but the relationship between these factors and the distribution of these subfamilies is unclear. The changes in mussel assemblages along the length of the St. Croix River seem related to macrohabitat characters, including the presence of fish migration barriers and stream gradient.

Acknowledgements. I would like to that James Perry from the University of Minnesota for prompting me to undertake this analysis. Also, thanks to Jan Serie of Macalester College and Mark Hove of the University of Minnesota for reviewing early drafts of this chapter. Funding for this project was provided by the St. Croix National Scenic Riverway, National Park Service, through a contract with James Perry. Access to primary data sources was graciously provided by the Minnesota Department of Natural Resources with assistance by Rich Baker and Bonita Eliason.

References

Baker FC (1928) The fresh water Mollusca of Wisconsin: Part II Pelecypoda Bull Wisc Geol Nat Hist Surv, 70 University of Wisconsin, Madison, Wisconsin

Baker SM, Hornbach DJ (1997) Acute physiological effects of zebra mussel (*Dreissena polymorpha*) infestation on two unionid mussels, *Actinonaias ligamentina* and *Amblema plicata*. Can J Fish Aquat Sci 54:512–519

Bauer G, Hochwald S, Silkenat W (1991) Spatial distribution of freshwater mussels: the role of host fish and metabolic rate. Freshwater Biol 26:377–386

Bogan AE (1993) Freshwater bivalve extinctions (Mollusca: Unionoida): a search for causes. Am Zool 33:599–609

Brönmark C, Malmquist B (1982) Resource partitioning between unionid mussels in a Swedish lake outlet. Holarct Ecol 5:389–395

Busacker GP (1988) A biological assessment of *Lampsilis higginsi* in the St Croix River at Stillwater, Minnesota: possible impacts associated with replacing the St Croix River crossing between Stillwater, Minnesota and Houlton, Wisconsin. A study by the Minnesota Department of Transportation, Technical Services Division St Paul, Minnesota

Cummings KS, Mayer CA (1992) Field guide to freshwater mussels of the Midwest. Illinois Natural History Survey Manual 5. Champaign, Illinois

DiMaio J, Corkum LD (1995) Relationship between the spatial distribution of freshwater mussels (Bivalvia: Unionidae) and the hydrological variability of rivers. Can J Zool 73:663–671

Doolittle TCJ (1988) Distribution and relative abundance of freshwater mussels in the Saint Croix National Scenic Riverway Report to the Wisconsin Division of Natural Resources, Bureau of Endangered Resources, Madison, Wisconsin

Doolittle TCJ, Heath DJ (1997) Second sampling of freshwater mussel communities for long-term monitoring of the St Croix National Scenic Riverway, Minnesota and Wisconsin. Report to the US Department of the Interior, National Park Service, St Croix National Scenic Riverway, St Croix Falls, Wisconsin

Dunn HL (1994) List of species relocated from the I-94 bridge. Letter to Macalester College from Ecological Specialist, October 6, 1994. St. Peters, Missouri

Fago D, Hatch J (1993) Aquatic resources of the St Croix River Basin. In: Hesse LW, Stalnaker CB, Benson NG, Zuboy JR (eds) Proceedings of the symposium on restoration planning for the rivers of the Mississippi River Ecosystem. Biological Report 19. US Department of the Interior, National Biological Survey, Washington, DC 20240, pp 23–56

Fish and Wildlife Service (1991) Endangered and threatened wildlife and plants 50 CFR 1711 & 1712 July 15, 1991. US Fish and Wildlife Service, Washington, DC
Fuller SLH (1974) Clams and mussels (Mollusca: Bivalvia). In: Hart CW, Fuller SLH (eds) Pollution ecology of freshwater invertebrates. Academic Press, New York, pp 215–260
Fuller SLH (1980) Freshwater mussels (Mollusca: Bivalvia: Unionidae) of the Upper Mississippi River: observations at selected sites within the 9-foot navigation channel project for the St Paul District, United States Army Corps of Engineers, 1977–1979, vol I, II. Academy of Natural Science, Philadelphia, Pennsylvania, pp 215–260
Graczyk DJ (1986) Water quality in the St Croix National Scenic Riverway, Wisconsin. Open-file report. US Geological Survey, Madison, Wisconsin, 85–4319
Graf DL (1997) Northern redistribution of freshwater pearly mussels (Bivalvia: Unionoidea) during Wisconsin deglaciation in the southern glacial Lake Agassiz region: a review. Am Midl Nat 138:37–47
Green RH (1971) Multivariate statistical approach to the Hutchsonian niche: bivalve molluscs of central Canada. Ecology 52:543–556
Haag WR, Warren ML (1998) Role of ecological factors and reproductive strategies in structuring freshwater mussel communities. Can J Fish Aquat Sci 55:297–306
Harman WN (1972) Benthic substrates: their effect on fresh-water mollusca. Ecology 53:271–277
Havlik ME (1993) Unionids and margaritiferids (Mollusca: Bivalvia) in the St Croix River Bordering Minnesota at Afton and Wild River State Parks, 8–17 June 1992. Report to the Minnesota Department of Natural Resources, Nongame Wildlife Program, St Paul, Minnesota
Heath DJ (1989) Saint Croix River US highway 10 bridge freshwater mussel relocation project at Prescott, Wisconsin Phase I: Mussel removal. Report prepared for Ayres Associates, Madison, Wisconsin
Hildrew AG, Giller PS (1992) Patchiness, species interactions and disturbance in the stream benthos. In: Giller PS, Hildrew AG, Raffaelli DG (eds) Aquatic ecology: scale, pattern and process. Blackwell, Oxford, pp 21–91
Holland-Bartels LE (1990) Physical factors and their influence on the mussel fauna of a main channel border habitat of the upper Mississippi River. J N Am Benthol Soc 9:327–335
Holmberg KL, Perry J, Ferrin RS, Sharrow DL (1997) Water resources management plan: St Croix National Scenic Riverway. Report to the National Park Service, St Croix National Scenic Riverway, St Croix Falls, Wisconsin
Hornbach DJ (1991) Factors influencing the distribution of unionid mussels in the St Croix River at Franconia, MN. Report to the Minnesota Department of Natural Resources, Nongame Wildlife Program, St Paul, Minnesota
Hornbach DJ (1992) An examination of the population structure, community relationships and habitat characteristics for the winged mapleleaf mussel (*Quadrula fragosa*) at Interstate Park, Saint Croix River, Wisconsin and Minnesota. Report to the Wisconsin Department of Natural Resources, Bureau of Endangered Resources, Madison, WI and Minnesota Department of Natural Resources, Nongame Wildlife Program, St Paul, Minnesota
Hornbach DJ (1995a) The effect of water depth and velocity on mussel distributions in the St Croix River, Interstate Park. A report prepared for the Winged Mapleleaf Recovery Team, Fish and Wildlife Service, US Department of the Interior, Twin Cities, Minnesota, July 27, 1995
Hornbach DJ (1995b) The effect of water depth and velocity on mussel distributions in the St Croix River, Interstate Park: an addendum. A report prepared for the Winged Mapleleaf Recovery Team, Fish and Wildlife Service, US Department of the Interior, Twin Cities, Minnesota, 15 August 1995
Hornbach DJ (1998) Temporal changes in the mussel population density at Interstate Park, St. Croix River, Minnesota and Wisconsin. Report to the National Park Service, St Croix National Scenic Riverway, St Croix Falls, Wisconsin
Hornbach DJ, Deneka T (1996) A comparison of a qualitative and a quantitative collection method for examining freshwater mussel assemblages J N Am Benthol Soc 15:587–596

Hornbach DJ, Miller AC, Payne BS (1992) Species composition of the mussel assemblages in the upper Mississippi River. Malcol Rev 25:119–128

Hornbach DJ, Troelstrup N, March JG, Deneka T (1994) The determination of the preferred mussel habitat for unionid mussel species inhabiting the St Croix River and the examination of factors influencing the distribution of *Quadrula fragosa* (Conrad). Final Project Report submitted to the National Park Service, Department of the Interior, St Croix Scenic Riverway, St Croix Falls, Wisconsin

Hornbach DJ, Baker P, Deneka T (1995a) Abundance and distribution of the endangered mussel, *Lampsilis higginsi* in the lower St Croix River, Minnesota and Wisconsin. Final Report to the US Fish and Wildlife Service, Contract No 14-48-000394-1009

Hornbach DJ, March JG, Deneka T (1995b) The potential factors influencing the distribution of freshwater mussel communities within the St. Croix and Mississippi Rivers and the examination of factors influencing the distribution of *Quadrula fragosa* (Conrad) and *Lampsilis higginsi* (Lea). Final Report Subpermit 93-50R, March 1995. Fish and Wildlife Service, US Department of the Interior, Twin Cities, Minnesota

Hornbach DJ, March JG, Deneka T, Troelstrup NH, Perry JA (1996) Factors influencing the distribution and abundance of the endangered winged mapleleaf mussel, *Quadrula fragosa* in the St Croix River, Minnesota and Wisconsin. Am Midl Nat 136:278–286

Hornbach DJ, Kemperman M, Kurth K (1997) Freshwater mussel ecology of the St. Croix River: comparison of juvenile and adult habitats. Report to the National Park Service, St Croix National Scenic Riverway, St Croix Falls, Wisconsin

Horne FR, McIntosh S (1979) Factors influencing distribution of mussels in the Blanco River of central Texas. Nautilus 94:119–133

Hove MC (1995) Phase II endangered species report: 7.0 Malacological (mussel) analysis. Report to Barton-Aschman Associates, Inc. 111 Third Ave. South, Suite 350 Minneapolis, Minnesota

Johnson RI (1970) The systematics and zoogeography of the Unionidae (Mollusca: Bivalvia) of the southern Atlantic Slope region. Bull Mus Comp Zool 140:263–450

Johnson RI (1980) Zoogeography of the North American Unionacea (Mollusca: Bivalvia) north of maximum Pleistocene glaciation. Bull Mus Comp Zool 149:77–189

Johnson SL (1995) Instream flow requirements of *Quadrula fragosa* and the aquatic community in the lower St Croix River downstream of the Northern States Power hydroelectric dam at St Croix Falls, Wisconsin. Report to the Bureau of Endangered Resources, Wisconsin Department of Natural Resources, Madison, Wisconsin

Kondo T (1989) Differences in clutch size and host recognition by glochidia between summer and winter breeders of Japanese unionid mussels. Venus 48:40–50

Kondo T (1993) Reproductive strategies of Japanese unionid mussels. In: Cummings KS, Buchanan AC, Koch LM (eds) Conservation and management of freshwater mussels. Proc UMRCC Symp, 12–14 October 1992, St Louis, MO. Upper Mississippi River Conservation Committee, Rock Island, Illinois, pp 142–146

Lee HL, DeAngelis DL (1997) A simulation study of the spatio-temporal dynamics of the unionid mussels. Ecol Model 95:171–180

McMahon RF (1991) Mollusca: Bivalvia. In: Thorp JH, Covich AP (eds) Ecology and classification of North American freshwater invertebrates. Academic Press, New York, pp 315–399

Morris TJ, Corkum LD (1996) Assemblage structure of freshwater mussels (Bivalvia: Unionidae) in rivers with grassy and forested riparian zones. J N Am Benthol Soc 15:576–586

Neves RJ (1993) A State-of-the-Unionids Address. In: Cummings KS, Buchanan AC, Koch LM (eds) Conservation and management of freshwater mussels. Proc UMRCC Symp, 12–14 October 1992, St Louis, MO. Upper Mississippi River Conservation Committee, Rock Island, Illinois, pp 1–10

Neves RJ, Widlak JC (1988) Occurrence of glochidia in stream drift and on fishes of the upper North Fork Holston River, Virginia. Am Midl Nat 119:111–120

Neves RJ, Weaver LR, Zale AV (1985) An evaluation of host fish suitability for glochidia of *Villosa vanuxemi* and *V. nebulosa* (Pelecypoda: Unionidae). Am Midl Nat 113:13–19

Raffaelli DG, Hildrew AG, Giller PS (1992) Scale, pattern and process in aquatic systems: concluding remarks. In: Giller PS, Hildrew AG, Raffaelli DG (eds) Aquatic ecology: scale, pattern and process. Blackwell, Oxford, pp 601–606
Salmon A, Green RH (1983) Environmental determinants of unionid clam distribution in the middle Thames River, Ontario. Can J Zool 61:832–838
SAS Institute (1994) JMP statistics and graphics guide, SAS Institute, Cary, North Carolina
Strayer DL (1983) The effects of surface geology and stream size on freshwater mussel (Bivalvia: Unionidae) distribution in southeastern Michigan, USA. Freshwater Biol 13:253–264
Stern EM (1983) Depth distribution and density of freshwater mussels (Unionidae) collected with Scuba from the lower Wisconsin and St Croix rivers. Nautilus 97:36–42
Strayer DL (1993) Macrohabitats of freshwater mussels (Bivalvia: Unionacea) in streams of the northern Atlantic slope. J N Am Benthol Soc 12:236–246
Strayer DL, Ralley J (1991) The freshwater mussels (Bivalvia: Unionidae) of the upper Delaware River drainage. Am Malacol Bull 9:21–25
Strayer DL, Ralley J (1993) Microhabitat use by an assemblage of stream-dwelling unionaceans (Bivalvia), including two rare species of *Alasmidonta.* J N Am Benthol Soc 12:247–258
Thioulouse J (1989) Statistical analysis and graphical display of multivariate data on the Macintosh Computer. Appl Biosci 5:287–292
Thioulouse J (1990) MacMul and GraphMu: two Macintosh programs for the display and analysis of multivariate data. Comput Geosci 16:1235–1240
Turgeon DD, Bogan AE, Coan EV, Emerson WK, Lyons WG, Pratt WL, Roper CFE, Scheltema A, Thompson FG, Williams JD (1988) Common and scientific names of aquatic invertebrates from the United States and Canada: mollusks. American Fisheries Society Spec Publ 16 Bethesda, Maryland
Van der Schalie H (1945) The value of mussel distribution in tracing stream confluence. Mich Acad Sci Arts Lett 30:355–373
Vannote RL, Minshall GW (1982) Fluvial processes and local lithology controlling abundance, structure, and composition of mussel beds. Proc Natl Acad Sci USA 79:4103–4107
Vannote RL, Minshall GW, Cummins KW, Sedell JR, Cushing CE (1980) The river continuum concept. Can J Fish Aquat Sci 37:130–137
Vaughn CC (1993) Can biogeographic models be used to predict the persistence of mussel populations in rivers? In: Cummings KS, Buchanan AC, Koch LM (eds) Conservation and management of freshwater mussels. Proc UMRCC Symp, 12–14 October 1992, St Louis, MO. Upper Mississippi River Conservation Committee, Rock Island, Illinois, pp 117–122
Wallace JB, Webster JR, Woodall WR (1977) The role of filter-feeders in flowing waters. Arch Hydrobiol 79:506–532
Watters GT (1992) Unionids, fishes, and the species area curve J Biogeogr 19:481–490
Watters GT (1993) Mussel diversity as a function of drainage area and fish diversity: management implications. In: Cummings KS, Buchanan AC, Koch LM (eds) Conservation and management of freshwater mussels. Proc UMRCC Symp, 12–14 October 1992, St Louis, MO. Upper Mississippi River Conservation Committee, Rock Island, Illimois, pp 111–116
Watters GT (1994a) Form and function of unionoidean shell sculpture and shape (Bivalvia). Am Malacol Bull 11:1–20
Watters GT (1994b) An annotated bibliography of the reproduction and propagation of the Unionoidea (Primarily of North America). Ohio Biol Surv Misc Contrib VI
Way CM, Miller AC, Payne BS (1989) The influence of physical factors on the distribution and abundance of freshwater mussels (Bivalvia: Unionidae) in the lower Tennessee River. Nautilus 103:96–98
Williams JD, Warren ML, Cummings KS, Harris JL, Neves RJ (1993) Conservation status of freshwater mussels of the United States and Canada. Fisheries 18:6–22
Zar JH (1984) Biostatistical analysis. Prentice-Hall, Englewood Cliffs, New Jersey

Part IV
Evolutionary Biology

13 Framework and Driving Forces for the Evolution of Naiad Life Histories

GERHARD BAUER

13.1 Introduction

So far the various taxa of the Unionoida have been introduced (Chap. 1, this Vol.), the variability of life-history traits has been shown (Chaps. 5, 7, 6, this Vol.) and it is evident that tremendous losses occur during the glochidial stage (Chap. 11, this Vol.), whereas in permanent waters, the habitat of most naiads, the adults suffer only low mortality under natural conditions (Chap. 9, this Vol.).

In combination with the common mode of feeding and brooding, tradeoffs, constraints and age-specific mortality patterns are the factors that guide naiad evolution in order to optimize life-time reproductive success. This chapter will try to demonstrate various options for maximizing fitness among naiads. Important problems to be addressed concern, for example, evolutionary trends, the tremendous differences in life span, the different host ranges etc. In order to elucidate the associations between fitness, life history, mode of life and habitat, a case study will be presented on four European species representing many of the life-history variations observed among naiads. Finally some general conclusions are drawn about plastic and canalized traits and about the relationships between life history and rate of evolution.

There is, however, a lack of knowledge about the life histories of the Etherioidea and the Hyriidae. Particularly the host relationships among these groups are largely unknown. There is no reason to assume that important selective factors, e.g. mortality during host infection, general evolutionary trends with respect to hydrodynamic effieciency etc., differ much from the Margaritiferidae and Unionidae; but this chapter deals mainly with the latter groups.

13.2 The Evolution of Major Life-History Traits

13.2.1 Evolutionary Trends: From Primitive to Derived

Evolutionary trends which are evident throughout a whole taxon evolve in response to selective pressures valid for this taxon. Among the naiads the common mode of feeding and brooding creates a selective pressure for improved hydrodynamic efficiency, e.g. separation of the in- and exhalent current and for improved gill morphology with respect to stability and marsupial functions. Another evolutionary trend is evident in the mode of host infection, where various mechanisms have evolved to decrease the tremendous mortality in free larvae (Chap. 11, this Vol.), i.e. to increase their chance of infecting a host.

Ecological Studies, Vol. 145
G. Bauer and K. Wächtler (eds.) Ecology and Evolution of the Freshwater Mussels Unionoida

13.2.1.1 Hydrodynamic Efficiency

Due to the mode of feeding and reproduction (Chap. 1, this Vol.), it is of central importance that water and food can be pumped effectively into the animal, and water, waste and glochidia can be expelled. This process involves structures at the posterior end, namely the so-called diaphragma and the mantle margin. These character states are least developed among the Margaritiferidae, where incurrent and excurrent water is separated very imperfectly. The weak pumping system correlates with a lack of papillae (which would impede exit of particles) at the excurrent mantle aperture.

The hydrodynamic system is highly developed in the lampsilids. It is nearly closed due to fusion of the mantle margins of in- and excurrent aperture and due to the posterior extension of the diaphragma. Most other taxa occupy intermediate positions (Davis and Fuller 1981).

13.2.1.2 Demibranch Morphology

Brooding the offspring within the gills interferes with other gill functions, namely filtration and oxygen uptake (Tankersley and Dimock 1992, 1993; Heino and Kaitala 1996). Thus, an evolutionary trend is evident to increase the number and organization of the interlamellar connectives (Kat 1984) and to separate gills responsible for respiration and filtration from those of marsupial functions (Davis and Fuller 1981).

Most workers have so far accepted the following hypothesis: the margaritiferids show the most primitive demibranchs with scattered interlamellar connectives which are not modified for marsupial functions (but see Chap. 14, this Vol.). Eggs are incubated in all four gills (tetragenous). In the other groups the gill structure is stabilized by increasing the number of connectives and by organizing them into continuous walls (Kat 1984). To facilitate respiration and food intake, frequently in some taxa only the outer (ectobranchous; e.g. *Anodonta, Unio*) or inner (endobranchous: Etherioidea) gills are used as marsupia. Finally, the lampsilids and hyriids reach the highest developmental stage with a portion of the respectively outer (lampsilids; Richard et al. 1991) or inner (hyriids; Chap. 2, this Vol.) gills being set aside as a permanently modified marsupium.

An interesting contradicting hypothesis is presented by Hoeh et al. (this Vol.), who state that the lack of interlammelar gill septa among the margaritiferids is already a derived character and that endobranchy is ancestral and ectobranchy developed via a tetragenous intermediate stage.

13.2.1.3 Mode of Host Infection

Many short-term breeding (tacytictic) species (Chap. 5, this Vol.) simply broadcast their larvae into the water as soon as they are ripe. This applies, for example, for *Margaritifera* and *Unio*, and results in very high mortality, since the chance of contacting a fish passively is extremely low (Chap. 11, this Vol.). Furthermore, glochidia are not able to select a proper host. Rather, they will attach to unsuitable fish species where they die (Kat 1984). If the mussel-fish relationship began as outlined by Watters (this Vol.), this mode of host infection must be considered primitive.

Mechanisms to improve host infection involve morphological adaptations and/or incorporation of behavioral factors into the reproductive process. Some unionids, lampsilids and hyriids release their glochidia within conglutinates which mimic fish items like worms and leeches (Kat 1984; Chap. 2, this Vol.) or dipteran larvae (Hartfield and Hartfield 1996; Watters 2000). When a fish takes them up, at least part of the glochidia will reach its gills in the ventilating current. Another subtle mechanism occurs in some lampsilids which attract hosts with mantle flaps resembling prey fish (Kat 1984). Both strategies may at least "select" a guild of host fish. Behavioral adaptations to increase glochidial survival are known from European *Anodonta* species (Jokela and Palokangas 1993; Trübsbach 1998; Welte 1999). These long-term breeders (bradytictic) perceive the presence of hosts and then release some of their glochidia (see Sect. 13.3.4.1). Release of small numbers of glochidia is aided by their being loosely packed within the marsupia (Davis and Fuller 1981). Since this feature occurs among all *Anodonata* species, the probability of perceiving hosts is probably widespread among this genus.

13.2.2 Responses to Environmental Variation: Constraints, Tradeoffs, Plasticity and Adaptations

13.2.2.1 Body Size

The evolution of this life-history trait among naiads is very little understood. On the one hand, there is an allometric constraint for fertility, such that fitness of large individuals is increased due to a high number of glochidia (Bauer 1994; Chaps. 7, 2, this Vol.). The development of parasitism, leading to tremendous mortalities of glochidia, certainly was a selective factor for large size and thus high fertility. Accordingly, most species of the Unionoida are much larger than other freshwater mussels, e.g. Dreissenidae, Crobiculidae, Sphaeriidae. (This does not mean that non-parasitic bivalves are not selected for high life-time reproductive success, but they may diversify also along lines other than reproduction). A tradeoff between growth and reproduction is suggested by the European *Unio* species: they are small but compensate for their low fertility by producing up to five consecutive broods per reproductive season (Chap. 7, this Vol.). Another selective advantage may be gained by large individuals of thick-shelled species, who escape the muskrat by exceeding its upper size limit (Chap. 10, this Vol.).

On the other hand, large individuals of thin-shelled species may be preferred by this predator (Chap. 10, this Vol.), such that there should be a selective advantage for slow growth, small size and early maturation. That no such effect has been found so far has been attributed to there being usually a large pool of individuals of normal growth which are protected from the muskrat, for example, because they live in deep water (Hanson et al. 1989; Jokela and Mutikainen 1995).

The available data so far suggest that within a species, maximum or asymptotic size show respectively a negative relationship to temperature, such that mussels in warm habitats tend to remain smaller. Presumably, in most cases this variation must be considered a passive phenotypic response (Bauer 1992; Chap. 7, this Vol.).

13.2.2.2 Life Span

An important physiological constraint is the negative relationship between life span and the growth constant k (derived from Ford-Walford plots, Walford 1946). This relationship (Fig. 13.1) cannot be explained on purely mathematical grounds, resulting from the Ford-Walford plot. It must rather reflect one common rule imposed on the whole group, although the ultimate factors are not well understood (Bauer 1992; why the regressions differ between Margaritiferids and Unionids will be discussed in Section 13.3.2.2). The growth constant does not only determine the rate at which the asymptotic size is approached, rather it is also considered a coefficient of catabolism, being related to metabolic activity (Bertalanffy 1938). As shown in Bauer (1992) and Hochwald (this Vol.), growth constant and maximum age are highly plastic (see particularly *M. margaritifera*, *A. anatina* and *U. crassus* in Fig. 13.1).

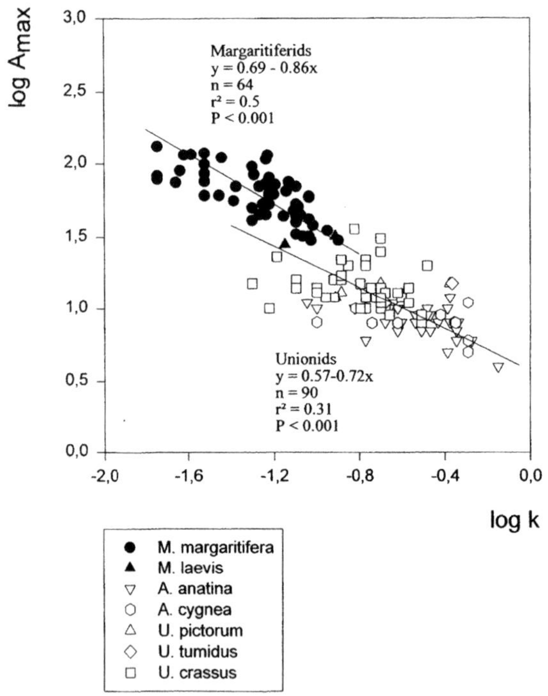

Fig. 13.1. Relationship between the growth constant k and the maximum life span among naiads. k is estimated by plotting shell length (length of annulus) at age t + 1 against shell length at age t. From the regression equation $L_{t+1} = a + b^*L_t$, k can be calculated as: $k = -\ln(b)$. Each value refers to one population. The difference of the regressions margaritiferids-unionids is highly significant (Ancova: effect of k: $F = 910$, $P < 0.000$; effect of the factor margaritiferid-unionid: $F = 138$, $P < 0.000$; the slopes of the regressions are not significantly different from one another). (Data from Hendelberg 1961; Negus 1966; Tudorancea and Gruia 1968; Tudorancea 1972; Grundelius 1987; Engel 1990; Nagel 1991; Valerius 1991; Franke 1993; Ziuganov et al. 1994; Müller 1995; Hochwald 1997; Lechner 1997; Teutsch 1997 and own investigations)

13.2.2.3 Host Range, Glochidial Size, Fertility

Glochidial size is species-specific and within a species there is a positive relationship between female size and number of glochidia in its marsupia (Hanson et al. 1989; Bauer 1994; Chap. 2, this Vol.). However, the degree to which fertility is constrained by allometry is different among naiad taxa (Bauer 1998).

Glochidia are small (0.05–0.45 mm) among margaritiferids and unionids (Fig. 13.2) and around 0.2 mm among most hyriids (Chaps. 2, 6, this Vol.), short-lived and only capable for a few days of attaching successfully to a host (Chap. 11, this Vol.). Above the species level there is a tradeoff between size and number such that fertility is high in species with small glochidia and vice versa (Fig. 13.3; Bauer 1994). Glochidial size is not constrained by allometry and also phylogenetic constraints seem to be of minor importance (Bauer 1994). However, hooked glochidia tend to be larger than unhooked ones (Fig. 13.2; Bauer 1994). An important physiological constraint is the duration of the parasitic stage, which is inversely related to glochidial size (Table 13.1). The largest glochidia are highly developed when released by the parent (Harms 1908, 1909). They are able to complete their metamorphosis on the host within a short period of time (a few weeks) compared to small glochidia, which are only poorly developed (Fig. 13.3).

Glochidial size increases as the host range increases (Fig. 13.3). This relationship also holds if it is controlled for phylogeny (Bauer 1994). Probably it is due to an interplay of the duration of the parasitic stage and the hosts immunological defences.

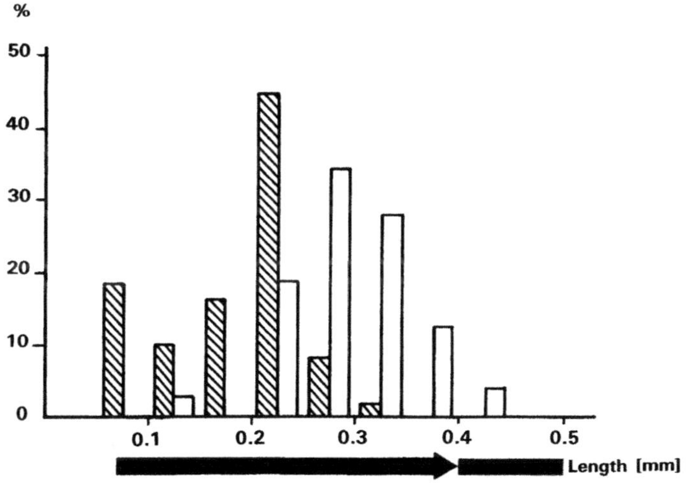

Fig. 13.2. Frequency distribution of shell lengths (0.05-mm intervals) of hooked (*white bars* 32 species) and unhooked (*hatched bars* 49 species, 4 of them margaritiferids) glochidia among Margaritiferidae and Unionidae. The *arrow below the abscissa* indicates growth of *M. margaritifera* in its parasitic stage from 0.07 up to 0.4–0.5 mm (*shaded bar*). (After Lefèvre and Curtis 1912; Surber 1912; Howard 1915; Awakura 1968; Yokley 1972; Fuller 1974; Clarke 1973, 1981; Heard 1975; Karna and Millemann 1977; Britton 1979; Buchanan 1980; Zale and Neves 1982; Yeager and Neves 1986; Claes 1987; Maaß 1987; Hochwald 1988; Hüby 1988; Nagel 1988; Buddensiek 1991)

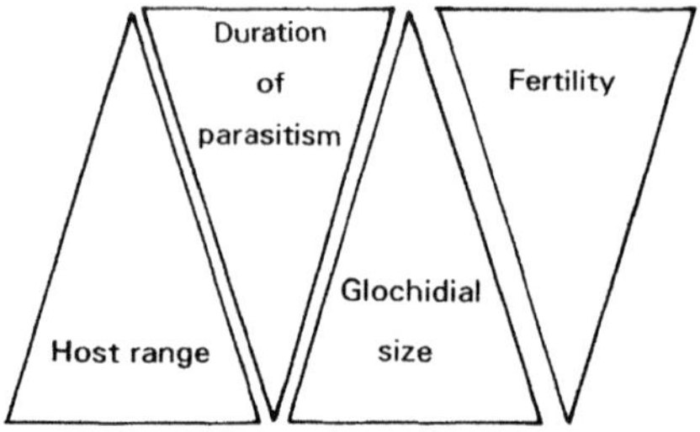

Fig. 13.3. Schematic presentation of the relationships between size and number of glochidia (*Fertility*), duration of the parasitic stage and size of the host range (above the species level)

Table 13.1. Glochidial size (shell length) and duration of the parasitic stage in degree-days of *Margaritifera margaritifera, M. auricularia, Unio tumidus, U. pictorum, Anodonta imbecilis, A. cygnea* and *A. grandis*

	Size (mm)	T(°C)	Degree-days	Reference
M. margaritifera	0.07	15.5–17	1300–1430	Hruska (1992)
M. auricularia	0.13	23–24	690	Araujo (this Vol.)
U. tumidus	0.2	20.5	410	Maass (1987)
U. pictorum	0.2	17	493	Maass (1987)
A. imbecilis	0.31	21	170	Trdan and Hoeh (1982)
A. cygnea	0.34	20	240	Claes (1987)
A. grandis	0.36	21	130	Trdan and Hoeh (1982)

Large glochidia have probably already left the host before its specific immune response is initiated and accordingly there is only little selective pressure to adapt to a particular host taxon. Support for this hypothesis was found by Terberger (1999). He showed that a first infection with *Anodonta* glochidia produced young mussels during a long period of time, compared to a second infection, where the duration of parasitism was shorter. However, in the latter case, the number of young mussels was lower. This indicates that only those which developed very quickly could tolerate the hosts' immune response. On the other hand, small glochidia are likely to be exposed to the antibodies of the host for a long period. In order to minimize the host response, they have to adapt very closely to the host, which apparently can only be achieved for a few closely related fish species (Bauer 1994). This is consistent with findings of Isom and Hudson (1984), who reared glochidia of various naiad species in artificial medium, in this way bypassing the fish host. They state "that there is no specifity for one host's blood by a given mussel species. This implies that there is no physiological plasma basis, other than possible immune response, for mussel host specifity."

13.3 A Case Study: Life-History Evolution Along a River and a Trophic Gradient

The considerations so far suggest that important options for a naiad life history are: imperfect or highly developed morphological organization, primitive or derived mode of host infection, large or small body size, long or short adult life span, broad or narrow host range with its consequences for fertility. Four European species, the freshwater pearl mussel *M. margaritifera* and the unionids *U. crassus*, *A. anatina* and *A. cygnea* exhibit a great many of these possibilities. This chapter will try to explain how their life-history patterns evolved.

13.3.1 Habitat Distribution of Four European Species

Stream order and trophic level are important parameters for freshwater communities. Many environmental factors change along the length of a river (Vannote et al. 1980). For example, the summer temperatures increase downstream and the size of detritus and substrate particles decreases. In waters of very low trophic level only few specialized species can establish, whereas more species are able to live in waters of higher nutrient concentration such that the diversity of the freshwater community increases. Furthermore, oxygen availability may decrease as stream order and trophic level increase.

With the exception of first-order streams, central European naiads inhabit all types of freshwaters, from small brooks to large streams and standing waters. They show a longitudinal zonation, with *M. margaritifera* occurring preferably in headwater streams, *U. crassus* extending further downstream and the *Anodonta* species preferring large streams and standing waters (Fig. 13.4).

A discriminant analysis with physicochemical parameters calculates the following equation whose scores are maximally different between the habitats of the four mussel species: score = $-3.31 + 0.0005(0.1)^{*}\text{conductivity} + 0.48(0.34)^{*}\text{BOD}_5 + 0.01(0.45)^{*}\text{Ca} + 2.1(0.69)^{*}\log(\text{Cl}) + 1.0(0.4)^{*}\log(\text{NH}_4)$; (standardized function coefficients in brackets). All these parameters are related to the trophic level and their signs in the equation are positive. Thus, the higher the score, the higher the trophic level of the habitat. According to Fig. 13.4, the pearl mussel occurs in streams poor in nutrients. *U. crassus* shows a slight overlap with the trophic level of pearl mussel habitats, which is consistent with its occurrence: it is the only unionid to be occasionally associated with pearl mussels (Bauer et al. 1991). The *Anodonta* species often inhabits waters of high trophic level.

13.3.2 Life Span

13.3.2.1 What Factor Determines Life Span?

According to Fig. 13.5a, there are tremendous differences in the maximum age attained by the four mussel species. The freshwater pearl mussel is one of the longest-lived invertebrates, whereas *Anodonta* is comparatively short-lived and *U. crassus* is in between. In accordance with the negative relationship between life span and k (Fig. 13.1), the distribution pattern of the growth constant k is opposite to that of the life span (Fig. 13.5b). Two factors, one intrinsic and one extrinsic, are responsible for these patterns.

As in other molluscs (Spärk 1936; Branch 1979; Bayne and Newell 1983), the rate of metabolic energy expenditure of naiads seems to be adjusted by evolution to correspond with the availability of food, i.e. the concentration of food particles in the water. Bauer et al. (1991) showed that under the same temperature conditions, the

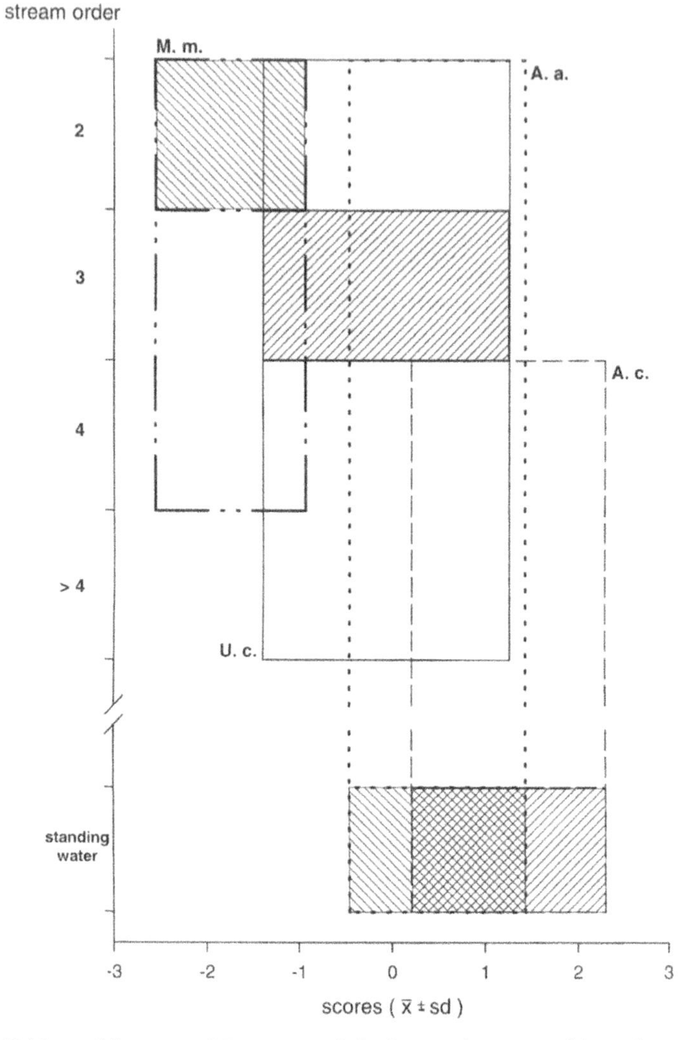

Fig. 13.4. Habitats of four central European naiads along a river or trophic gradient. *Ordinate* The distribution pattern along the length of a stream (up to standing water) is based on 33 *M. margaritifera* (*M.m.*), 23 *U. crassus* (*U.c.*), 75 *A. anatina* (*A.a.*) and 46 *A. cygnea* (*A.c.*) habitats; *hatched area* habitat where more than 40% of the recorded populations occur; *abscissa* values are based on a discriminant analysis with physicochemical factors in 13 *M. margaritifera*, 9 *U. crassus*, 27 *A. anatina* and 4 *A. cygnea* populations. Only healthy populations are considered. The trophic level increases *to the right*, the range occupied by each species refers to the mean ± SD of the calculated discriminant scores. (Engel 1990; Nagel 1997; Teiber 1991; Valerius 1991; Franke 1993; Niemeyer 1993; Hochwald 1997; Lechner 1997; Lang 1998; Schneidt 1998; Trübsbach 1998 and our own data)

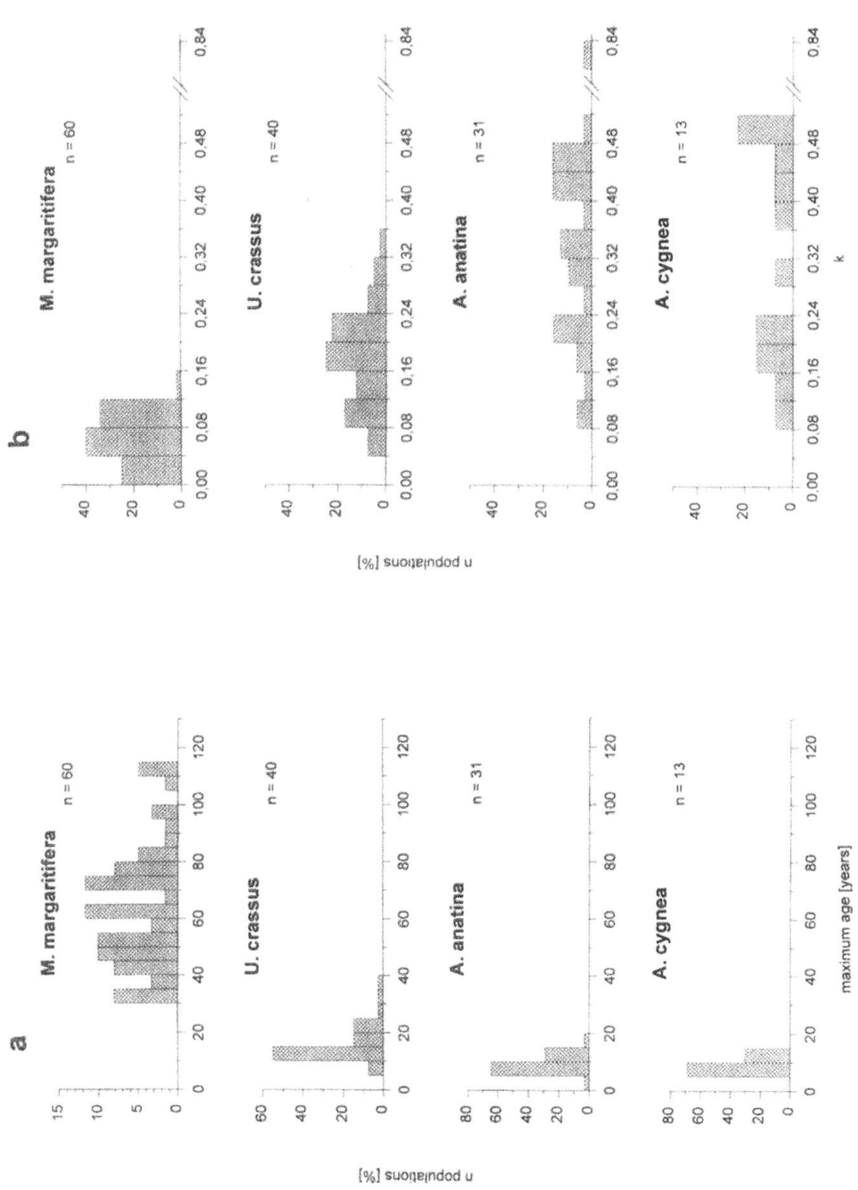

Fig. 13.5a,b. Frequency distribution of growth parameters among *M. margaritifera*, *U. crassus* and *Anodonta* in central Europe. *n* Number of populations. **a** Life span. **b** Growth constant. (Sources as in Fig. 13.4)

metabolic rate of *M. margaritifera* is lower than that of unionids. Species with a high metabolic rate, such as *Anodonta*, require a rich food supply, whereas pearl mussels are adapted to rivers of low productivity by means of a low metabolic rate (Bauer et al. 1991). A similar explanation is supposed by Hornbach (this Vol.), for the different distribution patterns of Lampsilinae and Ambleminae, since he found higher metabolic rates in a lampsilid compared to *Amblema plicata* (Baker and Hornbach 1997).

This might explain why the growth constants of pearl mussels generally are lower, such that they attain a longer life than unionids (Fig. 13.1;5). However, within each species a considerable plasticity of life span and k is evident (Fig. 13.1;5; Chap. 7, this Vol.). Furthermore, according to Fig. 13.5, the life span of *U. crassus* is significantly higher ($\chi^2 = 30.5$; $P < 0.001$) and k significantly lower ($\chi^2 = 18.4$; $P < 0.001$) than in the case of *A. anatina*, despite no significant differences in metabolism between both species under the same temperature conditions (Bauer et al. 1991). To investigate the mechanisms underlying this pattern, water temperature was recorded in 1998 from 1 June till 31 August in five pearl mussel populations, five *U. crassus* populations and six *A. anatina* populations. For most of these rivers, long-term data on hydrochemical factors (conductivity, BOD_5, Ca, Cl, NH_4, NO_3, PO_{4diss}) were also available. Partial correlation analysis showed that water temperature (expressed as degree-days) was the only parameter explaining a significant amount of the variation in k (Fig. 13.6a). This indicates, in accordance with Hochwald (this Vol.), that the metabolic rate is in fact the underlying mechanism of the relationship between life span and k (Fig. 13.1).

Anodonta lives in waters of moderate or high productivity (Fig. 13.4) at high temperatures during the summer months (Fig. 13.6). Its metabolic rate is high (Bauer et al. 1991). It grows rapidly and therefore is a priori short-lived (Fig. 13.1). *U. crassus* lives further upstream (Fig. 13.4). The lower summer temperatures lead to a slower growth and a longer life. The pearl mussel is highly specialized for living in streams poor in nutrients. The reduced metabolism, an adaptation to the low food supply, in combination with low summer temperatures in its headwater habitats (Fig. 13.6) lead to low growth rates and thus provide the basis for a very long life.

13.3.2.2 A Strategy for Maximizing Life Span: *Margaritifera margaritifera*

13.3.2.2.1 Evidence

If the pearl mussel attains the maximum possible life span, a female may pass through as many as 40 to 50 reproductive periods (Bauer 1987), whereas, for example, *Anodonta*, which is a priori short-lived, reproduces less than ten times. Accordingly, for the pearl mussel, the value of immediate self-preservation is high in relation to immediate reproduction, which should lead to a strong selective pressure for extended longevity.

This becomes evident in the difference of the regressions in Fig. 13.1: at the same growth constant, k, margaritiferids on the average reach a higher life span than unionids. Furthermore, it is evident in Fig. 13.6b: margaritiferids reach a higher life span than predicted by the relationship between temperature and life span among unionids and mortality rates are very low (Chap. 9, this Vol.).

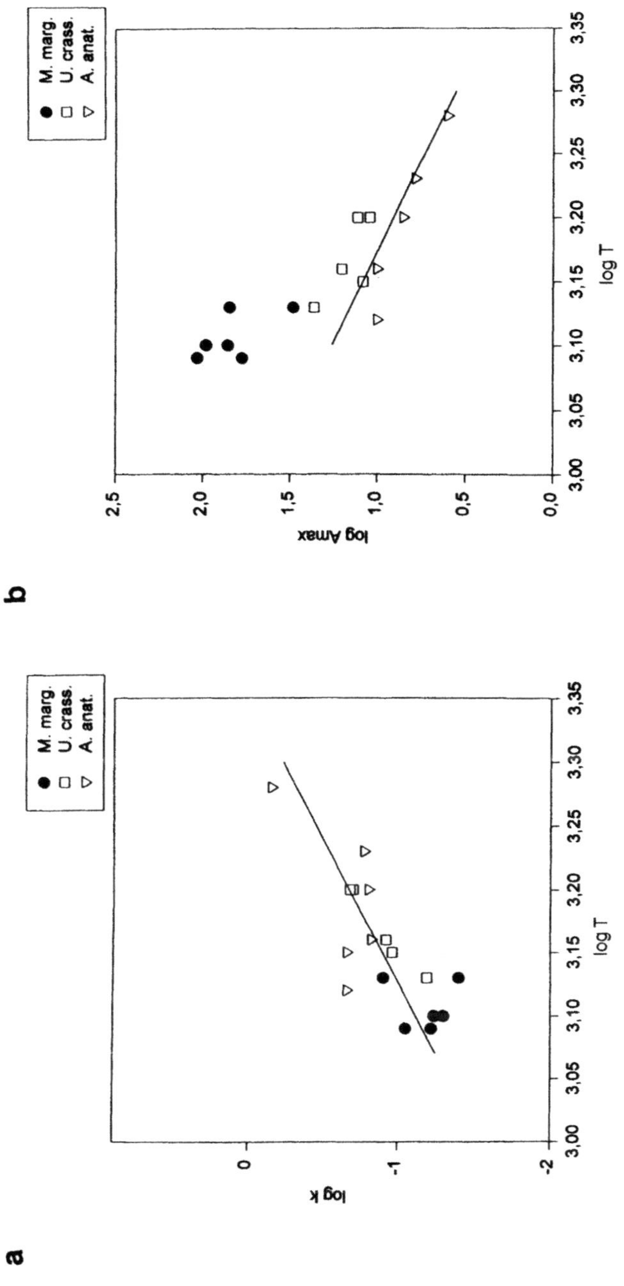

Fig. 13.6a,b. Growth parameters and water temperature (T = degree-days = sum of daily mean temperatures from 1.6. till 31.8 1998. Temperature was recorded at 36-min intervals). *Each value* represents one population. **a** Relationship between temperature and growth constant. ($y = 14.85 + 4.43x$; $r^2 = 0.68$; $P < 0.001$). **b** Relationship between temperature and maximum individual life span. The regression is calculated only for *U. crassus* and *A. anatina*. ($y = 12.2 - 3.53x$; $r^2 = 0.64$; $P < 0.01$)

13.3.2.2.2 Mechanisms

The resource allocation strategy must be considered an important adaptation of pearl mussels for an extended life span. Approximately one third of the organic substance is invested in the shells as periostracum and conchiolin layers (Chap. 5, this Vol.), such that they can resist the acid water for more than 100 years. Furthermore, reproductive effort of females is very low. Whereas per reproductive season *Anodonta* allocates 5–40% of its body weight to egg production (Haukioja and Hakala 1978; Niemeyer 1993), *M. margaritifera* invests only between 0.3 and 5.3%. A low reproductive effort probably exerts only a low impact, if any, on somatic functions through absorption of resources, and a low reproductive effort should hardly interfere with filtration efficiency, since frequently only small parts of the gills contain glochidia. Additionally, this allocation system is highly plastic. Only those females which are in good physiological condition reproduce, and reproductive effort is not simply a function of age or size as predicted by life-history theory (Pianka and Parker 1975; Kozlowski 1991) and shown for many molluscs (e.g. Hughes and Roberts 1980; Vahl 1981; Bayne and Newell 1983; Tuomi et al. 1983). Rather, it is adjusted to the amount of surplus resources. Thus, the resource allocation policy of pearl mussels favours survival over current reproduction (Bauer 1998).

13.3.3 Host Range, Glochidial Size, Fertility

13.3.3.1 Impact of the Fish Fauna

According to Fig. 13.7, the number of fish species is very low in pearl mussel rivers. In central Europe the brown trout (*Salmo trutta*) is the most common species, comprising ca. 80% of all fish individuals (Bauer 1988). The two other functions in Fig. 13.7 start at higher values and they quickly approach high species numbers. That means that (1) species richness generally is higher in unionid habitats, and (2) the fish fauna is more heterogenous in unionid habitats, i.e. different habitats may contain different fish faunas.

Thus, the freshwater pearl mussel occurs together with a fish fauna low in diversity, where brown trout is by far the most common species. In unionid habitats many more fish species live and the species composition may differ considerably between habitats. Therefore, for these unionids a broad host range should be advantageous. Survival of glochidia should increase if they are able to utilize many fish species as hosts. On the other hand, a broad host range has only little advantage for the pearl mussel, since its glochidia usually meet brown trout. Figure 13.8a shows that the host range of the pearl mussel is, in fact very narrow, comprising only the four native Salmoninae species (Bauer 1997). By contrast, *Anodonta* has a very broad host range. *U. crassus* is in between, a contradiction to the diverse fish fauna (Fig. 13.7) which will be discussed below.

In accordance with the mechanisms described in Section 13.2.2.2 (Fig. 13.3) glochidial size shows the same pattern as the size of the host range (Fig. 13.8b). Since there is a tradeoff between size and number of glochidia (Fig. 13.3), size-specific fertility (number of glochidia in the marsupia of a gravid female at a given shell length) is highest in pearl mussels, lowest in *Anodonta* and in between in *Unio* (Bauer 1994). Calculating mean fertility (regressions given in Bauer 1994) yields for

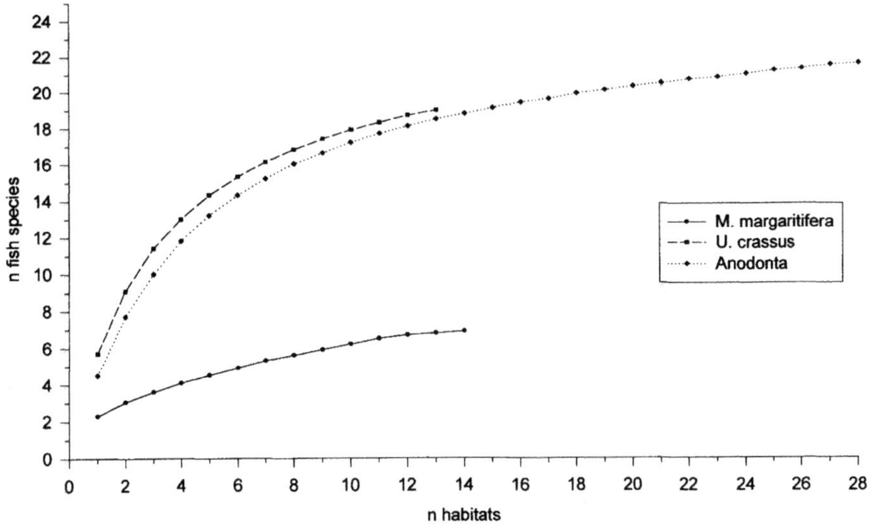

Fig. 13.7. Species richness of the fish fauna in pearl mussel rivers and in habitats of *U. crassus* and *Anodonta*. Relationship between the total number of recorded fish species (*ordinate* recalculated according to the Shinozaki rarefaction method, Shinozaki 1963; Achtziger et al. 1992) and the number of investigated habitats (*abscissa*). (Data from Engel 1990; Bauer et al. 1991; Franke 1993; Niemeyer 1993; Lechner 1997; Lang 1998)

a 10-cm-long pearl mussel 2 500 000, but only 140 000 glochidia for an *Anodonta* of same size.

13.3.3.2 A Strategy for Maximizing Fertility as Well as Young Mussel Size: *Margaritifera margaritifera*

Figure 13.2 shows that the glochidium of *M. margaritifera* with a shell length of 0.07 mm is definitely at the lower boundary of the size range of naiad glochidia, suggesting that it is minimized and, due to the tradeoff between offspring size and number, fertility is maximized. However, whereas most naiad glochidia do not increase in size during parasitism, pearl mussel glochidia grow considerably on their host and, although unhooked, at the end of parasitism reach even the maximum size of hooked glochidia (Fig. 13.2), suggesting that young mussel size is maximized. As far as we know (Chap. 9, this Vol.), mortality in the postparasitic stage is still high until maturity. It is therefore advantageous to outgrow this dangerous stage as soon as possible. A large size when dropping off the host might shorten this period of time.

13.3.4 Brooding and Host Infection

The pearl mussel is a short-term breeder, retaining its offspring only for 4 to 6 weeks during late summer in the marsupia (Fig. 13.9). *U. crassus* also is a short-term breeder, but it produces several times eggs (up to five times) during its spawning season in spring (Lechner 1997; Lang 1998; Chap. 7, this Vol.). On the other hand, *Anodonta* is

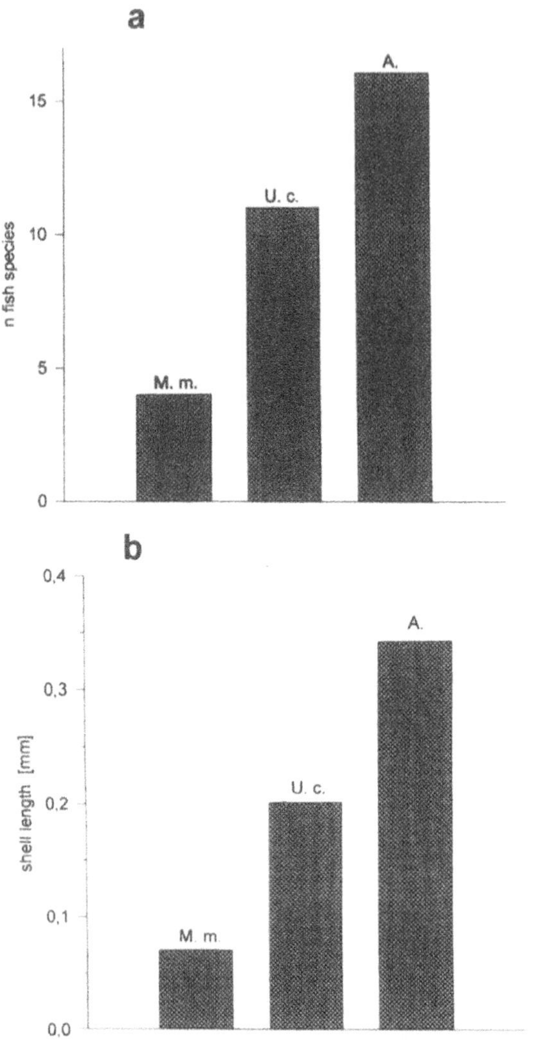

Fig. 13.8a,b. a Number of fish species in the host range of the freshwater pearl mussel, *U. crassus* and *Anodonta* in Central Europe. (As far as we know, the two *Anodonta* species do not differ with respect to their host range). **b** Size of glochidia (shell length). (Data from Maaß 1987; Engel 1990; Fleischauer-Rössing 1990; Franke 1993; Niemeyer 1993; Lechner 1997; Bauer 1997; Nagel 1997; Teutsch 1997)

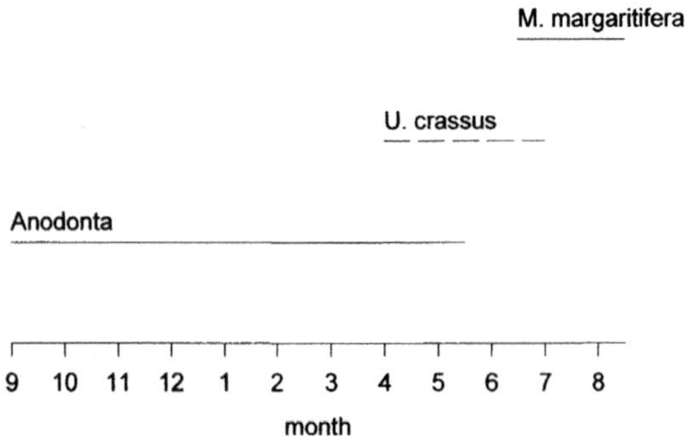

Fig. 13.9. Periods of gravidity

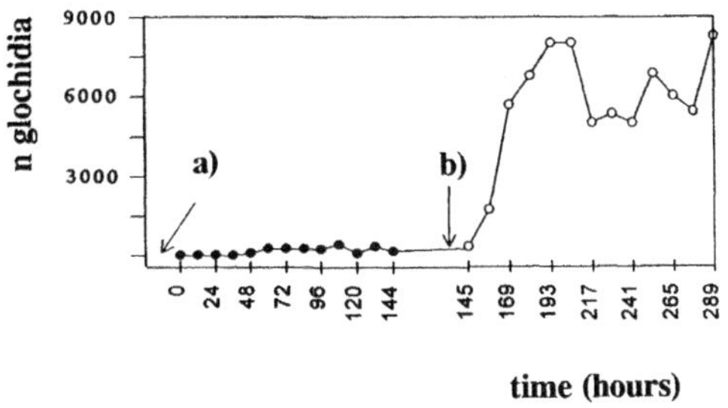

Fig. 13.10a,b. Number of glochidia released by 15 *Anodonta cygnea* females when water was added. **a** From an aquarium containing non-host fish, bitterling (*Rhodeus sericeus*). **b** From an aquarium containing hosts, rudd (*Scardinius erythrophthalmus*). (Trübsbach 1998)

a long-term breeder; the marsupia contain ripe (infectious) glochidia from October until the end of the next May.

A considerable difference exists in the mode of host infection between both modes of breeding. The short-term breeders just release their glochidia into the water as soon as they are ripe. On the other hand, *Anodonta* has a long period of time available where it could release the glochidia if conditions are favourable. Indeed, it is able to perceive the presence of hosts and preferably then releases some of its offspring (Jokela and Palokangas 1993; Trübsbach 1998; Welte 1999).

13.3.4.1 A Strategy for Improving Survival of Glochidia: *Anodonta*

Davis and Fuller (1981) state that *Anodonta* glochidia are very loosely packed within the marsupia such that they can be released in small numbers. They attribute this feature to the advantage that the adult does not have to overcome the inertia of a large mass during expulsion of marsupial contents. Now an even greater advantage is evident. *Anodonta* is not only able to perceive the presence of fish, rather it distinguishes between presence of non-hosts and presence of hosts and only in the latter case releases some of its glochidia (Fig. 13.10; Trübsbach 1998; Welte 1999).

13.3.5 Conclusions: a Sequence of Strategies

Along a river or trophic gradient we find the pearl mussel in headwater streams extremely poor in nutrients (Fig. 13.4). Here it fares well with its simple organization. The low hydrodynamic efficiency is compensated for by exposing the inhalent aperture right against the current, such that the water is blown into the animal (Fig. 13.11). The tetragenous mode of breeding should not interfere much with filtration, since (1) reproductive effort is very low (Sect. 13.3.2.2) and thus frequently only small parts of the gills are occupied by eggs, and (2) the period of gravidity is short (Fig. 13.9). Finally, the risk of food or oxygen shortage is diminished by the reduced metabolism.

Low metabolic rates, low temperature and a resource allocation in favour of somatic functions lead to a high life expectancy and thus to a high number of reproductive periods. The very small glochidia and the large body size (up to 140 mm) result in a high fertility, and in combination with the high number of reproductive periods, in a high life time glochidial production, despite the low reproductive effort. Thus the pearl mussel's strategy maximizes the number of glochidia but also leads to

Fig. 13.11. Pearl mussels typically direct their inhalent aperture against the current (*arrows*)

large young mussels (Sect. 13.3.3.2). The very high host specifity certainly must be viewed in relation to the minimized glochidial size (Fig. 13.3) and to the prevalence of the Salmoninae in its habitat (Fig. 13.7).

By contrast, the strategy of *Anodonta* maximizes the size of the host range, which requires large glochidia leading inevitably to low fertility (Sect. 13.2.2.3) despite the large (up to 200 mm) body size. Additionally, the number of reproductive periods is low, since *Anodonta* is short-lived (Sect. 13.3.2.1). The low life-time fertility is compensated by a behavioral adaptation to increase the chance of glochidia contacting a host (Sect. 13.3.4.1).

In many respects, *U. crassus* is intermediate. Here host range, glochidial size and fertility probably are a compromise between two opposing factors, one for a fertility as high as possible (requiring small glochidia) due to the inefficient mode of host infection, and one for a host range as broad as possible (requiring large glochidia) due to the diverse fish fauna (Fig. 13.7). A further factor lowering fertility is the small body size (40–80 mm; Chap. 7, this Vol.). However, *U. crassus* produces several broods per reproductive season.

13.4 Plastic and Canalized Traits

According to the relatively few available data, the following scenario can be suggested.

Above the species level, life span and fertility are independent of each other. Both traits may respond to selection, the former to the food supply via metabolism (Sect. 13.3.2), the latter to body size (Sect. 13.3.3.1) and to the fish fauna (Sect. 13.3.3.2) via glochidial size (Fig. 13.12a).

On the species level, however, the relationships between these traits are fundamentally different (Fig. 13.12b). The temperature dependence of metabolism (Sect. 13.3.2.1) leads to considerable plastic reactions of life span and body size, and due to the allometric constraint, fertility (Bauer 1992; Chap. 7, this Vol.).

On the other hand, traits related to the host relationship are highly canalized. This is particularly evident in *Margaritifera* sensu stricto (*M. margaritifera, M. falcata,*

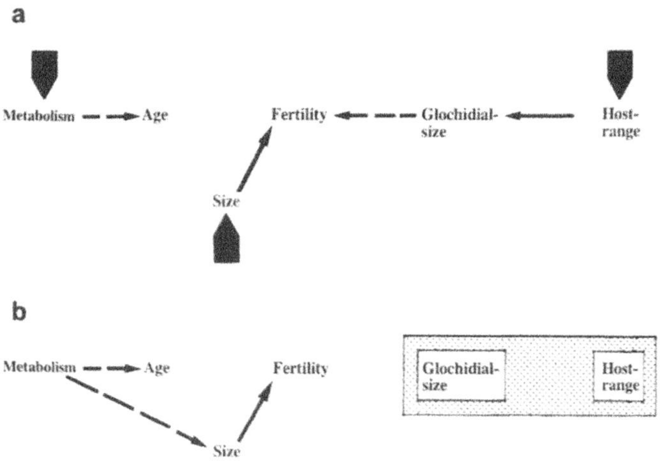

Fig. 13.12a,b. Relationships between naiad life-history traits. *Stippled arrows* negative relationships; *Thick arrows* traits which are likely to be modified by selection. **a** Above the species level. **b** On the species level (*Margaritifera margaritifera*). The *hatched block* indicates that glochidial size and host range are strongly canalized

M. laevis), where, on both sides of the northern Atlantic and the northern Pacific, glochidial size, size at the end of the parasitic stage and host range (only the native Salmoninae) are nearly identical (Bauer 1997).

This might indicate that these host-related traits are under strong selective control (Berven and Gill 1983; Smith-Gill 1983). They are optimized by selection and in their optimized form buffered against environmental variation.

13.5 Life Histories and Rates of Evolution

The high species diversity of North American naiads is treated extensively in this book (Chaps. 2, 14, 15, this Vol.). Particularly the Lampsilinae show high rates of speciation, which correlate with an efficient hydrodynamic system, and reduced amounts of the outer demibranch that is marsupial (Davis and Fuller 1981). Species-specific glochidial size is highly variable (from 0.06 to 0.31 mm shell length; Surber 1912; Fuller 1974), indicating specialized and opportunistic host relationships. Furthermore, morphological and (or) behavioral adaptations to increase host infection are widespread. Thus, a variety of different strategies has been adopted by this group. The rates of evolution are very high; some species are thought to have evolved in less than 150000 years (Kat 1984). Another, apparently recent, radiating taxon is *Elliptio* (Unioninae), where many species are believed to be of post-Pleistocene origin (Davis et al. 1981). Nagel and Badino (this Vol.) estimate the late Pleistocene as the main period of differentiation among the recent *Anodonta*.

On the other hand, an explanation is required for the extremely low rates of evolution among the Margaritiferidae. Despite their large distribution areas, less than 15 species are recognized worldwide (Chap. 3, this Vol.; Chesney et al. 1993), some of them sharing almost the same life history (Bauer 1997). A striking example is *M.*

margaritifera. Though its distribution areas in both the New and the Old World have been separated since the Eocene (60 million years ago), hardly any difference in their life histories has evolved on either side of the Atlantic. Thus, some factors must act on the margaritiferids to prevent genetic divergence (Chesney et al. 1993; Chap. 4, this Vol.), some of which I will try to outline.

1. The weak hydrodynamic system reduces the types of suitable habitats considerably. The animals apparently use the current to support water pumping (Fig. 13.11). Thus, they are confined to fast-flowing waters. There is no margaritiferid living in standing water.
2. Most species attain a long life span (Vannote and Minshall 1982; Bauer 1992; Ziuganov et al. 1994; Chap. 8, this Vol.). The longer the life span, the longer the generation time and the slower the rates of evolution (Freeland 1986).
3. All margaritiferids so far investigated are capable of becoming hermaphroditic (and probably self-fertile) at low population densities (Heard 1970; Bauer 1987; Chap. 8, this Vol.). Thus, a small number of founder animals might establish a population by the production of large numbers of glochidia, however, at the expense of genetic variability.
4. Important fitness traits are already maximized such that they have reached "design barriers" (Stearns 1977). This holds true particularly for the three *Margaritifera* species sensu stricto, i.e. *M. margaritifera, M. laevis, M. falcata*, with fairly identical glochidial and young mussel size (Bauer 1997). As shown in Section 13.3.3.2 (Fig. 13.2), glochidial size must be considered minimized (and thus fertility maximized) and young mussel size must be considered maximized. Glochidia of similar small size are also reported for other margaritiferids: *M. dahurica* 0.08-mm egg diameter (Miculic 1954), *Cumberlandia monodonta* 0.06-mm shell length (Howard 1915; Gordon and Smith 1990). Even the very large *M. auricularia* has a glochidium which, compared to most other naiads, is very small (0.13-mm shell length, Chap. 8, this Vol.).
5. As far as we know, margaritiferids are very specialized parasites. Their degree of adaptation to a narrow host range is very high. It comprises only salmonids among the northern species (*Margaritifera* sensu stricto, Bauer 1997). The southern species use other hosts, but the host range is very narrow as well (Johnson and Brown 1998; Chap. 8, this Vol.). Additionally, a high degree of habitat specificity is shown by *Margaritifera* sensu stricto (Bauer 1997) and *Margaritifera hembeli* (Johnson and Brown 1998).

Thus, a number of factors select for a high degree of adaptation to a specific niche where some important life-history traits can be either minimized or maximized (for example, glochidial size of *M. margaritifera* can be minimized, and thus fertility maximized, due to the prevalence of the salmonids in the habitat; see Sect. 13.3.3; Fig. 13.7). It seems as if the margaritiferids had arrived at the end of an evolutionary one-way blind alley.

This does not mean that these strategies are not successful. On the contrary, the high degree of fitness attained by the life histories of *Margaritifera* sensu stricto (probably the margaritiferids with the most accomplished strategy) is evident in their high phylogenetic age (Chap. 15, this Vol.), their large distribution areas and the high population densities (Murphy 1942; Karna and Milleman 1978; Bauer 1991). Frequently, they are so abundant and closely packed that they completely obscure the stream bed (Israel 1913; Clarke 1973).

13.6 Summary

Evolutionary trends among the naiads comprise increased hydrodynamic efficiency, advanced demibranch morphology, spatial reduction of marsupia (but see Chap. 14, this Vol., for different hypotheses) and morphological and/or behavioral adaptations to improve host infection.

The allometric constraint for fertility probably was a factor for the evolution of the large size of naiads. Maximum life span increases at lower metabolic rates and thus at lower temperatures. There is a tradeoff between glochidial size and fertility; however, small glochidia require adaptation to a narrow host range.

The life histories of four central European species represent nearly all of these options: the large freshwater pearl mussel (*Margaritifera margaritifera*) lives in head-water streams poor in nutrients. Here, the tetragenous mode of breeding (interfering with food and oxygen uptake) and the weak hydrodynamic system are compensated for by the strong current, the high oxygen availability and particularly the reduced metabolism. This latter factor, combined with low summer temperatures, leads to a very long life. The prevalence of salmonids in these rivers makes a narrow host range possible and thus a reduction in glochidial size, leading to an increase in fertility. This strategy leads to a high life-time fertility through a high number of reproductive periods and a high number of glochidia per reproductive period.

On the other hand, two *Anodonta* species live in large rivers or standing waters of high trophic level. They are a priori short-lived due to high summer temperatures, and they cooccur with a diverse fish fauna. Their strategy maximizes the size of the host range (i.e. the number of fish species where glochidia can develop). This reqires large glochidia, which are kept for many months in highly developed marsupia. The low fertility is compensated for by a behavioral adaptation to increase glochidial survival. They perceive the presence of host fish and then release some glochidia.

Unio crassus, which prefers intermediate river sections, also occupies an intermediate position with respect to life span, size of the host range, glochidial size and fertility. Due to the small body size, fertility is low, which is compensated for by producing some consecutive broods per reproductive season.

The growth process, i.e. life span, growth rate, body size, shows a great deal of phenotypic plasticity. On the other hand, traits closely related to the host relationship, i.e. glochidial size and host range, are strongly canalized.

Rates of evolution are extremely low among the margaritiferids. This can be explained by a number of factors selecting for a high degree of adaptation to a specific niche where important life-history traits can be either minimized or maximized. Some traits seem to have already reached a "design barrier" such that no further accomplishment is possible.

Acknowledgements. Comments by R. Hoeh, D.J. Hornbach, G.T. Watters and H. Zwölfer improved this manuscript considerably. Mrs. S. Röck provided most of the graphs.

References

Achtziger R, Nigmann U, Zwölfer H (1992) Rarefaction-Methoden und ihre Einsatzmöglichkeiten bei der zooökologischen Zustandsanalyse und Bewertung von Biotopen. Z Ökol Naturschutz 1:89–105

Awakura T (1968) The ecology of parasitic glochidia of the fresh-water pearl mussel, *Margaritifera laevis* (Haas). Hokkaido Fish Hatch Sci Rep 23:1–17

Baker SM, Hornbach DJ (1997) Acute physiological effects of zebra mussels (*Dreissena polymorpha*) infestation on two unionid mussels, *Actinonaias ligamentina* and *Amblema plicata*. Can J Fish Aquat Sci 54:512–519

Bauer G (1987) Reproductive strategy of the freshwater pearl mussel *Margaritifera margaritifera*. J Anim Ecol 56:691–704

Bauer G (1988) Threats to the freshwater pearl mussel in central Europe. Biol Conserv 45: 239–253

Bauer G (1991) Plasticity in life history traits of the freshwater pearl mussel - consequences for the danger of extinction and for conservation measures. In: Seitz A, Loeschcke V (eds) Species conservation: a population-biological approach. Birkhäuser, Basee, pp 103–120

Bauer G (1992) Variation in the life span and size of the freshwater pearl mussel. J Anim Ecol 61:425–436

Bauer G (1994) The adaptive value of offspring size among freshwater mussels (Bivalvia; Unionoidea). J Anim Ecol 63/4:933–944

Bauer G (1997) Host relationships at reversed generation times. Margaritifera and salmonids. In: Dettner K, Bauer G, Völkl W (eds) Vertical food web interactions: evolutionary patterns and driving forces. Ecological Studies, vol 130, Springer, Berlin, Heidelberg New York, pp 69–79

Bauer G (1998) Allocation policy of female freshwater pearl mussels. Oecologia 117:90–94

Bauer G, Hochwald S, Silkenat W (1991) Spatial distribution of freshwater mussels: the role of host fish and metabolic rate. Freshwater Biol 26:377–386

Bayne BR, Newell RC (1983) Physiological energetics of marine molluscs. In: Saleuddin ASM, Wilbur KM (eds) The Mollusca, vol 4. Academic Press, New York, pp 407–415

Bertalanffy VL (1938) A quantitative theory of organic growth. Hum Biol 10:181–213

Berven KA, Gill DE (1983) Interpreting geographic variation in life history traits. Am Zool 23:85–97

Branch GM (1979) Respiratory adaptations in the limpet *Patella granatina*. Comp Biochem Physiol 62:641–647

Britton JC (1979) The freshwater bivalve Mollusca of the Savannah River plant. Academy of Natural Sciences, Philadelphia

Buchanan AC (1980) Mussels of the Meramec River Basin, Missouri. Missouri Department of Conservation, Jefferson City

Buddensiek V (1991) Untersuchungen zu den Aufwuchsbedingungen der Flußperlmuschel *Margaritifera margaritifera* Linnaeus (Bivalvia) in der frühen postparasitären Phase. PhD Thesis, University Hannover, 288 pp

Chesney HCG, Oliver PG, Davis GM (1993) *Margaritifera durrovensis* Phillips, 1928: taxonomic status, ecology and conservation. J Conchol Lond 34:267–299

Claes M (1987) Untersuchungen zur Entwicklungsbiologie der Teichmuschel *Anodonta cygnea*. Ms Thesis, Tierärztliche Hochschule Hannover

Clark AH (1973) The freshwater molluscs of Canada. National Museum of Canada, Ottawa

Clarke AH (1981) The freshwater molluscs of the Canadian Interior Basin. Malacologia 13:495

Davis GM, Fuller SLH (1981) Genetic relationships among Recent Unionacea (Bivalvia) of North America. Malacologia 20(2):217–253

Davis GM, Heard WH, Fuller SLH, Hesterman C (1981) Molecular genetics and speciation in *Elliptio* and its relationships to other taxa of North American Unionidae (Bivalvia). Biol J Linn Soc 15:131–150

Engel H (1990) Untersuchungen zur Autökologie von *Unio crassus* (Philipsson) in Norddeutschland. PhD Thesis, University Hannover, 213 pp
Fleischauer-Rössing S (1990) Untersuchungen zur Autökologie von *Unio tumidus* Philipsson und *Unio pictorum* Linnaeus (Bivalvia) unter besonderer Berücksichtigung der frühen postparasitären Phase. PhD Thesis, University Hannover, 192 pp
Franke G (1993) Zur Populationsökologie und Geschlechtsbiologie der Teichmuscheln *A. anatina* L. und *A. cygnea* L. MS Thesis, University Bayreuth, 84 pp
Freeland B (1986) Arms races and covenants: the evolution of parasite communities. In: Kikkawa J, Anderson DJ (eds) Community ecology: pattern and process. Blackwell, Melbourne, pp 289–303
Fuller SLH (1974) Clams and mussels. Pollution ecology of freshwater invertebrates. In: Hart CW, Fuller SLH (eds) Academic Press, New York, pp 215–273
Gordon ME, Smith DG (1990) Autumnal reproduction in *Cumberlandia monodonta*. Trans Am Microsc Soc 109:407–411
Grundelius E (1987) Flodpärlmusslans tilbakagang i Dalarna. Information fran Sötvattens-Laboratoriet Drottningholm 4, 72 pp
Hanson JM, Mackay WC, Prepas EE (1989) Effect of size-selective predation by muskrats on a population of unionid clams. J Anim Ecol 58:15–28
Harms W (1908) Die postembryonale Entwicklung von *Unio pictorum* und *Unio tumidus*. Zool Anz 32:693–703
Harms W (1909) Postembryonale Entwicklungsgeschichte der Unioniden. Zool Jahrb 28: 325–386
Hartfield P, Hartfield E (1996) Observations on the conglutinates of *Ptychobranchus greeni*. Am Midl Nat 135:370–375
Haukioja E, Hakala T (1978) Life-history evolution in *Anodonta piscinalis* (Mollusca, Pelecypoda). Oecologia 35:253–266
Heard WH (1970) Hermaphroditism in *Margaritifera falcata*. Nautilus 83:113–114
Heard WH (1975) Sexuality and other aspects of reproduction in *Anodonta* (Pelecypoda: Unionidae). Malacologia 15(1):81–103
Heino M, Kaitala V (1996) Optimal resource allocation between growth and reproduction in clams: why does indeterminate growth exist? Funct Ecol 10:245–251
Hendelberg J (1961) The freshwater pearl mussel. Rep Inst Freshwater Res Drottningholm 41: 149–171
Hochwald S (1988) Untersuchungen zur Populationsökologie und Fortpflanzungsbiologie der Bachmuschel *Unio crassus* (Phil.) 1788. MS Thesis, University Bayreuth, 86 pp
Hochwald S (1997) Das Beziehungsgefüge innerhalb der Größenwachstums- und Fortpflanzungsparameter bayerischer Bachmuschelpopulationen (*Unio crassus* Phil. 1788) und dessen Abhängigkeit von Umweltfaktoren. PhD Thesis, University Bayreuth, 172 pp
Howard AD (1915) Some exceptional cases of breeding among Unionidae. Nautilus 29:4–11
Hruska J (1992) The freshwater pearl mussel in South Bohemia. Arch Hydrobiol 126:181–191
Hüby B (1988) Zur Entwicklungsbiologie der Fließgewässermuschel *Pseudanodonta complanata*. PhD Thesis, Tierärztliche Hochschule, Hannover, 111 pp
Hughes RN, Roberts DJ (1980) Reproductive effort of winkles with contrasted methods of reproduction. Oecologia 47:130–136
Isom BG, Hudson RG (1984) Freshwater mussels and their fish hosts; physiological aspects. Am Soc Parasitol 70(2):318–319
Israel W (1913) Biologie der europäischen Süßwassermuscheln. KG Lutz, Stuttgart
Johnson PD, Brown KM (1998) Intraspecific life history variation in the threatened Lousiana pearlshell mussel, *Margaritifera* hembeli. Freshwater Biol 40:317–329
Jokela J, Mutikainen P (1995) Phenotypic plasticity and priority rules for energy allocation in a freshwater clam: a field experiment. Oecologia 104:122–132
Jokela J, Palokangas P (1993) Reproductive tactics in *Anodonta* clams: parental host recognition. Anim Behav 46:618–620
Karna DW, Millemann RE (1977) Glochidiosis of salmonid fishes. III. J Parasitol 63:728–733

Karna DW, Millemann RE (1978) Glochidiosis of salmonid fishes. III. Comparative susceptibility to natural infection With *Margaritifera margaritifera* (L.) (Pelecypoda: Margaritanidae) and associated histopathology. J Parasitol 64(3):528–537
Kat PW (1984) Parasitism and the Unionacea (Bivalvia). Biol Rev 59:189–207
Kozlowski J (1991) Optimal energy allocation models – an alternative to the concepts of reproductive effort and cost of reproduction. Acta Oecol 12:11–33
Lang K (1998) Die Bachmuschel *Unio crassus* (Philipsson 1788) im Niederbach: Bestandssituation, Fortpflanzungsbiologie und Wanderverhalten. MS Thesis, University Freiburg, 76 pp
Lechner S (1997) Populationsökologische Untersuchungen an der Bachmuschel *Unio crassus* (Philipsson 1788) im Einzugsgebiet der hessischen Kinzig. MS Thesis, University Freiburg, 91 pp
Lefevre G, Curtis W (1912) Studies on the reproduction and artificial propagation of freshwater mussels. Bull Bur Fish 756:1–190
Maaß S (1987) Untersuchungen zur Fortpflanzungsbiologie einheimischer Süßwassermuscheln der Gattung Unio. PhD Thesis, Tierärztliche Hochschule, Hannover, 107 pp
Mikulic LV (1954) Plodovitost daurskoj zemcuznicy. Bull Fish Ocean 39:357–358
Müller D (1995) Populationsökologie der grossen Teichmuschel *Anadonta cygnea* (L.) in Seen des Salzburger Alpenvorlandes. MS Thesis, University Salzburg, 107 pp
Murphy G (1942) Relationship of the fresh-water mussel to trout in the Truckee River. Calif Fish Game 28/2:89–102
Nagel K-O (1988) Anatomische, morphologische und biochemische Untersuchungen zur Taxonomie und Systematik der europäischen Unionacea. PhD Thesis, Gesamthochschule Kassel
Nagel K-O (1991) Gefährdete Flußmuscheln in Hessen. Wachstum, Reproduktionsbiologie und Schutz der Bachmuschel (Bivalvia: Unionidae: *Unio crassus*. Z Angew Zool 78(2):205–218
Nagel K-O (1997) Flußmuscheln: Bestandsentwicklung und Schutzmöglichkeiten in Hessen. Mitt Auenzentrum Hessen 1:61–72
Negus C (1966) A quantitative study of growth and production of unionid mussels in the River Thames at Reading. J Anim Ecol 35:513–532
Niemeyer B (1993) Vergleichende Untersuchungen zur bionomischen Strategie der Teichmuschelarten *Anodonta cygnea* L. und *Anodonta anatina* L. PhD Thesis, University Hannover, 143 pp
Pianka ER, Parker WS (1975) Age-specific reproductive tactics. Am Nat 109:454–464
Richard PE, Dietz TH, Silverman H (1991) Structure of the gill during reproduction in the unionids *Anodonta grandis*, *Ligumia subrostrata*, and *Carunculina parva texasensis*. Can J Zool 69:1744–1754
Schneidt A (1998) Sympatrische und allopatrische *Anodonta*-Vorkommen in Südbaden: Bestandssituation, Fortpflanzungsstrategie und Konkurrenz. MS Thesis, University Freiburg, 75 pp
Shinozaki K (1963) Note on the species-area curve. Proc 10th Annu Meet Ecol Soc Jpn Tokyo, p 5
Smith-Gill SJ (1983) Developmental plasticity: developmental conversion versus phenotypic modulation. Am Zool 23:47–55
Spärk R (1936) On the relation between metabolism and temperature in some marine lamellibranches and its zoogeographical significance. Kongelige Danske Videnskabernes Selskabb. Biol Medd K Dan Videns Selsk 13:1–27
Stearns SC (1977) The evolution of life history traits; a critique of the theory and a review of the data. Annu Rev Ecol Syst 8:145–171
Surber T (1912) Identiffication of the glochidia of freshwater mussels. Bureau of Fisheries Document 771. Government Printing Office, Washington, DC
Tankersley RA, Dimock RV (1992) Quantitative analysis of the structure and function of the marsupial gills of the freshwater mussel *Anodonta cataracta*. Biol Bull 182:145–154

Tankersley RA, Dimock RV (1993) The effect of larval brooding on the filtration rate and particle-retention efficiency of *Pyganodon cataracta* (Bivalvia: Unionidae). Can J Zool 71: 1934–1944

Teiber P (1991) Die Vorkommen der Teichmuschel-Arten *Anodonta cygnea* (L.) und *Anodonta piscinalis* (Nilsson) in ausgewählten oberschwäbischen Seen in Abhängigkeit von Lebensraumfaktoren. MS Thesis, University Ulm, 73 pp

Terberger J (1999) Wirtsbeziehungen bei *Anodonta* (Bivalvia; Unionidae). MS Thesis, University Freiburg

Teutsch S (1997) Populationsbiologische Untersuchung an *Anodonta*: Fortpflanzungsstrategien und Charakterisierung von Teilpopulationen innerhalb eines Stillgewässers. MS Thesis, University Freiburg, 82 pp

Trdan RJ, Hoeh WR (1982) Eurytopic host use by two congeneric species of freshwater mussels. Am Midl Nat 108:381–388

Trübsbach J (1998) Fortpflanzungsbiologie und Mortalität der Gemeinen Teichmuschel *Anodonta anatina* (Linnaeus 1758). MS Thesis, University Freiburg, 58 pp

Tudorancea C (1972) Studies on Unionidae populations from the Crapani-Jijila complex of pools. Hydrobiologia 39:527–561

Tudorancea C, Gruia L (1968) Observations on the *Unio crassus* Philipsson population from the Nera River. Trav Mus Hist Nat Grigore Antipa 8:381–394

Tuomi J, Hakala T, Haukioja E (1983) Alternative concepts of reproductive effort, costs of reproduction, and selection in life-history evolultion. Am Zool 23:25–34

Vahl O (1981) Age-specific reproductive value and reproductive effort in the Iceland Scallop. Oecologia 51:53–56

Valerius K (1991) Populationsbeschreibungen und Dispersionsmessung bei Fluß- und Teichmuscheln (Bivalvia: Unionidae) der Haffenschen Landwehr am Niederrhein. MS Thesis, University Köln, 109 pp

Vannote RL, Minshall GW (1982) Fluvial processes and local lithology controlling abundance, structure, and composition of mussel beds. Proc Natl Acad Sci USA 79:4103–4107

Vannote RL, Minshall GW, Cummins KW, Sedall JR, Cushing CE (1980) The river continuum concept. Can J Fish Aquat Sci 37:130–137

Walford LA (1946) A new graphic method of describing the growth of animals. Biol Bull Mar Biol Lab Woods Hole 90:141–147

Watters GT (2000) Morphology of the conglutinate of the kidneyshell freshwater mussel, *Ptychobranchus fascionaris*. Invertebr Biol (in press)

Welte R (1999) Wirtserkennung bei der Teichmuschel *Anodonta cygnea* und der Bachmuschel *Unio crassus*. Ms Thesis, University Freiburg

Yeager BL, Neves RJ (1986) Reproductive cycle and fish hosts of the rabbits foot mussel in the upper Tennessee River drainage. Am Midl Nat 116:329–340

Yokley P (1972) Life history of *Pleurobema cordatum*. Malacologia 11:351–364

Zale AV, Neves RJ (1982) Reproductive biology of four freshwater mussel species (Mollusca: Unionidae) in Virginia. Freshwater Invertebr Biol 1(1):17–28

Ziuganov V, Zotin A, Nezlin L, Tretiakov V (1994) The freshwater pearl mussels and their relationships with salmonid fish. VNIRO, Moscow, 104 pp

14 A Phylogenetic Perspective on the Evolution of Morphological and Reproductive Characteristics in the Unionoida

WALTER R. HOEH, ARTHUR E. BOGAN, and WILLIAM H. HEARD

14.1 Introduction

Unionoid bivalves have a long history of classification and reclassification. Pre-Darwin classifications were deductive and artificial, reflecting character selection and weighting. More recent classifications, which should be natural and inductive to reflect phylogeny, reveal a diversity of methodologies and results.

The nonphylogenetic classifications of Simpson (1900), Modell (1942), and Starobogatov (1970) included the establishment of numerous new higher taxa, viz., genus-group, family-group, and genus- and family-group taxa, respectively, based largely on shell characters. (For a description of the "Russian school" of bivalve systematics, see Shikov and Zatravkin 1991 and Korniushin 1998).

Ortmann (1912, 1921) characterized family-group taxa of unionoideans (Margaritiferidae; Unionidae: Unioninae, Anodontinae, Lampsilinae; Mutelidae: Hyriinae, Mutelinae) on the basis of anatomical, reproductive, and developmental features. Parodiz and Bonetto (1963) later stressed larval differences in dividing Ortmann's Mutelidae into Unionoidea: Hyriidae and Muteloidea: Mutelidae and Mycetopodidae. [However, Kabat (1997) recently made some nomenclatural corrections: Iridinidae Swainson has priority over Mutelidae Gray, and Etherioidea Deshayes has priority over Muteloidea Gray].

Heard and Guckert (1970) stressed anatomical and reproductive differences in characterizing Unionoidea: Margaritiferidae and Amblemidae (both tetragenous, i.e., all four demibranchs are used for brooding), Hyriidae (endobranchous, i.e., only the inner two demibranchs are used for brooding), and Unionidae (ectobranchous, i.e., only the outer two demibranchs are used for brooding), and they presented dendrograms to suggest affinities of these higher groupings.

The aforementioned studies were conducted by nonphylogenetic systematists and evolutionary taxonomists, which led to inconsistencies. It remained for Davis and Fuller (1981) to apply a more objective philosophy and methodology based on a form of biochemical systematics, viz., a phenetic analysis of immunoelectrophoretic data. In examining some North American unionoideans, they reported the following family-level groups: Unionidae: Margaritiferinae, Anodontinae, and Ambleminae: Amblemini, Gonideini, Pleurobemini, and Lampsilini. In interpreting relationships based on an analysis of mitochondrial DNA (16S ribosomal gene), Lydeard et al. (1996) elevated Margaritiferinae to Margaritiferidae but found the same other taxa as did Davis and Fuller (1981). All of the unionoid reproductive characters evaluated in Davis and Fuller (1981) and Lydeard et al. (1996) were found to be homoplasious. However, a limitation of the latter two studies was their focus on only North American unionoids. These studies' evaluations of morphological character state

Ecological Studies, Vol. 145
G. Bauer and K. Wächtler (eds.) Ecology and Evolution of the Freshwater Mussels Unionoida

transitions were based on estimates of phylogeny for a nonmonophyletic group (i.e., North American unionoids). The presence of anodontine and margaritiferine higher taxa in Eurasia suggests that North American unionoids are at best a paraphyletic group (i.e., not all of the descendants of the ancestral lineage are represented as part of the group). The analyses of nonmonophyletic groups to estimate phylogeny and character state transitions can produce misleading inferences (e.g., see Eldredge and Cracraft 1980; Brooks and McLennan 1991). Recently, the relationships of North American unionoids to the faunas of other continents were clarified as Hoeh et al. (1998a) revised some conclusions on the basis of mitochondrial DNA analysis of species from Africa, South America, and Australasia.

The above brief discussion suggests that there is little congruence amid current estimates of evolutionary relationships for unionoid bivalve higher taxa. Furthermore, this plethora of hypotheses regarding unionoid evolutionary relationships leads to multiple discordant inferences regarding unionoid morphological character evolution. Two relatively recent treatments of unionoid evolutionary relationships (Heard and Guckert 1970; Hoeh et al. 1998a) will be used herein to illustrate this point. Heard and Guckert (1970: Fig. 1), somewhat following Hannibal (1912), explicitly diagrammed the following set of evolutionary relationships for unionoid families: (Margaritiferidae, (((Iridinidae, Mycetopodidae), Hyriidae), Unionidae)) (Fig. 14.1A).

In contrast, Hoeh et al. (1998a: Fig. 2) presented a phylogeny that indicates the following evolutionary relationships: ((((Unionidae [excluding *Coelatura*], Margaritiferidae), *Coelatura*), (Iridinidae, Mycetopodidae)), Hyriidae) (Fig. 14.1B). *Coelatura* is an African freshwater mussel genus, described by Conrad in 1853, that has tradi-

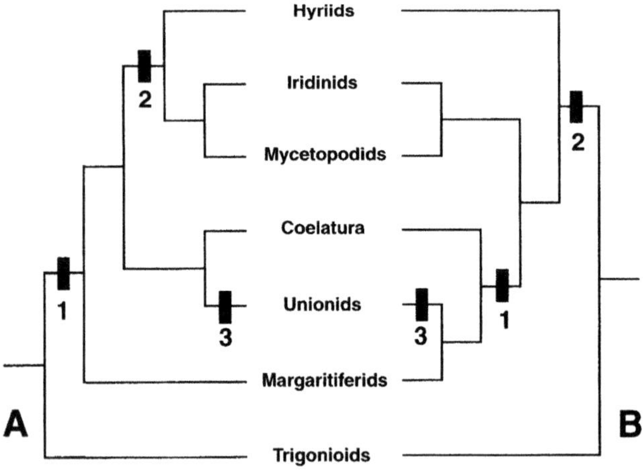

Fig. 14.1. Unionoid relationships, based on morphological characteristics, after Heard and Guckert (1970) (*A*) and, based on DNA sequences, modified from Hoeh et al. (1998a) (*B*). Each of the two hypotheses indicates a single origin for the Unionoida, but differs in the implications for the evolution of larval brooding location. Hypothesis *A* indicates that tetrageny (*1*) is ancestral for the Unionoida with endobranchy (*2*) and ectobranchy (*3*) as derived character states. Hypothesis *B* suggests that endobranchy (*2*) is ancestral for the Unionoida with tetrageny (*1*) and ectobranchy (*3*) as derived character states

tionally been placed within the Unionidae (e.g., Heard and Guckert 1970). Notwithstanding, the former study was based on evaluations of select anatomical and phenological characteristics while the latter was based on an analysis of mitochondrial DNA sequences obtained from the cytochrome c oxidase subunit I (COI) gene. Not surprisingly, the two studies produced very different estimates of unionoid evolutionary relationships. This comparison points out the high degree of incongruence among current hypotheses of unionoid evolutionary relationships, particularly with respect to the placement of the Hyriidae and Margaritiferidae. This incongruence impedes the understanding of unionoid character state transitions. For example, character mapping (e.g., Maddison and Maddison 1997) on the Heard and Guckert (1970) hypothesis (Fig. 14.1A) most parsimoniously suggests that tetragenous brooding, as in the Margaritiferidae, is the ancestral character state for unionoid bivalves. Alternatively, the Hoeh et al. (1998a) hypothesis (Fig. 14.1B) implies that endobranchous brooding is the ancestral condition for this taxon. Furthermore, the latter study also suggests that (1) the glochidium is the ancestral larval type, (2) the ventrally hooked glochidium evolved at least twice within the Unionoida, and (3) some of the other presumed ancestral character states found in the Margaritiferidae, such as the lateral muscle scars, anatomically incomplete diaphragm, and lack of interlamellar gill septa, are actually derived characteristics.

The current phylogenetic uncertainty within the Unionoida does not facilitate the testing of evolutionary hypotheses. Rigorous analyses of multiple data sets, individually and in combination, have led to the construction of more robust estimates of evolutionary history (e.g., Kluge 1989, 1998; de Queiroz et al. 1995; Hillis 1995; Miyamoto and Fitch 1995; Whiting et al. 1997). Therefore, the establishment of a robust estimate of phylogenetic relationships for the higher taxa within the Unionoida worldwide, based on all of the available data, will enable critical evaluations of evolutionary hypotheses concerning unionoid bivalve (1) origins, (2) larval characteristics, and (3) marsupial and other anatomical character state transitions. Furthermore, robust phylogenetic hypotheses will facilitate the construction of a more stable classification of unionoid bivalves that denotes monophyletic groups and their evolutionary relationships. To these ends, a morphological data matrix was analyzed alone and in combination with previously utilized DNA sequences (Hoeh et al. 1998a) to generate estimates of phylogeny for the major lineages within the Unionoida. The use of both morphological and molecular systematic analyses has been helpful in situations where analyses of single data sets were inconclusive (e.g., Whiting et al. 1997).

14.2 Materials and Methods

14.2.1 Organisms

The 31 bivalve species examined in this study are as follows:

Out-group
Order Trigonioida
  Superfamily Trigonioidea
    Family Trigoniidae
      Neotrigonia margaritacea (Lamarck, 1804)

In-group
Order Unionoida
Superfamily Etherioidea
Family Iridinidae
Mutela dubia (Gmelin, 1791)
Mutela rostrata (Rang, 1835)
Family Mycetopodidae
Anodontites guanarensis Marshall, 1927
Anodontites trigonus (Spix, 1827)
Monocondylaea minuana d'Orbigny, 1835
Superfamily Unionoidea
Family Hyriidae
Castalia stevensi (H.B. Baker, 1930)
Diplodon deceptus Simpson, 1914
Hyridella menziesi (Gray, 1843)
Lortiella rugata (Sowerby, 1868)
Velesunio angasi (Sowerby, 1867)
Family Margaritiferidae
Cumberlandia monodonta (Say, 1829)
Margaritifera margaritifera (Linnaeus, 1758)
Family Unionidae
Actinonaias ligamentina (Lamarck, 1819)
Amblema plicata (Say, 1817)
Anodonta cygnea (Linneaus, 1758)
Coelatura aegyptiaca (Cailliaud, 1827)
Cyrtonaias tampicoensis (Lea, 1838)
Elliptio dilatata (Rafinesque, 1820)
Fusconaia flava (Rafinesque, 1820)
Glebula rotundata (Lamarck, 1819)
Gonidea angulata (Lea, 1838)
Ligumia recta (Lamarck, 1819)
Pleurobema clava (Lamarck, 1819)
Potamilus alatus (Say, 1817)
Pyganodon grandis (Say, 1829)
Quadrula quadrula (Rafinesque, 1820)
Strophitus undulatus (Say, 1817)
Toxolasma lividus (Rafinesque, 1831)
Unio pictorum Linnaeus, 1758
Unio tumidus Retzius, 1788

14.2.2 Methods

Unionoid morphological character state data, representing attributes of the shell, soft tissue anatomy, and larval structures, were obtained from direct observation of specimens and/or from the literature (Tables 14.1, 14.2).

The DNA sequence data (630 base pairs of COI, with only transversions coded at third positions) are those analyzed in Hoeh et al. (1998a). The suitability

Table 14.1. Morphological characters and character states based on literature and specimen examination

1. Supra-excurrent opening
   0 = absent, 1 = present
2. Posterior end of mantle sheets
   0 = unfused, with simple incurrent and excurrent openings, 1 = with a simple incurrent opening, but fused to provide a short, muscular excurrent siphon, 2 = fused to provide short, muscular incurrent and excurrent siphons
3. Incurrent papillae
   0 = absent, 1 = present, simple, 2 = present, branched, 3 = present, arboresecent, 4 = present, reduced, crenulations
4. Mantle margin of females with specialized structures (e.g., flaps, caruncles, etc.) anteroventral to the incurrent opening
   0 = not applicable, 1 = absent, 2 = present
5. Position of anus on the posterior adductor muscle
   0 = dorsal edge of posterior adductor muscle, 1 = posterior edge of posterior adductor muscle, 2 = posteroventral margin of posterior adductor muscle, 3 = ventral margin of posterior adductor muscle
6. Intestine complexity
   0 = simple, undifferentiated, 1 = complex, with three compartments
7. Attachment of the dorsal margin of the outer lamella of the outer demibranchs to the inner surface of the mantle
   0 = except at the posterior end of those demibranchs, 1 = for the entire length of those demibranchs
8. Attachment of the dorsal margin of the inner lamella of the inner demibranchs to the visceral mass
   0 = only at the anterior region of those demibranchs, 1 = for the entire length of the visceral mass
9. Diaphragm (tissue separation of the suprabranchial and branchial components of the mantle cavity)
   0 = complete, formed in part by the dorsal margin of the inner lamella of the inner demibranchs and in part by the siphonal musculature, 1 = incomplete, with a single perforation in the siphonal musculature, 2 = incomplete, formed only by the inner demibranchs
10. Number and location of marsupial demibranchs
    0 = not applicable, 1 = endobranchous (only the inner two), 2 = tetragenous (all four), 3 = ectobranchous (only the outer two)
11. Marsupial region of marsupial demibranchs seasonally extends below the ventral margin of the filaments
    0 = not applicable, 1 = no, 2 = yes
12. Interlamellar space of all marsupial and nonmarsupial demibranchs divided into vertical water tubes by vertical, transverse interlamellar septa
    0 = not applicable, 1 = absent, 2 = present
13. Relative number and spacing of transverse (primary) marsupial septa
    0 = not applicable, 1 = greater in marsupial than in nonmarsupial regions of marsupial and in nonmarsupial demibranchs, 2 = similar in marsupial and nonmarsupial regions of marsupial and in nonmarsupial demibranchs
14. Completeness of transverse (primary) septa
    0 = not applicable, 1 = vertically perforated in marsupial but imperforate in nonmarsupial regions of marsupial demibranchs and throughout nonmarsupial demibranchs, 2 = vertically imperforate throughout marsupial and nonmarsupial demibranchs, 3 = vertically perforated in marsupial and nonmarsupial demibranchs

Table 14.1. *(Continued)*

15. Tripartite water tubes
    0 = absent, 1 = present, 2 = not applicable
16. Lateral ridges on primary septa approximating secondary vertical septa
    0 = absent, 1 = present, 2 = not applicable
17. Larval type
    0 = veliger, 1 = glochidium, 2 = lasidium, 3 = haustorium
18. Medioventral glochidial hooks
    0 = absent, 1 = present, 2 = not applicable
19. Glochidial shape (lateral view)
    0 = not applicable, 1 = subtriangular, 2 = subcircular, 3 = subovate, 4 = celtiform
20. Shell disk sculpture
    0 = absent, 1 = plications, 2 = zigzag, 3 = radial, 4 = pustules
21. Lateral teeth dentition
    0 = absent, 1 = reduced, 2 = present, 3 = not applicable, 4 = pseudotaxodont
22. Pseudocardinal teeth dentition
    0 = absent, 1 = reduced, 2 = present, 3 = not applicable
23. Lamellar-ligament fossette
    0 = v-shaped, 1 = otherwise
24. Beak sculpture
    0 = absent, 1 = double-looped, 2 = radial, 3 = concentric bars, 4 = zigzag
25. Lateral muscle scars
    0 = absent, 1 = present
26. Labial palp shape
    0 = triangular, 1 = semicircular to kidney-shaped, 2 = falciform
27. Anterior adductor muscle shape
    0 = elongate, 1 = round
28. Marsupium restriction
    0 = absent, 1 = middle, 2 = posterior, 3 = not applicable

of the morphology dataset for phylogenetic analyses at this hierarchical level was evaluated using the g_1 statistic of a random tree distribution (from 10000000 random trees; e.g., see Hillis 1991; Hillis and Huelsenbeck 1992) and the PTP (permutation tail probability) test (Archie 1989; Faith and Cranston 1991) as implemented in PAUP* (Swofford 1998). Prior findings from plots of substitution pattern, g_1 statistics, and PTP testing are consistent with the hypothesis that significant phylogenetic signal exists in the transformed COI nucleotide data matrix utilized herein and validate its use in this particular phylogenetic context (Hoeh et al. 1998a,b).

Phylogenetic analyses were carried out on the morphology matrix and on the combined morphology + COI sequence matrix using the maximum parsimony (MP) algorithm in PAUP*. Based on previous morphological (e.g., see Atkins 1937; Taylor et al. 1969, 1973; Tevesz 1975; Popham 1979; Tevesz and Carter 1980; Smith 1986; Healy 1989; Waller 1990, 1998) and molecular (Hoeh et al. 1998b) systematic analyses, which indicated that the Trigonioida was the sister taxon to a monophyletic Unionoida, *Neotrigonia margaritacea* was used to root the resulting topologies. One thousand random terminal taxa addition order runs, combined with global branch rearrangement options, were employed to generate topologies (all changes received equal weight and were unordered). These options increased the probability of finding the best topology under the parsimony optimality criterion (e.g., see Maddison 1991).

Table 14.2. Data matrix based on 28 morphological characters. The out-group is listed first followed by all of the unionoid taxa in alphabetical order. Characters are numbered right to left and correspond to the characters and character states listed in Table 14.1

	1111111111222222222 1234567890123456789012345678
Neotrigonia	00312?1020000022020333121003
Actinonaias	1011101123121200103022130210
Amblema	1011101022121200102122110210
Anodonta	1011101023121210111000130210
Anodontites gua.	001111101122201 2??000000100
Anodontites tri.	0011111011222012??000000100
Castalia	0111101101121100111322120011
Coelatura	101110102212110010222214021?
Cumberlandia	00410000 2211????102011131210
Cyrtonaias	10111010232212001 0?022110210
Diplodon	0111201101121100111011120211
Elliptio	1011101023121200103022130210
Fusconaia	1021101022121200102022130210
Glebula	1011?01123221200103022100212
Gonidea	1021?01022122300102022130210
Hyridella	0111??1111121100111011140011
Ligumia	1012101123221200103022110112
Lortiella	???1??1??1121100 11?011100???
Margaritifera	0011000 02211????102011131210
Monocondylaea	011121110112 2?012??002020100
Mutela dub.	0201??11011222003??04000011?
Mutela ros.	0201??11011222003??04000011?
Pleurobema	1021101023121200103022130210
Potamilus	1011101123221200104022130012
Pyganodon	1011101023121210111000110210
Quadrula	1031101022121200103422130210
Strophitus	1011101023121210111000130210
Toxolasma	1012?01023221200103022130012
Unio pic.	1011101023121100111022110210
Unio tum.	1011101023121100111022110210
Velesunio	0?11??1111121100111022100011

The robustness of the resulting topologies was evaluated by bootstrap analyses (1000 replicates).

In order to evaluate the level of incongruence between the topologies generated from analyses of the morphology and the DNA sequence matrices (1) an overall strict consensus tree was constructed from the two strict consensus trees obtained from the separate analyses and (2) the incongruence length difference test (ILD, Farris et al. 1994) was performed (using the partition homogeneity test as implemented in PAUP*). In addition, character mapping, using MacClade (Maddison and Maddison 1997), was carried out on the topologies generated to investigate their implications for morphological character evolution within the Unionoida.

14.3 Results

PTP testing ($P = 0.001$, 1000 replicates) and g_1 statistics ($g_1 = -0.655$, $P < 0.01$; from an evaluation of 10000000 random trees) suggest that the morphology data matrix contains significant hierarchical structure and thus is suitable for phylogenetic analysis. The parsimony analysis of the morphology matrix produced 240 equally parsimonious trees of 94 steps (retention index = 0.8121). The strict consensus tree of the 240 equally parsimonious trees is displayed in Fig. 14. 2.

The ILD test (i.e., the partition homogeneity test in PAUP*) determined that the morphology and DNA sequence matrices were significantly incongruent with one another ($P = 0.016$ 500 replicates). This finding was reinforced by the relative lack of resolution displayed in the overall strict consensus tree (Fig. 14.3).

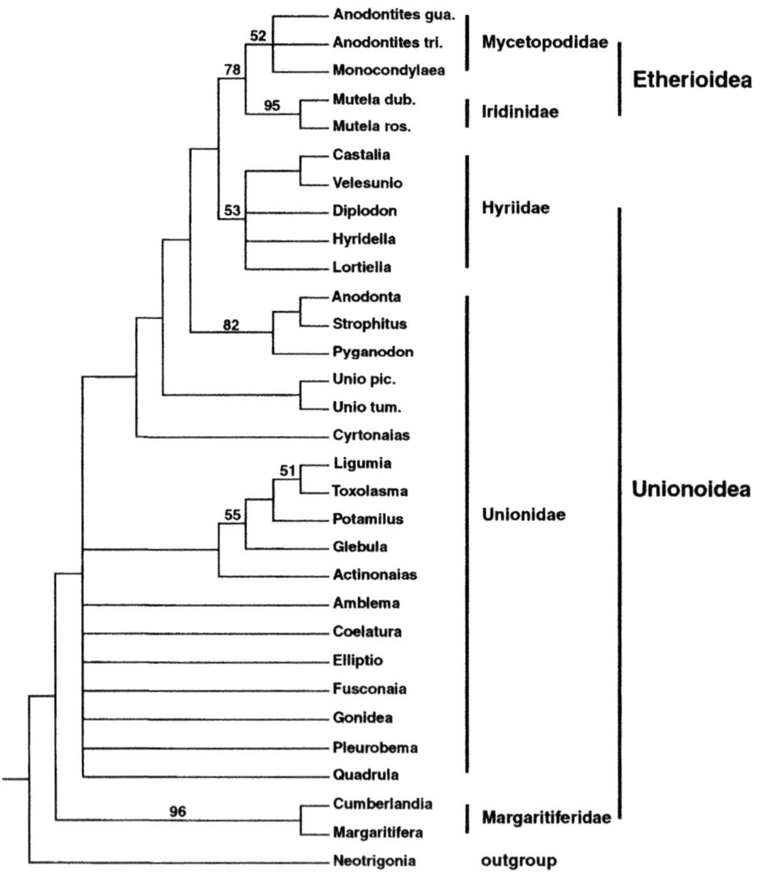

Fig. 14.2. Strict consensus tree topology derived from 240 equally parsimonious trees produced by maximum parsimony (MP) analysis of the morphology matrix. Numerals are bootstrap percentages for the MP analysis (1000 replicates). Only bootstrap values greater than 50% are shown

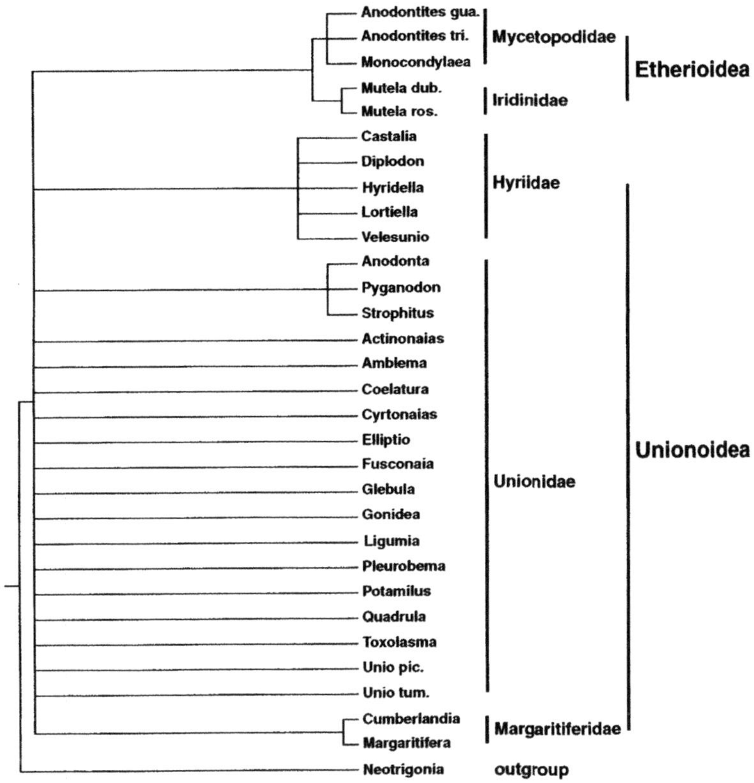

Fig. 14.3. Overall strict consensus tree topology showing the clades in agreement between the strict consensus tree obtained from the morphological analysis presented herein and the strict consensus tree obtained from analyses of DNA sequences presented in Hoeh et al. (1998a)

However, this same strict consensus tree does indicate six clades (i.e., Anodontinae, Hyriidae, Margaritiferidae, Iridinidae, Mycetopodidae, Etherioidea) that are common to the best trees obtained from the two separate analyses. Notwithstanding, a parsimony analysis of the combined morphology + DNA sequence matrix was conducted which produced two equally parsimonious trees of 873 steps (retention index = 0.5864). The strict consensus tree of the two equally parsimonious trees is displayed in Fig. 14.4.

14.3.1 Higher-Level Evolutionary Relationships Within the Unionoida

14.3.1.1 Morphology Analysis

All 240 equally parsimonious trees resulting from analysis of the morphology matrix supported the monophyly of the Hyriidae, Margaritiferidae, Iridinidae, Mycetopodidae, and Etherioidea (e.g., Fig. 14.2). The evolutionary relationships among these higher taxa resemble those portrayed by Heard and Guckert (Fig. 14.1A) in that (1)

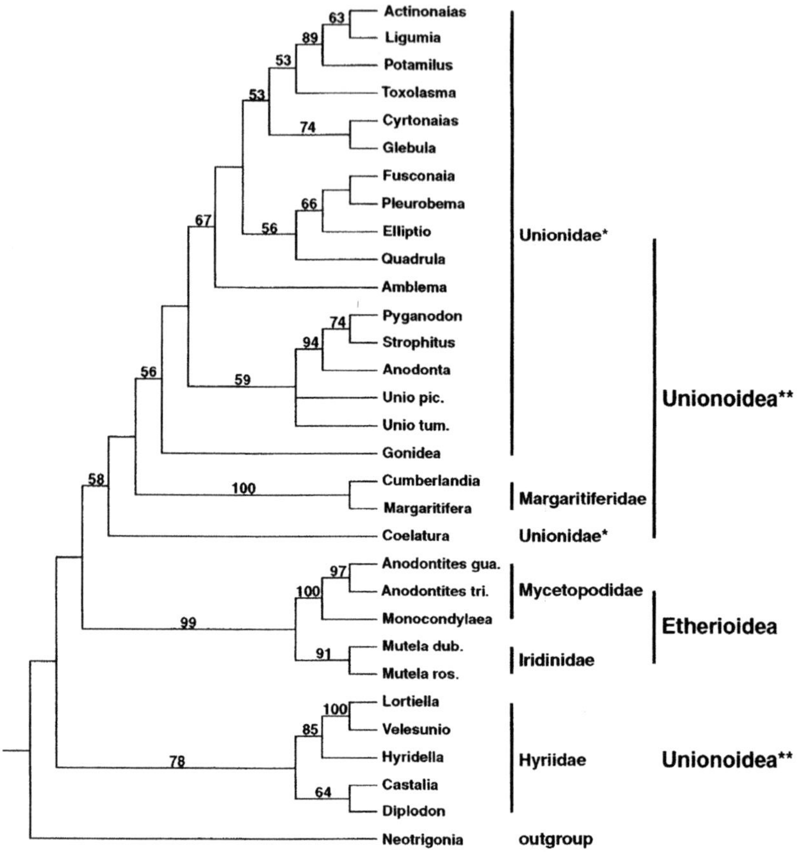

Fig. 14.4. Strict consensus tree topology derived from two equally parsimonious trees produced by maximum parsimony (MP) analysis of the combined morphology + transformed COI nucleotide matrix. Numerals are bootstrap percentages for the MP (1000 replicates) analyses. Only bootstrap values greater than 50% are shown. The paraphyly of the Unionidae (*) and Unionoidea (**) is indicated

the Etherioidea (i.e., iridinids + mycetopodids) and Hyriidae are sister taxa and (2) the margaritiferids represent a lineage descended from the basal-most cladogenic event within the ancestral unionoid line. However, questions regarding the monophyly of the Ambleminae (sensu Davis and Fuller 1981) as well as the higher-level relationships within this subfamily are largely unresolved (except for a clade containing five lampsiline genera). Support for most clades in the morphology analysis was relatively weak in that only the Anodontinae, Etherioidea, Margaritiferidae, and Iridinidae had bootstrap percentages above 70 (Fig. 14.2).

14.3.1.2 Combined Analysis

Both equally parsimonious trees from the combined morphology and COI sequence matrix (Fig. 14.4) support the monophyly of the Hyriidae, Margaritiferidae,

Iridinidae, Mycetopodidae, and Etherioidea, as did the morphology analysis. Based on bootstrap percentages, support for the monophyly of these taxa was typically greater in the combined analysis than in the morphology analysis (except for the slight decline in the bootstrap percentage for the Iridinidae). Furthermore, the strict consensus tree from the combined analysis (Fig. 14.4) supports the monophyly of the Lampsilini (represented herein by the following six genera: *Actinonaias*, *Cyrtonaias*, *Glebula*, *Ligumia*, *Potamilus*, and *Toxolasma*) and Ambleminae (sensu Davis and Fuller 1981 [excepting the position of *Gonidea*]; represented herein by the above-listed lampsiline genera plus the following five genera: *Amblema*, *Elliptio*, *Fusconaia*, *Pleurobema*, and *Quadrula*). *Gonidea* is a freshwater mussel genus, described by Conrad in 1857, that is found in western North America. Nevertheless, the evolutionary relationships among the higher taxa depicted in this analysis resemble those portrayed by Hoeh et al. (1998a) (Fig. 14.1B) in that (1) the Etherioidea (i.e., iridinids + mycetopodids) is the sister taxon to the "Unionidae + Margaritiferidae" clade and (2) the Hyriidae is the sister taxon to the "Etherioidea + Unionidae + Margaritiferidae" clade.

However, the placement of *Gonidea* varied between the combined analysis presented herein (Fig. 14.4) and that of Hoeh et al. (1998a). The latter's parsimony analysis of COI sequences placed *Gonidea* as the sister taxon to a clade composed of anodontine bivalves + *Unio*. The combined parsimony analysis presented in Fig. 14.4 indicates instead that *Gonidea* is the sister taxon to an "amblemine + anodontine + *Unio*" clade. This latter phylogenetic placement of *Gonidea* is similar to that depicted in Heard and Guckert (1970: Fig. 1). Nevertheless, the evidential support for either of these two placements of *Gonidea*, as judged by bootstrap values, is weak. In addition, two non-North American unionid genera are represented in the combined analysis: *Coelatura* and the afore-mentioned *Unio*. The Old World genus *Unio* appears closely related to anodontine bivalves while *Coelatura* represents one of the two lineages resulting from the earliest cladogenic event in the lineage ancestral to African, Eurasian, and North American unionoideans. This topology thus contradicts, as did Hoeh et al. (1998a), the perceived ancestral nature of the Margaritiferidae with respect to the Unionoida as well as to the Unionidae (e.g., see Ortmann 1911, 1919; Heard and Guckert 1970; Davis and Fuller 1981; Smith 1986). Within the Hyriidae, the Australasian (i.e., *Hyridella*, *Lortiella*, and *Velesunio*) and South American (i.e., *Diplodon* and *Castalia*) genera are contained in two geographically delimited clades which is consistent with the hyriid subfamilial classifications of Parodiz and Bonetto (1963) and Bonetto (1997). However, the combined analysis indicates that the Unionidae and Unionoidea (= Unionacea, sensu Parodiz and Bonetto 1963) are paraphyletic taxa (Fig. 14.4). That is, regarding the Unionoidea, the Unionidae + Margaritiferidae clade and the Etherioidea are depicted as sister taxa in Fig. 14.4 with the Hyriidae as the sister taxon to the Unionidae + Margaritiferidae + Etherioidea clade [rather than to the Unionidae + Margaritiferidae as traditionally believed (e.g., see Parodiz and Bonetto 1963)]. Regarding the Unionidae, the margaritiferids are represented as a clade with in the Unionidae (Fig. 14.4) rendering the latter paraphyletic. However, if the margaritiferids were treated as a subfamily within the Unionidae (following Davis and Fuller 1981), then the Unionidae, as represented herein, would be a monophyletic taxon.

14.4 Discussion

14.4.1 Unionoid Character Evolution

MacClade was used to evaluate competing hypotheses regarding the evolution of morphological and reproductive characteristics in unionoids. This goal was accomplished by determining the most parsimonious order of character state transitions based on both the morphology (Fig. 14.2) and total-evidence topologies (Fig. 14.4).

14.4.1.1 Morphology Analysis

The morphology-based topology presented in Fig. 14.2, while largely unresolved, is nevertheless suggestive regarding the order of several transitions in unionoid reproductive characteristics. Among malacologists, there has been a general belief that many of the morphological character states found in margaritiferid bivalves represent the ancestral condition for the Unionoida (e.g., see Ortmann 1912; Walker 1917; Heard and Guckert 1970; Davis and Fuller 1981; Smith 1986). Thus, the ancestral unionoid brooding arrangement has typically been regarded as tetrageny based on this presumed ancestral nature of the tetragenous Margaritiferidae (e.g., see Hannibal 1912; Ortmann 1912; Walker 1917; Heard and Guckert 1970; Davis and Fuller 1981). Nonetheless, the hypothesized evolutionary relationships presented by Parodiz and Bonetto (1963), Modell (1964), Morrison (1973), and Bonetto (1997) are consistent with the hypothesis that the endobranchus condition (i.e., only the inner demibranchs are used for brooding) is ancestral in unionoids.

Mapping the location of brooding onto the strict consensus tree topology derived from the morphology analysis (Fig. 14.2) produces a hypothesis consistent with tetrageny as the ancestral unionoid condition (Fig. 14.5).

Furthermore, this instance of character mapping suggests that the endobranchous brooding observed in etherioidean and hyriid bivalves was derived a single time from the ectobranchous condition. These findings are inconsistent with those obtained from character mapping onto topologies generated from analyses of DNA sequences which indicated endobranchy as the ancestral character state for unionoids (Hoeh et al. 1998a). The hypothesized transition from ectobranchy to endobranchy, without a tetragenous intermediate stage, seems rather unlikely based on functional arguments (Hoeh et al. 1998a). However, tetragenous intermediates could have existed in taxa, phylogenetically interposed between the extant etherioids+hyriids and anodontines (Fig. 14.5), that have subsequently become extinct.

A second contentious point in the literature regarding transitions in unionoid reproductive characteristics is the polarity of the shift between the glochidium and the lasidium/haustorium larval types. These two (or three) unionoid larval types are so distinct morphologically as to have prompted some researchers to suggest that these types could not have descended from a common ancestor [= unionoid bivalves are polyphyletic (= multiple, independent origins), Parodiz and Bonetto 1963]. However, a large body of work suggests that unionoids are indeed a monophyletic assemblage (e.g., see Atkins 1937; Taylor et al. 1969, 1973; Tevesz 1975; Popham 1979; Tevesz and Carter 1980; Smith 1986; Healy 1989; Waller 1990, 1998; Hoeh et al. 1998b). Thus, investigations to evaluate the polarity of evolutionary transitions in unionoid larval type are valid.

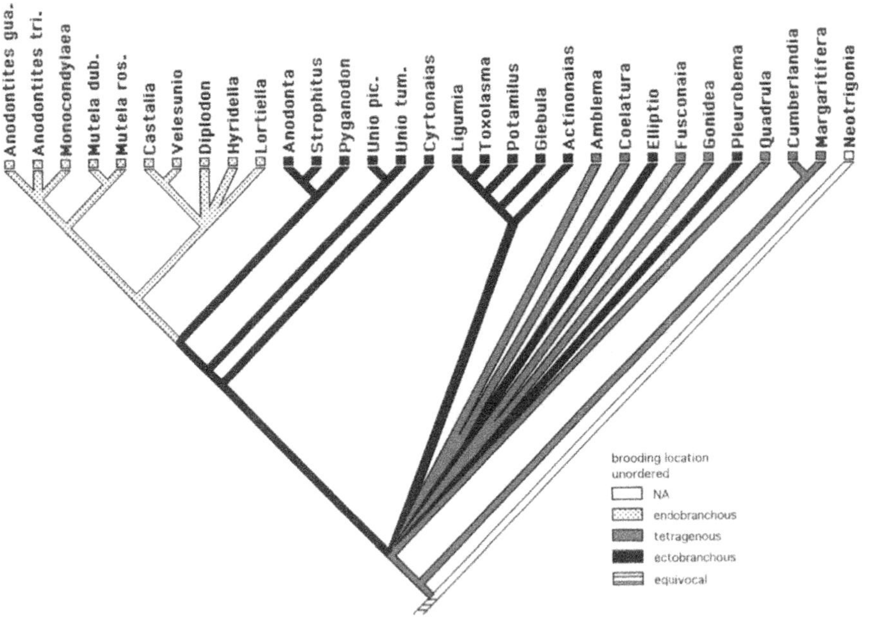

Fig 14.5. Morphological character mapping on the strict consensus tree topology produced from maximum parsimony analysis of the morphology matrix. The most parsimonious estimate for brooding location transition order suggests that tetrageny was the ancestral unionoid character state from which arose ectobranchy and, subsequently, a single origin of endobranchy in the lineage ancestral to the etherioidean + hyriid clade (*NA* = not applicable)

Given that margaritiferids possess glochidial larvae, many malacologists believe that the glochidium is the ancestral larval type for the Unionoida (e.g., see Ortmann 1912; Heard and Guckert 1970; Davis and Fuller 1981; but see Modell 1964 and Morrison 1973 for disparate opinions). Mapping unionoid larval type onto the topology represented in Fig. 14.2 corroborates the hypothesized ancestral nature of the glochidial larvae and further suggests that the derived lasidium/haustorium larval type arose a single time in the lineage ancestral to the Etherioidea (Fig. 14.6).

This particular hypothesis of larval morphology transition has recently been corroborated by a molecular systematic study (Hoeh et al. 1998a).

A third disagreement concerning the evolutionary dynamics of unionoid reproductive characteristics is the question of homology regarding medioventrally hooked glochidia. The subsequent four unionoid taxa possess medioventrally hooked glochidia: anodontines, the genera *Unio* and *Cristaria*, and most of the species within the Hyriidae. The distinct morphology of the hyriid hook led Ortmann (1921) to hypothesize independent origins for the anodontine and hyriid glochidia. Nonetheless, the morphological similarity between the hooked glochidia of anodontine bivalves and those of the genus *Unio* has prompted differing ideas regarding the evolutionary relatedness of these two groups. Hannibal (1912), Heard and Guckert (1970), and Morrison (1955, 1973) suggested a relatively close evolutionary relationship between anodontines and the genus *Unio* which necessarily indicates the possi-

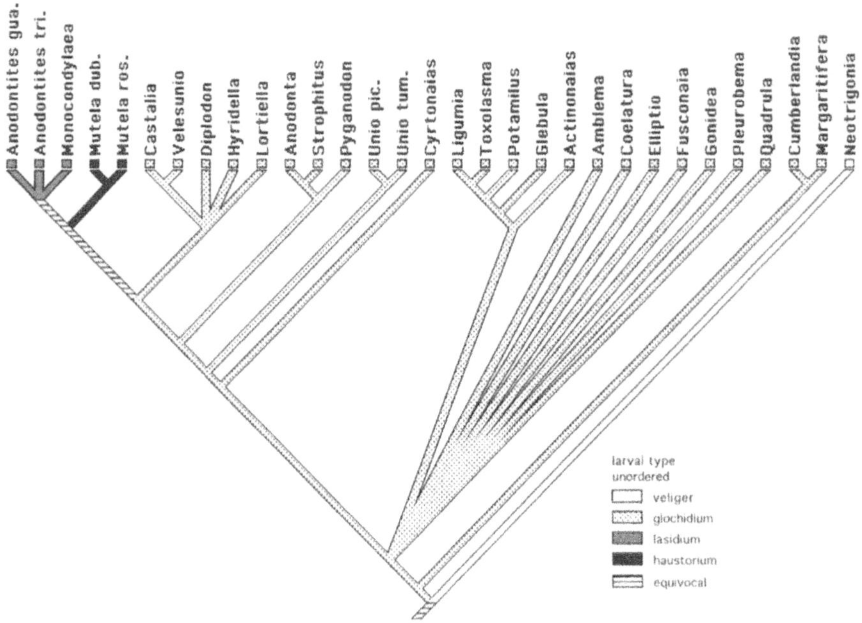

Fig. 14.6. Morphological character mapping on the strict consensus tree topology produced from maximum parsimony analysis of the unionoid morphology matrix. The most parsimonious estimate of transition order suggests that the glochidium was the ancestral unionoid larval type from which the lasidium and haustorium larvae arose in the lineage ancestral to the extant etherioideans

bility of a single origin of hooked glochidia in an ancestor common to both groups. Alternatively, both Ortmann (1912) and Davis and Fuller (1981) hypothesized a relatively distant evolutionary relationship regarding anodontines and the genus *Unio*, which suggests independent origins of hooked glochidia in the anodontine and *Unio* ancestral lineages. Mapping the presence/absence of ventrally hooked glochidia onto the topology in Fig. 14.2 indicates a single origin for this glochidial type and thus suggests that the ventral hooks are best interpreted as homologous structures (Fig. 14.7).

This single-origin hypothesis runs counter to the two-origin hypothesis corroborated recently (Hoeh et al. 1998a).

14.4.1.2 Methodological Concerns

The separate analyses of morphology (Fig. 14.2) and DNA sequences (Fig. 14.1B; Hoeh et al. 1998a) provided significantly different estimates of unionoid phylogeny and, in two of the three instances outlined above, morphological character evolution. There are a number of factors that could be responsible for the topological discrepancy between the morphological and molecular analyses. One possibility is that an insufficient number of characters have been examined for the morphological and/or molecular phylogenetic analyses. In general, larger datasets are believed to provide

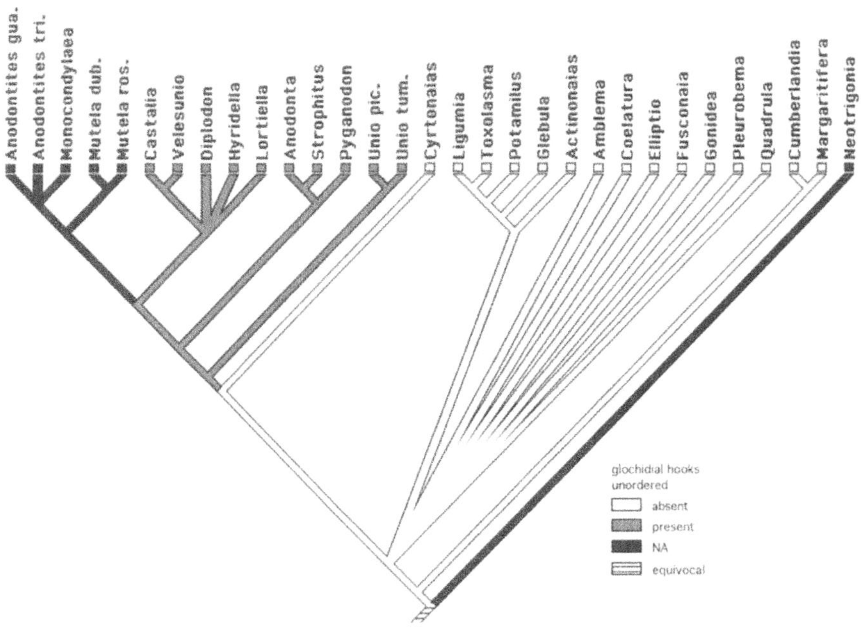

Fig 14.7. Morphological character mapping on the strict consensus tree topology produced from maximum parsimony analysis of the unionoid morphology matrix. The most parsimonious estimate for the evolution of medioventrally hooked glochidia suggests that they arose a single time (*NA* = not applicable)

more accurate estimates of phylogenetic history than smaller datasets (e.g., Hillis et al. 1994; Graybeal 1998). Given the relatively small number of unionoid characters analyzed to date, either the morphology- or molecule-based tree (or both!) could be a highly inaccurate representation of phylogeny. Additionally, it has been demonstrated that the topologies of phylogenetic trees based on analyses of gene sequences do not always accurately represent organismic evolutionary history (e.g., Pamilo and Nei 1988; Doyle 1992; Bull et al. 1993). Factors that can be responsible for this phenomenon include horizontal gene transfer (mediated via bacteria or hybridization events), the sorting of ancestral gene polymorphism, and highly variable within- and among-gene rates of molecular evolution (e.g., Avise 1994). The latter phenomenon can produce distinct topologies from phylogenetic analyses of different genetic loci for a given set of taxa (e.g., Reed and Sperling 1999). Acknowledging this potential for disparate phylogenetic outcomes, how should malacologists view the current state of affairs regarding unionoid phylogeny and character evolution?

One viewpoint might be to say that we are currently at an impasse regarding these topics and that further illumination/discussion is not possible or warranted until additional datasets are obtained. For instance, one could anticipate that the analysis of a third data set might provide a tree congruent with that produced by one of the first two datasets, thus enabling a majority rule view of unionoid phylogeny. Alternatively, this third dataset might lead to a third distinct estimate of unionoid evolutionary history rather than congruence with a prior estimate! Notwithstanding, the viewpoint of an "impasse until more data are obtained" would inhibit hypothesis tests regarding unionoid phylogeny and character evolution and is relatively nihilistic in that it ignores the hierarchical structure contained within the original two data matrices (especially since there are six clades in common between the separate morphology and DNA sequence analyses: Fig. 14.3).

A second viewpoint to deal with the disparity in topologies is to apply the principle of taxonomic congruence to the problem (e.g., Miyamoto and Fitch 1995). Taxonomic congruence requires that the best estimate of evolutionary history be the clades in absolute agreement between the topologies produced by analyses of "independent" data partitions (e.g., morphology and DNA sequences). This may be interpreted as producing a conservative hypothesis (e.g., Swofford 1991). In the present case, invoking taxonomic congruence would suggest that the overall strict consensus tree (Fig. 14.3) of the best trees produced by the separate morphology and DNA sequence analyses is the best current hypothesis of unionoid evolutionary relationships. However, the strict consensus tree produced by the taxonomic congruence method is much less well resolved than either of the trees from the original analyses and is not the globally most parsimonious estimate of evolutionary history for the group. As such, this hypothesis of evolutionary relationships has less explanatory power [i.e., less ability to ascribe similar, congruent-derived attributes to common history (e.g., Farris 1983; Miyamoto 1985; Kluge 1989)] and fails to inform regarding the historical distribution of multiple, independently evolved character states [i.e., homoplasy (e.g., Kluge 1998)]. Moreover, the rationale underlying data partitioning has been soundly criticized (e.g., Kluge 1989, 1998), as has the use of relatively unresolved consensus trees as estimates of phylogeny (e.g., Swofford 1991; Hillis 1995).

A third viewpoint to accommodate the two disparate views of unionoid evolutionary history provided by the separate phylogenetic analyses of morphology and DNA sequences is to employ instead a total-evidence (= global parsimony) approach that entails an analysis of a combined morphology + DNA sequence matrix and emphasizes character congruence rather than taxonomic congruence [i.e., it evaluates the potential agreement across all characters simultaneously to determine the best estimate of evolutionary history (e.g., Miyamoto 1985; Kluge 1989, 1998; Eernisse and Kluge 1993; Jones et al. 1993; Kluge and Wolf 1993)]. Since the current morphological and DNA character state distributions in unionoids are products of a single evolutionary history, a simultaneous analysis of all characters seems justified. The advantages of this approach to phylogeny estimation include offering (1) an explicit and testable phylogenetic hypothesis, (2) a maximization of the explanatory power of the resulting phylogenetic hypothesis, and (3) insight on the historical distribution of homoplasy. Furthermore, this technique has precedence in that it has been used in prior phylogenetic analyses of North American unionoids (Hoeh 1990; Lydeard et al. 1996). Given these factors, the phylogenetic hypothesis obtained from the combined analysis of morphology and DNA sequences (Fig. 14.4) is offered as the best current estimate of unionoid evolutionary relationships and, as such, will be used henceforth herein to evaluate character evolution within the taxon.

14.4.1.3 Combined Analysis

The morphology + COI sequence-based topology presented in Fig. 14.4 suggests that a significant reinterpretation of unionoid morphological character evolution is required. This point is illustrated in Fig. 14.8, where morphological character changes have been plotted on the morphology + COI sequence-based topology.

Notable reinterpretations of morphological character evolution include the following: (1) the presence of interlamellar septa is symplesiomorphic rather than synapomorphic for the Unionidae sensu lato (= including margaritiferids) since the present reconstruction suggests that they evolved once in the unionoid common ancestor and were subsequently lost in the ancestral margaritiferid (see Fig. 14.8: character 12) and (2) the supra excurrent opening evolved once in the common

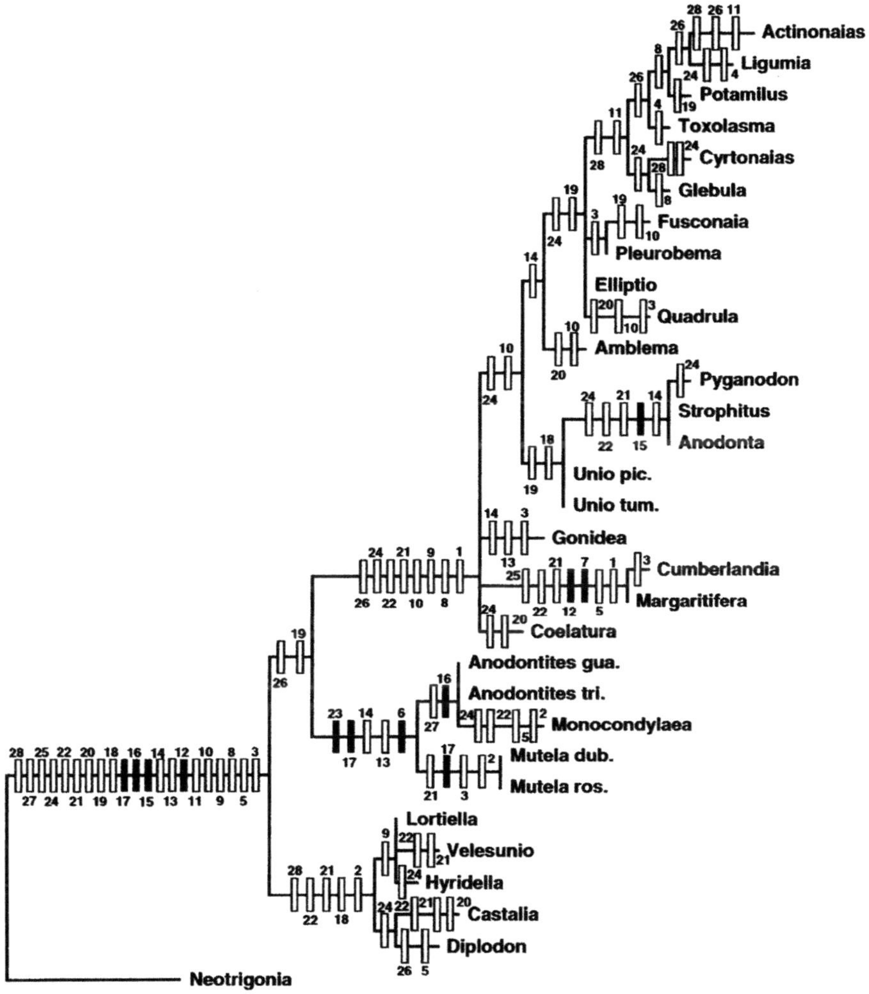

Fig. 14.8. The strict consensus tree topology from the maximum parsimony analysis of the combined morphology + transformed COI nucleotide matrix presented as a phylogram (i.e., branch lengths are proportional to the estimated number of changes on a branch) representing only the morphological changes. The numbers refer to the characters listed in Table 1 and indicate both unique (*dark bars*) and homoplasious (*open bars*) morphological synapomorphies as determined by the ACCTRAN character-state optimization algorithm in PAUP*

ancestor of the Unionidae s.l. and was subsequently lost in margaritiferids (as illustrated in Fig. 14.8: character 1) or, alternatively, the supraexcurrent opening evolved twice, once in the ancestor of *Coelatura* and once in the ancestor of the amblemine + anodontine + *Unio* + *Gonidea* clade. These two examples run counter to the claim that the presence of both interlamellar septa and the supraexcurrent opening represent unique and unreversed synapomorphies for the Unionidae sensu stricto (Lydeard et al. 1996). Assuming that this topology is a reasonable representation of unionoid phylogeny, the often-reported homoplasious nature of morphological character evolution in the Unionoida is corroborated by the analysis presented in Fig. 14.8.

The topology presented in Fig. 14.4 is also consistent with the following hypotheses: (1) endobranchy is the ancestral unionoid brooding condition and ectobranchy evolved from a tetragenous intermediate state (Fig. 14.9), (2) the glochidium is the ancestral unionoid larval condition (Fig. 14.10), and (3) the medioventrally hooked glochidium evolved at least twice; once in the lineage ancestral to the Hyriidae and once in the lineage ancestral to the *Unio* + anodontine clade (Fig. 14.11).

The ancestral nature of endobranchous brooding for the Unionoida is most likely the result of functional constraints. The paired unionoid gonopores open into the inner suprabranchial canals above the inner demibranchs (e.g., Howes 1902). Ciliary

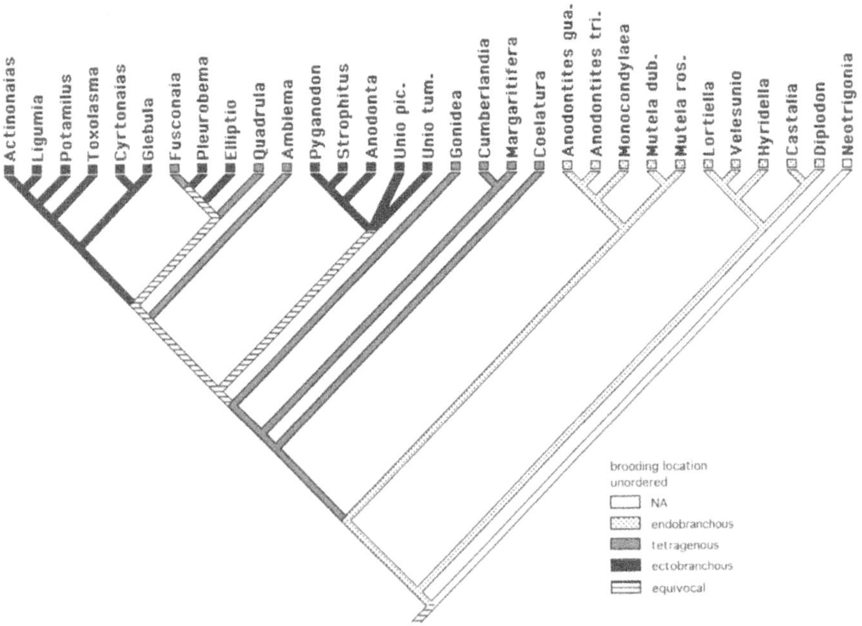

Fig. 14.9. Morphological character mapping on the strict consensus tree topology produced from maximum parsimony analysis of the combined morphology + transformed COI nucleotide matrix. The most parsimonious estimate for brooding location transition order suggests that endobranchy was the ancestral unionoid character state from which tetrageny arose in the lineage ancestral to the margaritiferid + unionid clade. Subsequently, ectobranchy arose at least once from a tetragenous ancestral lineage

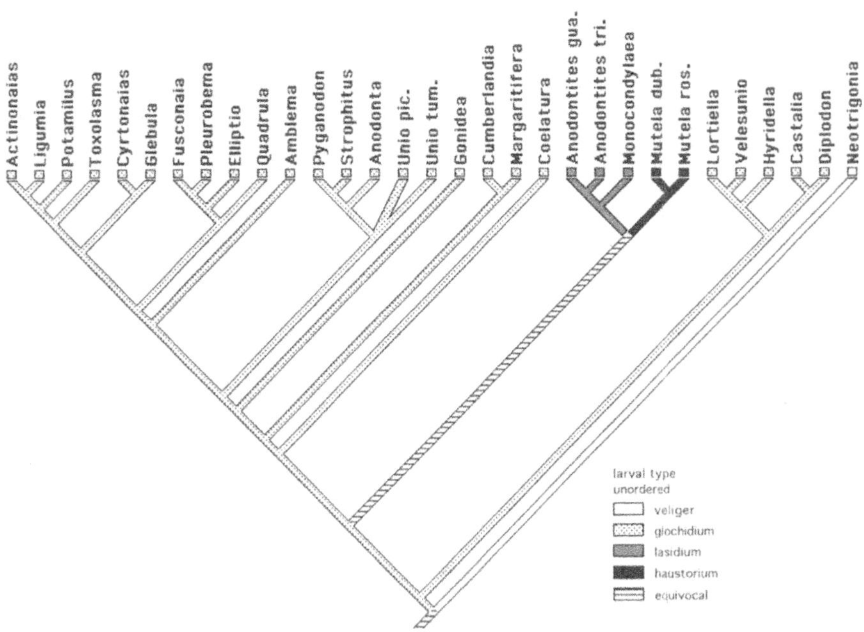

Fig. 14.10. Morphological character mapping on the strict consensus tree topology produced from maximum parsimony analysis of the combined morphology + transformed COI nucleotide matrix. The most parsimonious estimate of larval type transition suggests that the glochidium was the ancestral unionoid larval type and that the lasidium/haustorium arose a single time in the lineage ancestral to the Etherioidea

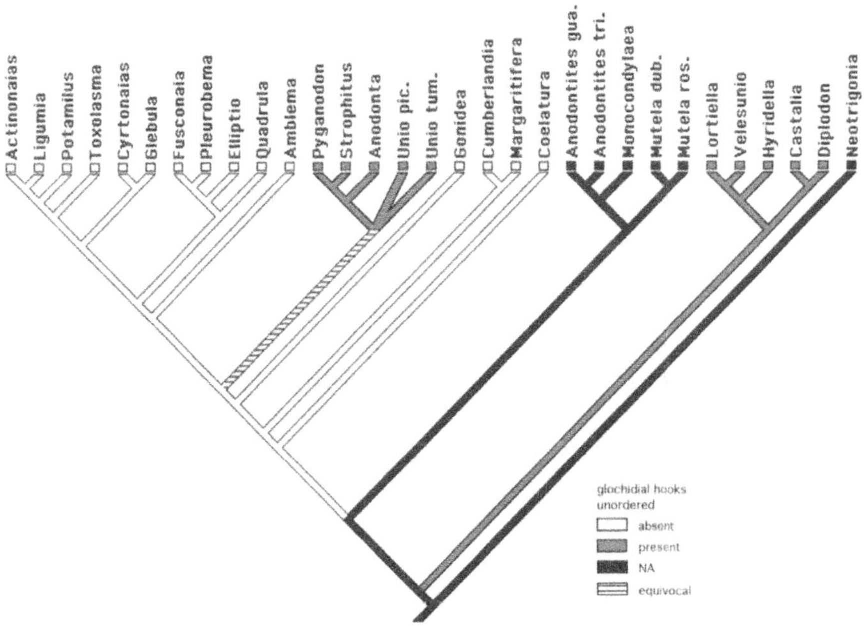

Fig. 14.11. Morphological character mapping on the strict consensus tree topology produced from maximum parsimony analysis of the combined morphology + transformed COI nucleotide matrix. The most parsimonious estimate for the evolution of medioventrally hooked glochidia suggests that hooked glochidia arose at least twice in the Unionoida (*NA* = not applicable)

tracts are most likely used to facilitate the transport of oocytes from the inner suprabranchial canals to the outer suprabranchial canals (and then into the outer demibranchs) in tetragenous and ectobranchus taxa (Latter 1891). Therefore, the initial transition to brooding in the ancestral unionoid lineage may have been aided by the use of only the inner demibranchs. Selection for increased brooding capacity may have subsequently acted to fix a mutant tetragenous condition in the common ancestral lineage of unionids + margaritiferids. Additional fitness benefits accrued from the specialization of demibranch function may have consequently potentiated the transition(s) to ectobranchy in the amblemine + anodontine + *Unio* clade.

The lack of medioventral glochidial hooks in some species of *Diplodon* (related to a transition from host-mediated to direct development?) most likely represents a loss of the ventral hook rather than an ancestrally hookless condition. However, this possibility should be evaluated with data. Notwithstanding, hypotheses #1 and #3 run counter to those suggested by the morphology-based analysis (above) while the findings on unionoid character evolution from the combined analysis here are identical to those presented in Hoeh et al. (1998a), which were based on analyses of COI sequences only.

14.5 Conclusions

The topology based on the total evidence approach (Fig. 14.4) should be used as the best current estimate of unionoid phylogeny. While the use of the total-evidence topology presented here is advocated to interpret broad-scale patterns of unionoid character evolution, we do not imply that the search for the best estimate of unionoid phylogeny has been completed. Rather, the total-evidence topology presented in this chapter should be viewed as a starting point for evaluating evolutionary patterns and processes within unionoids. At least two factors suggest that additional work is required: (1) the significant discrepancies between the morphology- and molecule-based topologies are unsettling given that both sets of characteristics evolved during a common unionoid phylogenetic history and (2) the relatively low bootstrap percentages for some of the nodes on the total evidence topology, which represent many of the fundamental cladogenic events within unionoids, suggest that additional data could substantially change the best current estimates of unionoid phylogeny and character evolution. The accretion of more morphological and molecular data to that analyzed herein, as well as adding ecological and behavioral characteristics, would bring additional evidence to bear on questions regarding early unionoid evolutionary history as well as enable a more rigorous evaluation of the degree of incongruence between the morphological and molecular datasets. Furthermore, we believe that a reevaluation of unionoid classification, while being justified by the total-evidence tree topology presented herein, would be somewhat premature at this time given the relatively small datasets included in the present analyses.

14.6 Summary

Uncertainty regarding the phylogenetic relationships within the Unionoida have hampered attempts to understand evolution within the group and establish a stable classification system. To test alternative hypotheses of unionoid phylogeny, a matrix of 28 morphological characters was developed for 30 species representing the higher taxa within the Unionoida (= in-group) and one species (*Neotrigonia margaritacea*) representing the Trigonioida (= out-group). Parsimony analyses were employed to generate phylogenetic trees from the morphology matrix alone and from a combined morphology + DNA sequence matrix. The parsimony analysis of the morphological data produced trees that supported the monophyly of the Hyriidae, Margaritiferidae, Iridinidae (= Mutelidae), Mycetopodidae, and Etherioidea. The evolutionary relationships among these higher taxa resemble those portrayed by Heard and Guckert (1970) in that (1) the Etherioidea (i.e., iridinids + mycetopodids) and Hyriidae are sister taxa and (2) the margaritiferids represent a lineage descended from the basalmost cladogenic event within the ancestral unionoid line. Furthermore, the phylogenetic analysis of the combined morphology + DNA sequence matrix also produced trees that supported the monophyly of the Hyriidae, Margaritiferidae, Iridinidae, Mycetopodidae, and Etherioidea. However, the evolutionary relationships among these higher taxa resembled those portrayed by Hoeh et al. (1998a): ((((Unionidae [excluding *Coelatura*], Margaritiferidae), *Coelatura*), (Iridinidae, Mycetopodidae)), Hyriidae). The latter topology suggests that endobranchy and glochidial larvae were the ancestral unionoid conditions and that medioventrally hooked glochidia evolved at least twice.

Acknowledgements. J.M.K. Berlocher, S. Boyd, J.B. Burch, R.S. Butler, A.H. Clarke, S.E. Cordoba, K.S. Cummings, R.V. Dimock, T.E. Dowling, A.G. Gerberich, M.E. Gordon, P.D. Hartfield, R.G. Howells, G.A. Long, D. Manning, M.C.D. Mansur, J.E. Maunder, C.A. Mayer, C.P. Morgan, R.G. Noseworthy, J.M. Olson, P.W. Parmalee, T.A. Pearce, W.F. Ponder, R.S. Prezant, D.G. Smith, G. Soliman, S.J. Stoehr, R.J. Trdan, J.D. Williams, F.R. Woodward, K.E. Wright, and R.G. Young are acknowledged for assistance in procuring specimens. We thank the volume editor, Dr. G. Bauer, S.L. Neale, and an anonymous reviewer for helpful comments on the manuscript. We also acknowledge the Illinois Natural History Survey for the loan of South American unionoid specimens from their collection.

References

Archie JW (1989) A randomization test for phylogenetic information in systematic data. Syst Zool 43:1796–1800

Atkins D (1937) On the ciliary mechanisms and interrelationships of lamellibranchs. Part II: sorting devices on the gills. Q J Microsc Sci 79:339–373

Avise JC (1994) Molecular markers, natural history and evolution. Chapman & Hall, New York

Bonetto AA (1997) The freshwater oysters (Muteloidea: Mutelidae). Their taxonomy and geographic distribution in the whole of the naiads of the world. Biociencias 5:113–142

Brooks DR, McLennan DA (1991) Phylogeny, ecology, and behavior: a research program in comparative biology. The University of Chicago Press, Chicago

Bull JJ, Huelsenbeck JP, Cunningham CW, Swofford DL, Waddell PJ (1993) Partitioning and combining data in phylogenetic analysis. Syst Biol 42:384–397

Davis GM, Fuller SLH (1981) Genetic relationships among recent Unionacea (Bivalvia) of North America. Malacologia 20(2):217–253

de Queiroz A, Donoghue MJ, Kim J (1995) Separate versus combined analysis of phylogenetic evidence. Annu Rev Ecol Syst 26:657–681

Doyle JJ (1992) Gene trees and species trees: molecular systematics as one-character taxonomy. Syst Bot 17:144–163

Eernisse DJ, Kluge AG (1993) Taxonomic congruence versus total evidence, and amniote phylogeny inferred from fossils, molecules, and morphology. Mol Biol Evol 10:1170–1195

Eldredge N, Cracraft J (1980) Phylogenetic patterns and the evolutionary process. Columbia University Press, New York

Faith DP, Cranston PS (1991) Could a cladogram this short have arisen by chance alone? On permutation tests for cladistic structure. Cladistics 7:1–28

Farris JS (1983) The logical basis of phylogenetic analysis. In: Platnick NI, Funk VA (eds) Advances in cladistics. Columbia University Press, New York, pp 7–36

Farris JS, Kallersjo M, Kluge AG, Bult C (1994) Testing significance of incongruence. Cladistics 10:315–319

Graybeal A (1998) Is it better to add taxa or characters to a difficult phylogenetic problem? Syst Biol 47:9–17

Hannibal H (1912) A synopsis of the Recent and Tertiary freshwater Mollusca of the Californian Province, based upon an ontogenetic classification. Proc Malacol Soc Lond 10(2):112–166, 10(3):167–211

Healy JM (1989) Spermiogenesis and spermatozoa in the relict bivalve genus *Neotrigonia*: relevance to trigonioid relationships, particularly Unionoidea. Mar Biol 103:75–85

Heard WH, Guckert RH (1970) A re-evaluation of the recent Unionacea (Pelecypoda) of North America. Malacologia 10(2):333–355

Hillis DM (1991) Discriminating between phylogenetic signal and random noise in DNA sequences. In: Miyamoto MM, Cracraft J (eds) Phylogenetic analysis of DNA sequences. Oxford University Press, New York, pp 278–294

Hillis DM (1995) Approaches for assessing phylogenetic accuracy. Syst Biol 44:3–16

Hillis DM, Huelsenbeck JP (1992) Signal, noise, and reliability in molecular phylogenetic analyses. J Hered 83:189–195

Hillis DM, Huelsenbeck JP, Cunningham CW (1994) Application and accuracy of molecular phylogenies. Science 264:671–677

Hoeh WR (1990) Phylogenetic relationships among eastern North American *Anodonta* (Bivalvia: Unionidae). Malacol Rev 23:63–82

Hoeh WR, Bogan AE, Cummings KS, Guttman SI (1998a) Evolutionary relationships among the higher taxa of freshwater mussels (Bivalvia: Unionoida): inferences on phylogeny and character evolution from analyses of DNA sequence data. Malacol Rev 31:121–140

Hoeh WR, Black MB, Gustafson RG, Bogan AE, Lutz RA, Vrijenhoek RC (1998b) Testing alternative hypotheses of *Neotrigonia* (Bivalvia: Trigonioida) phylogenetic relationships using cytochrome c oxidase subunit I DNA sequences. Malacologia 40:267–278

Howes GB (1902) Atlas of practical elementary zootomy. Macmillan London

Jones TR, Kluge AG, Wolfe AJ (1993) When theories and methodologies clash: a phylogenetic reanalysis of the North American ambystomatid salamanders (Caudata: Ambystomatidae). Syst Biol 42:92–102

Kabat AR (1997) Correct family names for the freshwater "muteloid" bivalves (Unionoida: Etherioidea). Occas Pap Mollusks Mus Comp Zool Harv Univ 5:379–392

Kluge AG (1989) A concern for evidence and a phylogenetic hypothesis of relationships among *Epicrates* (Boidae, Serpentes). Syst Zool 38:7–25

Kluge AG (1998) Total evidence or taxonomic congruence: cladistics or consensus classification. Cladistics 14:151–158

Kluge AG, Wolfe AJ (1993) Cladistics: what's in a word? Cladistics 9:183–199

Korniushin AV (1998) Review of the studies on freshwater bivalve mollusc systematics carried out by the Russian taxonomic school. Malacol Rev Suppl 7:65–82

Latter OH (1891) Notes on *Anodon* and *Unio*. Proc Zool Soc Lond 1891:52–59

Lydeard C, Mulvey M, Davis GM (1996) Molecular systematics and evolution of reproductive traits of North American unionacean mussels (Mollusca: Bivalvia) as inferred from 16S rRNA gene sequences. Philos Trans R Soc Lond 351B:1593–1603

Maddison DR (1991) The discovery and importance of multiple islands of most-parsimonious trees. Syst Zool 40:315–328
Maddison WP, Maddison DR (1997) MacClade: analysis of phylogeny and character evolution. Sinauer Associates, Sunderland, Massachusetts
Miyamoto MM (1985) Consensus cladograms and general classification. Cladistics 1:186–189
Miyamoto MM, Fitch WM (1995) Testing species phylogenies and phylogenetic methods with congruence. Syst Biol 44:64–76
Modell H (1942) Das natürliche System der Najaden. Arch Molluskenkd 74(5/6):161–191
Modell H (1964) Das natürliche System der Najaden. 3. Arch Molluskenkd 93:71–129
Morrison JPE (1955) Family relationships in the North American freshwater mussels. Am Malacol Union Annu Rep 21:16–17
Morrison JPE (1973) The families of the pearly freshwater mussels. Bull Am Malacol Union 1972:45–46
Ortmann AE (1911) The anatomical structure of certain exotic naiades compared with that of the North American forms. Nautilus 24(9):103–108, 24(10):114–120, 24(11):127–131
Ortmann AE (1912) Notes upon the families and genera of the naiades. Ann Carnegie Mus 8:222–365
Ortmann AE (1919) A monograph of the naiades of Pennsylvania. Part III: Systematic account of the genera and species. Mem Carnegie Mus 8(1):xvi–384
Ortmann AE (1921) South American naiades: a contribution to the knowledge of the freshwater mussels of South America. Mem Carnegie Mus 8(3):451–668
Pamilo P, Nei M (1988) Relationships between gene trees and species trees. Mol Biol Evol 5:568–583
Parodiz JJ, Bonetto AA (1963) Taxonomy and zoogeographic relationships of the South American naiades (Pelecypoda:Unionacea and Mutelacea). Malacologia 1(2):179–213
Popham JD (1979) Comparative spermatozoon morphology and bivalve phylogeny. Malacol Rev 12:1–20
Reed RD, Sperling FAH (1999) Interaction of process partitions in phylogenetic analysis: an example from the swallowtail butterfly genus *Papilio*. Mol Biol Evol 16:286–297
Shikov EV, Zatravkin MN (1991) The comparative method of taxonomic study of Bivalvia used by Soviet Malacologists. Malakol Abh 15(17):7–10
Simpson CT (1900) Synopsis of the naiades, or pearly fresh-water mussels. Proc US Nat Mus 22:501–1044
Smith DG (1986) The stomach anatomy of some eastern North American Margaritiferidae (Unionoida: Unionacea). Am Malacol Bull 4:13–19
Starobogatov YI (1970) Mollusc fauna and zoogeograhical partitioning of continental water reservoirs of the world. Zool Inst Akad Nauk SSSR, Leningrad
Swofford DL (1991) When are phylogeny estimates from molecular and morphological data incongruent? In: Miyamoto MM, Cracraft J (eds) Phylogenetic analysis of DNA sequences. Oxford University Press, New York, pp 295–333
Swofford DL (1998) PAUP. Phylogenetic analysis using parsimony (and other methods). Sinauer, Sunderland, Massachusetts
Taylor JD, Kennedy WJ, Hall A (1969) The shell structure and mineralogy of the Bivalvia. Introduction. Nuculacea-Trigonacea. Bull Br Mus (Nat Hist) Zool Suppl 3:1–125
Taylor JD, Kennedy WJ, Hall A (1973) The shell structure and mineralogy of the Bivalvia II. Lucinacea-Clavagellacea. Conclusions. Bull Br Mus (Nat Hist) Zool 22:256–294
Tevesz MJS (1975) Structure and habits of the "living fossil" pelecypod *Neotrigonia*. Lethaia 8:321–327
Tevesz MJS, Carter JG (1980) Environmental relationships of shell form and structure of unionacean bivalves. In: Rhoads DC, Lutz RA (eds) Skeletal growth of aquatic organisms: biological records of environmental change. Plenum Press, New York, pp 295–322
Walker B (1917) The method of evolution in the Unionidae. Occas Pap Mus Zool Univ Mich 45:10

Waller TR (1990) The evolution of ligament systems in the Bivalvia. In: Morton B (ed) The Bivalvia-Proceedings of a memorial symposium in honour of Sir Charles Maurice Yonge, Edinburgh. Hong Kong University Press, Hong Kong, pp 49–71

Waller TR (1998) Origin of the molluscan class Bivalvia and a phylogeny of major groups. In: Johnston PA, Haggart JW (eds) Bivalves: an eon of evolution. University of Calgary Press, Calgary, pp 1–45

Whiting MF, Carpenter JC, Wheeler QD, Wheeler WC (1997) The Strepsiptera problem: phylogeny of the holometabolous insect orders inferred from 18S and 28S ribosomal DNA sequences and morphology. Syst Biol 46:1–68

15 The Evolution of the Unionacea in North America, and Its Implications for the Worldwide Fauna

G. Thomas Watters

15.1 Introduction

"The distribution in space of the uniones of the continental interior has been shown to be in many respects very peculiar" Keyes 1894:26.

The Recent North American unionacean fauna is unsurpassed in the world. Nearly 300 species occurred there in modern times, and even small creeks may have more species than the entire continents of Europe and Australia combined. Yet the origins of this fauna are shrouded in the veil of time. The fossil record shows us glimpses of forms very similar to our modern species, but 70+ Ma old. We find tantalizing pieces of shells that belong to families that no longer occur there. Evidence suggests the existence of an earlier Mesozoic fauna that may have been as diverse in species and morphology as that of Recent times. This chapter will discuss the evolution of the North American unionacean fauna as illustrated by the fossil record, and compare it to similar faunas in Europe and Asia.

Over 200 fossil unionaceans have been named from North America. As with Recent species, early describers of fossil mussels separated them based on the presence or absence of hinge teeth. Those that had teeth were placed in the great melting-pot known as *Unio*, and those without teeth were placed in *Anodonta*. Both genera are based on Recent species. As the systematics of Recent unionaceans became more refined, numerous new generic names were introduced to handle the diversity. Curiously, as fossil mussel groups were split from *Unio* and *Anodonta*, some workers advocated placing Mesozoic, and even Paleozoic, species into Recent genera. In particular, Russell (1934) and Modell (1957) "revised" the fossil mussels and, almost without exception, assigned them to Recent genera. White's (1883) review of the fossil freshwater mussels of North America described and illustrated many of the taxa but continued to use Recent generic assignments. Given the great amount of time such genera would have to exist for these assignments to be true, and in light of the high degree of shell-shape convergence among freshwater bivalves (Watters 1994), these generic pigeon holes cannot be maintained. Later workers realized this and began to adopt new genera. Henderson's (1935) monumental catalogue of the nonmarine fossil mollusks of North America was the first attempt to organize the plethora of fossil forms and his nomenclature was a compromise of fossil and Recent generic names. Haas's (1969) chapter in the *Treatise on Invertebrate Paleontology* put the fossil unionacean groups into a modern perspective. None of these important works attempted to outline the evolution of the group in a systematic way. White (1907) briefly addressed the evolution of the group, but he believed that the Mesozoic species were the direct ancestors of our modern fauna, ignoring any convergence in shell shapes and sculpture.

Ecological Studies, Vol. 145
G. Bauer and K. Wächtler (eds.) Ecology and Evolution of the Freshwater Mussels Unionoida

This chapter is not meant to be a revision or a catalogue of the fossil mussels. No attempt is made to mention every species, or even every genus, of fossil mussel. However, even a brief study of these fossils reveals that further refinements and new genera are needed to capture the phylogeny of the group. This fact makes a review such as this difficult. This is not the place to introduce new taxa, yet identifiers must be given to unnamed groups to facilitate discussion. I have adopted several vernacular names for higher groups, such as proparryesians and vetulonaians, and referred to some undescribed genera with the admittedly clumsy monikers of Elliptio A, etc.

15.2 What Is a Unionacean?

Our Recent freshwater bivalves are the result of independent, heterochronic invasions by marine groups. In the streams and lakes of the present day we find unrelated groups side by side: corbiculids, dreissenids, donacids, unionids, etc. Similar melting pots may be found throughout geologic history. Because many groups have convergent shell features, the question becomes: what is a unionacean?

All unionaceans are presumed to have a parasitic larval stage, but exceptions have been found in the Mycetopodidae (Parodiz 1969; Mansur and Campos-Velho 1990), the Mutelidae (Kondo 1990), and the Unionidae (Barfield and Watters 1998; Lellis and King 1998). This trait ignores the fossil taxa, for, as discussed later in this chapter, we cannot be sure if fossil species were parasitic or when parasitism may have arisen. Undoubtedly, the Unionacea arose before the subsequent evolution of parasitic larvae.

The families of freshwater mussels recognized today fall into two groups based on larval type. The Unionidae, Margaritiferidae, and Hyriidae possess a glochidium. The Mycetopodidae have a lasidium and Mutelidae a haustorial larva. These latter two larval types are similar to each other, but very different from the glochidium. The Unionidae/Hyriidae/Margaritiferidae may represent a separate freshwater invasion from the Mycetopodidae/Mutelidae group, or an early divergence from that group. Both groups have converged on shell shapes and both have evolved parasitic larvae, but the similarities may end there. Based on fossil shell characteristics, all five families were present at least by the Cretaceous, and all five once existed in North America as part of a Pangaean fauna.

Recent unionaceans have a nacreous inner shell layer and an outer covering of periostracum, neither of which are useful characteristics for identifying fossil species. Dentition usually consists of anterior cardinals and posterior laterals, but the laterals may be absent, all teeth may be absent, or in some groups, even pseudotaxodont teeth are present. The umbos are often, but not always, sculptured differently from the remainder of the shell. This gives us few criteria for recognizing fossil unionaceans. Nevertheless, some freshwater groups that superficially resembled the Unionacea have been identified: the Archanodontacea, Anthracosiacea, Pachycardiidae, and *Murraia*. These are briefly discussed below in the context of their relationships to the unionaceans.

The superfamily Archanodontacea was a Paleozoic freshwater group with large (to 22 cm), *Anodonta*-like shells. Weir (1969) gave few distinctions to separate archanodontaceans from unionaceans. He believed their sporadic distribution indicated a lack of a parasitic stage, but this in itself is not sufficient to eliminate them from the

Unionacea. Most species were edentulous, but some apparently possessed claustra, longitudinal ridges that acted like lateral teeth, which also occur in anodontine unionids (although as a product of convergence). As mentioned by Weir, there is a gap in time between the last archanodontaceans in the Lower Permian and the first presumed unionaceans in the Triassic. Based on this admittedly meager evidence, it appears that the archanodontaceans were an early independent invasion by marine bivalves into the freshwater enivironment that disappeared before the origin of the unionaceans.

The superfamily Anthracosiacea also was a freshwater Paleozoic group resembling unionaceans. They lacked lateral teeth and the cardinals (when present) may be subdivided into pseudotaxodont teeth. These bivalves lived in the Paleozoic forest swamps and are particularly common in the Carboniferous deposits of England. Like the archanodontaceans they were an independent and ultimately unsuccessful invasion into freshwater apart from the unionaceans.

The Pachycardiidae is an enigmatic group of Permo-Triassic bivalves that closely resembled unionaceans such as the Mesozoic vetulonaians (see below). Most were marine or brackish-water inhabitants, but some are believed to have lived in freshwater. They possessed anterior cardinals and posterior laterals (rarely with anterior laterals), a simple pallial line, and at least some had a nacreous inner layer (Cox 1969). Conchologically, the pachycardiids are much more similar to the unionaceans than are the Trigoniacea, the purported ancestors of the unionaceans. Hoeh et al. (1998) found the trigoniaceans to be the closest living relatives of the unionacaeans, but that study could only look at the one living trigoniacean genus (*Neotrigonia*), and of course no living pachycardiids. Where (and if) the pachycardiids fit into the evolution of the unionaceans deserves further study.

The genus *Murraia* is well represented in the same Lower Cretaceous deposits as many unionaceans. These were small (2 cm) species with thin shells, divergent cardinal teeth, no lateral teeth, and a shallow pallial sinus. They are another extinct group of freshwater bivalves independent of the unionaceans.

Finally, the "Carboniferous" unionids from the Coal Measures of Marietta, Ohio, are probably based on Quaternary specimens. Two "*Unio*" species from the Silurian of Ohio are marine modiolopsids (Henderson 1935).

15.3 A Brief Review of North American Unionaceans Through Time

15.3.1 Triassic

Fossil records of unionaceans for the Triassic are not particularly numerous, but are widespread. Specimens are known from Arizona, Massachusetts, New Mexico, Pennslyvania, and Texas. During this time, the present continents were united in the supercontinent of Pangaea II, and the interior of the continent was warm and humid with stable river systems and lakes (Dubiel et al. 1989, 1991). In North America this Period contains members of several important groups such as the proparreysians, hyriids, and vetulonaians.

Hyriid-like species were widely distributed in the Triassic of North America, with species occurring in Texas, Arizona, Pennslyvania, and perhaps Massachusetts. Assignation of fossil taxa to the Hyriidae is based on the shared characteristic beak sculpture found in Recent species. This may be a convergent feature in the fossil forms, but the radial beak sculpture rarely occurs in other living mussel families and, with only the meager shell morphological evidence to guide us, these species seem most similar to the Hyriidae. They are elongate, *Elliptio*- or *Unio*-like species. However, Marshall (1929) clearly illustrated their hyriid beak sculpture, which consisted of well-defined radiating ribs (Fig. 15.1), and Parodiz (1969) compared them to modern species of *Diplodon*. I concur with Pilsbry (1921) and Parodiz (1968) that this now South American family once occurred in North America.

Elsewhere, hyriids date back to at least the Jurassic of Uruguay (Martinez and Figueiras 1991; Martinez et al. 1993), Australia (McMichael 1957), and Antarctica (Camacho 1957; Morton 1984). If the North American species actually belong to the Hyriidae, then this group experienced a very wide distribution in Pangaea early on. Whether they ultimately originated in Antarctica (Herbst and Camacho 1970), South America, or North America (Parodiz 1969) cannot be ascertained without additional records. The Uruguayan Jurassic? species, *Tacuaremboia caorsii* Martinez, Figueiras and Da Silva, 1993 has defied familial placement, having qualities of unionids and archanodontaceans. If it was a unionid, it would be one of the first recorded from South America. It also is perhaps the largest freshwater bivalve known, measuring nearly 35 cm in length.

The Triassic species *"Diplodon?" haroldi* Reeside, 1927 (Fig. 15.2) from the Dockum Group of Texas probably represents an early member of the important group referred to here as proparreysians. Proparreysians are trigonal to oval forms characterized by course chevron beak sculpture that often persists onto the shell disk as ribs and pustules. Some species with concentric sculpture are placed here as well. They show an amazing similarity to Recent species due to convergent shell sculpture (Watters 1994) and many have been erroneously placed in Recent genera by other workers. Their affinities are uncertain. The beak sculpture is reminiscent of Recent hyriid genera

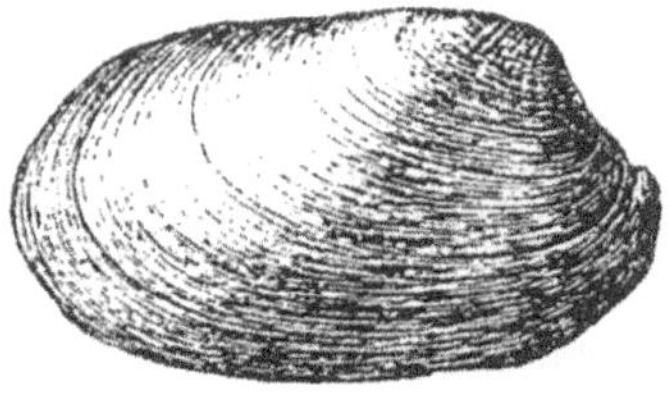

Fig. 15.1. *Unio dockumensis* Simpson, 1896. Dockum beds, Triassic, Texas. 60 mm. (Simpson 1896)

Fig. 15.2. *Diplodon? haroldi* Reeside, 1927. Dockum beds, Triassic, Texas. 35 mm (fragment). (Reeside 1927)

Fig. 15.3. *Unio subplanatus* Simpson, 1896. Dockum beds, Triassic, Texas. 85 mm. (Simpson 1896)

such as *Triplodon* and *Chevronaias* but also resembles unionid genera such as *Scabies* and *Rhombuniopsis*. Ortmann (1910), after examining the anatomy of a single living species of *Parreysia*, argued that the radial sculpture by itself was not sufficient grounds to place a species in the Hyriidae. Of course, this observation offers us little assistance with fossil forms. Haas (1969) placed the proparreysians in the Unionidae and Modell (1934) rather arbitrarily assigned them to Recent genera such as *Quincuncina* and *Orthonymus*. Henderson (1935) placed them alongside *Parreysia* in the Unionidae.

The name vetulonaians is here applied to a group of Mesozoic unionids that all have a distinctly "modern" look about them despite their age. They resemble members of the Recent genera *Pleurobema* and *Fusconaia*. Vetulonaians are trigonal and unsculptured, generally having heavy hinges, and concentric or undulating beak sculpture. The earliest representative may be *Unio subplanatus* Simpson, 1896, from the Triassic of Texas (Fig. 15.3), although Parodiz (1969) included this species under his "*Diplodon*-like forms." Branson (1935) created the generic name *Vetulonaia* for *V. whitei* Branson, 1935, a typical form from the Jurassic Morrison Formation of Wyoming.

Other Triassic species do not seem to be assignable to any nominal genera or larger groups: *Unio cristonensis* Meek, 1875, *U. thomasi* Henderson, 1934, *U. emersoni* Troxell, 1914, and *Anoplophora wilbrahamensis* Emerson, 1900. Parodiz (1969) included the last two species as "*Diplodon*-like forms." These include some poorly preserved species from Pennsylvania and Massachusetts – quite distant from the known Mesozoic centers of unionacean diversity in western North America.

15.3.2 Jurassic

Our North American Jurassic mussel records come mainly from the western half of the continent and late in that Period. During the Late Jurassic, the Sundance Sea occupied most of this region. As that sea regressed it left behind the area that became the Morrison Formation. Deposits and paleontological evidence suggest that this area was relatively dry and that many of the available water bodies may have been saline (Stanley 1986). Aquatic organisms in general seem to have been scarce. Nevertheless, most of the unionacean records from this Period are from the Morrison Formation, and the group seems to have prospered during this time.

Jurassic unionaceans are characterized by the appearance of a *Unio*-like group and an increase in the diversity of the vetulonaians, both of which persisted into the Cenozoic and may be ancestral to Recent forms. A third group, *Hadrodon*, is limited to the

Fig. 15.4. *Hadrodon lateralis* Yen, 1952. Morrison Formation, Jurassic, Colorado. 67 mm. (Yen 1952)

Fig. 15.5. *Vetulonaia whitei* Branson, 1935. Morrison Formation, Jurassic, Colorado. 50 mm. (Branson 1935)

Fig. 15.6. *Vetulonaia faberi* Holt, 1942. Morrison Formation, Jurassic, Colorado. 55 mm. (Holt 1942)

Jurassic. These groups are discussed below. Proparreysians have not been found in the Jurassic. Their presence (though rare) in the Triassic and their great diversity in the Cretaceous indicate that they were present at this time, but have not yet been identified or found.

The peculiar genus *Hadrodon* is known from Colorado. These were squarish mussels with a prominent, angular, posterior ridge and a wing-like posterior slope that often was twisted or folded over laterally (Yen 1952; Fig. 15.4). They seem to represent a localized, short-lived, divergent group of unionids. They are important as reminders that much of the diversity of fossil unionaceans may lie with these kinds of endemic taxa, scattered and isolated over the continent, most of which we may never find.

Several species of vetulonaians appear in the Morrison Formation in Montana, Utah, Colorado, and Wyoming (Figs. 15.5, 15.6). Taxa such as *V. faberi* Holt, 1942 and *V. utahensis* Yen, 1952 seem to be direct ancestors of Cretaceous forms like *V. mammillaris* (Yen 1952). Their shell shape is that of a Big River morph, a suite of shell characteristics found in Recent species adapted for life in large rivers (Watters 1994).

Fig. 15.7. *Unio felchii* White, 1886. *Atlantosaurus* beds, Jurassic, Colorado. 80 mm. (White 1886)

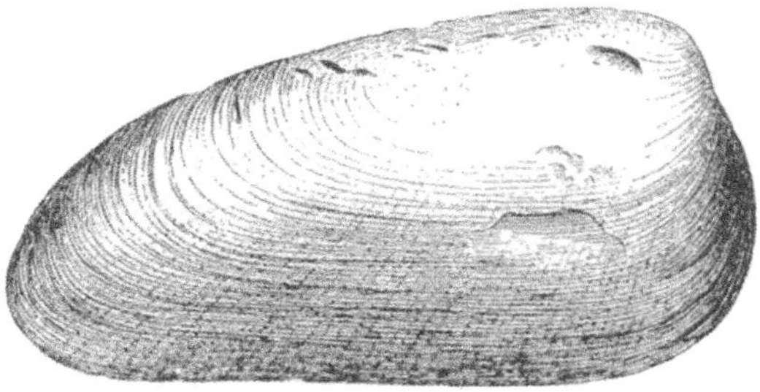

Fig. 15.8. *Unio subspatulatus* Meek & Hayden, 1857. Judith River Formation, Jurassic, Montana. 70 mm. (White 1883)

There is no reason to believe that these adaptations did not perform the same function in vetulonaians, indicating a riverine habitat for these species as well.

The most problematic group of fossil unionids is the *Unio*-like species (Figs. 15.7, 15.8). Morphologically they blend into the vetulonaians on the one hand and the elliptioids on the other. As a group they are heavy-shelled, often quite elongate, with prominent anteriorly placed umbos. Beak sculpture is absent or reduced. Generally, the hinges were much more massive than those of Recent *Elliptio* and *Unio*. The Jurassic species, mostly from the *Atlantosaurus* beds of Colorado, were rather homogeneous and nondescript.

15.3.3 Cretaceous

The Cretaceous witnessed the first great explosion of diversity of unionaceans, in terms of both species and morphological variety. The proparreysians in particular are well represented, and many other new groups appear and prosper; but most groups do not survive the transition to the Cenozoic. Like the Jurassic records, our knowledge of unionaceans in the North American Cretaceous is known almost exclusively from the western region. During the Early Cretaceous this area was partially isolated from the east by the Mowry Sea. Beginning in the Albian, 105 Ma implie, an interior seaway formed, completely cutting off western North America from the east (Smith et al. 1994). Portions of this seaway persisted into the Cenozoic. The fossil unionaceans come from the dinosaur-rich deposits along the western shore of that seaway, apparently from rivers flowing eastward into the seaway.

Cretaceous vetulonaians are found in Alberta and Montana. By this time they have a decidedly *Pleurobema*-like aspect and continue to display Big River morphological characteristics. Two Cretaceous species belonging to an undescribed genus are related to the vetulonaians, but have single or double rows of knobs on the posterior half of the shells: *Unio biaesopoides* Whitfield, 1907 (Fig. 15.15), and *Unio aesopiformis* Whitfield, 1903. Russell (1976) placed them in the Recent genus *Plethobasus*. While they are not related to that group, they are analogous. *Plethobasus* species are sculptured relatives of *Pleurobema*, and *U. biaesopoides* and *U. aesopiformis* are sculptured relatives of vetulonaians.

Also related to the vetulonaians is a group referred to here as the cylindricoids from Wyoming, Montana, and New Mexico. They were heavy, trigonal to cylindrical forms with an anteriorly placed umbo and an occasionally corrugated posterior slope. The group is noteworthy because of *Unio cylindricoides* Whitfield, 1907 (Fig. 15.16), which is strikingly similar to the Recent *Quadrula cylindrica* (Say, 1817) (Fig. 15.17), itself a singular and remarkable species. *Unio cylindricoides* is linked morphologically with the other members of the group through *Unio aldrichi* White, 1878, a slightly less elongated species. At least one species, *Unio* sp. cf. *proavitus* Tozer, 1956 survived to the Paleocene.

Other Cretaceous unionaceans appear to be modified vetulonaians as well. *Unio gonionotus* White, 1876 (Fig. 15.18) from Utah had an expanded, strongly ribbed posterior portion of the shell reminiscent of the female shells of modern *Epioblasma*. *Unio goniambonatus* White, 1878 (Fig. 15.19) and *U. nebrascensis* Meek, 1870 have strong posterior ridges resembling modern *Amblema*.

The proparreysians also were widespread in the Cretaceous. Like the vetulonaians of the same time, morphologically this group converged on shell shapes and sculpture found in Recent unionaceans, though undoubtedly unrelated. Some of these convergences are striking. *Proparreysia letsoni* (Whitfield 1907; Fig. 15.9) has the same row of knobs for substrate-anchoring and posterior corrugations for antiscouring as does the Recent *Obliquaria reflexa* Rafinesque, 1820 (Fig. 15.10). *Proparreysia verru-*

Fig. 15.9. *Unio letsoni* Whitfield, 1907. Hell Creek Formation, Cretaceous, Montana. 44.3 mm. (Russell 1976)

Fig. 15.10. *Obliquaria reflexa* Rafinesque, 1820. Recent Mississippi River drainage. Average 50 mm. (Burch 1973)

Fig. 15.11. *Unio verrucosiformis* Whitfield, 1903. Hell Creek Formation, Cretaceous, Montana. 52.5 mm. (Russell 1976)

Fig. 15.12. *Quadrula pustulosa* (Lea 1831). Recent Mississippi River drainage. Average 60 mm. (Burch 1973)

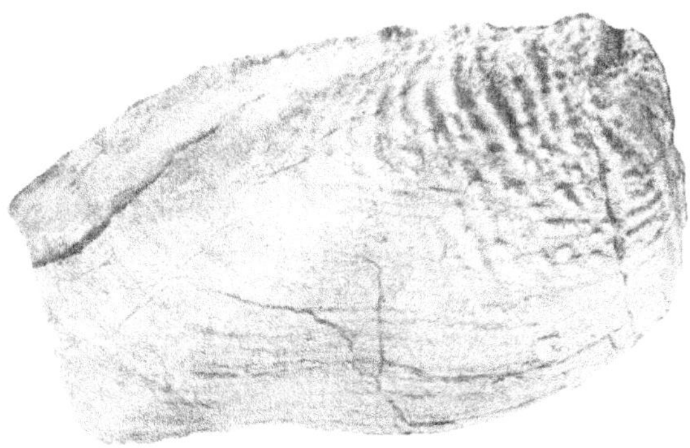

Fig. 15.13. *Unio natosini* McLearn, 1929. Blairmore Formation, Cretaceous, Alberta. 68 mm. (Russell 1967)

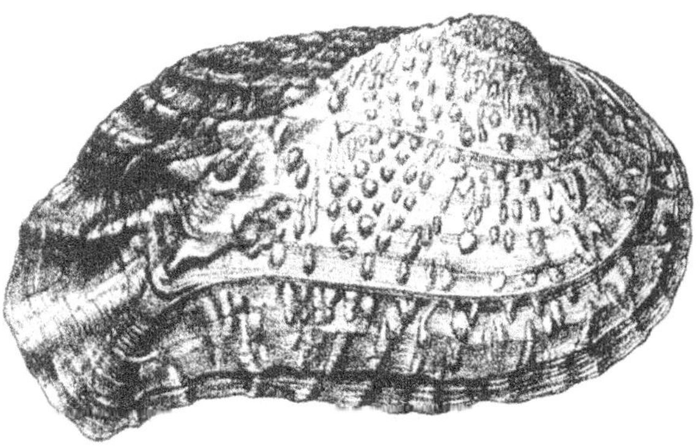

Fig. 15.14. *Tritogonia verrucosa* (Rafinesque 1820). Recent Mississippi River drainage. Average 150 mm. (Burch 1973)

cosiformis (Whitfield 1907) (Fig. 15.11) is the dopplegänger of *Quadrula pustulosa* (Lea 1831) (Fig. 15.12), having the same anchoring pustules scattered over the shell disk. *Unio natosini* McLearn, 1929 (Fig. 15.13) is the Cretaceous analogue of the modern *Tritogonia verrucosa* (Rafinesque 1820) (Fig. 15.14). These convergences are all the more interesting when we remember that the proparreysians may have been hyriids.

The first *Anodonta*-like forms appear in the Upper Cretaceous. The sole characteristic that unites these species is the lack of hinge teeth. Because hinge teeth have been lost independently in Recent unrelated forms, great care must be taken in assigning these species to genera, or even families. Early on, Keyes (1894) doubted the place-

Fig. 15.15. *Unio biaesopoides* Whitfield, 1907. Lance Formation, Cretaceous, Montana. ~70 mm. (Whitfield 1907)

Fig. 15.16. *Unio cylindricoides* Whitfield, 1907. Lance Formation, Cretaceous, Montana. ~40 mm (juvenile). (Whitfield 1907)

Fig. 15.17. *Quadrula cylindrica* (Say 1817). Recent Mississippi River drainage. Average 90 mm. (Burch 1973)

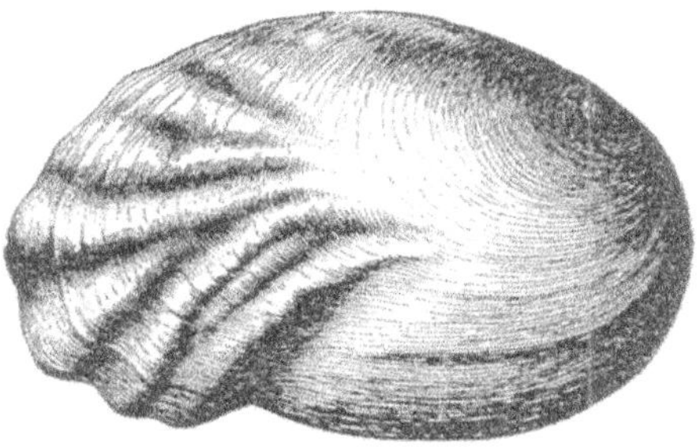

Fig. 15.18. *Unio gonionotus* White, 1876. Upper Kanab, Cretaceous, Utah. 60 mm. (White 1883)

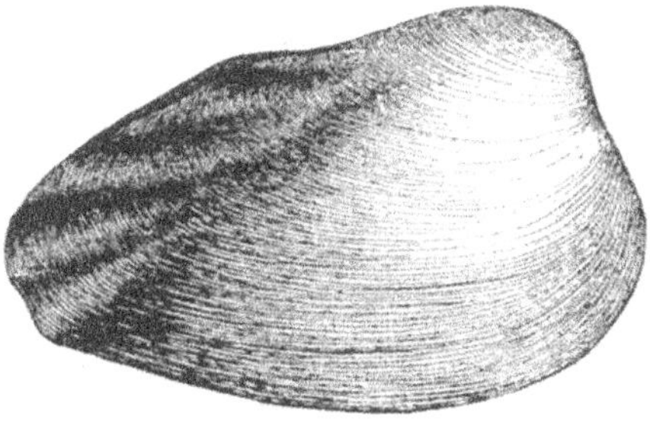

Fig. 15.19. *Unio goniambonotus* White, 1878. Black Buttes, Cretaceous, Wyoming. 58 mm. (White 1883)

ment of these species in *Anodonta*. I cannot agree with Davis and Fuller (1981) that the "Anodontinae are known with certainty from the Cretaceous." There is no evidence that these edentate fossils are related to each other or to any modern forms. Undoubtedly, the adaptation to soft substrates that seems to be the basis for losing the hinge and teeth arose independently in several groups.

Elliptioids is a general term used here to apply to groups of elongate *Elliptio*-like species. They first appear in the Cretaceous and persist at least into the Eocene and may be ancestral to Recent *Elliptio*. Elliptioids occur in three groups: protelliptios, elliptioid A, and the Greenland elliptioids.

The protelliptios were elongate to oval, had pronounced concentric beak sculpture, a well-defined dorsal ridge, and occasionally a corrugated posterior slope. They include the genera *Protelliptio* and *Plesielliptio* between which I can discern no important differences. The highly sculptured *Unio belliplicatus* Meek, 1870, from Wyoming seems to be a specialized offshoot of this group. Protelliptios also occur in Nebraska, Montana, and Alberta. They persisted into the Eocene.

Elliptioid A has been placed with the protelliptios, but differs by lacking pronounced beak sculpture (Fig. 15.20). They may be an unrelated group that has converged upon the same overall shell shape as the protelliptios.

The Greenland elliptioids consist of the species *Unio thulei* and *U. sewardi*, both Woodward, 1971. They represent one of the few records of Cretaceous unionaceans from the eastern North American continent. During the Cenomanian, the presumed age of these species, Greenland had not yet separated from the rest of eastern North

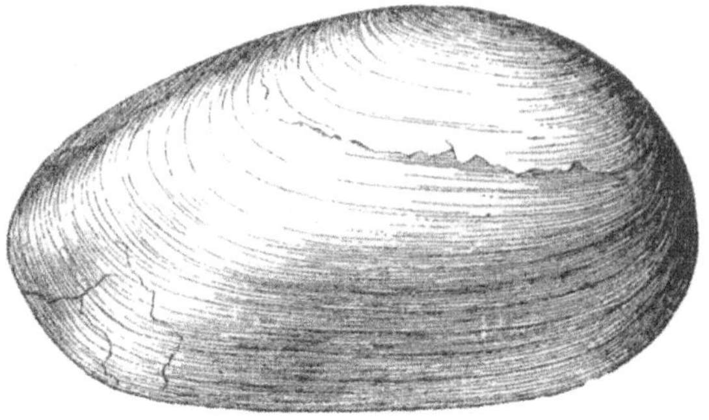

Fig. 15.20. *Unio stantoni* White, 1905. Black Buttes, Cretaceous, Wyoming. 83 mm. (White 1883)

America (Smith et al. 1994). Of all of the elliptioids, these species most closely resemble Recent forms. Unfortunately, the beak sculpture of these species is unknown. Whether Greenland species like these are derived from Europe or North America is uncertain.

Unio-like forms occur in Montana and Wyoming. Like their Jurassic ancestors they were heavy-shelled, elongate forms with anterior beaks. They may represent ancestors of true *Unio*, now so widespread in Europe, in North America.

In 1988 Morris and Williamson reported on a pseudotaxodont hinge fragment from the Hell Creek Formation of Montana. The hinge is strikingly similar to Recent mutelids, and the authors placed the species in *Pleiodon*, a Recent African mutelid genus. I cannot agree with their generic placement for the same reasons I cannot agree with Modell's pigeonholing of other fossil species into Recent genera – it is unlikely that *Pleiodon* has been in existence for 65+ Ma. The fact that most of the shell is missing does not help their claim. Nevertheless, the specimens raise important questions. Mutelids are known from the Cretaceous and Cenozoic of Africa, but no mutelids had ever been found in North America before these fragments were discovered. If their specimens are indeed mutelids, then when and where did this group originate? By the Cretaceous, North America and Africa had long since been separated by the breakup of Pangaea in the Jurassic. This suggests that mutelids, like hyriids and unionids, may once have had a much wider distribution across Pangaea. Indeed, a Cretaceous mussel from Brazil has been, on occasion, assigned to the Mutelidae (Ihering 1912). If true, then North American mutelids remain to be discovered in much earlier deposits. The alternate explanation is that the fragments represent convergence to a pseudotaxodont condition by a then existing North American group; but if this is the case, no intergrades or similar species have been found.

The enigmatic species *Anodonta parallela* White, 1878 (Fig. 15.21) occurs in Colorado and Wyoming. It resembles the mycetopodid *Mycetopoda*, a Recent genus from South America. Whether this is a true mycetopodid or the result of convergence from another group cannot be ascertained; but if mutelids and hyriids existed in Mesozoic North America, then perhaps so did mycetopodids. [Henderson (1935) placed the Triassic *Mycetopoda diliculi* Pilsbry, 1921, in this family but Parodiz (1969) referred it to the Hyriidae]. The arguments for and against mycetopodids in North America are the same as those for the mutelids: where are the earlier representatives? and if not a mycetopodid, then what?

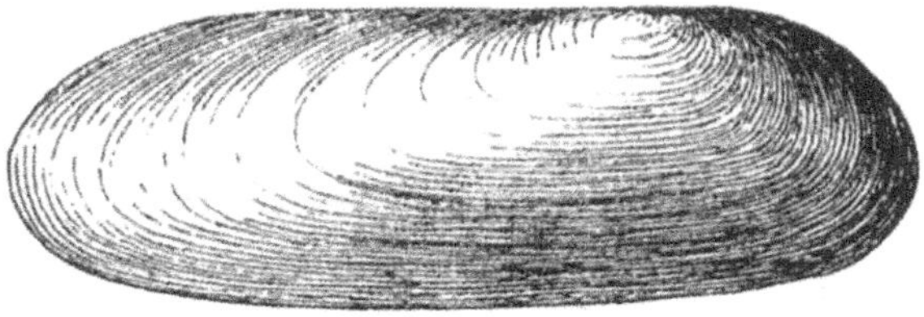

Fig. 15.21. *Anodonta parallela* White, 1878. Laramie Formation, Cretaceous, Colorado. 62 mm. (White 1883)

Fig. 15.22. *Nippononaia asinaria* Reeside, 1957. Burro Canyon Formation, Cretaceous, Colorado. 26 mm (mold). (Reeside 1957)

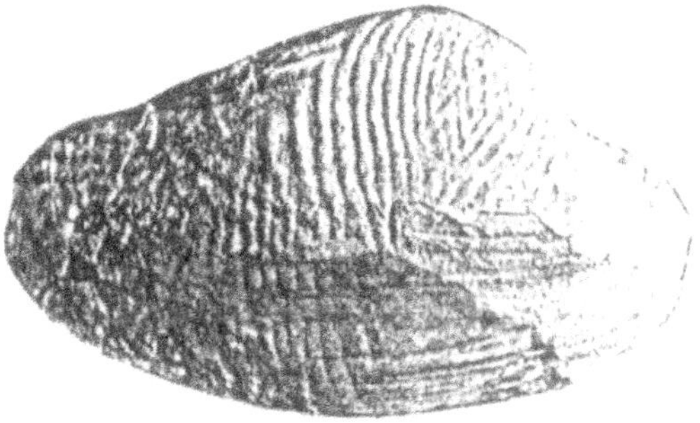

Fig. 15.23. *Unio paraholmesianus* Yen, 1954. Mount Garfield Formation, Cretaceous, Colorado. 48 mm. (Yen 1954)

Nippononaia is a problematic group. Based on *Nippononaia ryosekiana* Suzuki, 1941, from the Lower Cretaceous of Japan, these are species with pronounced chevron sculpture over the entire shell. *Nippononaia asinaria* Reeside, 1957 (Fig. 15.22), described from the Lower Cretaceous of Colorado, is very similar to *N. ryosekiana*. Although Haas (1969) placed this genus in the Unionidae, the dentition is not typical of that group. The "cardinal" teeth are long and lamellar, very similar to the lateral teeth, suggesting anterior and posterior laterals rather than the typical unionid dentition. *Unio paraholmesiana* Yen, 1954 (Fig. 15.23) from the Late Cretaceous of Wyoming also is a member of *Nippononaia*, and may be an earlier name for *N. asinaria*. The hinge of *U. paraholmesiana* is unknown (Yen 1954). The sculpture of these species is very similar to *Trigonioides*, a group of Lower Cretaceous freshwater mussels from Asia that possessed schizodont dentition (Jingeng 1992). Reeside hinted at corrugations on the teeth of *N. asinaria*, which suggested that perhaps *Nippononaia* and *Trigonioides* were closely related. If *N. asinaria* is congeneric with *N.*

ryosekiana, then Asia and North America shared faunal elements as far back as the Lower Cretaceous. This is all the more interesting when we realize that Japan and Colorado were at opposite ends of the Pangaean supercontinent. *Nippononaia* and *Trigonioides* may have been a separate, perhaps nonunionacean group of freshwater mussels that were distributed over at least Laurasia in Cretaceous time.

Lea (1868) and Whitfield (1885) described "Cretaceous" unionids from the Fish House or Camden Clay beds of New Jersey. These beds are now known to be Pleistocene, and the species still extant. This fauna was thoroughly studied by Kat (1983) and Bogan et al. (1989).

15.3.4 Cenozoic

By the early Eocene the great intercontinental seaway had closed, uniting the eastern and western continents into present-day North America (Smith et al. 1994). The Bering land bridge intermittently linked western North America and eastern Asia. To the east, North America was similarly linked to Europe via Greenland. These changes dramatically altered the composition of freshwater mussels.

Tertiary unionaceans appear in three areas. The first is a remnant of the Mesozoic fauna that persisted in dwindling numbers in Wyoming, Montana, and North Dakota. These few survivors of the K-T boundary are true descendants of the Cretaceous fauna. The second area is northwestern North America – California, Oregon, Alberta, Saskatchewan, Alaska. Many species here appear without any obvious Cretaceous ancestors in the region. It is likely that they are immigrants from Asia across the Bering land bridge. The third area is the Mississippi Embayment of southern North America. Here we see species that have obvious ties to our modern fauna. Whether they are derived from the great Cretaceous fauna of the western North American continent or the product of an unknown eastern Cretaceous fauna remains to be seen. However, based on the evidence of pollen (Tschudy 1971) and freshwater snails (Taylor 1988), the western Rocky Mountain area did not have any appreciable faunal exchange with the Mississippi Embayment, suggesting that the latter fauna was derived from elsewhere.

Most of the Mesozoic unionacean groups did not survive into the Cenozoic, or are represented by only a few species. It is tempting to ascribe some catastrophic event to their demise in the fashion of the dinosaurs, but as stated by Taylor (1988), the data do not support this hypothesis. Certainly most of the groups, particularly the speciose vetulonaians and proparreysians, reached their zenith in the Upper Cretaceous, but whether they disappeared all at once or were declining previous to the K-T boundary needs to be established. The loss of these groups is probably attributable to the general climatic changes and orogeny during the Maastrichian stage in western North America. Some snail genera, now confined to tropical areas, disappeared from this region as the climate became progressively colder (Taylor 1988). Apparently, the mussels themselves and/or their hosts also did not survive these changes. This regional extinction, rather than catastrophic extinction, is supported by the persistence of some "Mesozoic" mussel faunas in other parts of the world well into the Cenozoic.

The vetulonaians, a once diverse and widespread group since the Triassic, may be represented in the North American Paleocene by *Pleurobema dyeri* Russell, 1967, from

Alberta. The cylindricoids, a probable sister group of the vetulonaians, became extinct in the Paleocene. Elliptio A may be represented by the Montana Paleocene *Elliptio silberlingi* Russell, 1934. No Greenland elliptios are known from the Paleogene of northeastern North America, but species inseparable from modern *Elliptio* taxa appear in Texas by the Miocene, thousands of kilometers away from the last known records of all of these groups. A group of small, streamlined species resembling miniature *Unio* briefly appeared in the Pliocene of Louisiana.

Protelliptios persisted into the Eocene of Wyoming, Montana, and North Dakota, and apparently invaded Japan across the Bering land bridge (MacNeil 1965; Taylor 1988). In the Green River and Wasatch Formations protelliptios were found in otherwise low-diversity riverine habitats dominated by these mussels (Hanley 1976) (In rare cases, habitat preferences have even been demonstrated for contemporaneous elliptioids: Hanley and Flores 1987).

Apparently true *Anodonta* forms, some perhaps representatives of modern species, appear in the Oligocene of Saskatchewan and the Pliocene of California and Oregon. It has been suggested that the western North American anodontines are related to eastern Asian species, and not to other anodontines in eastern North America. Walker (1910) and Heard (1974) considered the Recent western anodontines to be immigrants from Asia via the Bering land bridge. Given the absence of good *Anodonta* in this region prior to the Oligocene, this hypothesis seems plausible. Eastern anodontines, on the other hand, may be derived from a separate invasion from Europe across Greenland.

Absent from this discussion till now are the Margaritiferidae. Today this is a not particularly speciose group, nearly circumpolar in distribution (but with no circumpolar species), with an isolated species in Laos. In North America it is represented by five isolated species: *M. margaritifera* (Linné 1758) in the northeast, *M. falcata* (Gould 1850) in the northwest, *M. hembeli* (Conrad 1838) from Louisiana, *M. mariannae* Joinnéhnson, 1983 from Alabama, and *Cumberlandia monodonta* (Say 1829) from the Mississippi River system. The distribution of these species clearly indicates a much wider distribution of the group in geological time. Unfortunately, identification of fossil margaritiferids from other groups is difficult. Species such as *Rhabdotophorus gracilis* Russell, 1935 from the Cretaceous of Alberta and *Unio nanaimoensis* Whiteaves, 1901 from the Cretaceous of Vancouver Island may represent early margaritiferids – or convergent unrelated taxa. *Unio senectus* White, 1877, placed by Tozer (1956) in the Paleocene of Alberta, seems the best candidate for a true fossil margaritiferid. As suggested by Walker (1910) the margaritiferids may have invaded North America from two directions – from Asia by the Bering land bridge, and from Europe by Greenland. One or both of these two immigrations apparently spread throughout much of North America only to contract in area to yield the remaining isolated species. As an aside, Taylor and Uyeno (1965) advanced the fascinating argument that the North American *M. falcata* and the Japanese *M. laevis* share common hosts that date from the common ancestor of the two species prior to inundation of the Bering land bridge.

Gonidea also is believed to be Asian in origin (Heard and Guckert 1971; Davis and Fuller 1981). Originally placed in its own subfamily under the Amblemidae by Heard and Guckert (1971) and its own tribe under the Ambleminae by Davis and Fuller (1981), a recent study was unable to place *Gonidea* within the framework of a classification of North American mussels (Lydeard et al. 1996). The appearance of this

Fig. 15.24. *Protamblema macneili* Modell, 1957. Bear River Formation, Cretaceous, Wyoming. 79 mm. (MacNeil 1935)

Fig. 15.25. *Megalonoidea porcata* MacNeil, 1935. Catahoula Sandstone, Miocene, Texas. 92 mm. (MacNeil 1935)

Fig. 15.26. *Megalonoidea rugicosta* MacNeil, 1935. Catahoula Sandstone, Miocene, Texas. 98 mm. (MacNeil 1935)

genus in time and space, and its dissimilarity to other North American species, supports this Asian origin. Four taxa [including the modern *G. angulata* (Lea 1838)] have been recorded from the Neogene of California.

Costanaia and *Megalonoidea* are extremely similar to Recent *Amblema* and *Megalonaias*, respectively, and probably represent ancestral forms of those modern groups. *Protamblema macneili* Modell, 1957 (Fig. 15.24) from the Cretaceous of Wyoming may be the earliest *Costanaia*. Later Miocene and Oligocene *Costanaia* differ from *P. macneili* and Recent species by having the massive ribs arranged radiating from the umbos, rather than obliquely to them. *Megalonoidea porcata* MacNeil, 1935 (Fig. 15.25) from the Texas Miocene is nearly identical with the Cretaceous *P. macneili* but has chevron beak sculpture. *Megalonoidea rugicosta* MacNeil, 1935 (Fig. 15.26) is so

Fig. 15.27. *Unio brimhallensis* Stanton, 1917. Fruitland Formation, Cretaceous, New Mexico. 93 mm (fragment). (Stanton 1917)

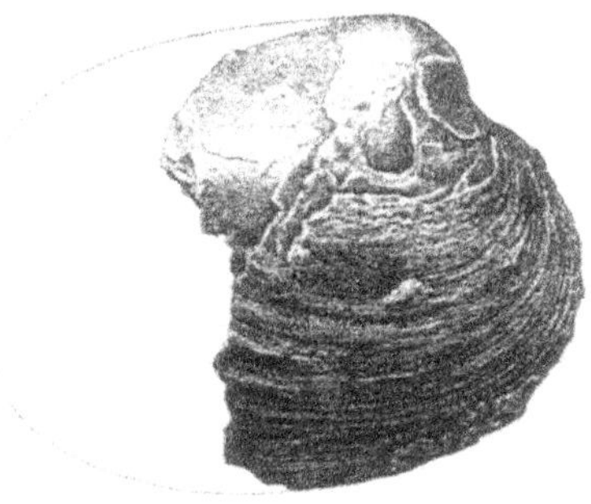

Fig. 15.28. *Lampsilis dalyi* Russell, 1952. Kishenehn Formation, Eocene, British Columbia. 99.6 mm. (Russell 1952)

similar to modern *Megalonaias nervosa* as to be placed in that modern genus. Although there are differences in reproductive timing and ctenidial morphology between *Amblema* and *Megalonaias*, Davis and Fuller (1981) considered them (and the Recent *Plectomerus*) so similar as to constitute a single genus, *Amblema*. This conclusion was refuted by Lydeard et al. (1996), who maintained the separation of these genera. Paleontologically, *Megalonaias* and *Amblema* are closely related. Both date from at least the Miocene, and probably have a Cretaceous ancestor. *Pseudohyria gobiensis* MacNeil, 1936, from Cretaceous Mongolia is similar to some members of *Costanaia*, suggesting a much wider original distribution for this group.

Unio brimhallensis Stanton, 1917 (Fig. 15.27) from the Cretaceous of New Mexico appears to be the earliest recognized lampsiline unionid. The species was very similar to modern *Lampsilis* such as *L. cardium* Rafinesque, 1820. In the Eocene, *Actinonaias*-like forms occur in British Columbia, Wyoming, and Colorado (Fig. 15.28). *Elliptio salissiensis* Russell, 1952 (Fig. 15.29) also from the British Columbia Eocene, although described as an *Elliptio*, seems more likely an ancestor of the *Lampsilis radiata* complex, a group widely distributed throughout eastern North America today. It also may have been sexually dimorphic. Some of these early lampsilines (*Lampsilis chadronensis* Mansfield, 1933, for instance) have corrugated posterior slopes, a feature found in many other Recent unionids but absent from most modern lampsilines. *Lampsilis marshalli* MacNeil, 1935 (Fig. 15.30) from the Texas Miocene is unmistakably a *Potamilus*.

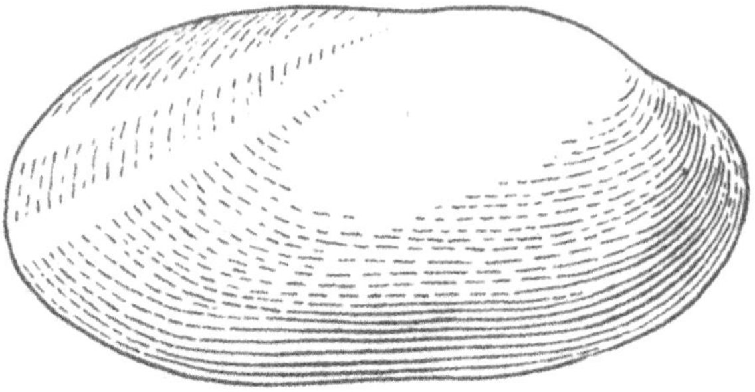

Fig. 15.29. *Elliptio salissiensis* Russell, 1952. Kishenehn Formation, Eocene, British Columbia. 67.5 mm. (Russell 1952)

Fig. 15.30. *Lampsilis marshalli* MacNeil, 1935. Catahoula Sandstone, Miocene, Texas. 76 mm. (MacNeil 1935)

15.4 Comparison of North America with Three European Faunas

15.4.1 The Swiss Oligocene

Modell (1934) briefly described the unionacean fauna of the Swiss Oligocene. He recognized only two genera, *Margaritifera* and *Unio*. I mention this depauperate fauna to call attention to three of Modell's *Margaritifera*: *M. geyeri* Modell, 1931, *M. flabellifera* Noul, 1846, and *M. subflabellata* Rollier, 1910. None of these resemble any living *Margaritifera*. However, they are virtually identical with the vetulonaians *Unio goniambonatus* White, 1878 from the Cretaceous of Wyoming, and to a lesser extent *U. nebrascensis* Meek, 1870, from the Cretaceous of Nebraska. Are the Swiss species holdovers of a Cretaceous fauna?

15.4.2 The Hungarian Tertiary

A thousand kilometers east and 30 million years later, the Pliocene Hungarian unionacean fauna was very different from the depauperate Oligocene Swiss fauna (Modell 1958). The Hungarian fauna is a surprising mixture of ancient Mesozoic survivors that bears little resemblance to today's species. As in Switzerland, in the Oligocene and Miocene we find *Unio* and representatives of the Swiss "*Margaritifera*" (mentioned above), both of which seem to be survivors of the Cretaceous; but in the

Pliocene the fauna takes an unexpected course. Rather than becoming more similar to the modern fauna, it becomes infused with numerous groups not found in Hungary today – but groups found in Asia.

15.4.3 The Romanian Pliocene

Wenz (1942) exhaustively described the rich freshwater molluscan fauna of the Romanian Pliocene. Here we clearly see a fauna much more related to the Recent Asian fauna than to the Recent European fauna. Included are elliptioids and *Unio*-like species seemingly taken straight from the Cretaceous of North America (Fig. 15.31). As members of *Cuneopsidea* and *Psilunio*, Wenz listed species that were nearly inseparable from Recent Asian *Rhombuniopsis* and *Lamprotula* on the one hand, and Cretaceous vetulonaians on the other (Figs. 15.32, 15.33). Spectacular species of *Hyri-*

Fig. 15.31. *Unio pristinus davilai* Porumbaru, 1881. Pliocene, Rumania. 84 mm. (Wenz 1942)

Fig. 15.32. *Unio doljiensis* Stephanescu, 1896. Pliocene, Rumania. 67 mm. (Wenz 1942)

Fig. 15.33. *Unio cymatoides* Brusina, 1874. Pliocene, Rumania. 59 mm. (Wenz 1942)

opsis, another Recent Asian group, occurred there as well. *Parunio*, described from the Pliocene of Mongolia (Ping 1931), are present as *Psilunio bielzi* (Czekelius 1864) and related taxa. None of these groups occur there today.

15.5 From Where Do North American Unionaceans Originate?

Banarescu (1971) stated that North American and European mussel faunas were derived in part from Asia. Certainly, it is apparent that during the Tertiary Europe and Asia shared a fauna composed of groups similar to North American Cretaceous taxa. South American, African, and Australian unionacean faunas were independent of North America by the Jurassic and play no further role in the mussel fauna of Laurasia.

Paleocene groups were able to disperse across the Bering land bridge from North America to Asia and vice versa. The presence of *Margaritifera* and *Anodonta* in eastern Asia and western North America proves that mussels moved across the Bering land bridge. A species of the Asian genus *Hyriopsis* in the Tertiary of Alaska indicates that this exchange worked both ways (Taylor 1988). To what extent the commonalties of the Asian/North American unionacean faunas are due to this bridge is not known. Cretaceous groups may have been distributed across Pangaea and the mussels of Europe and Asia therefore derive their faunas from the indigenous taxa that were present prior to the breakup of that supercontinent. Although climatic conditions led to their near extirpation in North America, they found refuge in Europe and Asia to varying degrees and times and persisted there until similar cooling events eliminated most of these species as well. The appearance of this fauna in the Hungarian Pliocene replacing an earlier depauperate fauna indicates several waves of colonization from Asia. Modern Europe is characterized by a depauperate fauna consisting principally of three genera, *Anodonta. Margaritifera*, and *Unio* - groups that were able to adapt to the new, cooler climates. The Mesozoic fauna fared better in Asia where modern *Rhombuniopsis* and *Lamprotula* probably represent descendants of the vetulonaians.

But where does this leave the Recent unionacean fauna of eastern North America, the most diverse on the planet? Were they derived from the Cretaceous fauna of the western North American continent as advocated by Davis and Fuller (1981), from Europe via Greenland, from Asia via Alaska, or from an endemic, preexisting eastern North American fauna?

The route from Greenland seems the least likely, as there is no fossil record in Western Europe of sufficient diversity to account for the North American fauna. Anodontas, margaritiferids, and even elliptioids may have crossed back and forth, and undoubtedly account for a portion of the modern North America diversity, but the main center of dispersal must lie elsewhere.

It is tempting to believe that Recent North American mussels are the refugees of that diverse Cretaceous fauna to the west of the intercontinental seaway; but if there was a genesis of sorts in the west, why could there not have been one in the east? After all, there *are* a few widely scattered fossils in the east, so at least some mussels were there during the Mesozoic.

If there was a Mesozoic eastern fauna, where is the evidence? Unlike the western fauna, the Mesozoic rocks of the eastern continent are either gone or overlain. Most outcrops of eastern North America are Paleozoic in age. A narrow strip of Triassic outcrops runs down the eastern seaboard, and indeed in it we find unionaceans; but most of the Mesozoic history of eastern North America is lost, including any eastern unionacean fauna. This eastern fauna would have been separated from its western counterpart by the seaway for approximately 30 Ma. How much did the two faunas diverge in this time, and what happened when they eventually were reunited? The answer to this question is lost as well. Forms very similar to modern species dominate the Tertiary of North America. Some have apparent ancestors in the western fauna, but others appear to form de novo. Perhaps these new lineages are descendants of the missing eastern fauna.

The answer to the question: from where did North American unionaceans originate is perhaps "a little of everywhere." During the existence of the supercontinent Pangaea, the ancestors of unionids, margaritiferids, hyriids, mycetopodids, and mutelids all may have lived in that section that became North America. Most seemed to have persisted into the North American Cretaceous, long after the other continents drifted away. During the latter part of the Mesozoic, unionaceans reached a zenith in diversity of species and morphology that was not equaled until recent times. This same fauna, or a version of it, persisted in Europe and Asia as well. As the Mesozoic drew to a close, North America began exchanging taxa with Asia to the west and Europe to the east. Cooler climates were inhospitable to this fauna and it died out at different times in different parts of the world. By the Pleistocene the stage was set for the survivors to begin recolonizing.

15.6 The Mussel-Fish Symbiosis: Thoughts on How the Relationship Began

Sphaeriids brood their young in their gills. They produce relatively few offspring, unlike marine groups, which often have high fecundity. I propose that ancestral unionaceans probably brooded their young and had low fecundity as well. The great fecundity of contemporary species is the result of the adoption of a parasitic life style.

It is perhaps not a great leap from free-living, brooded veligers to parasitic glochidia. Like the Sphaeriidae, the ancestral unionaceans probably had few larvae. These veligers may have been released freely or bound in temporary mucous packages. Fishes may have attempted to eat the veligers, or otherwise encountered them on the substrate. By simply closing their valves, some veligers may have attached themselves to these fishes as accidental passengers. The selective advantage imparted by the increased dispersal abilities of these veligers may have led to a true phoretic symbiosis. As veligers evolved methods to increase the success of attachment (hooks, increased adductor muscle size, etc.), damage to the tissues of the fish necessarily resulted. The host responded with a wound reaction. The "encystment" of glochidia is, after all, nothing more than the encapsulation of the larvae by the host's tissue in response to damage. What was needed to complete the transition was the ability of the glochidium to survive the immunological system of the host. In retrospect, the move from phoretic veliger to parasitic glochidium seems almost passive. Few glochidia grow while on the host, and although they apparently feed on their host's cellular fluids, they seem to be parasites for the sole purpose of transportation. When compared to other parasites, such as cestodes, *Plasmodium*, or rhizocephalan cirripedes, glochidia are not particularly sophisticated, and perhaps may be better described as embellished hitchhikers.

If unionaceans were parasites in the Mesozoic, what were they using as hosts? Many species of agnathans, acanthodians, crossopterygiians, and placoderms also lived in freshwater, but most became extinct before the earliest records of the Unionacea (Long 1995). Survivors, such as lampreys, have been implicated only once as hosts of Recent mussels (Wilson and Ronald 1967), and it is unlikely that the agnathans were ever important hosts for mussels. Mesozoic chondrichthyans also had freshwater representatives, such as *Xenacanthus*, but it is improbable that these large swimming predators would have had the opportunity to encounter glochidia in any numbers. Rays did not invade freshwater until the Tertiary. The widespread Dipnoi had lived in freshwater since the Carboniferous and may have been available as hosts. However, the near extinction of the subclass over geologic time certainly would have necessitated a shift to other fish groups or else face a similar fate.

The best candidate for Mesozoic hosts are the actinopterygian teleosts – the same group parasitized today. Indeed, Dubiel et al. (1991) reported an Upper Triassic freshwater fauna in western North America consisting of numerous mussels cooccurring with a wide variety of actinopterygians and rarer hybodontid sharks, coelacanths, and dipnoans. Teleosts appeared in the Triassic, the same Period in which unionaceans are first encountered, and had freshwater representatives at least by the Jurassic (Long 1995). Many of the modern groups of teleosts appeared by the early Cenozoic, as do many of the modern groups of mussels (*Costanaia* and lampsiline-like species). If mussels had early on evolved the ability to parasitize fishes, then the first great radiation of freshwater teleosts by the Cretaceous may have been directly responsible for a similar radiation of unionaceans.

Of course, these ideas are conjecture. We have no direct way of knowing if the earliest mussels were parasites or not. Fossil glochidia are indeed known, but only from the Quaternary (Brodniewicz 1968, 1969). Ctenidial calcium concretions that today act as sources of glochidial calcium have been found in Jurassic fossil unionaceans (Richard et al. 1991). However, the parallel explosion in diversity of Mesozoic unionaceans with the teleosts and the presence of glochidia-bearing species on nearly every continent suggests that unionaceans developed parasitic larvae early in their evolution.

15.7 Summary

North American unionaceans evolved from an as yet unidentified marine group by at least the Triassic. At that time all of the present-day continents were assembled into a single supercontinent, Pangaea. The presence of taxa resembling hyriids, mycetopodids, mutelids, margaritiferids, and unionids in North America in the Mesozoic suggests that the modern families once enjoyed a much wider distribution than they do today. Upon the fragmentation of Pangaea, the hyriids, mycetopodids, and mutelids persisted in North America until the Late Cretaceous, but then became extirpated, leaving only the margaritiferids and unionids. It is possible that similar scenarios occurred on other continents as well (Fig. 15.34).

The Mesozoic North American unionacean fauna approached the Recent fauna in terms of diversity of taxa and shell shapes. Many unrelated forms, including

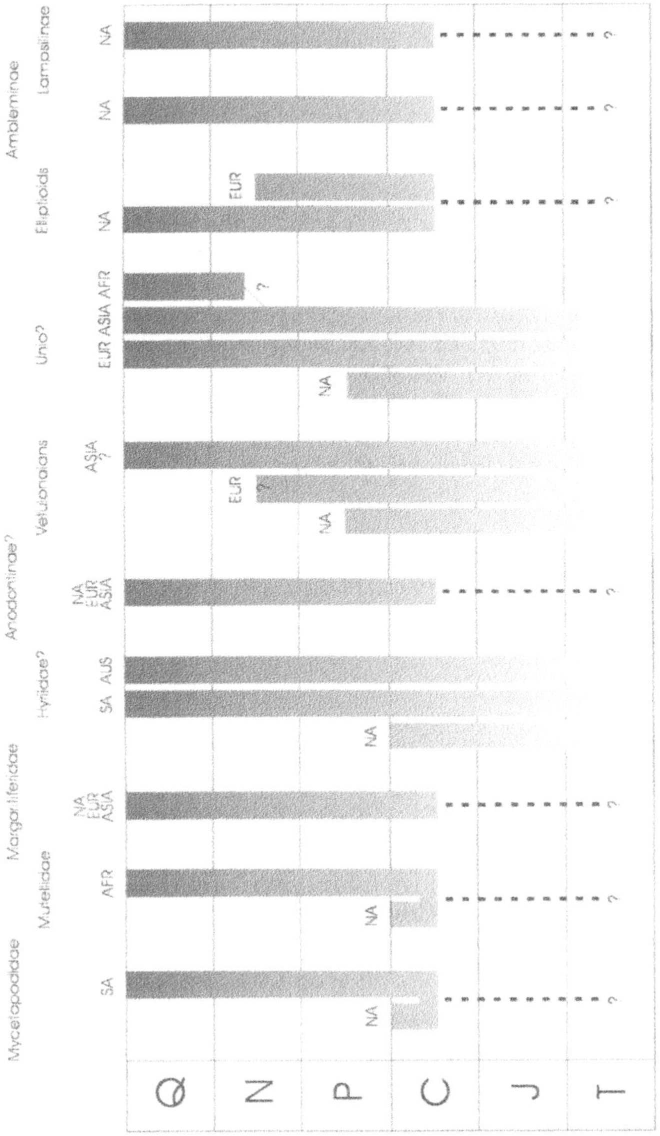

Fig. 15.34. Diagrammatic representation of chronological extent of the important fossil mussel groups on each continent. Periods: *Q* Quaternary; *N* Neogene; *P* Paleogene; *C* Cretaceous; *J* Jurassic; *T* Triassic. Continents: *NA* North America; *SA* South America; *AFR* Africa; *EUR* Europe; *ASIA*; *AUS* Australia. Durations are approximate and periods are not to scale

probable hyriids, converged on shell shapes that occur in the Recent fauna. These shapes have been shown to perform specific functions, such as anchoring and reducing scour, and similar shapes arose independently for similar functions (Watters 1994).

The Mesozoic fauna is known primarily from western North America. Eastern North America has few Mesozoic outcrops and the evidence of any eastern fauna that may have existed has been lost. By the close of the Mesozoic, climatic conditions caused a massive decline in the diversity of this fauna in the west. Modern forms appear by the Eocene, widely scattered across the continent. They probably had multiple origins. Faunal exchange across the Bering land bridge to the west and Greenland to the east undoubtedly occurred. Portions of the western fauna adapted to the cooler climates and eventually migrated east and south. Other Recent taxa may be derived from an undiscovered eastern Mesozoic fauna.

The worldwide distribution of glochidia-bearing mussels indicates that the evolution of parasitic larvae occurred in the Mesozoic and was probably tied to the parallel rise of the teleosts. This life style led directly to the success of this group in modern times.

Acknowledgements. I thank Andrew K. Rindsberg, Geological Survey of Alabama, and David Campbell, University of North Carolina at Chapel Hill, for suggesting valuable references on paleogeography. Portions of this study were funded by a grant from Conchologists of America.

References

Banarescu P (1971) Competition and its bearing on the fresh-water faunas. Rev Roum Biol 16:153–164

Barfield M, Watters GT (1998) Non-parasitic life cycle in the green floater, *Lasmigona subviridis* (Conrad, 1835). Trian Unionid Rep (16):22

Bogan AE, Spamer EE, Manville GC, Gallagher WB, Cain AJ (1989) Preliminary reexamination of the Fish House local fauna and flora (Pleistocene), Pennsauken, Camden County, New Jersey. Mososaur 4:111–126

Branson CC (1935) Fresh-water invertebrates from the Morrison (Jurassic?) of Wyoming. J Paleontol 9:514–522

Brodniewicz I (1968) On glochidia of the genera *Unio* and *Anodonta* from the Quaternary freshwater sediments of Poland. Acta Palaeontol Polon 13:619–628

Brodniewicz I (1969) Les glochidia et leur importance pour l'étude des dépots quaternaires d'eau douce. Mitt Int Ver Theor Angew Limnol 17:315–318

Burch JB (1973) Freshwater unionacean clams (Mollusca: Pelecypoda) of North America. US EPA Biota Freshwater Ecosyst Ident Man (11):1–176

Camacho HH (1957) Observaciones geológicas y petrográficas en Bahía Esperanza, Puerto Paraíso e Islas Shetland del Sur y Melchior. 2: Observaciones en Bahía Esperanza. Publ Inst Ant Argentino 4:25–30

Cox LR (1969) Family Pachycardiidae Cox, 1961. In: Moore RC (ed) Treatise on invertebrate paleontology, part N, vol 1, Mollusca 6, Bivalvia. Geological Society of America and University of Kansas Press, Lawrence, pp N467–N468

Davis GM, Fuller SLH (1981) Genetic relationships among recent Unionacea (Bivalvia) of North America. Malacologia 20:217–253

Dubiel RF, Good SC, Parrish JM (1989) Sedimentology and paleontology of the Upper Triassic Chinle Formation, Bedrock, Colorado. Mount Geol 26:113–126

Dubiel RF, Parrish JT, Parrish JM, Good SC (1991) The Pangaean megamonsoon – evidence from the Upper Triassic Chinle Formation, Colorado Plateau. Palaios 6:347–370
Haas F (1969) Superfamily Unionacea. In: Moore RC (ed) Treatise on invertebrate paleontology, part N, vol 1, Mollusca 6, Bivalvia. Geological Society of America and University of Kansas Press, Lawrence, pp N411–N467
Hanley JH (1976) Paleosynecology of nonmarine Mollusca from the Green River and Wasatch Formations (Eocene), southwestern Wyoming and northeastern Colorado. In: Scott RW, West RR (eds) Structure and classification of paleocommunities. Dowden, Hutchinson & Ross, Stroudsburg, pp 235–261
Hanley JH, Flores RM (1987) Taphonomy and paleoecology of nonmarine Mollusca: indicators of alluvial plain lacustrine sedimentation, upper part of the Tongue River Member, Fort Union Formation (Paleocene), northern Powder River basin, Wyoming and Montana. Palaios 2:479–496
Heard WH (1974) Anatomical systematics of freshwater mussels. Malacol Rev 7:41–42
Heard WH, Guckert RH (1971) A re-evaluation of the recent Unionacea (Pelecypoda) of North America. Malacologia 10:333–355
Henderson J (1935) Fossil non-marine Mollusca of North America. Geol Soc Am Spec Pap (3):1–313
Herbst R, Camacho HH (1970) Sobre el hallazgo de bivalvos de agua dulce (Unionidae y Mycetopodidae) en el Terciario Superior de Empedrado, Provincia de Corrientes, Argentina. Ameghiniana 7:335–340
Hoeh WR, Black MB, Gustafson R, Bogan AE, Lutz RA, Vrijenhoek RC (1998) Testing alternative hypotheses of *Neotrigonia* (Bivalvia: Trigonioida) phylogenetic relationships using Cytochrome C Oxidase Subunit I DNA sequences. Malacologia 40:267–278
Holt EL (1942) A new *Unio* from the Morrison Formation of the Grand River Valley, Colorado. J Paleontol 16:459–460
Ihering H von (1912) Analysen der Südamerikanischen Heliceen. J Acad Nat Sci Philadelphia 2:475–500
Jingeng S (1992) Ontogenetic variations of the Early Cretaceous non-marine bivalve *Trigonioides* (*T.*) *heilongjiangensis* and the concept of fossil species. Paläontol Z 66:241–264
Kat PW (1983) Fossil evidence from Fish House Clays for the origin and changes in species composition through time of the northern Atlantic Slope unionid fauna (Mollusca: Bivalvia). Proc Acad Nat Sci Philadelphia 135:85–101
Keyes CB (1894) Derivation of the unione fauna of the northwest. Proc Iowa Acad Sci 1(5):25–29
Kondo T (1990) Reproductive biology of a small bivalve *Grandidieria burtoni* in Lake Tanganyika. Venus 49:120–125
Lea I (1868) Descriptions of Unionidae from the Lower Cretaceous Formation of New Jersey. Proc Acad Nat Sci Philadelphia 20:162–164
Lellis WA, King TM (1998) Release of metamorphosed juveniles by the green floater, *Lasmigona subviridis* (Conrad 1835). Tran Unionid Rep (16):23
Long JA (1995) The rise of fishes. Johns Hopkins University Press, Baltimore
Lydeard C, Mulvey M, Davis GM (1996) Molecular systematics and evolution of reproductive traits of North American freshwater unionacean mussels (Mollusca: Bivalvia) as inferred from 16S rRNA gene sequences. Philos Trans R Soc Lond B 351:1593–1603
MacNeil FS (1935) Fresh-water mollusks from the Catahoula sandstone (Miocene) of Texas. J Paleontol 9:10–17
MacNeil FS (1936) *Pseudohyria gobiensis* MacNeil. In: Morris FK (ed) Central Asia in Cretaceous time. Bull Geol Soc Am 47:1514–1515
MacNeil FS (1965) Evolution and distribution of the genus *Mya*, and Tertiary migrations of the Mollusca. US Geol Surv Prof Pap 483G:1–51
Mansur MCD, Campos-Velho MNR de (1990) Técnicas para o estudo dos gloquídios de Hyriidae (Mollusca, Bivalvia, Unionoida). Acta Biol Leopold 12:5–18
Marshall WB (1929) New fossil land and fresh-water mollusks from the Reynosa Formation of Texas. Proc US Nat Mus 76(2798):1–6

Martinez S, Figueiras A (1991) Two new species of Mesozoic *Diplodon* (Bivalvia: Hyriidae: Hyriinae) from Uruguay. Walkerana 5:217–223

Martinez S, Figueiras A, Da Silva JS (1993) A new Unionoidea (Mollusca, Bivalvia) from the Tacuarembó Formation (Upper Triassic – Upper Jurassic), Uruguay. J Paleontol 67:962–965

McMichael DF (1957) A review of the fossil freshwater mussels (Mollusca, Pelecypoda) of Australia. Proc Linn Soc N S Wales 81:222–244

Modell H (1934) Oligocaene Najaden der Schweiz. Arch Molluskenkd 66:207–216

Modell H (1957) Die fossilen Najaden Nordamerikas. Arch Molluskenkd 86:183–200

Modell H (1958) Die tertiären Najaden des ungarischen Beckens. Geol Jahrb 75:197–250

Morris PJ, Williamson PG (1988) *Pleiodon* (Conrad) (Bivalvia: Mutelidae: Pleiodoninae) from the Late Cretaceous of Montana: a first North American record for the Mutelidae. J Paleontol 62:758–765

Morton LS (1984) *Diplodon baqueroensis* nov. sp. (Unionidae – Pelecypoda) de la Formation Baquero (Cretacico Medio) de Santa Cruz, Argentina. Facena 5:89–96

Ortmann EA (1910) The systematic position of the unionid-genus *Parreysia*. Nautilus 23: 139–141

Parodiz JJ (1968) Annotated catalogue of the genus *Diplodon* (Unionacea – Hyriidae). Sterkiana (30):1–22

Parodiz JJ (1969) The Tertiary non-marine Mollusca of South America. Ann Carnegie Mus 40:1–242

Pilsbry HA (1921) Report of fossil Mollusca. In: Wanner HE (ed) Some fossil remains of the Trias of York County, Pennslyvania. Proc Philadelphia Acad Nat Sci 73:25–37

Ping C (1931) On some new fossil mollusks of Mongolia. Bull Geol Soc China 10:229–247

Reeside JB (1927) Two new unionid pelecypods from the Upper Triassic. J Wash Acad Sci 17: 476–478

Reeside JB (1957) Nonmarine pelecypod (*Nippononaia asinaria*) from the Lower Cretaceous of Colorado. J Paleontol 31:651–653

Richard PE, Dietz TH, Silverman H (1991) Structure of the gill during reproduction in the unionids *Anodonta grandis*, *Ligumia subrostrata*, and *Carunculina parva texasensis*. Can J Zool 69:1744–1754.

Russell LS (1934) Reclassification of the fossil Unionidae (fresh-water mussels) of western Canada. Can Field Nat 48:1–4

Russell LS (1952) Molluscan fauna of the Kishenehn formation, southeastern British Columbia. Natl Mus Can Bull (126):120–133

Russell LS (1967) Unionidae from the Cretaceous and Tertiary of Alberta and Montana. J Paleontol 41:1116–1120

Russell LS (1976) Pelecypods of the Hell Creek Formation (Uppermost Cretaceous) of Garfield County, Montana. Can J Earth Sci 13:365–388

Simpson CT (1896) Description of four new Triassic unios from the Staked Plains of Texas. Proc US Natl Mus 18:381–385

Smith AG, Smith DG, Funnell BM (1994) Atlas of Mesozoic and Cenozoic coastlines. Cambridge University Press, Cambridge

Stanley SM (1986) Earth and life through time. WH Freeman, New York

Stanton TW (1917) Contributions to the geology and paleontology of San Juan County, New Mexico. 3. Nonmarine Cretaceous invertebrates of the San Juan Basin. US Geol Surv Prof Pap (98-R):309–319

Taylor DW (1988) Aspects of freshwater mollusc ecological biogeography. Palaeogeogr Palaeoclimatol Palaeoecol 62:511–576

Taylor DW, Uyeno T (1965) Evolution of host specificity of freshwater salmonid fishes and mussels in the North Pacific region. Venus 24:199–209

Tozer ET (1956) Uppermost Cretaceous and Paleocene non-marine molluscan faunas of western Alberta. Geol Surv Can Mem (280):1–125

Tschudy RH (1971) Palynology of the Cretaceous-Tertiary boundary in the northern Rocky Mountains and Mississippi Embayment regions. Geol Soc Am Spec Pap 127:65–111

Walker N (1910) The distribution of *Margaritana margaritifera* (Linn.) in North America. Proc Malacol Soc Lond 9:126–145
Watters GT (1994) Function and form of unionoidean shell shape and sculpture. Am Malacol Bull 11:1–20
Weir J (1969) Superfamily Archanodontacea. In: Moore RC (ed) Treatise on invertebrate paleontology, Part N, Vol. 1, Mollusca 6, Bivalvia. Geological Society of America and University of Kansas Press, Lawrence, pp N402–N404
Wenz W (1942) Die Mollusken des Pliozäns der rumänischen Erdöl-Gebeite als Leitversteinerungen für die Aufschluß-Arbeiten. Senckenbergiana 24:1–293
White CA (1883) A review of the non-marine fossil Mollusca of North America. US Geol Surv 3rd Annu Rep:403–550
White CA (1886) On the fresh-water invertebrates of the North American Jurassic. Bull US Geol Surv (29):1–24
White CA (1907) The ancestral origin of the North American Unionid, or fresh-water mussels. Smithson Misc Collect 48:75–88
Whitfield RP (1885) Brachiopoda and Lamellibranchia of the Raritan clays and Greesland marls of New Jersey. Monogr US Geol Surv (9):1–338
Whitfield RP (1907) Remarks on and descriptions of new fossil Unionid from the Laramie clays of Montana. Bull Am Mus Nat Hist 23:623–628
Wilson KA, Ronald K (1967) Parasite fauna of the sea lamprey (*Petromyzon marinus* von Linné) in the Great Lakes region. Can J Zool 45:1083–1092
Yen T-C (1952) Molluscan fauna of the Morrison formation. US Geol Surv Prof Pap (233-B):21–49
Yen T-C (1954) Nonmarine mollusks of Late Cretaceous age from Wyoming, Utah, and Colorado. US Geol Surv Prof Pap (254):45–66

Part V
Mussels and Environment

16 Environmental Relationships of Naiads: Threats, Impact on the Ecosystem, Indicator Function

GERHARD BAUER and KLAUS WÄCHTLER

16.1 Introduction

In the first part of this chapter the threats to naiad populations will be considered. There is growing awareness that these mussels are particularly sensitive to threats. More than 70% of the North American species are declining, some of them being already extinct (Master 1990; Bogan 1993; Johnson and Brown 1998; Chap. 12, this Vol.). However, in many countries still little attention is being paid to the situation of naiads (Pfeiffer 1999; Chap. 2, this Vol.). On the other hand, a very sensitive species whose decline is well documented in some of its distribution areas is the freshwater pearl mussel (*Margaritifera margaritifera*). Since it produces valuable pearls, its occurrence was frequently recorded a long time ago (e.g. v. Hessling 1859). An extensive summary on its status is given by Young et al. (this Vol.). Meanwhile, a number of conservation measures are being taken for various endangered species (e.g. Chaps. 2, 8, 19, this Vol.).

The second point is the impact of naiads on the environment. This concerns mainly their significance for removing seston, a topic which is discussed controversely.

A third point finally is the role of naiads as environmental monitors. In this respect, two features of naiads are of particular importance, namely the long life of some species and the annual formation of shell layers (at least in cold and temperate climates).

16.2 Threats

16.2.1 Adult Stage

Habitat changes through flow regulation (Bogan 1993; Chap. 2, this Vol.), commercial exploitation (Lefevre and Curtis 1912, Chaps. 19, 20, this Vol.) and pollution (Bauer 1988; Michaelson and Neves 1995; Chaps. 18, 19, 20, this Vol.) are important factors for reducing the density of adult mussels. This is most detrimental in strictly dioecious species, as there may be a minimum viable population below which fertilization of eggs is not ensured. This has been shown for *Unio crassus* (Hochwald 1988) and *Elliptio complanata* (Downing et al. 1993). Furthermore, small populations may be so isolated that genetic variability is reduced considerably. Nagel and Badino (this Vol.) attribute the absence of genetic variability in their sample of *Microcondylea bonelli* to this factor. We still have no data about the effect of the introduction of alien genetic material, (e.g. through glochidia-infected fish). However, as Nagel and Badino

Ecological Studies, Vol. 145
G. Bauer and K. Wächtler (eds.) Ecology and Evolution of the Freshwater Mussels Unionoida

(this Vol.) show for European species, geographically defined subgroups do exist, indicating adaptation to local conditions.

A severe recent threat, particularly in North America, is the introduced zebra mussel (*Dreissena polymorpha*), which (among other hard substrates) colonizes unionid mussel shells. Due to stress and starvation, mortality may be increased up to 100% (Ricciardi et al. 1996), particularly in species with high metabolic rates (Baker and Hornbach 1997).

On the other hand, a threat introduced in Europe is the muskrat. Its severe impact here, particularly in small rivers, is attributed to the absence of large predators controlling muskrat populations (Chap. 11, this Vol.).

16.2.2 Glochidial and Parasitic Stage

To our knowledge, there is no record that a naiad species is threatened due to a reduction of glochidial vitality by any environmental factor. This stage, (the free glochidium), seems to be remarkably resistant to pollution of any kind. However, crucial for this stage is the availability of a sufficient number of suitable hosts. Decreased host densities are a threat for a number of species (Hochwald 1988; Johnson and Brown 1998; Chaps. 19, 20, this Vol.). The decline of a highly host-specific mussel species due to the complete absence of its hosts is strikingly documented for *Margaritifera auricularia* (Chap. 8, this Vol.), which is now the rarest naiad species in Europe. However, host availability may also be lowered due to a decreased reproductive rate of host fish, because it may be the youngest age class of fish where glochidia preferably can develop, since older fish are resistant due to acquired immunity (Bauer 1987). Furthermore, host suitability may be decreased by other mussel species having infected the fish previously. For example, Hochwald (1997) could show that mortality of *U.crassus* glochidia increases on hosts which had already been infected by *Anodonta anatina* glochidia. The significance of this interspecific competition for hosts on mussel population dynamics has been very little investigated so far. A very subtle threat is reported for *Ptychobranchus greeni*, whose conglutinates, mimicking dipteran larvae, attach to hard surfaces, where they are taken up by fish. However, due to eutrophication, hard substrate is covered by mud and algae, such that the conglutinates cannot attach and are washed away (Hartfield and Hartfield 1996).

16.2.3 Postparasitic Stage

This seems to be the stage most sensitive to eutrophication among species adapted to headwater conditions (Bauer 1988; Hochwald 1997; Chaps. 6, 18, 19, this Vol.). Apparently, at this stage, the naiads living within the substrate, die due to some factor related to oxygen depletion when the substrate is enriched with organic matter (Buddensiek et al. 1993).

16.3 Significance of Naiads for Seston Removal

The high biomass frequently attained by naiad populations has led to the assumption that naiads are important for reducing seston concentrations (Kasprzak 1986; Libois and Hallet-Libois 1987, Chaps. 2, 18). Indeed, Pusch et al. (this Vol.) can show that naiads may remove on the average 15% of the seston in a eutrophic river from May to October.

On the other hand, it is argued that resource utilization rates are so low, that even at high densities there is no competition for food (Hanson et al. 1988). Also the high densities attained by freshwater pearl mussels argue against food limitation. Frequently, small (and clean) rivers were plastered for many kilometres with mussels (Israel 1913). However, it may be due to the low metabolism of this species (Chap. 13, this Vol.) that food availability apparently is no problem.

16.4 Naiads as Environmental Monitors

Concerning the role of mussels as environmental monitors, three aspects will be considered.

One point of importance is the age structure of the population. Naiad population dynamics depends on many environmental factors (see above) and therefore habitat perturbations may be easily reflected in the population's reproductive rate. Since many najads are long-lived, such perturbations may be read off from the age structure even if they occurred long ago. For example, most populations of the freshwater pearl mussel in Central Europe are overaged (Chaps. 18, 19, this Vol.). Some age structures show a continuous decline in reproduction and its complete cessation many decades ago (Bauer 1988). Another example is an overaged population of *M.auricularia* (Chap. 8, this Vol.), showing that its host, the sturgeon, disappeared from the rivers in Spain decades ago.

Similar information may be gained from analyses of growth rings. In some species the food supply seems to be related to eutrophication or temperature, such that periods with high or low food availability are documented in growth rings being above or below average (Negus 1966; Chaps. 18, 21, this Vol.).

Mussels store toxicants and trace elements. Therefore, one can use them as biomonitors by analyzing these substances in the soft parts (Chap. 2, this Vol.). An even more subtle method to use najad shells as time archives is provided by a Swedish group (Chap. 21, this Vol.). By means of a newly developed nuclear microscope, registrating the X-ray emission from the interaction between charged particles from an accelerator and the target atoms, the stored information on element dynamics is analyzed in shells of the freshwater pearl mussel. Since this species is very long-lived and since shells are available which were collected more than 100 years ago, they have been able to study interactions between environmental changes (e.g. pollution, acidification) and naiad life conditions in the past covering the last 2 centuries. Surprisingly, time resolution is as short as 1 week.

16.5 Summary

In this chapter the threats to naiad populations are analyzed, the role naiads play for organic matter dynamics is investigated and their value as environmental monitors is demonstrated.

All developmental stages are subject to threats. Particularly, strictly dieocious species are very sensitive to reduced density since fertilization rate of eggs decreases.

Among headwater species the postparasitic stage seems to be most sensitive, particularly to eutrophication.

Usually, resource utilization is low among naiads, such that there is no competition for food. However, in some cases, significant reduction of seston by naiads has been demonstrated.

With respect to their monitoring function, analysis of the age structure of a naiad population may provide information on past conditions. Furthermore, naiads store various substances and respond with different growth rates to environmental variation. Also the shell layers may be analyzed. This method allows a time resolution even as short as 1 week.

References

Baker SM, Hornbach DJ (1997) Acute physiological effects of zebra mussel (*Dreissena polymorpha*) infestation on two unionid mussels, *Actinonaias ligamentina* and *Amblema plicata*. Can J Fish Aquat Sci 54:512–519

Bauer G (1987) The parasitic stage of the freshwater pearl mussel. II. Susceptibility of brown trout. Arch Hydrobiol Suppl 76/4:403–412

Bauer G (1988) Threats to the freshwater pearl mussel in central Europe. Biol Conserv 45:239–253

Bogan AE (1993) Freshwater bivalve extinctions (Mollusca: Unionida): a search for Causes. Am Zool 33:599–609

Buddensiek V, Engel H, Fleischauer-Rössing S, Wächtler K (1993) Studies on the chemistry of interstitial water taken from defined horizons in the fine sediments of bivalve habitats in several northern German lowland waters II. Microhabitats of *Margaritifera margaritifera* L., *Unio crassus* (PHILIPSSON) and *Unio tumidus* PHILIPSSON. Arch Hydrobiol 127,2:151–166

Downing J, Rochon Y, Pérusse M, Harvey H (1993) Spatial aggregation, body size, and reproductive success in the freshwater mussel *Elliptio complanata*. J N Am Benthol Soc 12(2): 148–156

Hanson JM, MacKay WC, Prepas, EE (1988) The effects of water depth and density on the growth of a unionid clam. Freshwater Biol 19:345–355

Hartfield P, Hartfield E (1996) Observations on the conglutinates of *Ptychobranchus greeni*. Am Midl Nat 135:370–375

Hessling T v (1859) Die Perlmuscheln und ihre Perlen. Verlag Wilhelm Engelmann, Leipzig

Hochwald S (1988) Untersuchungen zur Populationsökologie und Fortpflanzungsbiologie der Bachmuschel *Unio crassus* (Phil.) 1788. MS Thesis, University Bayreuth, 86 pp

Hochwald S (1997) Das Beziehungsgefüge innerhalb der Größenwachstums- und Fortpflanzungsparameter bayerischer Bachmuschelpopulationen (*Unio crassus* Phil.1788) und dessen Abhängigkeit von Umweltfaktoren. PhD Thesis, University Bayreuth, 172 pp

Israel W (1913) Biologie der europäischen Süßwassermuscheln. KG Lutz, Stuttgart

Johnson PD, Brown KM (1998) Intraspecific life history variation in the threatened Lousiana pearlshell mussel, *Margaritifera hembeli*. Freshwater Biol 40:317–329

Kasprzak K (1986) Role of Unionidae and Sphaeriidae (Mollusca, Bivalvia) in the eutrophic Lake Zbechy and its outflow. Int Rev Ges Hydrobiol 71,3:315–334

Lefevre G, Curtis W (1912) Studies on the reproduction and artificial propagation of freshwater mussels. Bull Bur Fish 756:1–190

Libios RM, Hallet-Libois C (1987) The unionid mussels (Mollusca, Bivalvia) of the Belgian Upper River Meuse: an assessment of the impact of hydraulic works on the river water self-purification. Biol Conserv 42:115–132

Master L (1990) The imperiled status of North American aquatic animals. Biodivers Net News 3:5–7

Michaelson DL, Neves RJ (1995) Life history and habitat of the endangered dwarf wedgemussel *Alasmidonta heterodon* (Bivalvia: Unionidae). J N Am Benthol Soc 14(2): 324–340
Negus C (1966) A quantitative study of growth and production of unionid mussels in the river thames at reading. J Anim Ecol 35:513–532
Pfeiffer M (1999) Die Najaden von Nepal: Verbreitung, Systematik und einige ökologische Aspekte. Ms Thesis, University Freiburg
Ricciardi A, Whoriskey FG, Rasmussen JB (1996) Impact of the *Dreissena* invasion on native unionid bivalves in the upper St. Lawrence River. Can J Aquat Sci 53:1434–1444

17 Filtration and Respiration Rates of Two Unionid Species and Their Impact on the Water Quality of a Lowland River

MARTIN PUSCH, JÖRG SIEFERT, and NORBERT WALZ

17.1 Introduction

As a consequence of their filter-feeding activity, mussels produce a zone around them that is depleted in suspended particles. As the inhalant and exhalant openings of freshwater mussels are generally situated adjacent to each other, this effect may deteriorate the efficiency of filter feeding in standing waters (O'Riordan 1993; Lenihan et al. 1996). In running waters, however, water is permanently mixed by turbulent flow, so that no vertical food gradients may be produced. Therefore, streams and rivers are favourable habitats for freshwater mussels, which may there dominate the total biomass of benthic invertebrates. In streams and small rivers in the uplands of Europe, the great populations of unionid mussels that once lived there are mostly extinguished. However, in the lowlands there are still plankton-rich lake outlets or larger rivers, where this rich nutritional resource is exploited by large populations of unionid mussels of the genera *Unio*, *Anodonta* and *Pseudanodonta* (Unionidae) (Kasprzak 1986). In these type of streams, a second aspect favours mussel colonization, which is the absence of catastrophic spates, which may affect mussel populations in the uplands through the mobilization of bed sediments, and physical abrasion or cracking of the shells.

Dense populations of mussels have the potential to reduce the seston load of the water body they inhabit significantly, and thus profoundly change ecosystem-scale fluxes of matter (Kasprzak 1986; Ostrovsky et al. 1993; Ogilvie and Mitchell 1995; Prins et al. 1998; Strayer et al. 1999). In a lake outlet section of a lowland river, the River Spree, Welker and Walz (1998) found that the concentration of suspended algae decreased exponentially with greater distance from the lake. Some aspects of their results indicated that this retention should be due to benthic filter feeders, which were dominated there by mussels. In order to verify this hypothesis, data were needed on the in situ clearance rate of unionid mussels in flowing waters. Also, we were interested in the question of whether food availability could be a limiting factor for the unionid mussel population in this river section, which shares this resource with a dense population of the zebra mussel *Dreissena polymorpha* (Pallas 1791).

However, we were not aware of measurements made under near-natural flow conditions. Stánczykowska et al. (1976) measured clearance rates with natural seston and under flow conditions, but mussels had not burrowed into the sediment, which reduces the filtration activity of the mussels (Kryger and Riisgård 1988). Other studies that measured the clearance rate of unionid mussels in still water (reviewed in Kryger and Riisgård 1988) were conducted under experimental conditions that differed considerably from the natural situation. This applied to the general setup (e.g. no sediment to dig in for mussels, short observation time that did not account for individual

Ecological Studies, Vol. 145
G. Bauer and K. Wächtler (eds.) Ecology and
Evolution of the Freshwater Mussels Unionoida

resting behaviour) and to the food supply (cultured monospecific algae (e.g. Kryger and Riisgård 1988), graphite or clay particles (e.g. Alimov 1965). Therefore, it seems doubtful if the results obtained in these studies may be applied to estimate natural clearance rates in a river (see also Kryger and Riisgård 1988). Hence, we measured the clearance rates of two unionid species (*Anodonta anatina* and *Unio tumidus*) occurring in the River Spree under simulated stream conditions. In order to gain more insight into the internal metabolism, we measured the respiration rates in parallel.

17.2 Study Sites and Methods

Anodonta anatina (L.) and *Unio tumidus* Philpsson (Unionidae) were sampled in the River Spree directly upstream of Berlin near the village of Freienbrink (52° 22′ N, 13° 50′ E) from April to October 1995. The River Spree is a eutrophic lowland river (Köhler 1994) with the water temperature exceeding 20 °C for about 2 months in summer. In 1995, seston biomass peaked at 19 $mg\,l^{-1}$ dry mass (DM) in April and early May, mainly consisting of pelagic algae (diatoms). From May to September, seston content averaged at 8 $mg\,l^{-1}$ DM. Sampled mussels were subsequently transferred to the laboratory, where they were held in an aerated tank filled with river water. Each second day, they were fed by replacing the tank water by fresh river water. During the clear water phase of the River Spree in June, the mussels were additionally fed by dried and pulverized stinging nettle, and by commercial fish fry food. Mussels that were held in the tanks did not deviate significantly in their mass/length ratio from mussels taken directly from the river.

Filtration and respiration rates were measured under stream conditions using recirculating chambers. These consisted of 25-cm sections of two Plexiglas tubes with diameters of 65 and 25 cm which were glued to a circular bottom (Fig. 17.1). One experimental chamber and one control chamber were filled with 40 l of fresh river water before each experiment. Thus, experiments were carried out under ambient conditions in the river in respect to temperature and seston content and composition. Temperature was controlled by cooling water flowing through the inner volume of each chamber. In the outer volume a recirculating current was produced by an adjustable electric step motor affixed above the experimental setup. Three driving wings connected to the wheel of the motor were joined to the top disc of the chamber by cords. The top disc had Styrofoam flippers on its lower face which transferred the rotation of the disc to the water volume in the chamber. Mean flow velocity in experiments was adjusted to 18 $cm\,s^{-1}$, which equalled that in mussel habitats in the river.

During the experiments, the water recirculating in the chambers was alternatingly sampled by a peristaltic pump via a port in each chamber. From the port, the water was pumped through two three-way valves to a turbidometer (model 512/TF10, Fa. Optek; forward scattering technique) and an oxygen probe with a temperature sensor (model 4002, Fa. Syland), and then back through a second valve to the respective chamber. Gas permeability of the tubing was very low (Tygon, 4-mm wall). Temperature, turbidity and oxygen content of the two chambers were recorded continuously by a PC.

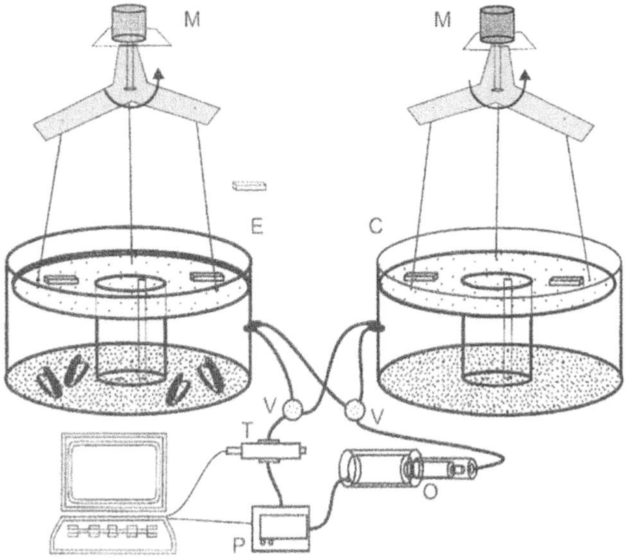

Fig. 17.1. Experimental setup with two recirculating chambers (experimental *E* and control chamber *C*) for the measurement of filtration and respiration rates with electric motor (*M*), valves (*V*), turbidity meter (*T*), peristaltic pump (*P*), oxygen probe (*O*) with temperature sensor and PC

For measurements, mussels were placed into several ca. 20 and 10-cm plastic trays filled with rinsed pebbles (1–2-mm particle size). Total mass of the mussels used ranged between 34 and 62 g. Adult mussel individuals of varying length were taken, representing the natural length composition of the population in the river (*A. anatina*: mean = 6.8 cm ± 0.83 SD; *U. tumidus*: mean = 5.0 cm ± 0.95 SD). Mussels were allowed to dig into the pebbles and to acclimatize to experimental conditions for 2 h before experiments were started. For measurements, the plastic trays were placed into the recirculating chambers. Three replicates were carried out, each lasting for 3.5 h on average. Then, more than 90% of the individuals had their syphons extended, as a sign for activity. Turbidity measurements were calibrated against the dry weight of the seston content before and after each measurement by filtration on glass fibre filters (Schleicher and Schüll no. 10). Measurements were terminated when oxygen content fell below 40% saturation. Mussels that had been used for experiments were measured in length, and a relationship between shell length and the dry mass of the soft parts was established. Thus, all results on mussel body mass are given as the dry mass of the soft parts. Measured turbidity data from experiments afterwards were modelled using a negative exponential regression. Using the regression equation, the slope in the seston concentration from the first to the second minute of the experiment was backcalculated. Thus, clearance rates could be given for conditions with nearly original seston concentration. Results on filter-feeding activity are given as of the retention rate for suspended matter (expressed in $\mathrm{mg\,DM\,g^{-1}\,DM\,h^{-1}}$), and as clearance rates of the water volume (expressed in $\mathrm{l\,g^{-1}\,DM\,h^{-1}}$), which gives the theoretical rate at which the water is cleared by the mussels. During the summer of 1995, 20 measurements were made with *U. tumidus*, and 22 with *A. anatina*.

The impact of a natural mollusc population on the seston content of the Spree during the vegetation period (April to October in 1995) was estimated for a lake outflow section of that river, the so-called Krumme Spree. There, water temperature and seston concentration were determined biweekly at two stations: station 1 near the village of Alt Schadow at the lake outflow (52° 07′ N, 13° 56′ E), and station 2 at a distance of 20 river km near the village of Trebatsch (Fig. 17.5a). Carbon content of riverine seston was 19.1% ± 3.9 SD during the vegetation period. Mussel density [Unionidae and *Dreissena polymorpha* (Pallas 1791)] was determined by repeated areal counts in various transects along the river. Average biomass of unionid mussels was 6 g m^{-2} for *A. anatina*, 52 g m^{-2} for *Unio* spp., and 140 g m^{-2} for *D. polymorpha*. The latter species colonizes most of the unionid mussel individuals as an epizoon, partially obstructing their water-inhalant syphon.

The scope for growth (SFG) was calculated as the difference between the supply of organic carbon to an individual mussel via filtration, and its carbon loss via respiration (Schneider et al. 1998). The SFG gives the surplus of organic matter that is available to the mussel, which may either be returned to the river via the production of pseudofaeces or faeces (ca. 70% Ostrovsky et al. 1993), or may be used for individual growth or reproduction. SFG was not calculated for the winter months, as our database did not cover the low levels of temperature and seston concentration present in the River Spree at that time.

17.3 Results

17.3.1 Filter-Feeding Activity

The individual dry mass of the soft parts of adult mussels averaged 3.5 g in *A. anatina*, and 2.5 g in *U. tumidus*. Retention rates of seston ranged between 0.16 and 7.4 mg DM $g^{-1} h^{-1}$ (mean: 2.85) for *A. anatina*, and between 0.81 and 7.4 mg DM $g^{-1} h^{-1}$ (mean: 2.60) for *U. tumidus* (Fig. 17.2). Retention rates were highly significantly correlated to the initial seston concentration in the measurements ($P < 0.0001$ in both species). In the range of seston concentrations studied, the correlation was linear, i.e. the filtration capacity of the mussels had not been exhausted to its full extent. Retention rates were lower in *U. tumidus* by 23% (Fig. 17.2).

Clearance rates ranged between 0.17 and 0.62 l $g^{-1} h^{-1}$ (mean: 0.36 at 19 °C) for *A. anatina*, and between 0.15 and 0.48 l $g^{-1} h^{-1}$ (mean: 0.34 at 19 °C) for *U. tumidus* (Fig. 17.3). In *A. anatina*, clearance rate correlated highly significantly to the respiration rate or temperature($P < 0.001$, respectively). These correlations were not significant for *U. tumidus*, possibly due to the narrower range of experimental temperatures (9.7–24.6 °C) in *A. anatina* versus 14.3–23.4 °C in *U. tumidus*.

17.3.2 Respiration Activity

Respiration rates ranged between 0.14 and 0.63 mg $O_2 g^{-1} h^{-1}$ (mean: 0.35) for *A. anatina*, and between 0.23 and 0.50 mg $O_2 g^{-1} h^{-1}$ (mean: 0.38) for *Unio tumidus* (Fig. 17.4). Respiration rates primarily depended on ambient temperature ($P < 0.001$ in *A.*

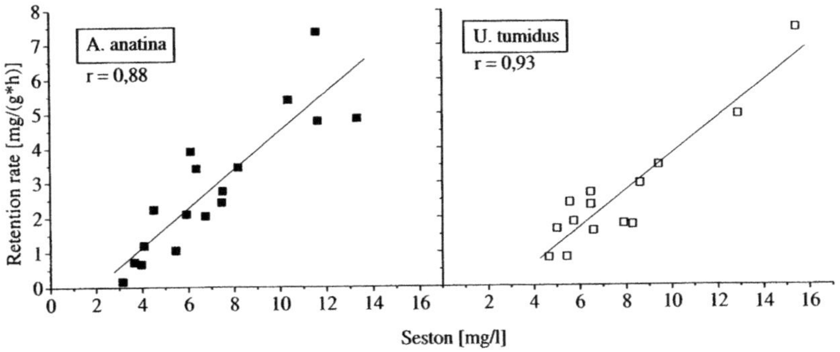

Fig. 17.2. Correlations between retention rates per unit dry mass of the soft parts and seston concentration for the two unionid species studied

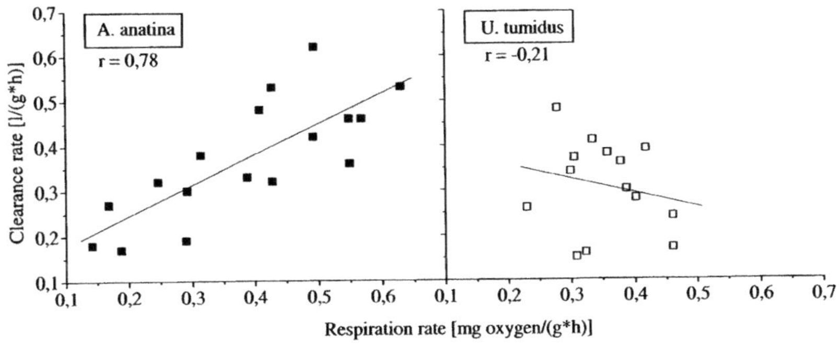

Fig. 17.3. Correlations between clearance rates per unit dry mass of the soft parts and respiration rates for the two unionid species studied

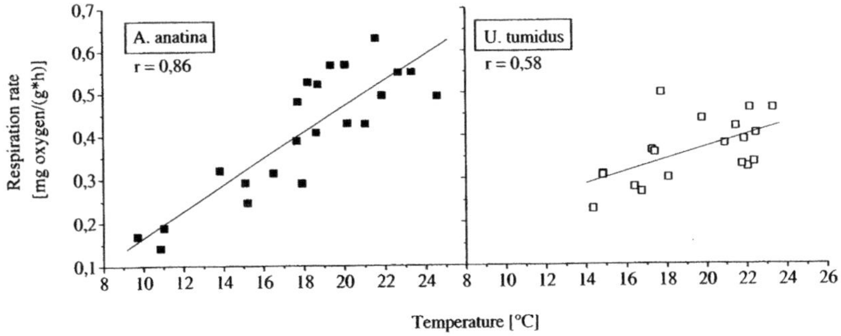

Fig. 17.4. Correlations between respiration rates per unit dry mass of the soft parts and temperature for the two unionid species studied

anatina, $P < 0.01$ in *U. tumidus*). The lower significance level in *U. tumidus* is possibly due to the narrower range of temperatures studied in this species. At temperatures exceeding 14°C, respiration rates were lower in *U. tumidus* (Fig. 17.4). The Q_{10} values were 2.35 for *A. anatina*, and 1.55 for *U. tumidus*.

17.3.3 Metabolic Balance of the Mussels

The scope for growth (SFG) was calculated on the basis of the laboratory measurements for the populations of *A. anatina* and *U. tumidus* in the River Krumme Spree for the period April to October 1995. The SFG showed a distinct seasonal pattern, with values exceeding 0.4 mg C g^{-1} h^{-1} from May to August (Fig. 17.5c). For *A. anatina*, SFG averaged at a seasonal mean of 0.45 mg C g^{-1} h^{-1}. The highest value of SFG reaching 0.8 mg C g^{-1} h^{-1} occured in July, and the lowest value with less than 0.1 mg C g^{-1} h^{-1} in October. For *U. tumidus*, SFG averaged at 0.30 mg C g^{-1} h^{-1}, ranging between 0.7 mg C g^{-1} h^{-1} and zero in the same months. This means that in autumn food supply and metabolic losses were just balanced. Thereby, it has to be kept in mind that the food supply in the downstream reaches of the Krumme Spree river section was clearly worse than in the upper lake outflow reach with its higher seston level. As *A. anatina* mainly occurred in the upper reaches, the above numbers cannot be compared directly between the two species. If SFG was calculated for *U. tumidus* only for the reaches where *A. anatina* also occurred, its mean SFG would be 0.41 mg C g^{-1} h^{-1}.

If it is assumed that 40% of filtered seston is assimilated (see review in Ostrovsky et al. 1993), which means that most of the organic content of the seston is used (see Kasprzak 1986; Hanson et al. 1988; Ostrovsky et al. 1993), the production/biomass (P/B) ratio of *Unio* sp. can be estimated at 0.8, and that of *A. anatina* at 0.4. These values are relatively high for mussels (see review in Kasprzak 1986), which might be an indication for habitat suitability. Besides, the number given here accounts not only for somatic but also reproductive production.

17.3.4 Effect on the River Spree Ecosystem

Seston concentrations in the Krumme Spree section showed a pronounced seasonal course (Fig. 17.5b), with highest concentrations during the algal spring bloom in April and May, a period of elevated discharge in June with lower concentrations, a summer peak during July and August, which was followed by a sharp autumnal decline. Retention of suspended matter occurred during May, and most markedly during the summer periods with low discharge (Fig. 17.5b). This gross retention effect may be generally attributed to the sum of the effects of sedimentation, retention by filter-feeding invertebrates on one side, and of phytoplankton growth on the other side.

In order to assess the contribution of benthic invertebrate filter feeders to this gross retention, the retention rates for the unionid mussels were calculated on the basis of the measurements reported above, and those for *Dreissena polymorpha* according to the results of Morton (1971) for the period April to October 1995. Measured gross seston retention and estimated retention by invertebrates showed a similar seasonal course, and were significantly correlated on a monthly basis (Pearson correlation, $r = 0.90$, $P < 0.015$, $n = 6$) (Fig. 17.5b). On the average, benthic filter feeders

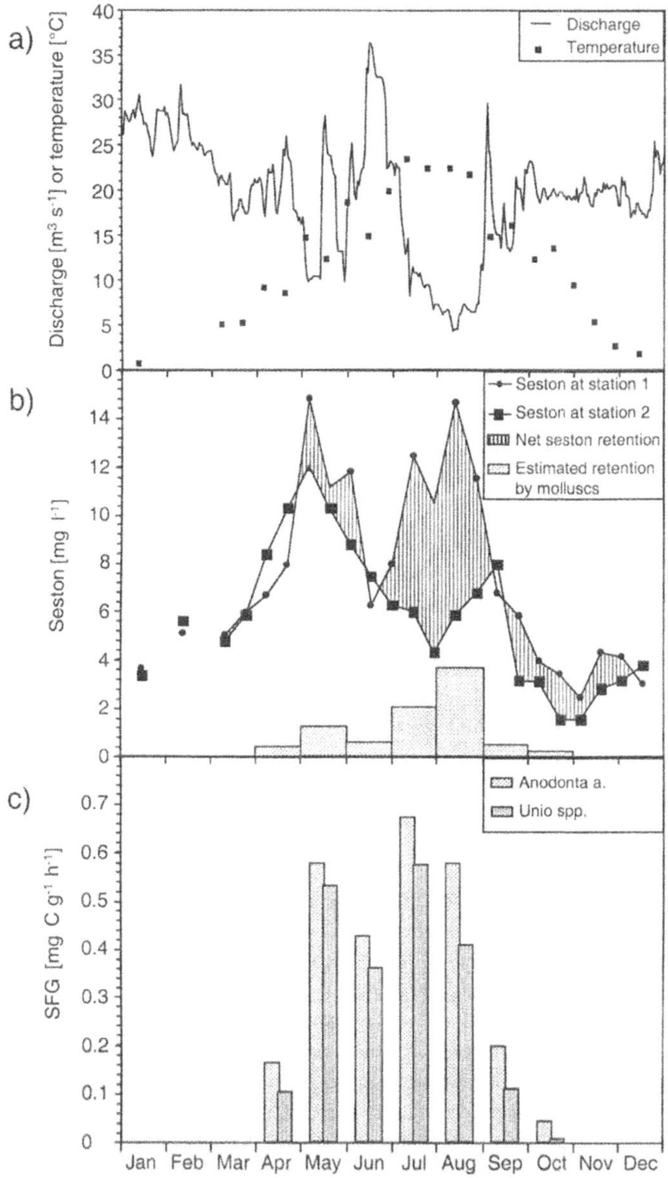

Fig. 17.5. **a** Water discharge and temperature in the Spree at station 1 (gauge at Leibsch UP) in 1995. Discharge data by Landesumweltamt Brandenburg. **b** Seston contents at the two stations on the River Spree with a flow distance of 20 km, and estimated seston retention by molluscs. **c** The scope for growth (SFG) calculated for April to October 1995 for the populations of *Anodonta anatina* and *Unio* spp. in the Krumme Spree river section

accounted for 47% of gross particle retention from May to October (Fig. 17.5b), which ranged between 17% in October and 70% in May and September (monthly means). This means that in that period 284 metric tons of seston were retained by invertebrates in that river section. Unionid mussels contributed by 32% to total filter-feeder retention, *D. polymorpha* by 64%, and the rest was mainly contributed by a filter-feeding snail. The prevalence of *D. polymorpha* can be attributed to their high abundance and feeding activity in the upper lake outlet reach. In most of the Krumme Spree river section (18 out of 20 km), filter-feeding activity was higher in the unionid population than in the *D. polymorpha* population.

17.4 Discussion

The experimental setup used in this study enabled measurements of the clearance rates of freshwater mussels under simulated river conditions. As the main goal of this study was the simulation of the mussels' activities in their habitat, experimental conditions were not as controlled as in other studies concerning food quality. Instead, the most important autecological (size, gravidity) and environmental (temperature, seston composition) factors influencing mussel activity were held at the respective level present in the river. As the respiration rate seems to be influenced by ambient oxygen concentration (Tankersley and Dimock 1993a), our measurements, which were run with undersaturated water, mimicked conditions that are frequently found in the field. Thus, our results are probably close to the activity levels that are realized in nature.

Clearance rates measured in this study approximately equalled those found by Alimov (1965) and Stánczykowska et al. (1976), but were considerably lower than those found by Kryger and Riisgård (1988). This was surprising to us, as the lower values reported by Alimov (1965) and Stánczykowska et al. (1976) were determined under conditions that were clearly adverse to the mussels. However, our values are means of long-term measurements (several hours) of the activity of several individuals, which included the normal resting behaviour of the mussels with reduced filtering activity, and included gravid females with reduced filtration efficiency (Tankersley and Dimock 1993b). On the other hand, the high values reported by Kryger and Riisgård (1988) were clearly influenced by the high nutritional quality of the cultured *Chlorella* algae that were fed, which probably enhanced filter-feeding activity considerably (own unpubl. results). Weight-specific retention rate was higher in *A. anatina* compared to *U. tumidus* under summer conditions. As activity was largely correlated to temperature, the latter species was more active at temperatures below about 15 °C due to its lower Q_{10} value. These autecological differences potentially contribute to different distributional patterns of mussels of the genera *Anodonta* and *Unio*, as found by Brönmark and Malmqvist (1982) and Libois and Hallet-Libois (1987). Respiration rates found in this study were close to those reported by Ornatowski (1967) and by Kasprzak (1986, recalculated from Alimov 1981).

A significant reduction in river plankton by mussels has been observed for *Dreissena polymorpha* by Effler et al. (1996), Basu and Pick (1997) and Findlay et al. (1998). The contribution of unionid mussels to plankton dynamics in lakes has also been estimated repeatedly (Kasprzak 1986), whereas it has rarely been quantified in rivers

(Libois and Hallet-Libois 1987). In the River Spree, unionid mussels are able to exert that influence on the river ecosystem despite of heavy epizoic colonization by *D. polymorpha*, which has reduced the populations of unionids in North America (Ricciardi et al. 1995; Strayer et al. 1999). However, no zoological data are available on our study site from the pre-*Dreissena* period.

Our data on the scope for growth (SFG) of the two species show that positive energy allocation by these mussels is mainly to be expected during summer, and that the numbers for SFG are greatly influenced by the seasonal dynamics of the riverine environment concerning temperature, discharge or algal concentration. Interannual variation in the seasonal course of these parameters may very likely influence their individual growth rate and reproductive effort (Jokela 1996). As the River Spree is impacted by a large-scale flow abstraction (Pusch 1997), the data presented in this study may contribute to a conservational strategy in order to manage the persistence of the mussel populations.

17.5 Summary

The filtration and respiration rates of two unionid mussel species were measured under stream conditions with natural seston. Seston clearance rates averaged $0.36 l g^{-1} h^{-1}$ for *Anodonta anatina* (L.) and at $0.34 l g^{-1} h^{-1}$ for *Unio tumidus* Philipsson. Respiration rates were mainly controlled by temperature and averaged 0.35 and $0.38 mg O_2 g^{-1} h^{-1}$, respectively, at 19 °C. From April to October, filter feeding supplied the mussels with more organic carbon than needed for respiration, with a high surplus from May to August. It was estimated that a natural mussel population in a plankton-rich lowland river (River Spree) accounts for 47% of the observed seston retention from May to October, of which 32% could be attributed to the two unionid species studied.

Acknowledgements. We are indebted to M. Böhme for the design of the recirculating chambers, to A. Hoffmann for mussel colonization density data, to J. Gelbrecht for providing water quality data and to the Landesumweltamt Brandenburg for providing discharge data.

References

Alimov AE (1965) Nekotorye obscie zakonomernosti processa filtracii u dvustvorcatych molljuskov (Some general regularities of the process of filtration in bivalved molluscs). Zh Obshch Biol 30:621–631

Alimov AF (1981) Funktsional'naya Ehkologiya Presnovodnykh Dvustvorchatykh Mollyuskov (Functional ecology of the freshwater bivalves). Trudy Zool Inst Akad Nauk SSSR 96, Nauka, Leningrad, 246 pp

Basu BK, Pick FR (1997) Phytoplankton and zooplankton development in a lowland, temperate river. J Plankton Res 19:237–253

Brönmark C, Malmqvist B (1982) Resource partitioning between unionid mussels in a Swedish lake outlet. Holarct Ecol 5:389–395

Effler SW, Brooks CM, Whitehead K, Wagner B, Doerr SM, Perkins M, Siegfried CA, Walrath L, Canale RP (1996) Impact of zebra mussel invasion on river water quality. Water Environ Res 68:205–214

Findlay S, Pace ML, Fischer DT (1998) Response of heterotrophic planktonic bacteria to the zebra mussel invasion of the tidal freshwater Hudson River. Microb Ecol 36:131–140

Hanson JM, Mackay WC, Prepas EE (1988) Population size, growth, and production of a unionid clam, *Anodonta grandis simpsoniana*, in a small, deep boreal forest lake in central Alberta. Can J Zool 66:247–253

Jokela J (1996) Within-season reproductive and somatic energy allocation in a freshwater clam, Anodonta piscinalis. Oecologia 105:167–174

Kasprzak K (1986) Role of unionidae and Sphaeriidae (Mollusca, Bivalvia) in the eutrophic Lake Zbechy and its outflow. Int Rev Ges Hydrobiol 71:315–334

Köhler J (1994) Dynamics of phytoplankton in the lowland river Spree. Verh Int Ver Limnol 25:1590–1594

Kryger J, Riisgård HU (1988) Clearance rate capacities of six species of European freshwater bivalves. Oecologia 77:34–38

Lenihan HS, Peterson CH, Allen JM (1996) Does flow speed also have a direct effect on growth of active suspension-feeders: an experimental test on oysters. Limnol Oceanogr 41:1359–1366

Libois RM, Hallet-Libois C (1987) The unionid mussels (Mollusca, Bivalvia) of the Belgian upper River Meuse: an assessment of the impact of hydraulic works on the water self-purification. Biol Conserv 42:115–132

Morton B (1971) Studies on the biology of *Dreissena polymorpha* Pall. V Proc Malacol Soc Lond 39, 289–301

Ogilvie SC, Mitchell SF (1995) A model of mussel filtration in a shallow New Zealand lake, with reference to eutrophication control. Arch Hydrobiol 133:471–482

O'Riordan CA, Monismith SG, Koseff JR (1993) A study of concentration boundary-layer formation over a bed of model bivalves. Limnol Oceanogr 38:1712–1729

Ornatowski Z (1967) Oxygen consumption in several species of freshwater mussels (Bivalvia). Bull Soc Amis Sci Lett Poznan Sér D 8:143–148

Ostrovsky I, Gophen M, Kalikhman I (1993) Distribution, growth, production, and ecological significance of the clam *Unio terminalis* in Lake Kinneret, Israel. Hydrobiologia 271:49–63

Prins TC, Smaal AC, Dame RF (1998) A review of feedbacks between bivalve grazing and ecosystem processes. Aquat Sci 31:349–359

Pusch M (1997) Ecological effects of flow rate reduction on a lowland river (Spree). Wasserwirtschaft 87:375–376

Ricciardi A, Whoriskey FG, Rasmussen JB (1995) Predicting the intensity and impact of *Dreissena* infestation on native unionid bivalves from *Dreissena* field density. Can J Fish Aquat Sci 52:1449–1461

Schneider DW, Madon SP, Stoeckel JA, Sparks RE (1998) Seston quality controls zebra mussel (*Dreissena polymorpha*) energetics in turbid rivers. Oecologia 117:331–341

Stánczykowska A, Lawacz W, Mattice J, Lewandowski K (1976) Bivalves as a factor effecting circulation of matter in Lake Mikolajskie (Poland). Limnologica 10:347–352

Strayer DL, Caraco NF, Cole JJ, Findlay S, Pace ML (1999) Transformations of freshwater ecosystems by bivalves. BioScience 49:19–27

Tankersley RA, Dimock RV Jr (1993a) The effect of larval brooding on the respiratory physiology of the freshwater unionid mussel *Pyganodon cataracta*. Am Midl Nat 130:146–163

Tankersley RA, Dimock RV Jr (1993b) The effect of larval brooding on the filtration rate and particle-retention efficiency of *Pyganodon cataracta* (Bivalvia:Unionidae). Can J Zool Rev Can Zool 71:1934–1944

Welker M, Walz N (1998) Can mussels control the plankton in rivers? – a planktological approach applying a Lagrangian sampling strategy. Limnol Oceanogr 43:753–762

18 Effects of Eutrophication on Unionids

Robert A. Patzner and Doris Müller

18.1 Introduction

Eutrophication is a process of nutrient enrichment in aquatic ecosystems. Under natural circumstances it occurs over long periods but is accelerated by human activities such as sewage disposal; the rapid increase in nitrate is caused by the application of nutrients and agrochemicals in cultivated areas, and phosphate nutrient levels in standing waters cause extended growth of phytoplankton. The ensuing bacterial decomposition may deplete oxygen levels severely. Muddy sediments are also often formed.

Human activities have caused pollution and eutrophication in most European and North American freshwaters. However, general recognition of the situation has led to some active management to help to alleviate the problem. Water purification plants, sewage works, drainage systems and restrictions on the use of fertilizers have made improvements in the quality of many inland waters in recent years.

Filter-feeding invertebrates such as mussels play a major role in the self-purification of many aquatic ecosystems (Strayer et al. 1994). As mussels are especially sensitive towards pollution, they are one of the first groups of the macrozoobenthos to be affected by decreasing water quality (Stansbery and Stein 1971; Fuller 1974; Horne and McIntosh 1979; Forester 1980).

The aim of the present chapter is to show the different influences of eutrophication on stenoecious and euryecious species of unionid mussels.

18.2 Methodological Approach

Growth rate, fertility and other parameters can vary between populations of mussels of the same species in dependency of their aquatic environment. Furthermore, the possibility of genetic differences should be considered (Paterson and Cameron 1985; Müller 1995). Therefore, it is not sufficient to make simple, direct comparisons of mussel populations from different waters in order to study the consequence of eutrophication. For the study of the effects of varying degrees of eutrophication on unionid mussels, a variety of methods have been employed in the literature:

1. Comparison of growth rates in subfossil and recent shells (Arter 1989).
2. Comparison of mussels in a river before and after inflow of eutrophicated water (Ostrovsky et al. 1993).
3. Comparison of mussels in a lake with different trophic zones (Zick 1998).
4. Comparison in connected waters with different trophic status (Müller 1995; Müller and Patzner, in prep.) (Fig. 18.1; Sect. 18.4).

Ecological Studies, Vol. 145
G. Bauer and K. Wächtler (eds.) Ecology and Evolution of the Freshwater Mussels Unionoida

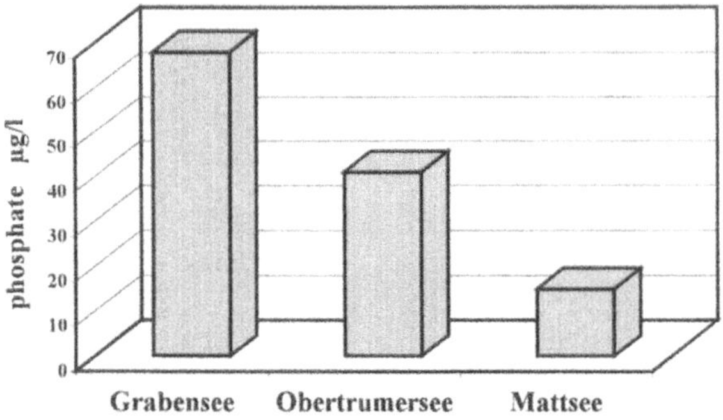

Fig. 18.1. Content of total phosphorous (reflecting the degree of eutrophication) in the water of three lakes of the Austrian Alpine foreland in 1988 (mean values). It is assumed that they contain genetically similar populations of unionids (see text). Values obtained from Amt der Salzburger Landesregierung

5. Comparison of chemistry in waters with living and extinct mussels (Hochwald 1997).
6. Transplantation of mussels into more eutrophic waters (Byrne 1998).

18.3 Effect of Eutrophication on a Stenoecious Species of Unionidae

In former times *Unio crassus* was abundant in many European running waters. Its distribution area stretches over northern and central Europe to the complete area of the Black Sea and Mesopotamia (Falkner 1990). Even in the beginning of the 20th century farmers shoveled *U. crassus* out of the streams to feed them to ducks and hens (Israel 1910). However, since then, the stocks have decreased by 90% as a result of pollution, eutrophication and impact of running waters (Engel and Wächtler 1990; Bauer et al. 1991). The remnant populations are mostly superannuated, with only few or even no juvenile mussels present (Bauer et al. 1991; Hochwald 1997). Although juvenile mussels are difficult to find because of their small size and their burrowing way of life, 6-year-old or older mussels should be found regularly. If this is not the case, the population is approaching superannuation. Several populations are known in which no reproduction have taken place for several years (Hochwald 1997).

The absence of juvenile mussels is often closely associated with increased levels of nitrate. Zettler (1996) and Hochwald (1997) measured concentrations of nitrate below $10\,mg\,l^{-1}$ in streams with normal juvenile mussel populations. Engel (1990) states values between $7.8\,mg\,l^{-1}$ and $11.9\,mg\,l^{-1}$. In streams with old-aged populations of *U. crassus*, nitrate concentrations have been noted between 16 and $22\,mg\,l^{-1}$. In streams where the mussel populations are already extinct, concentrations up to $25\,mg\,l^{-1}$ of nitrate were measured (Hochwald 1997; Fig. 18.2). It is presumed that in processes of nitrification and denitrification within the sediment nitrite is released. This may have a toxic effect on the juvenile mussels buried in the sediment (Buddensiek 1991; Buddensiek et al. 1993). Eutrophication of waters often causes an increased production of mud, which fills the interstitial spaces of the sediment. This

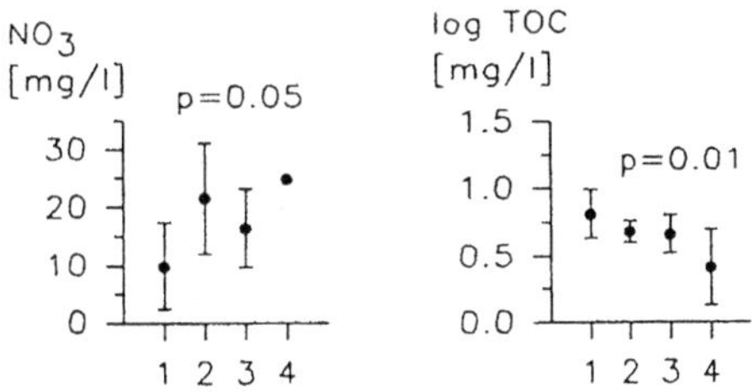

Fig. 18.2. Relation between the content of nitrate (NO_3) and total organic carbon (TOC) in the river water (mean ± SD)and the vitality of four population classes of *Unio crassus*. **1** *U. c.* with normal juveniles; **2** senile *U. c.* with few juveniles; **3** senile *U. c.* without juveniles; **4** extinct *U.c.* (After Hochwald 1997)

was found to be one of the main reasons for the decrease and/or extinction of the sensitive *Margaritifera margaritifera* (Bauer and Thomas 1980; Strecker et al. 1990; Bauer 1991, 1992). This might also be true for *U. crassus*. For both processes the content of nitrate could serve as an indicator (Hochwald 1997).

The content of total organic carbon (TOC) also seems to play an important role for the upcoming of *U. crassus*. In streams with well-developed populations of *U. crassus*, Hochwald (1997) found the highest values of organic carbon (Fig. 18.2). However, she could not find significant relations between the existence of the mussel populations and values of oxygen, sulfate, calcium, magnesium, chloride and pH.

Eutrophication and pollution of waters not only affect the mussels directly but may also cause changes within the fish fauna. As only a limited series of species serve as hosts for the glochidia larvae of *U. crassus* (Engel and Wächtler 1989; Hochwald and Bauer 1990; Hochwald 1997), this fact may also have a strong influence on the existence of the mussels.

18.4 Effect of Eutrophication on Euryoecious Species of Unionidae

When waters become more eutrophic, stenoecious species, such as *Unio crassus*, disappear, while the density of more tolerant species such as *Unio pictorum* and most species of the genus *Anodonta* may increase (Agrell 1948; Forsyth 1978; Franke 1993; Strayer et al. 1994). Adult *Anodonta* spp. and tolerant *Unio* species can utilize the increase in nutrients (Arter 1989; Franke 1993; Müller 1995).

18.4.1 Chemical Factors and Substrate

At a slightly raised content of nitrate ($18\,mg\,l^{-1}$) Franke (1993) found high population densities of *A. anatina*. This author states that the species is tolerant of water pollution, even though she found a lower population density in habitats with a higher content of chloride and low oxygen concentrations.

Low oxygen concentrations during the summer stagnation in the hypolimnion is the most important factor for limiting the depth distribution of adult mussels (Ornatowski 1967; Forsyth 1978; Arter 1989; Müller 1995). Lukacsovics (1966) found that *A. cygnea* had a mortality of 95% within 3 days under hypoxial conditions. Grantham (1969) found no mussels alive where oxygen occasionally dropped below $3\,mg\,l^{-1}$. *Anodonta imbecillis* can respire normally at down to concentrations of about $0.73\,mg\,l^{-1}$ (Histand 1938). Ornatovski (1967) found a demand for 0.52 mg oxygen h^{-1} per $10\,g^{-1}$ weight for *Anodonta cellensis* (= *A. cygnea*), indicating a higher oxygen consumption in juvenile animals. Eutrophic lakes have a high part of muddy sediment in which the oxygen content is low or even lacking. This has a strong negative influence on the juveniles, usually buried in the substrate (Arter 1989). Additionally, oxygen deficiency can stimulate a premature release of glochidia larvae (Tankersly and Dimock 1993).

Some species of mussels are adapted to live in the muddy substrates that are usually associated with eutrophication in standing or very slow-flowing waters. These species are more tolerant to eutrophication. Thus, Parmalee and Hughes (1993) found higher abundances of *Anodonta* but a strong reduction in the more sensitive mussels from 50 to 6 species with increased mud levels in the North American Tellico lake. Ostrovsky et al. (1993) observed highest densities of *Unio terminalis* on muddy sand in the River Jordan inlet zone which are probably related to highest food availability. The importance of sedimentation for juvenile mussels is discussed in Section 18.4.6.

18.4.2 Effects of Eutrophication on the Abundance of Mussels

Agrell (1948) found increasing abundances of *Anodonta piscinalis* (= *A. anatina*) and *Unio pictorum* in eutrophic waters, while *Unio tumidus* decreased.

In the early 1990s the abundance of unionids was investigated in some lakes of the Alpine foreland of Austria (Patzner et al. 1993). In the preceding 20 years the water quality had improved and the trophic situation changed from strongly eutrophic to mesotrophic. It has been shown that the abundances of *A. cygnea* and of *A. anatina* decreased dramatically during that time. The reason for this is not clear. It may be that the mussels have been accustomed to a high food supply and now actually starve to death in the clean water or that there are other reasons, not connected with the degree of eutrophication (Patzner and Müller 1996). A change of nutrient might be another reason. A. E. Bogan (pers. comm.) points out that the composition of bacteria and phytoplankton may have changed by the purification of the water. Thus, mussel populations adapted to eutrophic water may not obtain enough of the correct food.

18.4.3 Effect of Eutrophication on Mussel Growth

Early discussions about the effect of increasing eutrophication on mussel growth can be found in Agrell's study of 1948. After measuring *Anodonta piscinalis* (= *A. anatina*), *A. cygnea*, *Unio pictorum* and *U. tumidus* from 20 areas located on 18 water courses, 15 lakes and 3 streams in Sweden, Agrell (1948) concluded that variations in shell parameters are essentially a function of the degree of eutrophication, such that the shell proportionality is altered. He observed an increase in relative weight, height and thickness above the middle line and a decrease in the relative posterior part as well as in the ridge distance as eutrophication increases.

Furthermore, a higher growth rate in *A. anatina* (Franke 1993) and *U. tumidus* (Arter 1989; Fig. 18.3) results in a shorter life span. This correlation appears not to hold for the swan mussel *A. cygnea* (Franke 1993; Müller 1995; Fig. 18.4). This species is growth-oriented and is adapted to optimal food conditions (Niemeyer 1993). However, it has to be mentioned that Arter used the slope of the curve as growth-rate

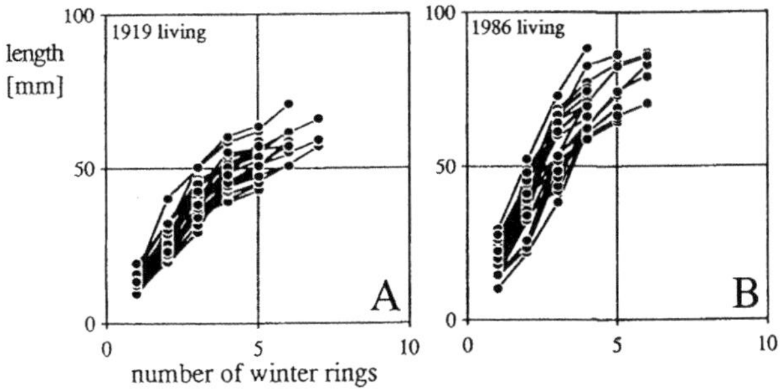

Fig. 18.3A,B. Individual growth curves of *Unio tumidus* in Lake Hallwil. **A** Living mussels collected in 1919 (low eutrophication). **B** Shells from living mussels collected in 1986 (high eutrophication). The size of the winter ring is plotted against the age estimate. A *line* connects the measurements of a single shell. The *highest point* gives the final size of the shell and *the slope* estimates the growth rate. (After Arter 1989)

Fig. 18.4. Comparison of maximal age against the growth constant k in *Anodonta cygnea* in the three lakes of the Austrian Alpine foreland with different populations of *A. cygnea* in Germany. (After Franke 1993; and Müller 1995), $r = 0.48$; $P = 0.1$. Both authors used the Bertalanffy growth constant

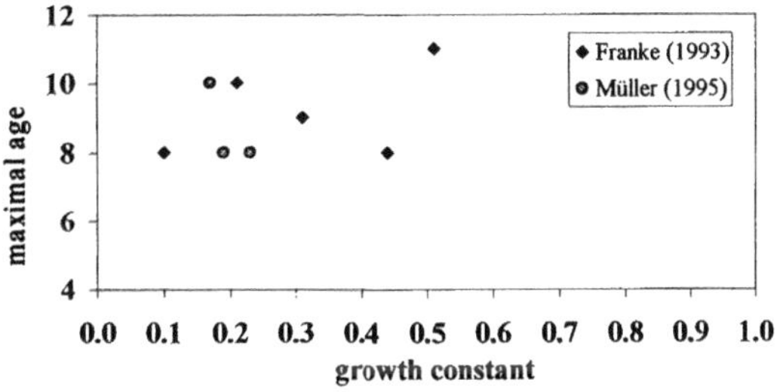

estimate (Fig. 18.3), whereas Franke (1993) and Müller (1995) used the Walford-Plot analysis to calculate a growth constant.

The influence of eutrophication on the age structure, growth and reproduction of the swan mussel *A. cygnea* has been studied in three lakes of the Austrian Alpine foreland with different levels of eutrophication (Müller 1995; see Fig. 18.1 and Sect. 18.2). After the glacial period, Lake Mattsee, Lake Grabensee and Lake Obertrumersee were combined to one single lake and today they are still connected by canals over short distances. Fish infected with the glochidia larvae of the mussels, can migrate between them. Nowadays the lakes show different degrees of eutrophication; Lake Grabensee is eutrophic, Lake Obertrumersee mesotrophic and Lake Mattsee moderately mesotrophic (Fig. 18.1). Growth constants k were similar (0.15 to 0.19) between these three localities. They reach their full length within the same period of time. This result may be related to similar water temperatures (23.5 to 24 °C up to a depth of 4 m) in the lakes. Furthermore, no correlation between the maximum age and the growth constant k could be observed ($P = 0.46$; Fig. 18.4).

18.4.4 Effect of Eutrophication on Body Condition

In the three Austrian lakes (see above) body condition of the swan mussel *Anodonta cygnea* correlates with the degree of eutrophication of the water (Müller 1995). Length

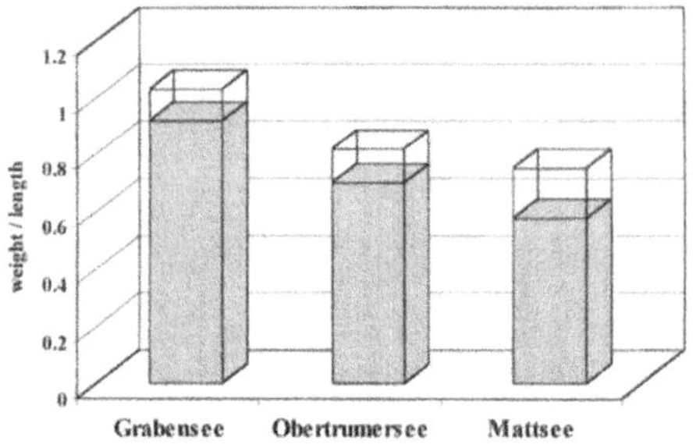

Fig. 18.5. Means of condition factors of *Anodonta cygnea* in lakes with different degrees of eutrophication (see Fig. 18.1). Lake Grabensee $n = 61$, Lake Mattsee $n = 129$, Lake Obertrumersee $n = 60$. *Transparent part* shows standard deviation (positive values). (After Müller 1995)

and width of shells were measured to the nearest 0.1 mm and wet weights of clams were taken to the nearest 0.1 g. Body condition factors such as wet weight/length and shell width/length were calculated. Clams from Grabensee showed the highest, those from Mattsee the lowest wet-weight/length index ($P < 0.0001$, $n = 250$; Fig. 18.5) at the same age. The same result was obtained for the shell width/length index ($P < 0.0001$, $n = 250$). Results of the Anova showed that weight differed significantly between sites ($F = 39.04$, $P < 0.0001$).

18.4.5 Effect of Eutrophication on Mussel Fecundity

A higher availability of food results not only in an increase in growth rate and a better condition factor (Fig. 18.5) but also in an increase in the production of glochidia. Müller (1995) studied the fertility of *Anodonta cygnea* in lakes with different degrees of eutrophication. The production of glochidia in mussels of the same shell length was lower in the moderately mesotrophic Lake Mattsee than in the eutrophic Lake Grabensee ($P < 0.0001$; Fig. 18.6). Additionally, in Lake Mattsee the number of glochidia did not increase with increasing length and therefore with increasing age. Normally, the number of glochidia is in direct correlation with the shell length and the size of the marsupium (Franke 1993; Downing et al. 1993; Bauer 1994), as is the case in the eutrophic Lake Grabensee. The decreasing production of glochidia with increasing age in the lakes with a relatively low degree of eutrophication may be connected to the condition factor and the low food supply (Müller 1995). We assume that the energy budget becomes worse with increasing age and thus not enough energy can be provided for a higher larval production. The mussels of Lake Grabensee have a better body condition due to the higher food supply. Therefore they have more energy, which can be diverted to tissue assessment and a higher production. In waters with sufficient nutrition the rate of filtration can be reduced. The energy thus saved can be used for a better condition and for reproductive purposes (Paterson and Cameron 1985). *A. cygnea* is a growth-oriented species (Niemeyer 1993), and therefore needs more energy for maintenance in localities with low food supply than species living in waters with high levels of eutrophication. The limited energy has to be divided for filtration, growth and shell production during summer and for filtration and production of glochidia in autumn. It can be assumed that this balance of energy changes with increasing age, with less energy available for the production of glochidia (Müller 1995).

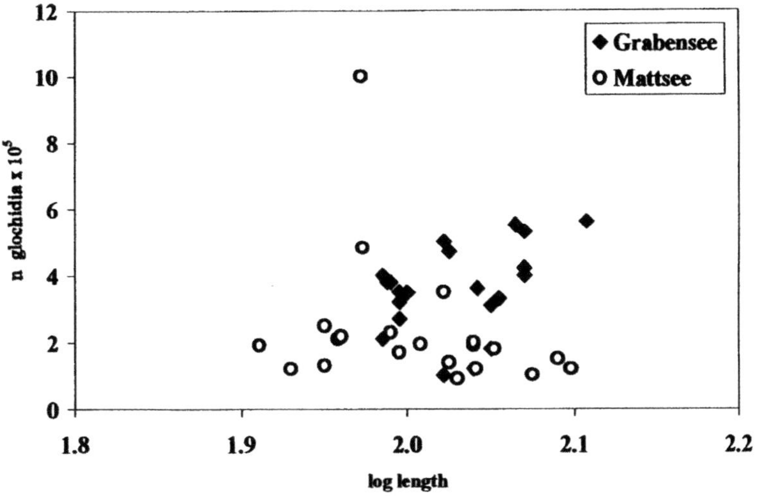

Fig. 18.6. Number of glochidia larvae of *Anodonta cygnea* in Lake Grabensee ($n = 21, r = 0.38$; $P = 0.9$) and in Lake Mattsee ($n = 20, r = 0, P = 0.9$). (After Müller 1995)

18.4.6 Effect of Eutrophication on Juvenile Mussels

In eutrophic waters the rate of mortality of juvenile *Unio tumidus* increases dramatically and can lead to a reduction in population size (Agrell 1948). The low oxygen contents in waters with high degrees of eutrophication mostly affect the mortality rate of juvenile (<1 year) mussels (see also Sect. 18.4.1).

Juvenile mussels are also extremely sensitive to the deposition of fine sediment, which can block their gills (Negus 1966). They need a stable substrate without anaerobic mud but soft enough to bury themselves (Isely 1911; Howard 1921; Negus 1966). Thus, Negus (1966) found juveniles of *A. anatina* especially in sand and clay with a low portion of mud.

18.5 Summary

Eutrophication caused by human activities generally results in changes to aquatic ecosystems. Unionid mussels react to several factors: (1) phytoplankton and bacteria as nutrition, (2) oxygen content, (3) sediment levels, especially relevant for juvenile mussels and (4) the fish fauna as host for the glochidia. It has been shown that eutrophication causes stenoecious species to decline or to become extinct, while euryecious species can – at least to a certain degree – profit from the rich availability of nourishment. Condition factor, growth and fecundity increase with increasing eutrophication degree, but the maximal age of the mussels decreases in some species.

References

Agrell I (1948) The shell morphology of some Swedish unionids as affected by ecological conditions. Ark Zool 41:1–30

Arter HE (1989) Effect of eutrophication on species composition and growth of freshwater mussels in Lake Hallwil (Aargau, Switzerland). Aquat Sci 51:87–99

Bauer G (1991) Plasticity in life history traits of the freshwater pearl mussel – consequences for the danger of extinction and for conservation measures. Species conservation: a population biological approach. Birkhäuser, Basel, pp 103–119

Bauer G (1992) Variation in the life span and size of the freshwater pearl mussel. J Anim Ecol 61:425–436

Bauer G (1994) The adaptive value of offspring size among freshwater mussels (Bivalvia; Unionidea). J Anim Ecol 63:933–944

Bauer G, Thomas W (1980) Die Ursachen für den Rückgang der Flußperlmuschel im Fichtelgebirge und Maßnahmen für ihren Schutz. Natur Landsch 55:100–103

Bauer G, Hochwald S, Schmidt C, Schmidt H, Reger RH (1991) Dauerbeobachtung von Muschelbeständen. Bayerische Akad Natursch Landschaftspflege 7:30–37

Buddensiek V (1991) Untersuchungen zu den Aufwuchsbedingungen der Flußperlmuschel *Margaritifera margaritifera* Linnaeus (Bivalvia) in ihrer postparasitären Phase. Thesis, Univ Hannover

Buddensiek V, Engel H, Fleischauer-Rössing S, Wächtler K (1993) Studies on the chemistry of interstitial water taken from defined horizons in the fine sediments of bivalve habitats in several northern German lowland waters. Arch Hydrobiol 127:151–166

Byrne M (1998) Reproduction of river and lake populations of *Hyridella depressa* (Unionacea: Hyriidae) in New South Wales: implications for their conservation. Hydrobiologia 389:29–43

Downing JA, Rochon Y, Perusse M (1993) Spatial aggregation, body size, and reproductive success in the freshwater mussel *Elliptio complanata*. J N Am Benthol Soc 12:148–156

Engel H (1990) Untersuchungen zur Autökologie von *Unio crassus* (Philipsson) in Norddeutschland. Thesis Univ Hannover

Engel H, Wächtler K (1989) Some peculiarities in developmental biology of two forms of the freshwater bivalve *Unio crassus* in northern Germany. Arch Hydrobiol 115:441–450

Engel H, Wächtler K (1990) Folgen von Bachentkrautungsmaßnahmen auf einen Süßwassermuschelbestand am Beispiel eines kleinen Fließgewässers des südlichen Drawehn (Lüchow-Daunenberg). Natur Landsch 65:63–65

Falkner G (1990) Binnenmollusken. In: Fechtner R, Falkner G: Weichtiere. Steinbachs Naturführer. Mosaik, München, pp 112–286

Forester AJ (1980) Monitoring of the bioavailability of toxic metals in acid-stressed shield lakes using pelecypod molluscs (clams, mussels). Proc Univ Mo Annu Conf Trace Subst Environ Health 14:142–147

Forsyth DJ (1978) Benthic macroinvertebrates in seven New Zealand lakes. NZ J Mar Freshwater Res 12:41–49

Franke G (1993) Zur Populationökologie und Geschlechtsbiologie der Teichmuschel *Anodonta anatina* L. und *Anodonta cygnea* L. MSc Thesis, Univ Bayreuth

Fuller SLH (1974) Clams and mussels (Mollusca: Bivalvia). In: Hart CW Jr, Fuller SLH (eds) Pollution ecology of freshwater invertebrates. Academic Press, New York, pp 215–273

Grantham BJ (1969) The fresh-water pelecypod fauna of Mississippi. Thesis, Univ Southern Mississippi, Hattiesburg

Hiestand WA (1938) Respiration studies with fresh-water molluscs: I. Oxygen consumption in relation to oxygen tension. Proc Indiana Acad Sci 47:287–292

Hochwald S (1997) Populationsökologie der Bachmuschel (*Unio crassus*). Bayreuther Forum Ökologie, vol 50, pp 1–166

Hochwald S, Bauer G (1990) Untersuchungen zur Populationsökologie und Fortpflanzungsbiologie der Bachmuschel *Unio crassus* (Phil. 1788). Schriftenr Bayer Landesamt Umweltsch 97:51–60

Horne FR, McIntosh S (1979) Factors influencing distribution of mussels in Blanco River of central Texas. Nautilus 93:119–133
Howard AD (1921) Experiments in the culture of freshwater mussels. Bull Bur Fish Wash 38: 63–89
Isely FB (1911) A preliminary note on the ecology of the early juvenile life of the Unionidae. Biol Bull 20:77–80
Israel W (1910) Die Najaden des Weidagebietes. Beiträge zur Kenntnis der mitteleuropäischen Najaden. Beil Nachrbl Dtsch Malakol Ges 4:49–64
Lukacsovics F (1966) Hypoxial examination of *Anodonta cygnea* L. on the O_2-consumption of gill tissues and the relation between body dimensions and the respiration of the gill tissue. Ann Inst Biol Hung Acad Sci 33:79–94
Müller D (1995) Populationsökologie der Großen Teichmuschel, *Anodonta cygnea* (L.), in den Seen des Salzburger Alpenvorlandes. MSc Thesis, Univ Salzburg
Negus C (1966) A quantitative study of growth and production of unionid mussels in the River Thames at Reading. J Anim Ecol 35:513–532
Niemeyer B (1993) Vergleichende Untersuchungen zur bionomischen Strategie der Teichmuschelarten *Anodonta cygnea* L. und *A. anatina* L. Thesis, Tierärztliche Hochschule Hannover
Ornatowski Z (1967) Oxygen consumption in several species of freshwater mussels (Bivalvia). Bull Soc Amis Sci Lett Poznan Ser D-8:143–148
Ostrovsky J, Gophen M, Kalikliman I (1993) Distribution, growth, production and ecological significance of the clam *Unio terminalis* in Lake Kinneret. Isr Hydrobiol 271:49–63
Parmalee PW, Hughes MH (1993) Freshwater mussels of Tellico Lake: 12 years after impoundment of the little Tennessee River. Ann Carnegie Mus 62:81–93
Paterson CG, Cameron IF (1985) Comparative energetics of two populations of the unionid, *Anodonta cataracta* (Say). Freshwater Invertebr Biol 4:79–90
Patzner RA, Müller D (1996) Gefährdung und Rückgang der Najaden-Muscheln (Unionidae, Bivalvia) in stehenden Gewässern. Bayerische Akad Natursch Landschaftspflege 20:177–196
Patzner RA, Loidl B, Glechner R, Hofrichter R (1993) Abundanz und Tiefenverteilung von Najaden (Mollusca: Bivalvia: Unionidae) in den Seen des Salzburger Alpenvorlandes. Natur Landsch 68:58–62
Stansbery DH, Stein CB (1971) Why naiades (pearly freshwater mussels) should be preserved. Stream channelization (Part 4). Hearings before a Subcommittee of the Committee on Government Operations, House of Representatives, Ninety-Second Congress, First Session, pp 2177–2179
Strayer DL, Hunter DC, Smith LC, Borg CK (1994) Distribution, abundance and roles of freshwater clams (Bivalvia, Unionidae) in the freshwater tidal Hudson River. Freshwater Biol 31:239–248
Strecker U, Bauer G, Wächtler K (1990) Untersuchungen über die Entwicklungsbedingungen junger Flußperlmuscheln. Schriftenr Bayer Landesamt Umweltsch 97:25–30
Tankersley RA, Dimock RV Jr (1993) The effect of larval brooding on the respiratory physiology of the freshwater unionid mussel *Pyganodon cataracta*. Am Midl Nat 130:146–163
Zettler M (1996) Populationen der Bachmuschel *Unio crassus* (Philipsson 1788) in den Einzugsgebieten der Elbe und Warnow in Mecklenburg-Vorpommern – ein Vergleich. Tagungsbericht Dtsch Ges Limnol, Krefeld
Zick D (1998) Die Molluskenfauna im Mattsee (Salzburg, Österreich). MSc Thesis, Univ Salzburg

19 The Extent of, and Causes for, the Decline of a Highly Threatened Naiad: *Margaritifera margaritifera*

MARK R. YOUNG, PETER J. COSGROVE, and LEE C. HASTIE

19.1 Introduction

Margaritifera margaritifera Linn., the freshwater pearl mussel, is distributed from the arctic and temperate regions of western Russia, westwards through Europe to the northeastern seaboard of North America (Fig. 19.1; Jungbluth et al. 1985). Throughout most of its range it has become scarce or threatened, and is listed as "vulnerable" by IUCN (Wells et al. 1983). It is widely believed that it has now become even more threatened (Chesney and Oliver 1998; Young 1991) and the purpose of this chapter is to review evidence for its current status and to identify the causes of its decline.

Accurate assessments of rates of decline of any species depend on reliable, comparable data from the past. However, although historical distribution records exist, it is only since about 1980 that standardized samples have been collected. Accounts of the occurrence of *M. margaritifera* are generally focused on specific countries, so we have tabulated information on a country-by-country basis.

19.2 Methods Used in the Collation of Data

We reviewed existing literature and contacted *M. margaritifera* researchers for data in broad date categories. Section 19.3 sets out the situation in each country separately, with a brief commentary on the occurrence of juveniles (here taken as mussels below 65 mm/12–15 years) and the causes of decline in each country. Finally, there follows a general assessment of trends in the status of *M. margaritifera* and a prediction for the future.

19.3 Trends in the Status of *M. margaritifera*

19.3.1 North America

M. margaritifera ranges from Pennsylvania north to Newfoundland and southern Labrador and westwards to the approximate level of Quebec. Walker (1910) lists various specific historical records, with most sites in New York State, Connecticut, Massachusetts, Vermont and Maine. There is no detailed account of the precise current status in individual rivers and streams, but Williams et al. (1993) record the overall status as Special Concern; that is not Endangered or Threatened but liable to enter those categories if the current damaging factors continue. They list it as still present in all areas of its historical range.

Ecological Studies, Vol. 145
G. Bauer and K. Wächtler (eds.) Ecology and Evolution of the Freshwater Mussels Unionoida

Fig. 19.1. The historical distribution of *Margaritifera margaritifera* (Linn.)

Fichtel and Smith (1995) and D. Smith (pers. comm.) report that it still persists in Massachusetts, Connecticut, Vermont, Maine and Rhode Island and that populations in Smith's study sites in Massachusetts are currently stable but sparse. However, in Pennsylvania and New York State it is declining severely (Strayer and Jirka 1997).

Clarke (1981) reviewed the Canadian distribution but did not provide detailed information on the size and status of individual populations. However, there are recent reports of large populations in New Brunswick, Canada (J.M. Hanson, pers. comm.), with an estimate of over a million adults in the catchment of the Petitcodiac and mussels still common in most other potentially suitable habitats. Hanson (pers. comm.) also records that at least one popluation survives in Nova Scotia and J. Maunder (pers. comm.) reports that it is "generally but not universally distributed in rocky streams.... in Newfoundland". There has been no coordinated survey of *M. margaritifera* in Canada, however, and records from other provinces, including southern Labrador, are scanty. Nevertheless *M. margaritifera* is not listed as threatened or endangered in Canada by the Canadian Wildlife Service (1999).

D. Smith (pers. comm.) reports that some juveniles are still present in the Massachusetts populations and that two host species, brown trout (*Salmo trutta*) and brook charr (*Salvelinus fontinalis*) are plentiful. At least one mussel of 25 mm length has been found recently in New Brunswick (J.M. Hanson, pers. comm.), but no specific search has been made for juveniles in Canada.

Many species of mussel are currently declining in northeast America, and Williams et al. (1993) note that this is being caused by a variety of factors. There has been physical habitat destruction, pollution (particularly including eutrophication) and commercial exploitation. These causes are not referred specifically to *M. margaritifera*.

M. margaritifera has official protected status in Vermont and Connecticut but not in other American States, nor in Canada.

19.3.2 Portugal

Originally, *M. margaritifera* was reported from tributaries of the river Douro, which flows to Porto, as listed by Nobre (1913). Bauer (1986) reports that he resurveyed the Rivers Paiva, Tamega, Sousa and Ferrera in June 1985 but failed to find any live or dead mussels. Nevertheless, he comments that, as judged by conductivity values of 44 µs, at least the River Paiva remains unpolluted and so potentially suitable. On Bauer's evidence, *M. margaritifera* must be regarded as extinct in Portugal but the cause of this extinction remains unknown.

19.3.3 Spain

Bauer (1986) (referring to Velado 1878) and Alvarez et al. (2000), note that *M. margaritifera* was once distributed in more or less all the rivers in northwestern Spain, ranging from the Rio Tambre in Galicia to the Rio Narcea in Asturias. Nevertheless, there are no data on the abundance of mussels in the past, although by implication they were relatively common in the last century.

Bauer (1986) revisited Galicia in 1985 and recorded mussels from Rio Tambre, Rio Mandeo and Rio Landro. Recently, J. Herrero, (pers. comm.) has also found mussels in the Rio Umia, near Vigo, but these are immediately threatened by a proposed hydro-

Table 19.1. Recent status of *M. margaritifera* in Spain

River and region	Date of survey	Density/relative abundance	Evidence of juveniles Life history characteristics	Source of record
Rio Tambre Galicia	1985	"Abundant" upstream of Ponte Carreira	Predominantly old specimens, up to 60 years	Bauer (1986)
Rio Mandeo	1985	"Abundant" upstream of Muniferal	Some juveniles present Life expectancy up to 50 years	Bauer (1986)
Rio Landro	1985	"Abundant" near Chavin	Some juveniles present Life expectancy up to 40 years	Bauer (1986)
Rio Umia	1998	"Common"	Some juveniles present	J. Herrero (pers. comm.)
Rio Esva, Asturias	1997	Dead shells only	–	Alvarez et al. (2000)
Rio Eo	1997	Dead shells only	–	Alvarez et al. (2000)
Rio Porcia	1997	Dead shells only	–	Alvarez et al. (2000)
Rio Narcea	1997	Present at 31 of 34 searched sites Densities range from 3–139 mussels m^{-2}	8 mussels from 22–36 mm long 22 mussels from 36–50 mm long	Alvarez et al. (2000)

electricity scheme. Since 1996 Alvarez and colleagues have been studying various rivers in Asturias and have found mussels in the Rio Narcea and Rio Esva. They also recorded many dead shells in Rio Esva, Rio Eo and Rio Porcia (Table 19.1).

Some juveniles were present in 1985 in two Galician rivers and in Rio Narcea live mussels were found in many sites in 1997, ranging up to 139 mussels m^{-2} but juveniles were very scarce, with only 30 found below 50 mm in length (Alvarez et al. 2000), so putting in question the long-term survival of *M. margaritifera* in Spain.

Bauer (1986) refers to canalization and the detrimental effect of a hydroelectric scheme in Spain. However, he believes that pollution, in the form of nutrient enrichment, is the main cause of decline. Nevertheless, in 1985 the Rio Mandeo and Rio Landro were still very clean and the populations were secure at that time. Alvarez et al. (2000) refers to river engineering and pollution as the main threats and the Rio Umia population is endangered by construction of a dam (J. Herrero pers. comm.).

M. margaritifera has no legal protection in Spain, beyond that provided by European designation.

19.3.4 France

Cochet (1996, 1998) summarizes the historical and recent distribution of *M. margaritifera* in France. Originally it was found in Finisterre and Brittany, Le Massif Amoricain, Les Vosges, Les Ardennes, Le Massif Central and Les Pyrenees Occidental and was also recorded, probably in error, from Les Alpes and Le Jura. This corresponds to almost all areas on calcifuge rocks. Cochet (1998) has resurveyed these areas but found only 38 small populations in streams of the Rhine, Seine, Loire, Garonne and Finisterre basins, with most confined to the upper reaches of the Loire. Bauer (1986) was unable to locate the Vosges mussels but Cochet (1998) found a few there, as well as in Le Massif Amoricaine and Le Morvan, although most are in Le Massif Central. None were found in the Pyrenees.

No data are available on whether juvenile mussels are still present in French populations.

Bauer (1986) suggests that pollution, as evidenced by high conductivity levels, may have caused local extinction in Les Vosges and Cochet (1996) also considers that pollution, as well as river engineering, has had an adverse effect. Remaining populations are in headwaters, generally above the influence of pollution.

M. margaritifera has been fully protected in France since October 1992.

19.3.5 Germany (Including Belgium and Luxembourg)

M. margaritifera was originally widely distributed in Germany, in the Rhine, Weser, Elbe and Danube basins. Jungbluth (1987) reports a comprehensive survey of 269 historical mussel sites with 93 populations remaining. Of these, 89 were in Bavaria and only singles in each of the other main areas of distribution. C. Schmidt and G. Wenz (pers. comm.) recommend that the records are grouped as follows for discussion of the current status.

1. Lower Saxony. Wellman (1938) refers to around 50 000 mussels in the Rivers Lachte and Lutter to the south of the Lüneburg Heath but none now remain in the River Lachte (Table 19.2). The dwindling population in the River Lutter has been the subject of extensive research (e.g. Wächtler et al. 1987) but at most 3000 individuals remain. Recently, attempts have been made to improve the habitat and to rear juveniles to restock the river (Buddensiek 1995).
2. Saxony and Thuringia. Baer (1981, 1995) reviews the status of mussels from Thuringia and the Vogtland mountains east of Dresden. It is apparent that *M. margaritifera* is now virtually extinct in Thuringia and the Vogtland mountains. In Saxonia there were originally 13 mussel rivers but by 1963 only approximately 25 000 mussels remained. Decline has continued and currently only around 800 mussels remain in one river in Vogtland.
3. Hessen. Jungbluth (1987) records a population of mussels surviving in the Vogelsberg mountains in the upper catchment of the Weser. By 1991 Nagel reported that only 20 individuals could be found in this stream. No historical records are available for the original extent or abundance of this population.
4. North Rhine – Westphalia. By 1987 only one population remained in this area, containing between 500 and 1000 mussels (Jungbluth 1987), although Bischoff et al. (1986) quote the population size as only 200. There seem to be no reliable estimates of previous population sizes but Jungbluth does refer to another stream, which previously held mussels but where they are now extinct.
5. Rhineland – Palatinate. Although at least three rivers held historical populations in this area (Jungbluth 1988), only two remain, with population sizes of around

Table 19.2. The historical and current occurrence of *Margaritifera margaritifera* in Germany. (Taken from references in text)

Region	Historical records	Current records
Lower Saxony	50000 in 1938	3000 in 1987
Saxony and Thuringia	At least 800 in Thuringia in 1969 Still present in Vogtland in 1969 25000 in 1963 in Saxonia	Extinct in Thuringia by 1995 Near extinction in Vogtland in 1995
Hessen	Only one population in 1987	20 in this river in 1991
North Rhine – Westphalia	Two populations – no estimates of numbers	500–1000 in one population in 1987
Rhineland – Palatinate	Three or more populations – no estimates of numbers	1800 in one population; 12 in another in 1995
Bavaria	– Rhon – 500 mussels in one river in 1985	250 survive by 1995
	– Originally 16 populations in Fichtelgebirge (1859) 70000 mussels in 1914	20000 in 8 rivers by 1988
	– 17 populations in Oberfalz in 1859	5 populations with 3000 by 1986
	– 98 populations in Bayerische Wald in 1859	60000 in 42 populations by 1986

1800 in one and as few as 12 in another (Groh 1995). No studies exist quoting reliable historical estimates.

6. Bavaria. This area is here taken to extend from Rhon in the northwest to the Bayerischer Wald in the southeast. Originally, at least 131 separate populations were found (Hessling 1859) but now many fewer survive. Jungbluth (1987) recorded 89 from the area; Schmidt (1990) found 61 populations and G. Wenz and C. Schmidt (pers. comm.) believe that at most 56 remain. These are divided between four subregions.

   At Rhon one population survives, with only 250 mussels, although 500 were present in 1985 (Bauer et al. 1991).

   In the Fichtelgebirge, where Hessling (1859) noted 16 rivers, only eight remain, although substantial numbers of mussels are still present in at least one of these. Meissner (1914) recorded 700000 mussels present here in 1914, whereas Bauer (1988) recorded fewer than 20000 overall. However, more intensive unpublished surveys since 1989 by Schmidt, Hochwald and Vogel-Bauer found that 60000 remained in the Mahringsbach and Hollbach, 40000 in the Zinnbach and another 40000 in the Südliche Regnitz. One of the most populous of these rivers has been the subject of an intensive programme of habitat improvement for mussels, with potentially polluting water being diverted and with sympathetic channel management (Bauer and Eicke 1986; Anonymous 1996).

   Of the 17 rivers that Hessling (1859) records from the Oberpfalz, five remain with up to 3000 mussels between them (C. Schmidt and G. Wenz, pers. comm.).

The largest numbers of mussel-bearing rivers remain in the Bayerischer Wald, at 42; but this is compared with Hessling's total of 98 in 1859. Steidle (1986) estimated the overall numbers of mussels present in 1850 at between 8 and 22 million, whereas C. Schmidt and G. Wenz (pers. comm.) believe that only around 60000 remain, with many rivers with very small populations.

Naturally occurring juveniles in Germany are now found only in Bavaria (Bauer et al. 1991) and Schmidt (1990) found mussels of less than 10 years in only ten populations there. However, G. Wenz and C. Schmidt (pers. comm.) believe that these may be in only two populations. R. Dettmer (pers. comm.) reports that juveniles resulting from artificial enhancement may now be found in the Lutter in Lower Saxony.

There is general agreement that the substantial decline in Germany has been caused by the following factors, arranged roughly in order of importance: (1) eutrophication (due to agricultural intensification, inadequate sewage treatment and the effects of fish farms) (Bauer and Zwölfer 1988; Foeckler et al. 1997); (2) habitat changes, including increased siltation (as a result of river engineering, the removal of bankside trees and erosion from exposed, sandy soils;K. Wächtler, pers. comm.); (3) illegal pearl fishing, (which has recently had a diminishing effect but may have been influential in the early stages of mussel decline); (4) industrial pollution (including acidification). In places, this has had only a limited impact, for in the Bayerischer Wald there are no major industries. G. Wenz and C. Schmidt (pers. comm.) do not believe that there has been a problem of declining fish stocks.

There is a general protection of *M. margaritifera* in Germany through European legislation but individual states also apply their own laws. The species has complete protection in Lower Saxony, for example, whereas in Bavaria it is still possible to obtain a licence for pearl fishing, but only with stringent conditions. The last pearl fisher mainly presses for conservation measures and water quality improvements (C. Schmidt and G. Wenz, pers. comm.).

Hannick (1980) reports that *M. margaritifera* was found in Belgium and Luxembourg, at least up to 1818, in four rivers and numerous streams nearest Germany. By implication, mussels were then abundant and supported a pearl fishery, but now no populations remain. The timing and causes of the decline to extinction are unknown.

19.3.6 Austria

M. margaritifera is found in Austria in the upper Danube and upper Elbe, on metamorphic rocks in Oberösterreich and Niederösterreich, and Riedl (1928) provided an inclusive list of all historic sites. Moog et al. (1993), during an extensive survey between 1990 and 1992, investigated 145 sites in 51 streams in Oberosterreich and, of these, 29 streams showed no evidence of mussels, 4 had only dead shells and live mussels were found only at 33 sites on 18 rivers. In Niederösterreich 86 sites on 37 streams were checked, with live mussels only at 24 sites in 9 rivers. Surviving populations are mainly in forested areas in East Mühlviertel and West Waldviertel.

Overall only three rivers held over 5000 mussels, but 20000 remain in the Waldaist and 5000, 3000 and 5400 in three tributaries of the Kamp. Most of the other streams now hold fewer than 1000 mussels. There are no precise estimates of the original population levels but there has been a very substantial decline, particularly in the Elbe catchment.

Age profiles were produced for all the remaining populations in Austria by Moog et al. (1993) but juveniles were present at only seven of these. In places, there were small dead shells but no live ones, indicating that adverse factors still apply. The numbers and proportion of juveniles show that only two Austrian populations can now be considered to be self-sustaining.

There is a long history of pearl fishing in Austria and this was an early cause of decline (Riedl 1928). However, fishing has now ceased and by 1950 industrialization and the intensification of agriculture had become important (Moog et al. 1993). This has caused direct toxic pollution, which completely eliminated some populations. Intensive agriculture, the effluent from aquaculture ponds and, to a lesser extent, sewage disposal have all contributed to eutrophication, which partly accounts for the failure of recruitment. Further problems have resulted from drainage, canalization and flow regulation.

M. margaritifera has no direct legal protection in Austria but several of the rivers which it inhabits have their own protection from fisheries authorities. Furthermore, some of the populations are within national nature conservation areas. Nevertheless, the status of the mussel remains highly vulnerable.

Recently, the condition of some of the old sites has improved and reinforcement and reestablishment programmes have been started.

19.3.7 Czech Republic

M. margaritifera originally had an extensive distribution throughout the rivers draining the metamorphic mountains of the western Czech Republic but now are restricted to the upper reaches of the rivers, especially in the Sumava and Jeseniky Mountains, where their upper altitudinal limit is 800 m (Dyk and Dykova 1974; Hruska 1992). Dyk and Dykova (1974) note that in past times they used to "pave the river bottoms" in places, but precise figures are difficult to obtain; nevertheless they report complete extinction in many rivers and dramatic reductions in all the rest. A unique population in the alkaline River Otava, near Zichovice, has been lost. Hruska has made an approximate census of the main remaining populations and his data for 1998 have been kindly provided by P. Trpak and S. Krivanek (pers. comm.) (Table 19.3).

It is clear that only two numerous populations remain but that one of these is both abundant and with juveniles present.

Table 19.3 shows that a substantial juvenile population exists in the Blanice River, with some in the Warm Vltava and Stropnice. Hruska (1992) notes that successful recruitment does not occur each year. For example, in the Blanice he records good numbers of juveniles produced in 1963, 1964 and 1976.

Dyk and Dykova (1974) blame decline on eutrophication and siltation, (caused by greatly increased agriculture, pond aquaculture and forest removal); general pollution, (caused by industrialization and urbanization of the lower catchments); stream regulation, damming and substrate extraction; overfishing for pearls; and timber transport.

Table 19.3. Numbers of *M. margaritifera* remaining in Czech rivers in 1998. (Data from J. Hruska, kindly provided by P. Trpak and S. Krivanek)

River	Population size
Warm Vltava (tributary of Vltava)	2000 adults plus 500 juveniles
Stropnice (tributary of Malse)	3000 adults plus 500 juveniles
Blanice	78 000 adults plus 42 000 juveniles
Luzni Brook (tributary of Rokytnice)	10 000 adults, no juveniles
Bystrina Brook (tributary of Rokytnice)	1700 adults, no juveniles

P. Trpak and S. Krivanek (pers. comm.) report that the River Vltava is in Sumava National Park and the River Blanice in Sumava Protected Landscape Area and Biogenetic Reserve. The River Malse is also in a proposed Nature Reserve. Otherwise,there is little effective protection of *M. margaritifera* in the Czech Republic.

19.3.8 Norway

M. margaritifera was originally distributed throughout Norway, except in the most mountainous regions (Okland 1976), extending up to ca. 70°N (Okland and Okland 1997). It has been found in 340 locations, especially in coastal areas, but has always been, and remains, most widespread in central Norway. Even by 1950 it had become extinct in many of its southern locations, as well as in a few in the north, but the decline has steepened, with only two remaining populations in streams in the southernmost part of Norway and greatly reduced occurrence elsewhere (Dolmen and Kleiven 1997). K. Wächtler (pers. comm.) records a substantial population (30 000–50 000) from Figjo, near Sandnes, West Norway, with some juvenile recruitment and healthy salmonid populations, but even here there have been considerable losses in the past 15 years, following siltation from gravel extraction and river engineering, and pearl fishing. Larsen (1997) comments that there is generally a lack of specific survey data, so that the full scale of the Norwegian decline is not known, but Dolmen and Kleiven (1997) collated what data exist.

Larsen (1997) implies that some recruitment occurs but the scale of this is unknown. He comments that in some places recovery schemes have allowed an "increase in recruitment" to occur.

Larsen (1997) suggests that past exploitation for pearls had a deleterious effect but that recently acidification, fish extinction, river engineering and hydroelectric regulation, erosion and siltation due to drainage and forest clearance, and eutrophication have all been major causes of decline, especially of juveniles. *M. margaritifera* is highly susceptible to acidification, because it lives only in unbuffered rivers, and will not survive below pH 5. In addition, acidification has reduced host fish populations, but in places liming has restored fish and has improved the reproductive success of the mussel (Larsen 1997).

M. margaritifera is protected fully in Norway and is placed on the country's Red List (Dolman and Kleiven 1997). Some recovery management has recently been started for large freshwater mussels by the Norwegian Directorate for Nature Management and extensive improvement work for salmonid fish may also benefit the mussel.

Historically, *M. margaritifera* also occurred in Denmark, and Björk (1962) notes that it was first reported in 1917 from the river Varde Aa in West Jutland, but that this river is now regulated and polluted by industrial waste, so that the population is extinct.

19.3.9 Sweden

M. margaritifera has occurred throughout Sweden on suitable rock types, extending north into Lapland, and much of the earliest work on the general biology and age structure of *M. margaritifera* was carried out on Swedish populations (e.g. Hendelberg 1961). However, even by the early 1900s, many populations had been seriously reduced by overfishing (e.g. Ekman (1905), as summarized by Björk (1962)). More recently, acidification, canalization and organic pollution have also reduced numbers greatly.

Grundelius (1987) conducted a detailed survey of the central province of Dalarna. She surveyed 54 potential locations but found *M. margaritifera* in only 20 sites and in only two of these was "an appreciable degree" of recruitment in evidence. Several sites held only a very few mussels, with only two having thousands present. Larsen (1997) records that *M. margaritifera* has vanished from at least one third of its pre-1900 localities in Sweden.

Originally pearl fishing was a significant cause of decline in Sweden, but latterly siltation from stream engineering and land drainage and acidification have applied. Grundelius (1987) attributes much of the failure of mussel recruitment to an absence of host fish, an indirect effect of acidification.

M. margaritifera is now protected in Sweden.

19.3.10 Finland

Valovirta (1998) reports that around 1900 there were about 200 rivers with populations of *M. margaritifera* in Finland, mostly in eastern Lapland, but that by 1998 only about 25% of these were still extant (Table 19.4). He records the greatest decline in southern areas. Since 1978 there has been an organized survey effort, including the use of subaqua gear, and over 60 rivers have been checked

Table 19.4. The decline of *M. margaritifera* in Finland up to 1995, referring to mapping units. (Valovirta 1998). Note that each mapping unit may include more than one mussel river

Original number of mapping units where *M. margaritifera* occurred	54
Units where reproductive populations remain	8
Units where a moderate decline is recorded	6
Units where a steep decline is recorded	20
Units where the mussel is extinct	20

and characterized. In Finland, populations of over 500 individuals per 500 m of river are regarded as worthy of conservation in the European Natura 2000 conservation network.

Glochidiosis is a problem in some Finnish trout farms, but there are young mussels in only a minority of rivers (Valovirta 1998).

According to Valovirta (1998), the main threat to *M. margaritifera* has been drainage of peatlands and wet forests, which flushes silt into the rivers and disrupts their hydrology. The artificial flow regulation that follows is also damaging. Further problems are acidification, causing direct loss and loss of host fish, and chemical input from anthropogenic sources. Siltation has also followed development of ski centres. He reports that *Salvelinus fontinalis* (American brook charr) have been introduced into many mussel rivers but they do not act as a host in Europe. Instead, they reduce the density of native salmonids, so reducing potential glochidial numbers.

Although *M. margaritifera* has had full legal protection in Finland since 1955, this has not been sufficient to prevent its habitat being disturbed. However, since 1995 a financial penalty, based on a confiscation value of around £500 per mussel, has helped reduce illegal mussel destruction. There have also been major attempts to restore natural hydrological regimes for salmonid rivers and in some cases the density of native juvenile salmonids has been increased to 0.5–1 m^{-2}. As a last resort, mussel translocation has been carried out. Within the same river system up to 90% survival has been achieved, but between-river transfer has resulted in less than 50% survival (Valovirta 1998).

19.3.11 Russia

According to Ziuganov et al. (1994), *M. margaritifera* occurs in northwest Russia only and is found in the Kola Peninsula, Karelia, the Arkhangel District and near St Petersburg. No precise figures exist for past population numbers, but there has undoubtedly been a substantial decline in many rivers.

Beletsky et al. (1995) record existing numbers from eight rivers as follows: Varzuga River (Kola) 100 million; Keret River (Karelia) 6 million; Umba River (Kola) 4 million; Gridina River (Karelia) 1 million; Torma River (Kola) 40 000; Olanga River (Karelia) 3000; Nuris River (Karelia) 1500; Vodopad Creek (Karelia) 300. This makes the populations from the Varzuga, Keret, Umba and Gridina Rivers of international importance, since they outnumber almost all other populations elsewhere.

Ziuganov et al. (1994) provide population age profiles indicating that some juvenile mussels are found in the Varzuga River catchment but that numbers are very variable from site to site. In limited areas recruitment seems to be reasonably good but in other areas very few juveniles are found. They comment that there are sufficient small salmonid fish there to sustain mussel reproduction, in contrast to other rivers. Enhancement of mussel stocks has been attempted by infecting wild salmon with glochidia, to provide a greater number of juveniles (Ziuganov et al. 1994).

Overfishing for pearls is reported by Ziuganov et al. (1994) as the major cause of current scarcity in the Kitsa River, a tributary of the Varzuga. However, the population in the main river has apparently recovered after a drastic decline due to fishing prior to 1939. Industrial pollution and mine wastes have also been detrimental in rivers near Murmansk, in Karelia and parts of the Kola Peninsula. A particular problem has been phenol pollution derived from rotting wood, following timber rafting in Karelia.

Ziuganov et al. (1994) also note that salmonid fish are now declining in many rivers and that this must be limiting mussel recruitment in many regions, including Karelia. Finally canalization, hydropower schemes and general physical disturbance have affected some rivers, such as the Niva River near Murmansk.

There seems to be no effective enforcement of conservation measures in Russia and so the species remains seriously endangered there. There is theoretical protection from fishing and the impacts of pollution but no practical protection (Zuiganov et al. 1994).

19.3.12 Eire

Originally, *M. margaritifera* was found throughout Eire, except for the central area and the eastern coastal section around Dublin. Beasley (1996) collated records of *M. margaritifera* from Eire, surveyed rivers in Co. Donegal and reviewed those of Lucey (1993) from southern Eire. There has also been intensive study of the river Nore catchment (Moorkens and Costello 1994) and so the least-known populations are now those on the central western seaboard.

Beasley and Roberts (1996) surveyed 131 sites on 30 river systems in Co. Donegal in 1993/1994 and found *M. margaritifera* at 31 sites on 7 river systems. They report the disappearance of the mussel from several historical sites and comment that numbers of mussels are now low. Ross (1988) estimated the total population of the River Owenea to be 200 000, based on extrapolation from small samples, whereas Beasley and Roberts (1996) found that this had been reduced to around 6000 (but they warn that the estimation methods differ, so that the results are only broadly comparable). Table 19.5 lists the mean densities of mussels found in a series of samples from the remaining seven rivers where living mussels were found.

Lucey (1993) surveyed 1812 sites on 398 rivers in southern Eire between 1987 and 1992, as well as reviewing the historical information. He reports that up to 1990 Kerney had collated 60 pre-1950, but only 15 post-1950, 10-km grid square records

Table 19.5. Densities of mussels from rivers in Co. Donegal. (Beasley and Roberts 1996)

River system	Recorded densities, no m^{-1}
Clady	0.20–8.45
Tullaghobegly	0.07–0.77
Owencarrow	0.0004–0.31
Leannan	0.11–1.16
Owenea	0.02–2.14
Oily	0.09–1.80
Eske	0.65–5.53

for *M. margaritifera*; however, this apparent decline could have been based on inadequate sampling. Lucey used systematic sampling and found either living or dead *M. margaritifera* from 58 rivers (14.6% of total surveyed), with living specimens at 124 sites. This greatly increased the recorded occurrence of the mussel, but Lucey believes this to be due solely to the increased sampling effort.

Costello et al. (1998) report that *M. margaritifera durrovensis* now occurs only in a 20-km stretch of the River Nore, and whereas Chesney et al. (1993) estimated a population of 5000 in 1988/1989, and 2500 in 1990, Costello et al. (1998) estimated 2000, 1000, 800, 500 and 360 overall in 1991–1995, respectively. However, they have noted (E.A. Moorkens, pers. comm.) that the numbers found are very dependent on viewing conditions and that the numbers visibly increased in 1996. In one 500-m stretch Costello et al. (1998) recorded 230, ca. 90, ca. 60, 52 and 110 from 1992 to 1996, respectively.

As far as we know, there is no reliable recent information on the status of *M. margaritifera* in central western Eire.

No *M. m. durrovensis* of less than 20 years old have been found in the latest surveys of the river Nore and the population is senescent, although *S. trutta* have been successfully infected with glochidia in the laboratory with a view to rearing juveniles. Lucey (1993) comments that juveniles of *M. margaritifera* were rarely encountered in southern Eire, although it is not clear whether a specific search was made for them. Beasley and Roberts (1996) found no mussels below 15 mm in length and only 27 of 15–35 mm at 11 sites in 4 rivers, despite regular searches for them. It must be concluded that few Irish populations are recruiting successfully.

Pearl fishing has been reported commonly in Eire, even in recent years, especially in Co. Donegal, and this remains a serious threat. However, Lucey (1993) and Beasley and Roberts (1996) agree that nutrient enrichment from farming and sewage disposal represents the main threat and that this is widespread. In local areas river engineering has also had an impact. Costello et al. (1998) found direct evidence of fouling due to cattle wastes and agricultural fertilizer use and regard this as the primary problem in the River Nore.

M. margaritifera has full legal protection throughout Eire.

19.3.13 England

Historical records exist for *M. margaritifera* from Devon and north Cornwall; the Rivers Wye and Severn; the North Yorkshire Moors; the North Tyne; near Lancaster; and from various rivers and streams in the Lake District (Kerney 1976). By 1995, when English Nature published a Species Action Plan, anecdotal and scattered records suggested that *M. margaritifera* had become scarce. Consequently, a survey was carried out covering the main historical sites by Oliver and Killeen (1996), who comment that the populations in the Rivers Clun, Irt and North Tyne are marginally viable, whereas the River Ehen population is both relatively abundant and has some limited recruitment occurring, at least in the upper sections (Table 19.6). It appears that this is now the only potentially viable population left in England, whereas N. McCormick (pers. comm.) noted "abundant mussels in the River Rede in 1972" and Woodward (pers. comm.) noted "relatively plentiful in North Tyne in 1977". Survey work in 1997 by Environment

Table 19.6. The numbers of *M. margaritifera* recorded in each of eight English localities in 1995. (Oliver and Killeen 1996)

River and region	Total no. found alive in survey	Size range (mm)	Comments
R. Torridge, Devon	409	79–124	Very scattered and scarce
R. Clun, Shropshire	2227	39–117	Still common only in one, short stretch
R. Esk, North Yorks	130	88–141	Very scattered and scarce
R. North Tyne, Northumberland	ca. 1000[a]	47–138	Still common only in one, short stretch
R. Rede, Northumberland	455[a]	61–112	Very scarce and scattered
R. Ehen, Cumbria	ca. 10 000 (see text below)	15–136	Only large population remaining. A few juveniles only in upper stretches
R. Irt, Cumbria	1346	52–139	Scattered, only common in one short stretch
Dubbs Beck, Cumbria	600	Not measured	Scattered population, coexisting with crayfish

[a] In 1997, Environment Agency staff found 10 600 mussels in North Tyne and 1500 in Rede (Rooksby 1997).

Agency staff (Rooksby 1997) confirmed 1500 mussels in River Rede and 10 600 in River North Tyne.

Killeen and Oliver (1997) and Oliver and Killeen (1997) proposed a management plan to conserve mussel stocks in the River Ehen. They confirmed and extended their population estimates to several hundred thousand adult mussels, in the upper section of the river, with a maximum density of $586\,m^{-2}$. However, they found only five juveniles below 15 mm and noted that many dead mussels were present.

Overall, English populations show very restricted recruitment and most are not recruiting and must be regarded as functionally extinct.

Eutrophication and deleterious agricultural practices, such as trampling and fouling at cattle-watering sites, affects the Torridge, Clun, Esk, Irt and Ehen and excess pearl fishing has applied to the Clun, Irt, Rede and North Tyne (Oliver and Killeen 1996). In lowland England, eutrophication is almost universal. The creation of a reservoir, and associated river engineering, have seriously affected the North Tyne.

M. margaritifera now has full protection under Schedule 5 of the 1981 Wildlife and Countryside Act.

19.3.14 Wales

There were once pearl fisheries in the River Conway (Jackson 1925) and, to a lesser extent, the River Wye in Wales. Scattered mussel records exist from streams in Pembroke, Carmarthenshire and Cardiganshire in the southwest; streams on the southern edge of Snowdonia; and one or two upper tributaries of the River Severn. In 1992

Countryside Council for Wales commissioned surveys revealing that *M. margaritifera* has become very scarce throughout Wales (e.g. Killeen et al. 1997). It is now virtually extinct in the Rivers Conway and Wye (Killeen et al. 1998); reduced to a few hundred old mussels in southwest Wales; and in similar numbers in most of Snowdonia. Only one river, at the southern edge of Snowdonia, holds a population which shows signs of recent recruitment (Killeen et al. 1997). In 1996 the population there was estimated at around 3000, with a very small number below 5 years old. Killeen et al. (1998) report the loss through dredging of a previously unknown population in Snowdonia estimated to have numbered 5000 individuals. They unsuccessfully translocated some of the few remaining live mussels to an apparently more suitable location downstream.

Mortality due to pearl fishing has been very heavy in the major Welsh rivers and some lowland populations may have been affected by organic enrichment, (for example the River Wye). In addition, stream engineering has had a localized impact.

M. margaritifera has recently been given full protection, under Schedule 5 of the Wildlife and Countryside Act (1981).

19.3.15 Northern Ireland

The early records of *M. margaritifera* in Northern Ireland have been summarized by Beasley (1996) and Beasley et al. (1998). Since 1987 Ross (1987), Mackie (1992) and Beasley and Roberts (1996) have conducted detailed surveys. The current position is presented in Beasley et al. (1998).

M. margaritifera was originally found in four catchment areas, namely the Foyle; central Loagh Neagh; the Bush; and the Erne. Since 1950, however, mussels have only been found in parts of the Foyle and Erne catchments and one river draining to Lough Neagh, with live mussels found at only 20 sites in 1996, out of over 250 surveyed. Beasley et al. (1998) protect the precise identity of these sites but provide evidence to show that there has been a substantial decline since 1990/1991 on the Foyle system (Table 19.7).

Virtually no mussels below 10 years in age were found in the 1996 surveys and pooled data from all rivers indicate a strongly modal size class of 91–120 mm (ca. 50+ years).

Table 19.7. Estimated numbers of mussels at sites in the Foyle catchment during 1990/1991 and 1996. (Beasley et al. 1998)

Site	No in 1990/1991	No in 1996
I	40	0
II	50	0
III	20	0
IV	30	20 (dead shells only)
V	50	<10
VI	200	0
VII	200	<10 (in 1995)
VIII	250	0

Some destructive pearl fishing continues on all catchments, with recent dead shells found in 1996. Beasley et al. (1998) also speculate that dredging and river alteration have adversely affected the eastern catchments, and that pollution, in the form of agricultural enrichment and increased sedimentation, has had most effect in the Foyle system.

M. margaritifera is currently listed on Schedule 7 of the Wildlife (NI) Order, which prohibits sale of the mussels, but there is no complete protection at present.

19.3.16 Scotland

M. margaritifera was originally present throughout the mainland, except in the River Tweed catchment and in the sandstone area of Caithness (Cosgrove 1997). It has also been recorded from a small number of the larger Western Isles but only very doubtfully from Orkney and Shetland.

Cosgrove and Young (1998) completed a standardised survey of almost all historical locations, as well as a number of previously unsurveyed areas (Table 19.8). It is evident that very few locations remain where mussels are both still common and actively recruiting and all of these are in the northern half of Scotland. In most, the numbers of juveniles were lower than should be expected in a fully recruiting population, with only four rivers where the numbers of specimens below 65 mm long exceeded 25% of the total. Although some of the declines occurred before 1970, it is also quite clear that the rate of population loss has been twice as rapid since then (Cosgrove and Young 1998). Nevertheless, Scotland retains at least nine rivers where juveniles are present and mussels are abundant, and these populations are of international significance.

Cosgrove and Young (1998) noted pearl fishing in 99% of populated rivers, including the most remote highland sites. In some cases dead mussels showed damage attributable to the use of opening tongs, which were supposed to allow 'safe' fishing. Other significant causes of mussel mortality were pollution (23% of sites) and hydrological engineering (16%); however, it is likely that pollution originally played a more significant role in the loss of mussels from lowland areas near agricultural land and settlements. More minor losses were associated with rubbish dumping, forestry operations, and acidification. Recently, host fish populations have collapsed in northern and western Scotland, leading to real concern for mussel recruitment.

Table 19.8. Numbers of Scottish rivers in each category of mussel occurrence found in 1996–1998. (Data after Cosgrove and Young 1998). Categories listed in text

	No.	Percentage of total
Total number of rivers surveyed	150	100
Extinct populations	61	40
Functionally extinct populations	42	28
Functional populations of which:	47	31
Populations with mussels below 65 mm	15	10
"Common" or "abundant" populations with mussels below 65 mm	9	6

M. margaritifera now has complete protection under Section 5 of the 1981 Wildlife and Countryside Act.

19.4 Overall Population Trends and the Current Status of *M. margaritifera*

In every part of its range, *M. margaritifera* has declined substantially and is now threatened with extinction or is highly vulnerable. Furthermore, the rate of decline has been at least as great recently as it was in earlier decades. Of particular concern is the general lack of juveniles. There is no agreement about precisely what constitutes a functional population, in terms of proportionate numbers of juveniles but, by any definition, very few functional populations remain. An approximate assessment of the status in each country is set out in Table 19.9.

Early causes of decline were pearl fishing and industrial pollution, but recently eutrophication has been of predominant importance, caused variously by intensive agriculture, aquaculture and sewage disposal.

Populations of international importance are now found only in Canada; Russia and northeast Scandinavia; Scotland; possibly Eire; and Bavaria and adjacent parts of the Czech Republic and Austria. Conservation legislation and practical policies apply in Bavaria, Austria, Finland, and Scotland, although so far there is little evidence of any positive benefit to the mussel populations. Regrettably there are no effective conservation measures in Russia, where some of the most abundant

Table 19.9. The current status of *M. margaritifera* in each country in which it occurs

Country	Current status
USA	Declining and of Special Concern; stable in New England
Canada	Some large populations remain and threats are few
Portugal	Presumed extinct
Spain	Serious decline; one (? two) potentially functional population
France	Still present, but scarce in most of former range
Belgium; Lux'g	Extinct
Germany	Still present but serious decline; some functional populations in Bavaria
Austria	Serious decline; one (? two) functional populations
Czech Republic	Serious decline; one functional population
Norway	Serious decline, especially in south; juvenile status unclear
Denmark	Extinct
Sweden	Serious decline; "appreciable recruitment" at two sites
Finland	75% populations lost; 11 "important" populations remain; ca. 8 functional populations
Russia	Serious decline; four populations of over 1 million remain; good recruitment in limited areas
Eire	Serious decline; little recruitment
England	Only one common population remains; limited recruitment
Wales	Only one population above 3000; very few juveniles
Northern Ireland	90% populations lost; almost no recruitment
Scotland	Serious decline; ca. 10 fully functional populations

populations remain, although Ziuganov and colleagues are attempting positive action. In the USA some sparse populations seem to be stable but serious decline has also occurred, whereas in Canada threats are low and important populations remain.

19.5 Summary

1. *Margaritifera margaritifera* ranges from western Russia to the eastern seaboard of North America, from the northern limit of ice-free areas, south to Austria, France and northern Spain. It has been considered to be declining substantially throughout this range.
2. This study uses published data and personal contact to establish the true status of *M. margaritifera* in each country, with emphasis on the continued presence of juvenile mussels and on the causes of decline and the extent of any conservation measures.
3. *M. margaritifera* has certainly declined substantially throughout its range and is now vulnerable or threatened with extinction almost everywhere, with few populations which still have juveniles present in significant numbers.
4. Causes of decline have included industrial pollution; agricultural and domestic effluent leading to eutrophication; river engineering; and pearl fishing. Recently reduced salmonid fish stocks have also posed a problem.
5. Several apparently viable populations still occur in each of Canada, Russia, northeast Scandinavia and Scotland; and Eire, Bavaria, Austria and Czech Republic each harbour at most five such populations. Otherwise populations typically consist of small numbers of old adults, with juveniles absent.
6. *M. margaritifera* seems to be least threatened in Canada, but there its populations are largely unsurveyed. Elsewhere it has universal legal protection but few conservation schemes are in place. The few remaining abundant populations in Russia are seriously threatened by neglect of the protective legislation.
7. Unless conservation measures are improved it seems certain that *M. margaritifera* will continue its decline, so that soon it will survive only in four isolated areas in the north of its range.

Acknowledgements. We have benefited greatly from the advice and knowledge of many colleagues, including C. Alvarez, O. Baer, G. Bauer, V. Buddenseik, R. Dettmer, T. Fowler, J.M. Hanson, J. Herrero, S. Krivanek, J. Lucey, J. Maunder, E. Moorkens, J. Okland, P.G. Oliver, D. Roberts, C. Schmidt, S. Schmidt, D. Smith, P. Trpak, G. Wenz and V. Zuiganov. We offer them our grateful thanks. Scottish Natural Heritage kindly funded the recent Scottish survey, whose results we quote with permission.

References

Alvarez C, Garcia-Roves P, Ocharan R, Cabal JA, Ocharan FJ, Alvarez MA (2000) A new record of the freshwater pearl mussel, *Margaritifera margaritifera* L. (Bivalvia, Unionoida) from the River Narcea (Asturias, north western Spain) Aquat Conserv Mar Freshwater Ecosyst 10:93–102

Anonymous (1996) Die Perlmuschel im Dreiländereck Böhmen – Bayern – Sachsen. Inova Press, Plzen, 67 pp

Baer O (1981) Zur früheren Verbreitung der Flußperlmuschel in der Westlausitz. Veröff Mus Westlausitz 5:53–70

Baer O (1995) Die Flußperlmuschel *Margaritifera margaritifera* (L.). Westarp Wissenschaften, Magdeburg, 118 pp

Bauer G (1986) The status of the freshwater pearl mussel *Margaritifera margaritifera* L. in the south of its European range. Biol Conser 37:1–9

Bauer G (1988) Threats to the freshwater pearl mussel *Margaritifera margaritifera* L. in Central Europe. Biol Conserv 45:239–253

Bauer G, Eicke L (1986) Pilotprojekt zur Rettung der Flußperlmuschel (*Margaritifera margaritifera* L.). Natur Landschaft 61(4):140–143

Bauer G, Zwölfer H (1988) 200 million offspring and scarcely any future. German Res 1:8–9

Bauer G, Hochwald S, Schmidt C, Schmidt H, Reger K-H (1991) Dauerbeobachtung von Muschelbeständen. Laufener Seminarbeitr 7(91):30–37

Beasley CR (1996) The distribution and ecology of the freshwater pearl mussel *M. margaritifera* L. 1758 in County Donegal, Ireland, and implications for its conservation. PhD Thesis, The Queen's University of Belfast

Beasley CR, Roberts D (1996) The current distribution and status of the freshwater pearl mussel *Margaritifera margaritifera* L. 1758 in north-west Ireland. Aquat Conserv Mar Freshwater Ecosyst 6:169–177

Beasley CR, Roberts D, Mackie TG (1998) Does the freshwater pearl mussel, *Margaritifera margaritifera* L., face extinction in Northern Ireland? Aquat Conserv Mar Freshwater Ecosyst 8:265–272

Beletsky V, Popkovich E, Ziuganov V, Zotin A (1995) Modern distribution of freshwater pearl mussel in Russia. 12th Int Malacological Congr, 3–8 Sept 1995, Vigo, Spain

Bischoff W-D, Dettmer R, Wächtler K (1986) Die Flußperlmuschel, Biologie und kulturelle Bedeutung einer heute vom Aussterben bedrohten Art. Staatliches Naturhistorisches Museum, Braunschweig

Björk S (1962) Investigations on *Margaritifera margaritifera* and *Unio crassus*. Acta Limnol 4:1–109

Buddensiek V (1995) The culture of juvenile freshwater pearl mussels *Margaritifera margaritifera* L. in cages: a contribution to conservation programmes and the knowledge of habitat requirements. Biol Conserv 74:33–40

Canadian Wildlife Service (1999) http://www.cws-scf.ec.gc.ca/es/endan_e.html (updated April 1999)

Chesney HCG, Oliver PG (1998) Conservation issues for Margaritiferidae in the British Isles and western Europe. J Conchol Spec Publ 2:231–242

Chesney HCG, Oliver PG, Davis GM (1993) "*Margaritifera durrovensis*" Phillips, 1928: taxonomic status, ecology and conservation. J Conchol 34:267–299

Clarke AH (1981) The freshwater molluscs of Canada. National Museum of Natural Sciences, Ottawa, Canada

Cochet G (1996) Situation de *Margaritifera margaritifera* et des autres nayads en France. Ministère de L'Environment, Paris

Cochet G (1998) Distribution map of *Margaritifera margaritifera* in France. Lettre Vertigo 6:6

Cosgrove PJ (1997) Distribution survey of freshwater pearl mussels *Margaritifera margaritifera*: observation records by Vice-county in Scotland. Confidential report to Scottish Natural Heritage, Edinburgh

Cosgrove P, Young M (1998) The status of the freshwater pearl mussel *Margaritifera margaritifera* in Scotland. A report to Scottish Natural Heritage, Edinburgh

Costello MJ, Moorkens E, Larkin M, Kurz I, Dowse J (1998) Management priorities for the river Nore (Ireland) to conserve the pearl mussel *Margaritifera margaritifera durrovensis* Phillips 1928. J Conchol Spec Publ 2:257–264

Dolman D, Kleiven E (1997) The freshwater pearl mussel in Norway 1. Vitenskapsmuseet Rapp Zool Ser 6:1–27

Dyk V, Dykova S (1974) The pearl oyster (*Margaritifera margaritifera* Linn.) a neglected indicator of the pollution of mountain and submontane water flows of the crystalline region of Czechoslovakia. Acta Vet Brno 43:287–304

Ekman TH (1905) Undersökningar öfver flodpärlmusslans förekomst och lefnadsförhållanden i Ljusnan och dess tillflöden inom Härjedalen. Medd K Landtbruksstyrelsen 110: 1–12

Fichtler C, Smith DG (1995) The Freshwater mussels of Vermont. Vermont Fish and Game Department, Technical Report 18. Leahy Press, Montpelier

Foeckler F, Schmidt H, Deichner O (1997) Naturschutzfachliche Analyse und Bewertung der Auswirkungen von Fischteichen auf die Gewässerfauna von Flußperlmuschelbächen Nordostbayerns. Z Ökologie Natursch 6:111–123

Groh K (1995) Erfahrung bei der Betreuung von Flußperlmuscheln-Artenschutzprojekten westlich des Rheins. Schriftenr Lindberger Hefte 5:59–76

Grundelius E (1987) Information fran Sotvattens Laboratoriet Drottningholm. Inst Swed Res Swed Nat Board Fish 4:1–72

Hannick P (1980) Un rapport sur la pèche de perles dans Le Luxembourg en 1818. Bull Inst Archéol Luxembourg 1–2:9–12

Hendelberg J (1961) The freshwater pearl mussel, *Margaritifera margaritifera*. On the localisation, life and growth of the individual and on the composition of the population according to an investigation in Parlalven in Arctic Sweden. Rep Inst Freshwater Res Drottningholm 41:149–171

Hessling T von (1859) Die Perlmuscheln und ihre Perlen. Wilhelm Engelmann, Leipzig

Hruska J (1992) The freshwater pearl mussel in South Bohemia: evaluation of the effect of temperatures on reproduction, growth and age structure of the population. Arch Hydrobiol 126:181–191

Jackson JW (1925) The distribution of *Margaritifera margaritifera* in the British Isles. J Conchol 17:195–211, 270–278

Jungbluth JH (1978) Die Bestandsentwicklung der Flußperlmuschel (*Margaritifera margaritifera* L. Mollusca: Bivalvia) an mitteleuropäischen Standorten. Verh Int Ver Limnol 20: 2435–2441

Jungbluth JH (1987) Flußperlmuschel-Erfassung in den Mittelgebirgen von Bayern 1985–1987. Univeröff. Gutachten im Auftrag des Bayer Landesamtes für Wasserwirtschaft, Heidelberg

Jungbluth JH (1988) Zur Situation der Flußperlmuschel *Margaritifera margaritifera* (L.) in der ehemaligen Preußischen Rheinprovinz (Mollusca: Bivalvia: Margaritiferidae). Decheniana (Bonn) 141:209–229

Jungbluth JR, Coomans HE, Groh H (1985) Bibliographie der Flussperlmuschel *Margaritifera margaritifera* (Linn. 1758). Verslagenen Technische Gegevens No 41 Institut voor Taxonomische Zoologie, Universiteit Amsterdam, 220 pp

Kerney MP (1976) Atlas of the non-marine Mollusca of the British Isles. Institute of Terrestrial Ecology, Cambridge

Killeen IJ, Oliver PG (1997) The freshwater pearl mussel (*Margaritifera margaritifera* L. 1758) in the river Ehen. Part 1 – Report on the 1996 survey. English Nature Research Reports Series English Nature, Peterborough, 24 pp

Killeen IJ, Oliver PG, Wood D (1997) The status of the freshwater pearl mussel (*Margaritifera margaritifera* L. 1758) in north-west Wales. Countryside Council for Wales Contract Science Report 182, Bangor, 26 pp

Killeen IJ, Oliver PG, Fowles AP (1998) The loss of a freshwater pearl mussel (*Margaritifera margaritifera*) population in NW Wales. J Conchol Spec Publ 2:245–250

Larsen BM (1997) Pearl mussel (*Margaritifera margaritifera* L.) A review of national and international knowledge – NINA Fagrapport 37:1–51

Lucey J (1993) The distribution of *Margaritifera margaritifera* (L.) in southern Irish rivers and streams. J Conchol 34:301–310

Mackie TG (1992) The distribution and current status of *Margaritifera margaritifera* in the North of Ireland. MSc Thesis, The Queen's University of Belfast

Meissner O (1914) Die Perlmuschel in Oberfranken. Ber Natur wiss Ges, Bayreuth 2

Moog O, Nesemann H, Ofenboeck T, Stundner C (1993) Grundlagen zum Schutz der Flussperlmuschel in Osterreich. Bristol-Stiftung, Forschungsstelle fur Natur- und Umweltschutz, Schaan, 240 pp
Moorkens EA, Costello MJ (1994) Imminent extinction of the Nore freshwater pearl mussel *Margaritifera durrovensis* Phillips: a species unique to Ireland. Aquat Conserv Mar Freshwater Species 4:363–365
Nagel KO (1991) Gefährdete Flußmuscheln in Hessen. Z Angew Zool 78(2):205–218
Nobre A (1913) Moluscos de Portugal, I, Babtista, Torres, Lisbon
Okland J (1976) Utbredelsen av noen ferskvannsmuslinger i Norge, og litt om Europea Invertebrate Survey. Fauna 29:29–40
Okland J, Okland KA (1997) Distribution limits of freshwater molluscs in northern Europe – a survey of edge populations in Norway and adjacent areas. Heldia 4(5):78–93
Oliver PG, Killeen IJ (1996) Report on the 1995 survey of the freshwater pearl mussel (*Margaritifera margaritifera* L. 1758). Rivers Torridge, Clun, Esk, Irt, Ehen, Rede and N Tyne. English nature research reports series 162, English Nature, Peterborough, 40 pp
Oliver PG, Killeen IJ (1997) The freshwater pearl mussel (*Margaritifera margaritifera* L. 1758) in the river Ehen. Part 3 – A management plan. English Nature Research Reports Series, English Nature, Peterborough, 33 pp
Ricciardi A, Neves RJ, Rasmussen JB (1998) Impending extinctions of North American freshwater mussels (Unionoida) following the zebra mussel (*Dreissena polymorpha*) invasion. J Anim Ecol 67:613–619
Riedl G (1928) Die Flußperlmuschel und ihre Perlen. Jb. oberöst. Mus.-Ver. 82:238–259
Rooksby M (1997) Report on the 1997 survey of the Freshwater Pearl Mussel (*Margaritifera margaritifera* L., 1758) in the River Rede and North Tyne. Environment Agency, Newcastle-upon-Tyne
Ross HCG (1987) Distribution of *M. margaritifera*, the freshwater pearl mussel, in the north of Ireland. Report to the Royal Irish Academy, Dublin (unpublished)
Ross HCG (1988) Aspects of the ecology of the freshwater pearl mussel *Margaritifera margaritifera* (L.) in NW Ireland, with special reference to life history strategy. MSc Thesis, The Queen's University of Belfast (unpublished)
Schmidt H (1990) Entwicklung eines Artenhilfsprogramms für die beiden Großmuschelarten Flußperlmuschel (*Margaritifera margaritifera* L. 1758) und Bachmuschel (*Unio crassus* PHIL 1788). Schriftenr Bayer Landesamt für Umweltschutz 97:5–12
Steidle L (1986) Die Flußperlmuschel und Perlfischerei in bayerischen Gewässern. Sonderdruck aus Informationsbericht des Bayer Landesamt für Wasserwirtschaft, München 1:13–47
Strayer DL, Jirka KJ (1997) The pearly mussels of New York State. Mem N Y State Mus 26:1–113
Valovirta I (1998) Conservation methods for populations of *Margaritifera margaritifera* (L.) in Finland. In: Killeen IJ, Seddon MB, Holmes AM (eds) Molluscan conservation: a strategy for the 21st century. J Conchol Spec Publ 2
Velado J (1878) *Unio margaritifer*. An Soc Esp de Hist Nat 7:244–245
Wächtler K, Dettmer R, Buddensiek V (1987) Zur Situation der Flußperlmuschel (*Margaritifera margaritifera* (L.) in Niedersachsen: Schwierigkeiten, eine bedrohte Tierart zu erhalten. Ber Naturhist Ges Hannover 129:209–224
Walker B (1910) The distribution of *Margaritifera margaritifera* (Linn.) in North America. Proc Malacol Soc Lond 9:126–145
Wellman G (1938) Untersuchungen über die Flussperlmuschel (*Margaritana margaritifera* L.) und ihren Lebensraum in Bächen der Lüneberger Heide. Zeitschrift für Fischerei und deren Hilfswissenschaften XXXVI; 4:32–39
Wells S, Pyle R, Collings N (1983) The IUCN Red Data Book. Gresham Press, Old Woking
Williams D, Melvin L, Warren KS Jr, Cummings KS, Harris JL, Neves RJ (1993) Conservation status of freshwater mussels of the United States and Canada. Fisheries 18(9):6–22
Young MR (1991) Conserving the freshwater pearl mussel (*Margaritifera margaritifera* L.) in the British Isles and Continental Europe. Aquat Conserv Mar Freshwater Ecosyst 1:73–77
Ziuganov V, Zotin A, Nezlin L, Tretiakov V (1994) The freshwater mussels and their relationships with salmonid fish, moscow, Vniro Publishing House, 104 pp

20 The Pearl Mussel-Salmon Community in the Varzuga River, Northwest Russia: Problems of Environmental Impacts

VALERY ZIUGANOV, SVIATOSLAV KALIUZHIN, VICTOR BELETSKY, and EKATERINA POPKOVITCH

20.1 Introduction

On a global scale, the freshwater mussels Unionoidea have declined dramatically. About 1000 species, from tropical to polar climates, are being subjected to serious habitat degradation through dredging, pollution, sedimentation and similar anthropogenic impacts (Neves 1997). In the Northern Hemisphere, the freshwater pearl mussel (*Margaritifera margaritifera* L.) possesses a holarctic distribution, having colonized coldwater rivers and streams cohabited by fishes of the Salmonidae.

The Varzuga River has the largest known population of both the freshwater pearl mussels (listed as particularly endangered by the IUCN) and the Atlantic salmon (*Salmo salar* L.). The density of the brown trout (*Salmo trutta* L.) is considerable as well. Both salmonids are host fish on which the pearl mussel glochidia develop. *M. margaritifera* is an integral component of the river ecosystem, filtering particles, serving as a substratum for algae, and providing habitat for benthic invertebrates that serve as food for salmon parr. Thus, the fragile relationship between salmonids and pearl mussels has been sustained for centuries, with large populations of each contributing to the secondary production of the river.

We initiated a study to evaluate anthropogenic factors affecting current ecosystem health, as human impact will undoubtedly increase in future years. Factors such as poaching, introduction of pink salmon (*Oncorhynchus gorbusha* Walbaum) and other factors threaten the ecological integrity of the river. The goal of this chapter is to analyze the impact of human disturbance on salmonid and mussel populations and to recommend the best management practices to promote resource conservation in this valuable river in northwest Russia.

20.2 Materials and Methods

20.2.1 The Study Area

The Varzuga River is one of the largest rivers in northwest Russia, located in the southeast portion of the Kola Peninsula (67°6′ N, 36°4′ E) (Fig. 20.1). The mean annual discharge is $76.5\,m^3\,s^{-1}$. The total length of the river is 254 km, with a drainage area of approximately $8000\,km^2$ consisting of bogs (50%) and forest (50%). Water temperature in July averages 16.5 °C and near 0 °C in January. Dissolved oxygen is near satu-

Ecological Studies, Vol. 145
G. Bauer and K. Wächtler (eds.) Ecology and Evolution of the Freshwater Mussels Unionoida

Fig. 20.1. Map of Northern Europe, with the location of the Varzuga River

ration and hardness is 40–50 mg l^{-1} (Kazakov et al. 1992). In general, the Varzuga River has a low ionic concentration due to low dissolution of crystalline minerals in the river bed. Comparison of our water chemistry data (Ziuganov et al. 1996) with those of Mosevitch and Sokolov (1939) and Kazakov et al. (1992) indicates that the principal hydrochemical parameters have remained essentially unchanged for more than 50 years.

Until recently, the watershed was not subjected to economic development, because of the rugged terrain. Favourable wind direction has reduced the deposition of industrial emissions such as sulphur and nitrogen oxides from nearby industrial complexes at Apatyti and Monchegorsk. Thus, the Varzuga River basin has not been degraded by acidification; hydrologic and hydrochemical conditions of the river have remained unaffected (Ziuganov et al. 1998).

The riparian zone is well vegetated with shrubs, conifers and birch trees, and meadows are common in the lower watershed. The river has a characteristic aquatic vegetation such as *Potamogeton*, *Fontinalis* and *Ranunculus*. The fish fauna is composed of the following common species: Atlantic salmon, brown trout, pink salmon (introduced), arctic grayling *Thymallus thymallus* (L.) and minnow *Phoxinus phoxinus* (L).

The total size of spawning and nursery areas for Atlantic salmon in the Varzuga River drainage is about 5.5 km^2, with spawning area of 1 132 200 m^2 (Kazakov et al. 1992). Adults spend 2–3 years at sea, depending on the specific run. Up to 60 000 adult salmon enter the Varzuga River to spawn. The mussels are located in the main channel and tributaries, at sites used as spawning grounds for Atlantic salmon and brown trout. For centuries, mussels from the Varzuga River served as a source of freshwater pearls for czars since the Middle Ages.

Table 20.1. Number of pearl mussels in the Varzuga River basin[a]

River reach (tributary to tributary)	Mean length (m)	Mean width (m)	Mean density (no m^{-2})	Population estimate (×1000)
Usia to Kichisara	22 000	25	1.8	990
Kichisara to Pana	12 000	30	1.25	450
Pana to Falaley	12 000	50	29.3	17 580
Falaley to Japoma	30 000	50	10.4	15 600
Japoma to Varzuga village	27 000	60	2.0	3 240
Pana tributary	80 000	50	2.0	8 000
Indel tributary	20 000	25	10.0	5 000

[a] Total length of the river occupied by the mussels is about 200 km. Total number of mussels is about 51 million.

20.2.2 Estimated Number of Pearl Mussel

The population size of pearl mussels was estimated in 1990–1995 by the following method. An iron frame (0.7 m^2) was thrown randomly onto the river bottom at least ten times per site. The number of mussels inside the frame was counted, and mean density per 1 m^2 was calculated. An estimate of total population size was obtained by multiplying mean density by the length and width of river reach occupied by mussels. Mean densities and population estimates were calculated at 25 sites along the Varzuga River and its Pana and Indel tributaries (Table 20.1).

To compare the number of mussels counted visually inside the frame to the actual number present (includes juveniles, burrowed adults, mussels under cobbles, etc.), an excavated sample within the frame was taken at periodic intervals.

Age analyses of adult specimens ($n = 44$) from the Varzuga River were done by counting annual rings according to Hendelberg (1961).

20.2.3 Density Estimates of Juvenile Salmon

A DC-equipment (LUGAB L1000s) providing 600–800 V was used for the electrofishing study. Eleven sites of the Varzuga River (1–11, Fig. 20.2) were fished quantitatively (three successive removals) during 4–13 August 1997. One site (12, Fig. 20.2) of the neighbouring Thorma River was fished quantitatively on 16 August 1997. Population estimates for sites subjected to three removals were calculated together for all age classes of parrs according to Zippin (1956).

Collected parr were released to the river after measurement of length. The area of each sampling site was usually 120–475 m^2, and the total area examined was 3000 m^2.

20.3 Results and Discussion

20.3.1 Number of Pearl Mussels

The difference between the visual estimate (Table 20.1) and actual density was considerable; true mean density was 2.76 times greater than the observed density. There-

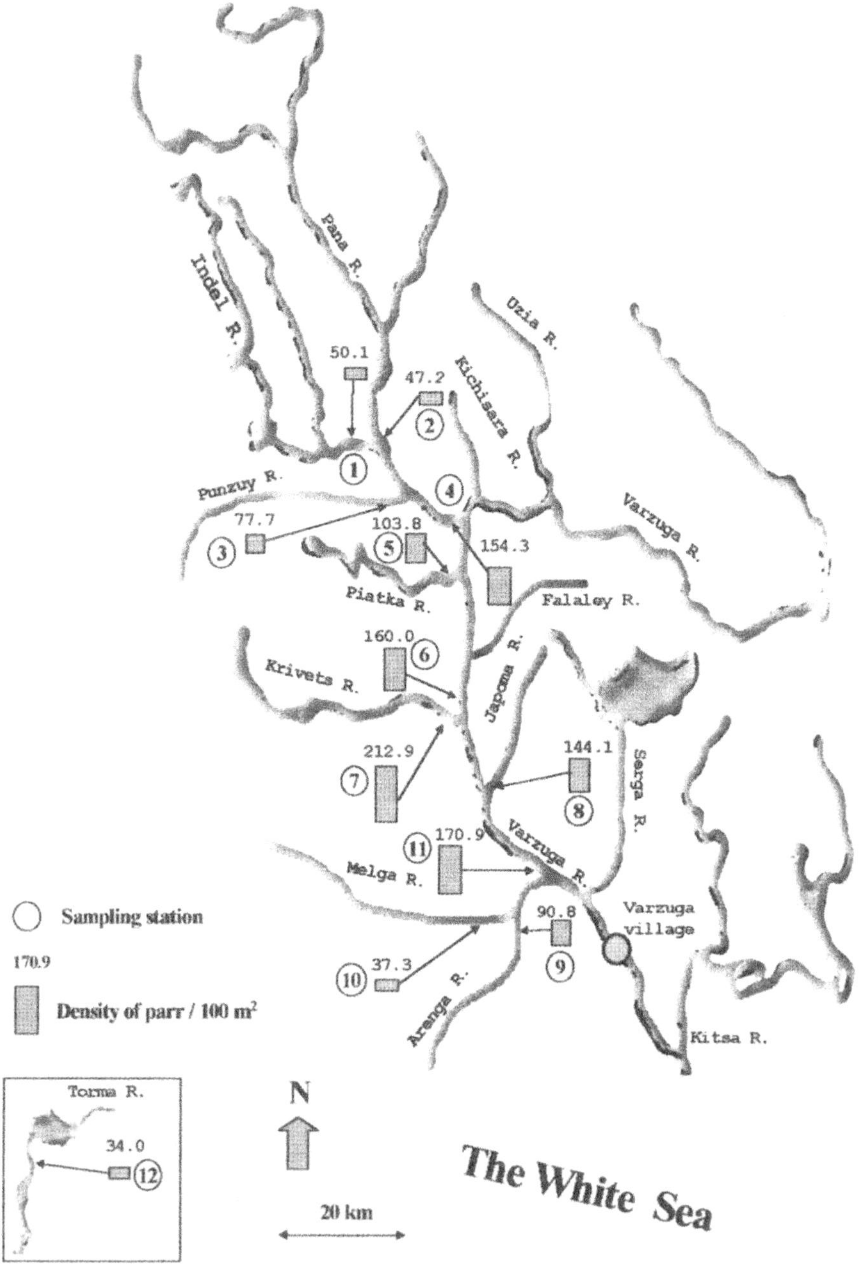

Fig. 20.2. Map of the Varzuga River basin with locations of sampling stations for presmoult Atlantic salmon. In the *left bottom corner* data on density of parrs in the Torma River (separate basin)

fore, the true population estimate of mussels greater than 1 cm in length was 140.76 million (51 million × 2.76). We found no mussels less than 5–10 mm in length because young mussels are burrowed deeply into the substrate. The size distributions of pearl mussels at three sites for 3 years were well distributed (Fig. 20.3). There was no deficiency of young mussels from 1990 to 1995; the proportions of the population less than 70 mm in length were as follows in 1990: 31% in the Pana River, 44% near the Falaley tributary and 26% near the Japoma tributary. The lowest proportion of young mussels near the Japoma tributary might be explained by the considerable turbidity of the water.

The population age structure of adult pearl mussels shows an exceptionally long life span in *M. margaritifera* inhabiting the Varzuga River (Fig. 20.4). When analyzing the age structure of a pearl mussel population we must take into account that in pristine rivers mortality increases in the second half of the maximum life span, whereas it is low in the first age classes (Bauer 1983, 1989). Thus, the age structure of a reproducing population should therefore demonstrate a predominance of specimens aged between 10 years and half the maximum life span (about 100 years for the Varzuga River) (Fig. 20.4).

20.3.2 Density of Parr

Average estimated density of Atlantic salmon parr (including fry 0+) in the Varzuga River was 113.6 ± 17.6 (SE) individuals $100\,m^{-2}$ in August, 1997. The highest densities were found at sites 4, 5, 6, 7 and 8 (Fig. 20.2). It is noteworthy that all these five sites are situated in the middle part of the Varzuga River and are practically untouchable by poachers due to absence of roads. The lowest densities were recorded at sites 1 and 2 (Indel-Pana junction) (50 and 47), perhaps due to high pressure by poachers, who can easily reach the Indel-Pana junction using the Umba-Indel road. However, the densities of parr at the latter site are considerably higher than in site 12 (Torma River, 34 parrs $100\,m^{-2}$). It is known that the stock of Atlantic salmon of the Torma River has been highly overfished during the past 10 years. Thus, one can see successful reproduction of salmon in the most investigated sites of the Varzuga River.

20.3.3 Threats to Survival

In spite of this reproductive success, threats to the population include illegal overfishing for mussels and salmon, introduction of non-native pink salmon, and invasion of *Gyrodactilus salaris* and its effect on Atlantic salmon.

20.3.3.1 Pink Salmon

Certainly the naturalization of ten populations of pink salmon in the White Sea basin has resulted in competitive use of spawning areas, as both species of salmon spawn in fall (September–October). Thousands of adult pink salmon enter the Varzuga River for spawning. Decomposition of pink salmon carcasses proceeds slowly in the cold, slightly acidic waters of Kola rivers. This change can alter the long-established ecological balance in these rivers. The mechanism of competitive interaction for spawning areas between pink salmon and Atlantic salmon, as well as the gradual

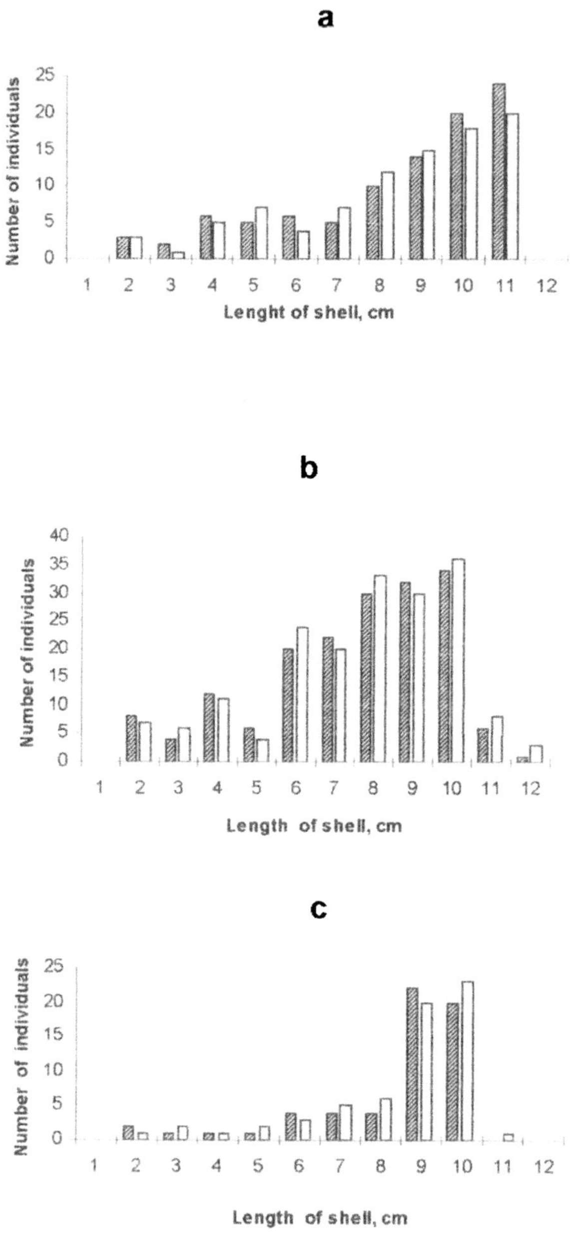

Fig. 20.3a–c. Size structure of *Margaritifera margaritifera* of the Varzuga River basin. **a** In the mouth of the Pana River. **b** In the Varzuga River near the tributary Falaley. **c** In the Varzuga River near the tributary Japoma. *Black column* data of 1990; *white column* data of 1995

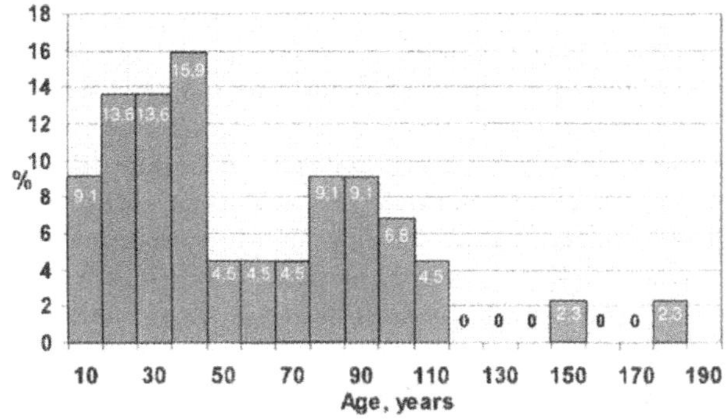

Fig. 20.4. Age structure of 44 adult pearl mussels from the Varzuga River

displacement of Atlantic salmon by pink salmon, will result in the decline and possible extinction of *M. margaritifera*, as the pink salmon is an unsuitable host (Ziuganov et al. 1994). Removal of the pink salmon from rivers in the White Sea basin would be advantageous to the native fauna.

20.3.3.2 Commercial Fishery and Poaching

The harvest of Atlantic salmon from the Varzuga River ranged from nearly 57 000 salmon in 1960 to 9000 fish in 1978 (Kazakov et al. 1992). In 1998, experts estimated that about 60 000 adult salmon entered the Varzuga River to spawn. About half of them were captured by poachers. Because of the closure and bankruptcy of many industrial and agricultural operations, high unemployment has resulted. Illegal fishing and hunting became the primary means of survival because of the high price for official licenses. Such economically hard times have resulted in an increased level of poaching for monetary gain. During this period, financial support for the Russian Environmental Protection Agency has decreased.

For the Varzuga River basin, fish inspectors record frequent poaching violations at the river mouth, within the Kitsa and Uzia tributaries, and along the entire Pana tributary (including Indel and Polisarka rivers). This illegal fishing is one of the most detrimental factors to population maintenance and reproduction of salmon and pearl mussels. In view of the increasing unemployment within the Murmansk Province, a greater level of poaching is to be expected in the future.

20.3.4 Measures to Conserve and Restore the Pearl Mussel-Salmon Community

The major steps to be taken to protect the salmon-pearl mussel community include preservation of their biotopes, conservation of salmonid populations and strict control over illegal pearl mussel harvest. To accomplish this, it is necessary to create a refuge or protection zone for pearl mussel colonies, especially from the junction of the Varzuga and Pana rivers downstream to Punzostrov Island. In this river reach, 80% of the population of pearl mussels and the main spawning ground for Atlantic salmon is to be found. In addition, a ban should be imposed on any economic development that would lead to environmental degradation such as forest cutting, mining, establishment of tourist camps etc. At the same time, monitoring of the pearl mussel population must be undertaken in this river system. From a regional perspective, the conservation of salmon-pearl mussel communities will require the establishment of protected areas (state reserves, sanctuaries, national parks) that should include entire watersheds when necessary. Closure of only the upper, middle or lower reaches of rivers will not guarantee the protection of salmon and pearl mussels, for obvious reasons. A source of industrial pollution in the upper river would stifle all efforts to protect pearl mussels in all downstream reaches. Overfishing of anadromous salmonids in the lower river would inhibit pearl mussel preservation in the upstream reaches.

20.4 Summary

Recent data on the pearl mussel-Atlantic salmon community in the pristine Varzuga River, Kola Peninsula, Russia, are presented and analyzed. Situated in a remote wilderness, the catchment area of the Varzuga River is as yet only weakly affected by pollution. This chapter reports the results of long-term investigations of the world's largest population of freshwater pearl mussels (more than 100 millian specimens), and one of the largest holarctic stocks of Atlantic salmon. Special attention is given to the prospects for conservation of the pearl mussel-salmon community.

References

Bauer G (1983) Age structure, age-specific mortality rates and population trend of the freshwater pearl mussel in North Bavaria. Arch Hydrobiol 98:532–532

Bauer G (1989) Die bionomische Strategie der Flussperlmuschel. Biol Unserer Zeit 3:69–75

Hendelberg J (1961) The freshwater pearl mussel *Margaritifera margaritifera* (L.). Rep Inst Freshwater Res Drottningholm 41:149–171

Kazakov R, Kuzmin O, Shustov J, Schurov I (1992) Atlantic salmon of the Varzuga River. Gidrometeoizdat, St.-Petersburg (in Russian)

Mosevitch N, Sokolov P (1939) Hydrochemical characteristics of some rivers of Kola peninsula with particular reference to geology of their basins. Izv VNIORCH 21:322–350 (in Russian)

Neves R (1997) Keynote address: a national strategy for the conservation of native freshwater mussels. In: Cummings K, Buchanan A, Mayer C, Naimo T (eds) Conservation and management of freshwater mussels II. Initiatives for the future. Proc Upper Mississippi River Conservation Committee Symp USA, 1992, pp 1–11

Zippin C (1956) An evaluation of the removal method of estimating animal populations. Biometrics 12:163–189

Ziuganov V, Zotin A, Nezlin L, Tretiakov V (1994) The freshwater pearl mussels and their relationship with salmonid fish. VNIRO, Moscow

Ziuganov V, Beletsky V, Mikhno I (1996) Influence of sport fishing of salmonid fish and other anthropogenic factors on biosystem salmon-pearl mussel of the White Sea basin. KP Polygraph, Moscow (in Russian)

Ziuganov V, Beletsky V, Neves R, Tretiakov V, Mikhno I, Kaliuzhin S (1998) The recreational fishery for atlantic salmon and the ecology of salmon and pearl mussels in the Varzuga River, northwest Russia. Virginia Tech Publ, Blacksburg, Virginia

21 How Environmental Information Can Be Obtained from Naiad Shells

HARRY MUTVEI and TORBJÖRN WESTERMARK

21.1 Introduction

The life conditions of naiads are governed by a multitude of environmental factors. Tolerance to some of these factors by the animal can be tested in aquaria, but such experiments necessarily have a limited duration, and laboratory conditions invariably differ from those in nature. Fortunately, some of the important environmental factors leave their record in the growth rate and chemistry of the shell.

As shown by several studies (Imlay 1982; Carell et al. 1995; Frantsevitch et al. 1996; Westermark et al. 1996), shell chemistry reflects the chemistry of the surrounding water. Elemental distribution in naiad shells has been analyzed by neutron activation at the Royal Institute of Technology, Stockholm, by nuclear microscopy (= proton microprobe or μ-PIXE) at the University of Uppsala, by ICP-MS (= inductively coupled plasma mass spectrometry) and by other methods. These studies elucidate acidification and oxygen depletion in the water, and their effects on shell growth rate. Shell growth rate also mirrors changes in food supply and, indirectly, reflects water temperature.

For neutron activation elemental analysis, shell samples of 10–50 mg were cut out, each representing 3–20 years growth record. The first growth stage, before sexual maturity, was omitted because of juvenile effects on elemental distribution. For proton-induced X-ray emission analysis, carried out with nuclear microscope (μ-PIXE), spatial resolution was 1–10 μm and time resolution as short as 1 week or even less (E. Dunca and H. Mutvei, unpubl.).

The naiad shells studied come from northern Europe: Scandinavia, Estonia and northern Russia. By using additional shells from museum collections, we have been able to study interactions between environmental changes and naiad life conditions in the past, covering the past two centuries, during which the agricultural and industrial revolutions have taken place. The following species of naiads were studied: *Margaritifera margaritifera* (Linné 1758), *Unio pictorum* (Linné 1758), *U. tumidus* (Philipsson 1788), *U. crassus* (Philipsson 1788), *Anodonta anatina* (Linné 1758), *A. cygnea* (Linné 1758) and *Pseudanodonta complanata* (Rossmaesler 1835). The methods of studies of shell growth rate and elemental distribution have been described in several papers (Carell et al. 1987; Mutvei et al. 1994; Westermark et al. 1996, 1997; Nyström and Dunca 1996; Nyström et al. 1996).

Ecological Studies, Vol. 145
G. Bauer and K. Wächtler (eds.) Ecology and Evolution of the Freshwater Mussels Unionoida

21.2 Annual Growth Increments and Growth Rate

Naiad shells consist of two principal aragonitic layers: outer prismatic and inner nacreous (Fig. 21.3B,C). The shell surface is covered by a comparatively thick periostracal layer composed of fibrous glycoprotein lamellae. Annual growth increments strongly resemble year rings in trees. They are best observed in vertical sections cut from umbo to ventral margin. To increase the distinctness of growth increments, the sections were chemically treated by etching the aragonite and simultaneously fixing the shell glycoproteins. Winter cessation of growth is marked by winter lines that are etch-resistant, opaque and separate the consecutive annual increments as narrow ridges (indicated by arrows in Fig. 21.2; see also Veinott and Cornett 1996).

Widths of annual growth increments were measured as the distance between two winter lines in the outer prismatic layer, at the border to the nacreous layer. The measurements were taken out using a light microscope equipped with a monitor, video camera and computer. If the environmental conditions are stable the accuracy of the measurements is high, even in *Margaritifera* shells which may attain an age of more than 200 years. Usually, however, growth is disrupted once or several times during a year. Growth disruptions show a certain similarity to the winter lines and constitute a source of error in counting and measuring annual growth increments.

When an animal is immature the annual growth rate is high and broad increments are formed. Once an animal has reached sexual maturity, the growth at the shell margin decreases more or less abruptly. In 100–200-year-old *Margaritifera* shells, the width of the growth increments at the shell margin is often less than 0.1 mm.

21.2.1 Variations in Growth Rates in a Single Population of *Margaritifera*

Individual variations of growth rates were studied in 25 shells from the River Vramsån (S Sweden), collected in 1989, 1991 and 1993. This river is only slightly influenced by acidification, and industrial and agricultural pollution. A mean population curve was produced by averaging the growth series of all collected individuals from the same locality. It shows four peaks of shell growth rates between 1930 and 1992 (I, II, III, IV, Fig. 21.1A). A positive correlation was found between the peaks and periods of high average summer temperatures. However, high temperature not only affects the metabolic activity of the animal but it also makes possible a longer feeding season, better availability of food, etc. Many environmental factors covary and it is therefore difficult to correlate growth rate to a single factor.

21.2.2 Shell Microgrowth Pattern in *Margaritifera margaritifera*

In transverse sections of the outer prismatic layer each annual growth increment shows numerous microgrowth lamellae. Impressions of these lamellae are also seen on thick glyco-protein sheets that encase individual prisms. In young individuals with wide annual growth increments, the thickness of the microgrowth lamellae is 2–3 µm. In old individuals with narrow annual growth increments, the thickness of the lamellae is about 0.1–0.2 µm. The number of microgrowth lamellae in wide annual increments is 160–180. This number roughly corresponds to the number of days during the growth season from April to November in southern Sweden. This indicates

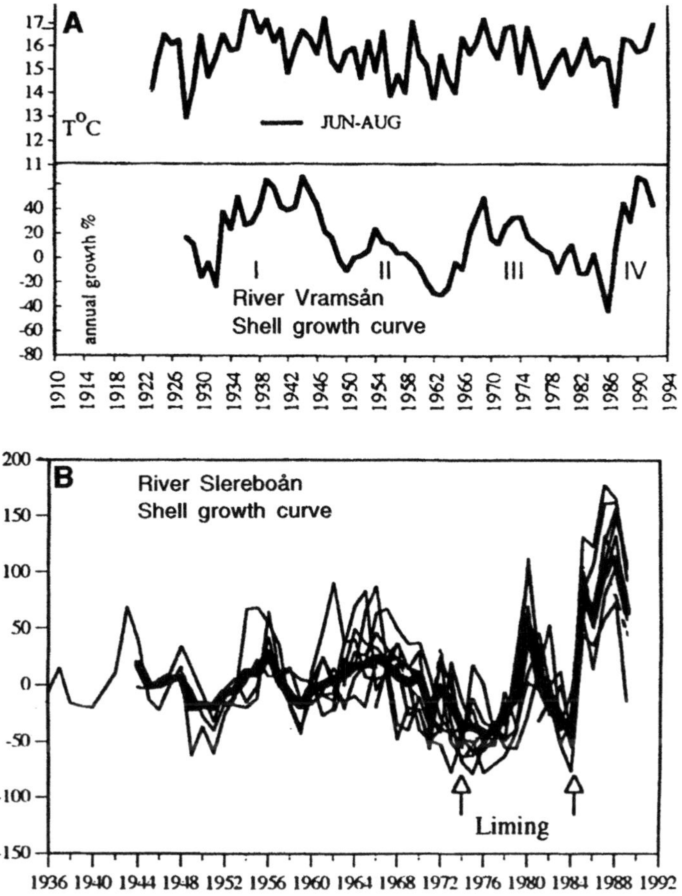

Fig. 21.1A,B. *Margaritifera margaritifera*. **A** Mean shell growth curve of the River Vramsån population showing four peaks of shell growth rates between 1924 and 1992; note positive correlation between the peaks (*I, II, III, IV*) and periods of high average summer temperatures. **B** Individual shell growth curves of the highly acidified River Slereboån population; note an abrupt increase in the growth rate after liming 1979 and 1984 (indicated by *arrows*). Percent of negative and positive annual growth rates designated on y-axis

that the microlamellae are formed diurnally like lamellae in fish otolites and in cephalopod statolites (E. Dunca and H. Mutvei, unpubl.).

Several periods of increase and decrease of thickness of microgrowth lamellae can be seen in most annual growth increments (Fig. 21.2). Thick lamellae indicate rapid growth and thin lamellae slow growth. In places, thin lamellae seem to form acid-resistant ridges which indicate a very slow growth or a growth cessation. The pattern of microgrowth lamellae and acid-resistant ridges is similar in the majority of young and middle aged individuals from the same year and same population. A certain similarity was also found in the pattern of acid-resistant ridges in individuals older than 50 years. In some annual growth increments the microgrowth lamellae and acid-

Fig. 21.2. *Margaritifera margaritifera*, River Sleréboån. Vertical section of the prismatic layer showing pattern of microgrowth lamellae, secreted during 1 year (1988); note acid-resistant ridges (*arrows*) that mark the winter lines and several periods of increase and decrease in thickness of the lamellae

resistant ridges show a distinct periodicity. Rapid growth periods are separated by slow growth periods. Up to eight such periods may occur in certain annual growth increments (Dunca and Mutvei 1996). The correspondence of the number and duration of such periods in shells of about the same age suggests that shell growth is governed by environmental factors and not by individual endogenous rhythms.

21.2.3 Shell Growth Rate in Relation to Availability of Food

21.2.3.1 *Margaritifera margaritifera*

Pearl mussels are considered as adapted to living in conditions with limited food resources. However, when food increases in abundance their shells show a considerable increase in growth rate, even in old individuals. This was observed in the River

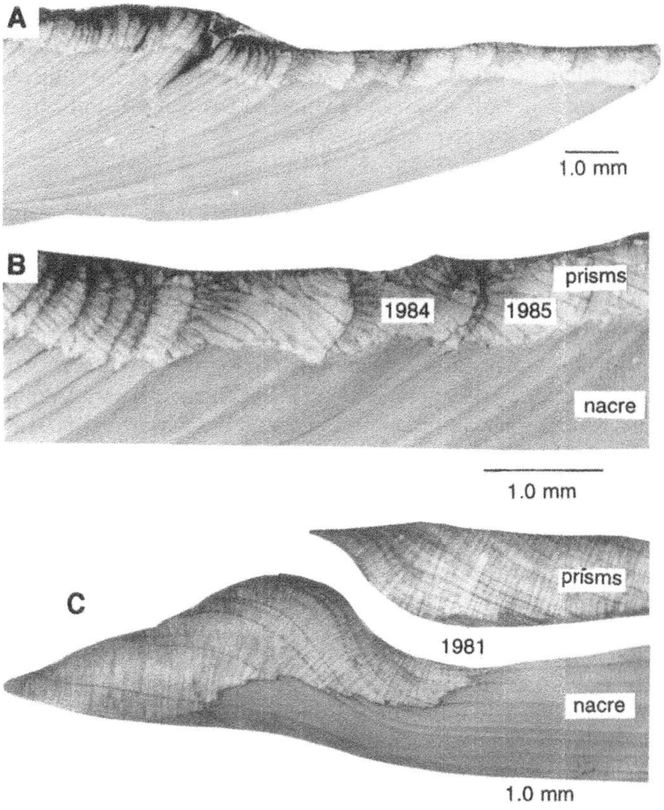

Fig. 21.3A–C. A, B *Unio crassus* from the Soodla River, N Estonia. **A** Vertical section at ventral shell margin showing an abrupt change in widths of annual growth increments after 1982 when the Soodla Reservoir was completed. **B** Closeup of **A. C** *Margaritifera margaritifera* from the River Kvarnbäcken; vertical section at ventral shell margin showing disruption of shell growth in 1981, marked by a crack; note that the widths of the annual growth increments increase considerably after 1981

Kvarnbäcken, County of Örebro, central Sweden. Shell growth was low from 1971 to 1981. In 1981 the sourcelake upstream was treated with lime, phosphorus and nitrogen in order to increase bioproduction and thereby eliminate the high mercury load in the lake. In all ten individuals from the river population examined by us, the shell growth was disrupted and a distinct crack was formed in 1981 (Fig. 21.3C). The crack indicates that the mantle retracted from the shell margin and ceased to secrete shell material. Then the mantle regained its secretory function and the annual growth rate increased several times from 1982. This was caused by increased pH after liming and increased food supply due to enhanced bioproduction in the sourcelake.

21.2.3.2 *Unio crassus*

Shell growth rate in *Unio crassus* from the River Soodla, N Estonia, showed a considerable increase after 1982 (Fig. 21.3A,B). This increase coincides with construction of the Soodla Reservoir in 1981, a short distance upstream. The growth increase seems to have been caused by an increase in the amount of phytoplankton used for feeding. Phytoplankton biomass in 1974–1976 was always below $1\,g\,m^{-3}$, but increased in the reservoir to $5.8\,g\,m^{-3}$ in 1982–1984, and fluctuated between 2.6 and $7.0\,g\,m^{-3}$ in 1990 (Timm and Mutvei 1993; Mutvei et al. 1994).

21.2.4 Shell Structure and Growth Rate in Relation to Acidification

SW Sweden has a high acidification load due to air-borne pollutants. Increase of acidification in rivers where *Margaritifera* lives is clearly indicated by changes in shell growth rate. Our material from the River Slereboån, ca. 30 km NW of Göteborg, comprises about 20 specimens 20–100 years old. Shell growth rate was extremely low during the periods 1970–1979, and 1982–1984, when the pH was around 5 or even lower. The shell surface shows dissolution pits and the periostracum extends deeply into the prismatic layer. Liming of the water system took place in 1974 and 1984, and a permanent liming device was installed upstream in 1986, after which the acidification decreased. The shell growth in all 20 mussels increased several hundred percent from 1985, and the growth remained constant until 1989 when the mussels were collected (Fig. 21.1B).

21.3 Elemental Distribution in Relation to Acidification

Studies of elemental distribution in naiad shells give us information about the environmental conditions under which the animals live. Altogether, 32 elements have been identified, and in most cases also quantified, in the shells by using neutron activation analysis, the nuclear microscope, ICP-MS and other methods. The concentration mechanism from the surrounding water to the shell has been shown for eight elements. An approximately linear relationship has been indicated for phosphorus, manganese, iron, cobalt, zinc, strontium and caesium, the latter from the study on radioactive nucleides by Frantsevitch et al. (1996), close to the Chernobyl reactor accident site. Indirect support for a linear relationship between elemental concentration and atmospheric fallouts has been found for nitrogen and sulphur.

21.3.1 Sulphur and Nitrogen

The emission of sulphur in Europe has decreased during the past two decades, after a heavy increase during the previous years of this century. On the other hand, the contribution of nitrogen compounds has increased and therefore total acidification has increased. Sulphur and nitrogen fallout causes lowering of pH and a higher solubility of manganese, cobalt, iron and zinc in water. The highest levels of sulphur in *Margaritifera* shells were found from SW Sweden. Northwards in Sweden, sulphur concentration in shells progressively decreases, which coincides with a decrease in acidification loads in the same direction. In N Sweden and NW Russia (South of the White Sea and central region of the Kola peninsula), close to the Arctic Circle, sulphur concentrations were found to be low in *Margaritifera* shells (Fig. 21.4). However, high sulphur levels occurred in *Margaritifera* shells in the 1950s in two rivers in N Sweden

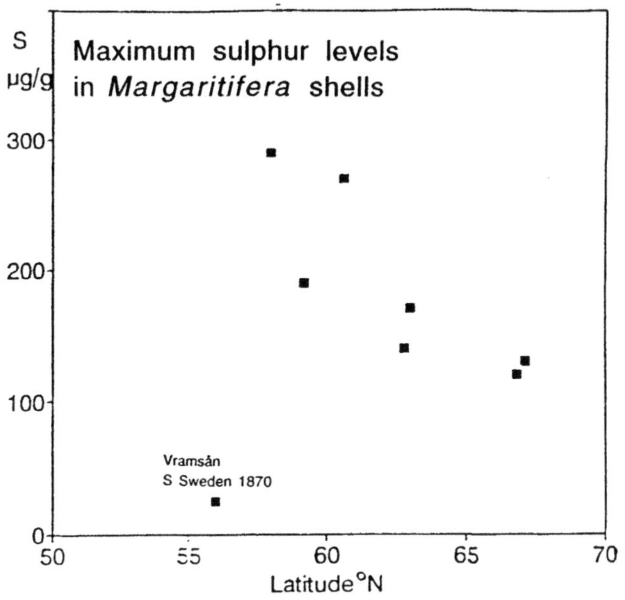

Fig. 21.4. Transect from S Sweden to N Sweden and N Russia showing decrease in maximum sulphur levels in *Margaritifera* shells using latitude as the coordinate. Note low sulphur level in the River Vramsån in 1870

and one in W Sweden, situated close to forest industries. The minimum sulphur concentrations in these rivers occurred during World War II when production rates of the forest industries were low.

The distribution pattern of nitrogen in *Margaritifera* shells is similar to that of sulphur. The highest nitrogen levels occurred in SW Sweden in 1970–1980, due to traffic, energy production and cultivation of land (Westermark et al. 1996). Low nitrogen levels were found in rivers near the Arctic Circle.

21.3.2 Solubility of Iron, Zinc and Cobalt in Relation to Acidification

The relationship between the solubility of iron, cobalt and zinc, and pH can be used for estimations of pH levels in the past. Zinc is to be preferred because its oxidation state is unique (+II) but for iron and cobalt it is not. In this way misinterpretations with regard to pH are avoided. The basis for this estimation is the minimum solubility at about pH 6.5, and the solubility increase at both lower and at higher pH, forming a "cradle" curve (Stumm and Morgan 1996).

One of the best examples of pH history was found in *Margaritifera* shells from the highly acidified River Slereboån, SW Sweden. Here neutron activation analyses show that iron concentration reached its minimum in 1960 when pH decreased to about pH 5. After liming in 1981 there was a dramatic increase of iron concentration, and the pH increased (Fig. 21.5A; Westermark et al. 1996). On the other hand, strontium concentration decreased significantly after liming (Malmberg 1998).

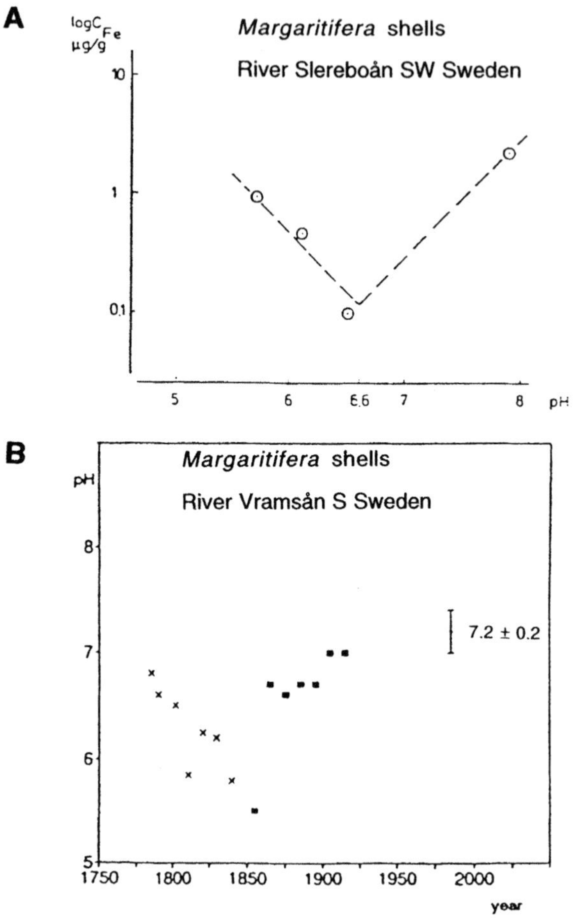

Fig. 21.5A,B. *Margaritifera.* **A** Tentative pH history based on the solubility of iron in a shell from the acidified and limed River Slereboån, SW Sweden; later pH estimations have been based on the more reliable zinc concentrations. **B** Tentative pH history of the River Vramsån, S Sweden, calculated from two shells

Neutron activation analyses of *Margaritifera* shells have also been carried out for estimations of pH changes in the River Vramsån, S Sweden, from 1770 to 1987 (Fig. 21.5B). A significant reduction in pH took place at the beginning of 19th century. This is assumed to have been caused by intense ditching and cultivation of new farming areas at that time. Soils from wetland and pastures were exposed to air and oxidized, causing nitrification and mineralization of nitrogen and sulphur (Mutvei et al. 1994; Westermark et al. 1996). After 1925 the pH stabilized at about pH 7.

21.3.3 Strontium Isotope Ratio in Relation to Acidification

Strontium isotope ratios in water can also be used to elucidate pH changes. This ratio is related to the weathering rate of primary minerals in soils and, consequently, mirrors the acidification (Åberg et al. 1995). Strontium isotope ratios for *Margaritifera* shells from two rivers from Central Sweden (Tansån and Vasslabäcken) are similar to the values for two source lakes in the vicinity of these rivers. These ratios show a decreasing trend from 1910 to 1965, indicating an increasing acidification of the environment. Shells from the highly acidified River Slereboån from SW Sweden had low, slightly decreasing, strontium ratios from 1945 to 1980, and then a sharp drop in strontium isotope ratios in 1984 when the source lake was limed and pH increased.

21.4 Manganese Concentration in Relation to Oxygen Depletion and Eutrophication

Concentration of manganese in bivalve shells is related to oxygen levels in the water and can therefore be used to discover oxygen variations and eutrophication in the past.

In Lake Krasnoe, Karelian Isthmus, Russia, eutrophication and oxygen depletion have been recorded since 1963 by the Institute of Limnology, St. Petersburg. In the past decades, a distinct increase in eutrophication of the lake has taken place, resulting in the increase of organic carbon, phosphorous and bacterioplankton. This caused oxygen depletion and a considerable decrease in the shell growth rate in *Unio tumidus* (Mutvei et al. 1996). Neutron activation analysis shows a high concentration of manganese in shells of these naiads (Fig. 21.6). Using high-resolution elemental mapping with nuclear microscope 20 analyses were carried out on a 1.5 mm-long-transect of

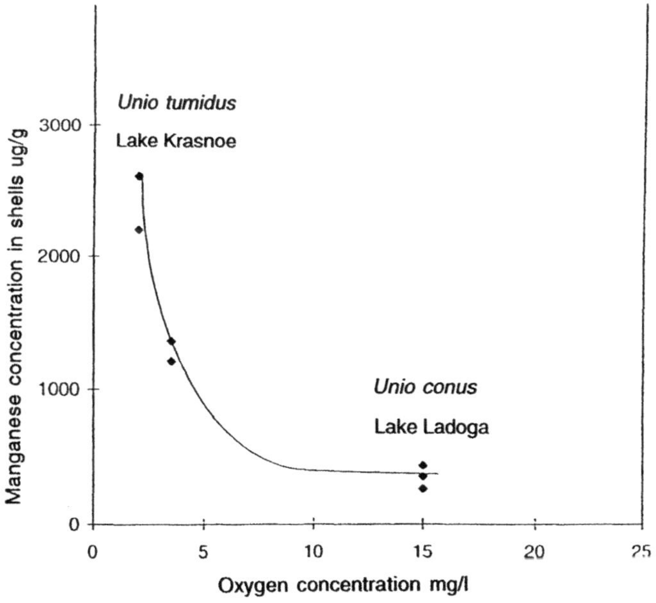

Fig. 21.6. Content of manganese in unionid shells in relation to mean values of oxygen in water; *Unio tumidus* from the eutrophicated Lake Krasnoe and *U. conus* from the Lake Ladoga, N Russia. (T. Westermark and T. Slepukhina, unpubl.)

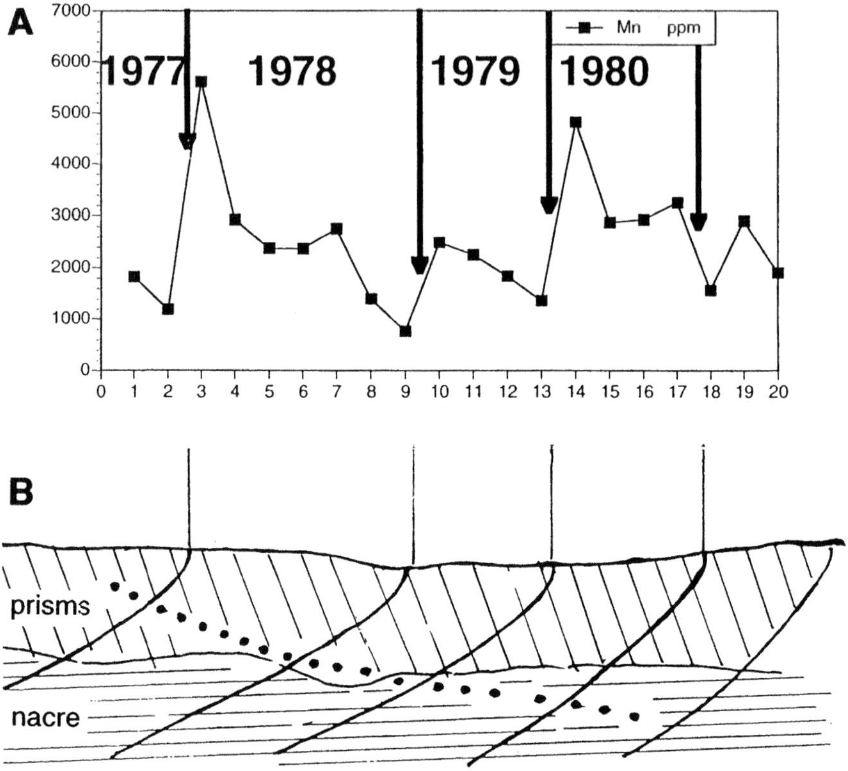

Fig. 21.7A,B. *Unio tumidus*, Lake Krasnoe, N Russia. **A** Seasonal variations of manganese in the shell, 1977–1980. **B** Schematic drawing of the prismatic and nacreous layers to show the position of points analyzed with a nuclear microscope. (E. Dunca and H. Mutvei, unpubl.)

the vertical shell section, representing the growth of 5 years from 1977–1981. Seasonal variation of manganese levels could be demonstrated in each year: manganese concentration was high in spring to early summer but low in the rest of the year (Fig. 21.7A,B; Westermark et al. 1997).

Similar seasonal variations of manganese and strontium content were found in *Margaritifera* shells from two other S Swedish rivers, Vramsån and Pauliström (Nyström et al. 1995, 1996; Nyström and Dunca 1996). The concentrations of these two elements in the River Vramsån obtained by analysis with a nuclear microscope are very similar to those recorded by neutron activation analysis (Carell et al. 1995).

21.5 Applications

Studies on shell growth rate, as reflected by variations in the number and thickness of microlamellae, coupled with elemental analysis with the nuclear microscope with high temporal resolution, will increase our understanding of the biology of freshwater bivalves.

Applications of elemental studies in freshwater bivalve shells can be summarized as follows: (1) Acidification history of freshwaters in acid-sensitive areas, and evaluation of measures against acidification: liming and/or reduction of acidifying agents. (2) Volcanic emission of sulphur oxides. (3) Eutrophication of freshwaters, and evaluation of measures against eutrophication: water purification by mechanical, biological and chemical means. (4) History of agricultural fertilizers. (5) Pollution history of process industry: iron-ore processing and metal production; wood and pulp industries; pollution by mining wastes. (6) Pollution by motor vehicle exhaust gases. (7) Pollution history of energy production units (coal, peat, oil etc; ash leaching). (8) Nuclear industry: mining and mine tailings; fuel element factories; reactor stations; reprocessing; reactor and radioactive waste accidents; nuclear weapon tests.

21.6 Summary

1. In the S Swedish river Vramsån, only slightly influenced by human activities, shell growth rates of *Margaritifera* population show three peaks between 1930 and 1992. A positive correlation was found between the peaks and periods of high average summer temperatures.
2. In transverse sections of the outer prismatic shell layer in *Margaritifera* each annual growth increment shows numerous microlamellae, which seemingly are formed diurnally like lamellae in fish otolites and cephalopod statolites. In some annual growth increments the lamellae vary in thickness, which indicates rapid growth periods separated by slow growth periods. Up to eight such periods may occur in certain annual growth increments.
3. In the central Swedish river Kvarnbäcken the source lake upstream was treated with lime, phosphorus and nitrogen in order to increase bioproduction. In the river population of *Margaritifera* the shell growth was disrupted and a distinct crack was formed after this treatment. Then the mantle regained its secretory function and the annual growth rate increased several times. Thus, when food increases in abundance the shell growth increases considerably, even in old individuals. A similar positive correlation between increase in annual growth rate and increase in amount of phytoplankton used for feeding was found in *Unio crassus* from an Estonian river.
4. In acidified rivers of SW Sweden the pH was around 5 or even lower during long periods. In these rivers the growth rate in *Margaritifera* shells was extremely low during these periods. After permanent liming of the water system the shell growth increased several hundred percent.
5. Sulphur and nitrogen fallout causes lowering of pH and higher solubility of manganese, cobalt, iron and zinc in water. The highest levels of sulphur and nitrogen in *Margaritifera* shells were found from SW Sweden. Northwards in Sweden, and in NW Russia, sulphur and nitrogen levels in shells progressively decrease.
6. The relationship between the solubility of iron, cobalt and zinc, and pH was used for estimations of pH levels in the past. The minimum solubility is at pH 5.5–6.5, and the solubility increases at both lower and higher pH.
7. Strontium isotope ratios in shells can also be used to elucidate pH changes. These ratios show a decreasing trend when the acidification of the environment increases.

8. Manganese concentration in naiad shells is related to oxygen levels in the water and can be used to discover oxygen variations and eutrophication in the past. High manganese levels in shells were found in the eutrophicated Lake Krasnoe, NW Russia. Seasonal variations in manganese levels in this lake could be demonstrated by using the nuclear microscope: manganese concentration was high in spring to early summer but low in the rest of the year.

Acknowledgements. We express our gratitude to numerous collaborators (see Carell et al. 1995), particularly to Dr. E. Dunca and Mrs. B. Carell. We are indebted to Dr. J. Dalingwater, Manchester, UK, for stylistic improvements.

References

Åberg G, Wickman T, Mutvei H (1995) Strontium isotope ratios in mussel shells as indicators of acidification. Ambio 25:265–268

Carell B, Forberg S, Grundelius E, Henrikson L, Johnels A, Lindh U, Mutvei H, Olsson M, Svärdström K, Westermark T (1987) Can mussel shells reveal environmental history? Ambio 16:2–10

Carell B, Dunca E, Gärdenfors U, Kulakowski E, Lindh U, Mutvei H, Nyström J, Seire A, Slepukhina T, Timm H, Westermark T, Ziuganov V (1995) Biomonitoring of pollutants in a historic perspective. Emphasis on mussel and snail shell methodology. Ann Chim 85:353–370

Dunca E, Mutvei H (1996) Periodic microgrowth patterns in shells of freshwater unionid bivalves. Bull Inst Océanogr Monaco Spéc 14:127–131

Frantsevitch L, Korniushin A, Pankov I, Ermakov A, Zachantjuk T (1996) Application of molluscs for radioecological monitoring of the Chernobyl outburst. Environ Pollut 94:91–100

Imlay MJ (1982) Use of shells of freshwater mussels in monitoring heavy metals and environmental stresses: a review. Malacol Rev 15:1–14

Malmberg C (1998) Mussel shells as sampler of our environment. Högskoletryckeriet, KTH, Stockholm, ISSN 0349-6465, 81 pp (in Swedish)

Mutvei H, Westermark T, Dunca E, Carell B, Forberg S, Bignert A (1994) Methods for study of environmental changes using structural and chemical information in molluscan shells. Bull Inst Océanogr Monaco Spéc 13:163–191

Mutvei H, Dunca E, Timm H, Slepukhina T (1996) Structure and growth rates of bivalve shells as indicators of environmental changes and pollution. Bull Inst Océanogr Monaco Spéc 14:65–72

Nyström J, Dunca E (1996) Nuclear microscopy of freshwater bivalves. Bull Inst Océanogr Monaco Spéc 14:121–125

Nyström J, Lindh U, Dunca E, Mutvei H (1995) A study of *M. margaritifera* shells from the River Pauliströmsån, S Sweden. Nucl Instr Meth Phys Res B 104:612–618

Nyström J, Dunca E, Mutvei H, Lindh U (1996) Environmental history as reflected by freshwater pearl mussels in the River Vramsån, southern Sweden. Ambio 15:350–355

Stumm W, Morgan JJ (1996) Aquatic chemistry, 3rd edn. John Wiley New York, 780 pp

Timm H, Mutvei H (1993) Shell growth of the freshwater unionid *Unio crassus* from Estonian rivers. Proc Estonian Acad Sci Biol 42:55–67

Veinott GI, Cornett RJ (1996) Identification of annually produced opaque bands in the shell of the freshwater mussel *Elliptio complanata* using the seasonal cycle of $\delta^{18}O$. Can J Fish Aquat Sci 53:372–379

Westermark T, Carell B, Forberg S, Mutvei H, Kulakowski E (1996) Freshwater unionid (Mollusca: Bivalvia) shells as environmental archives: methodology and observations. Bull Inst Océanogr Monaco Spéc 14:73–81

Westermark T, Mutvei H, Carell B, Dunca E, Bignert A, Kulakowski E, Nyström J, Olsson M, Seire A, Slepukhina T, Ahlgren I, Wahlberg O (1997) How can mussel shells contribute to environmental history information? Development of paleoceanography as a new field of science, Commemorating 50th Anniv Albatross Exp 1947–48, R Swedish Acad Sci, Stockholm, Abstr Vol

Part VI
Synopsis

22 Ecology and Evolution of the Naiads

Gerhard Bauer and Klaus Wächtler

22.1 Introduction

"Parasites present a fascinating array of life histories, from simple, straightforward life cycles . . . , to complex life cycles consisting of episodes of both sexual and asexual reproduction and several different developmental stages using a series of hosts. This range of life cycles would seem to provide an excellent opportunity to test some of the predictions of life-history theory. However, recent texts on life-history evolution (e.g. Roff 1992; Stearns 1992) have not included discussions of parasite examples . . ." (Poulin 1996).

Compared to most parasites, the life cycle of naiads is simple. It comprises only one larval stage, which is parasitic for a more or less short time (Chap. 13, this Vol.), immobile during this time and it does not reproduce. The longest-lasting stage is the free-living, semisessile and filter-feeding adult (Chap. 9, this Vol.). However, even this slight deviation from a free-living type of animal leads to a very peculiar parasitic strategy and to life-history patterns which contradict familiar theories. This strategy certainly contributes to the considerable evolutionary success of naiads.

22.2 Naiads as Parasites

The parasitism of naiads leads to tremendous mortality rates in free larvae since host transmission is highly ineffective, even in species with adaptations for improved host infection (Chaps. 9, 11, this Vol.). Wood (1974) showed that the adult *Anodonta cygnea* supplies its glochidia with nutrients and meanwhile there can be no doubt that there are some cases where indeed glochidia metamorphose directly into young mussels within the marsupia (Chap. 6, this Vol.). Thus, the question arises why the overwhelming majority of naiads retain their parasitic mode of reproduction, leading inevitably to tremendous glochidial mortality.

Prevention of an osmotic shock is frequently supposed (Kat 1984), but to our knowedge this hypothesis is not supported by data. (As *Dreissena* and *Corbicula* show, a veliger is well able to cope with freshwater).

Another advantage, cited by many authors (e.g. Davis and Fuller 1981; Kat 1984; Chap. 15, this Vol.) concerns the distribution. Indeed, distribution ability seems to be considerably high, as indicated by the large areas recolonized by naiads after the last ice age (Davis and Fuller 1981; Bauer 1987).

Also the food supply of the larvae might be of importance. In all cases so far investigated at least the carbon content of the larva increases during parasitism (Chap. 6, this Vol.); but there are also some groups where considerable growth is evident,

Ecological Studies, Vol. 145
G. Bauer and K. Wächtler (eds.) Ecology and Evolution of the Freshwater Mussels Unionoida

namely among the Mutelidae (Chap. 6, this Vol.), among the Margaritiferidae (Chaps. 8, 13, this Vol.) and among some Lampsilidae (Surber 1912). Thus, the larvae are not only hitchhikers, but drain considerable resources from their hosts. This facilitates a strategy to produce very many but "cheap" glochidia (i.e. glochidia largely without storage energy). As shown in Chapter 11 (this Vol.), the life span of free glochidia is only a few days, regardless of their size. The large *Anodonta* glochidia (0.34 mm shell length) are as short-lived as the intermediate (0.2 mm) *Unio* and the small (0.07 mm) *Margaritifera* glochidia.

Apparently, unlike other propagules (e.g. Roff 1992), size has nothing to do with storage energy, but only with the degree of development and thus the duration of the parasitic stage (Chap. 13, this Vol.). This strategy to produce a large number of larvae, while keeping investments into the single larva very low, and to shift the energy supply for the few successful ones to the host is also evident in the Mutelidae (Falk 1995; Herrmann 1995). It may have been a reason why the veliger's energy-consuming ability to swim was lost and has never been reinvented again. As the variety of glochidial forms (Chaps. 5, 6, 11, this Vol.) and particularly the highly derived lasidium and haustorium (Chaps. 6, 14, this Vol.) indicate, naiad larvae are subject to evolution; but all these larval types are not particularly sophisticated parasites. There is no case where an active swimming and active host-selecting larva was developed. Accordingly, since host infection occurs largely passively, the size of the host range should be adjusted to the diversity of the fish fauna in the preferred habitats. There are some data to support this hypothesis (Chap. 13, this Vol.): the host range is narrow in species coocurring with a fish fauna low in diversity and it is broad in species coocurring with a species-rich fish fauna. A striking exception is *Margaritifera auricularia*. Though it lives in large streams together with a diverse fish fauna, only Acipenseridae are hosts (Chap. 8, this Vol.). This argues for a strong phylogenetic constraint.

On a finer scale, the evolution of the host range is poorly understood. Frequently, closely related fish species coocurring with a particular naiad show tremendous differences with respect to their host suitability. The coho- (*Onchorhynchus kisutch*) and chinnook salmon (*O. tshawytscha*) are very closely related (Smith and Stearly 1989) and cooccur with *Margaritifera falcata*. The chinnook is an excellent host but the coho is nearly completely resistant (Karna and Millemann 1978). Among the cyprinids coocurring with *Unio crassus*, the chub (*Leuciscus cephalus*) is a good host, whereas the dace (*L. leuciscus*) is resistant (Hochwald 1997). This phenomenon is even more strange in species with a very broad host range. Glochidia of *Anodonta cygnea* develop on many cyprininids and other fish species (Chap. 13, this Vol.) and even on tadpoles (Terberger 1999), but the cyprinid belica (*Leucaspius delineatus*), which frequently cooccurs with *Anodonta* in large numbers, is unsuitable (Teutsch 1997).

There is growing evidence that in the evolution of the host-parasite relationship an active part is played by the fish, i.e. it is not the biochemical similarity of related species that is important, but the immune response (Isom and Hudson 1984). This may be different, rendering one species a good host and a related one unsuitable.

Some of these host-parasite relationships are insofar unique, as due to the long life of the adult mussels, the generation time of the parasite largely exceeds that of its hosts. This "reversal of generation times" should have profound effects on the evolution of the host-parasite relationship. This holds particularly true for *Margaritifera* sensu stricto, (i.e. *M. margaritifera*, *M. falcata*, *M. laevis*), exceeding the generation times of their hosts by a factor between 10 and 20 (Bauer 1997; Chap. 13, this Vol.). However, the shorter an organism's generation time, the faster its potential rate of evolution, and thus the faster the rate to evolve strategies against a parasite (Freeland 1986). This should render the mussels inferior to their fast-evolving hosts, but in spite of that, these host-parasite systems have been stable for at least 60 million years (Bauer 1997).

22.3 Naiads and Life-History Theory

Many basic assumptions of life-history theory are violated by the naiads. The following list contains only some examples; it could be extended considerably.

A number of allometric constraints, frequently encountered in the animal kingdom (Blueweiss et al. 1978), do not hold for naiads. There is no relationship between body size and life span, or between body size and offspring size (Bauer 1994; Chap. 13, this Vol.).

There is no relationship between offspring size and storage energy and thus starvation time of offspring (review in Roff 1992, p. 348 ff.). The life span of large glochidia does not exceed that of small ones (Chap. 11, this Vol.).

Kirkwood and Holliday (1979) developed a plausible theory that life span (1) evolves in response to mortality and (2) ends in a phase of senescence. Neither of these holds for naiads (Chaps. 9, 13, this Vol.).

Most models predict an increase in reproductive effort with age (Pianka and Parker 1975; Kozlowski 1991). However, in many naiads reproductive output is mainly determined by the physiological condition (Tuomi et al. 1983; Bauer 1998; Chaps. 2, 7, 13, 18, this Vol.), in this way favouring survival over current reproduction.

Accordingly, general models on life-history evolution which are based on such assumptions are not applicable to this group. The most familiar of these models is the r-K concept (Pianka 1972; Southwood et al. 1974). If one considers, for example, *Anodonta*, the following r-selected features are evident: low degree of specialization, short life span, fast development and early maturation. On the other hand, *Anodonta* exhibits the following K-selected features: large body size, highly developed brooding (Wood 1974), low fertility and large offspring size (Chap. 13, this Vol.). Clearly, the selective factors operating on life-history evolution of naiads are different from those underlying most general models. (In this case, the stability of resource availability in space and time).

22.4 Naiads and Evolutionary Success

No matter what criterion is used to define evolutionary success, be it the number of species within a taxon, the distribution area, the population density or the phylogenetic age, naiads provide examples (and even extreme examples) for either.

22.4.1 The Number of Radiating Species

North American naiads show one of the most diverse radiations among freshwater invertebrates. Particularly the heterogenous lampsilids and the Ambleminae are marked by a high species richness (Davis and Fuller 1981).

22.4.2 Distribution Area

Large distribution areas are found among many *Anodonta* species (Clarke 1973; Nagel 1988). Also *Margaritifera* sensu stricto, i.e. *M. margaritifera*, *M. falcata*, *M. laevis*, occupy large areas, particularly *M. margaritifera*, which occurs on both sides of the Northern Atlantic (Chaps. 3, 19, this Vol.).

22.4.3 Population Density

Extremely high densities are reported for some margaritiferids. In these cases where the rivers are plastered with mussels, as is reported for *M. margaritifera* (Israel 1913) and *M. falcata* (Clarke 1973), densities of up to 1000 individuals m^{-2} must be assumed (pers. observ.). For *M. hembeli* peak densities of up to 100 individuals m^{-2} are recorded (Johnson and Brown 1998). Probably these margaritiferids, though living exclusively in rivers extremely poor in nutrients, attain among the highest biomasses of freshwater invertebrates. Densities of unionids usually are lower, but exceptionally may also approach such values (Chap. 12, this Vol.). For hyriids, maximum densities of 814 individuals m^{-2} are reported, but usually densities are much lower (Chap. 2, this Vol.).

22.4.4 Phylogenetic Age

Due to the few conchological criteria for recognizing fossil Unionacea, the age of most taxa is discussed controversially (Chaps. 3, 15, this Vol.). However, according to Watters (this Vol.), Unionacea evolved from an as yet unidentified marine group by at least the Triassic and the modern families once enjoyed a much wider distribution than they do today.

At least 60 million years must be the phylogenetic age of *M. margaritifera*. It occurs on both sides of the Northern Atlantic (Chaps. 3, 19, this Vol.), areas which had been linked during the Eocene via the Greenland bridge.

22.5 Summary

The parasitic strategy of naiads is marked by largely inactive larvae, little resource investment into the single larva and by shifting the energy supply for the few successful ones to the host. Frequently, closely related fish species differ considerably with respect to their susceptibility for a particular mussel species, indicating that the host's immune response plays a major part in the evolution of the host range.

The naiads violate many basic assumptions of life-history theory. Accordingly, general models like the r-K concept are not valid for this taxon.

The evolutionary success of naiads is evident in their high species number, and among some taxa, large distribution areas, high population densities and a high phylogenetic age.

References

Bauer G (1987) Reproductive strategy of the freshwater pearl mussel *Margaritifera margaritifera*. J Anim Ecol 56:691–704
Bauer G (1994) The adaptive value of offspring size among freshwater mussels (Bivalvia; Unionoidea). J Anim Ecol 63/4:933–944
Bauer G (1997) Host relationships at reversed generation times. *Margaritifera* and salmonids. In: Dettner K, Bauer G, Völkl W (eds) Vertical food web interactions: evolutionary patterns and driving forces. Ecological Studies, vol 130. Springer, Berlin Heidelberg New York, pp 69–79
Bauer G (1998) Allocation policy of female freshwater pearl mussels. Oecologia 117:90–94
Blueweiss LH, Fox V, Nakashima D, Peters R, Sams S (1978) Relationships between body size and some life history parameters. Oecologia 37:257–272
Clarke AH (1973) The freshwater molluscs of Canada. National Museum of Canada, Ottawa
Davis GM, Fuller SLH (1981) Genetic relationships among recent Unionacea (Bivalvia) of North America. Malacologia 20(2):217–253
Falk L (1995) Das ektoparasitische Larvenstadium der afrikanischen Süßwassermuschel *Mutela bourguignati* Bourguignat 1885. PhD Thesis, Tierärztliche Hochschule Hannover, 128 pp
Freeland B (1986) Arms races and covenants: the evolution of parasite communities. In: Kikkawa J, Anderson DJ (eds) Community ecology: pattern and process. Blackwell, Oxford, pp 289–303
Herrmann F (1995) Untersuchungen zum Lebenszyklus der afrikanischen Süßwassermuschel *Mutela bourguignati* Bourguignat 1885. PhD Thesis, Tierärztliche Hochschule Hannover, 96 pp
Hochwald S (1997) Das Beziehungsgefüge innerhalb der Größenwachstums- und Fortpflanzungsparameter bayerischer Bachmuschelpopulationen (*Unio crassus* Phil.1788) und dessen Abhängigkeit von Umweltfaktoren. PhD Thesis, University Bayreuth, 172 pp
Isom BG, Hudson RG (1984) Freshwater mussels and their fish hosts; physiological aspects. Am Soc Parasitol 70(2):318–319
Israel W (1913) Biologie der europäischen Süßwassermuscheln. KG Lutz, Stuttgart
Johnson PD, Brown KM (1998) Intraspecific life-history variation in the threatened Lousiana pearlshell mussel (*Margaritifera hembeli*). Freshwater Biol 40:317–329
Karna DW, Millemann RE (1978) Glochidiosis of salmonid fishes. III. Comparative susceptibility to natural infection with *Margaritifera margaritifera* (L.) (Pelecypoda: Margaritanidae) and associated histopathology. J Parasitol 64(3):528–537
Kat PW (1984) Parasitism and the Unionacea (Bivalvia). Biol Rev 59:189–207
Kirkwood TBL, Holliday FRS (1979) The evolution of ageing and longevity. Proc R Soc Lond 205:531–546
Kozlowski J (1991) Optimal energy allocation models – an alternative to the concepts of reproductive effort and cost of reproduction. Acta Oecol 12:11–33
Nagel K-O (1988) Anatomische, morphologische und biochemische Untersuchungen zur Taxonomie und Systematik der europäischen Unionacea. PhD Thesis, Gesamthochschule Kassel
Pianka ER (1972) r- and K-selection or b- and d-selection? Am Nat 106:453–463
Pianka ER, Parker WS (1975) Age-specific reproductive tactics. Am Nat 109:453–463
Poulin R (1996) The evolution of life-history strategies in parasitic animals. Adv Parasitol 37:107–134
Roff DA (1992) The evolution of life histories; theory and analysis. Chapman and Hall, New York
Smith G, Stearly R (1989) The classification and scientific names of rainbow and cutthroat trout. Fisheries 14:4–10
Southwood TRE, May R, Hassel MP, Conway GR (1974) Ecological strategies and population parameters. Am Nat 108:791–804
Stearns SC (1992) The evolution of life histories. Oxford University Press, Oxford

Surber T (1912) Identification of the glochidia of freshwater mussels. Bureau of fisheries document 771. Government Printing Office, Washington, DC

Terberger J (1999) Wirtsbeziehungen der Teichmuscheln *Anodonta*. Ms Thesis, University Freiburg

Teutsch S (1997) Populationsbiologische Untersuchung an *Anodonta*: Fortpflanzungsstrategien und Charakterisierung von Teilpopulationen innerhalb eines Stillgewässers. MS Thesis, University Freiburg, 82 pp

Tuomi J, Hakala T, Haukioja E (1983) Alternative concepts of reproductive effort, costs of reproduction, and selection in life-history evolution. Am Zool 23:25–34

Wood EM (1974) Development and morphology of the glochidium larva of *Anodonta cygnea*. J Zool Lond 173:1–13

Subject Index

Ecological Studies

Volumes published since 1994

Volume 106
Minimum Animal Populations (1994)
H. Remmert (Ed.)

Volume 107
The Role of Fire in Mediterranean-Type Ecosystems (1994)
J.M. Moreno and W.C. Oechel (Eds.)

Volume 108
Ecology and Biogeography of Mediterranean Ecosystems in Chile, California and Australia (1995)
M.T.K. Arroyo, P.H. Zedler, and M.D. Fox (Eds.)

Volume 109
Mediterranean-Type Ecosystems. The Function of Biodiversity (1995)
G.W. Davis and D.M. Richardson (Eds.)

Volume 110
Tropical Montane Cloud Forests (1995)
L.S. Hamilton, J.O. Juvik, and F.N. Scatena (Eds.)

Volume 111
Peatland Forestry. Ecology and Principles (1995)
E. Paavilainen and J. Päivänen

Volume 112
Tropical Forests: Management and Ecology (1995)
A.E. Lugo and C. Lowe (Eds.)

Volume 113
Arctic and Alpine Biodiversity. Patterns, Causes and Ecosystem Consequences (1995)
F.S. Chapin III and C. Körner (Eds.)

Volume 114
Crassulacean Acid Metabolism. Biochemistry, Ecophysiology, and Evolution (1996)
K. Winter and J.A.C. Smith (Eds.)

Volume 115
Islands. Biological Diversity and Ecosystem Function (1995)
P.M. Vitousek, L.L. Loope, and H. Adsersen (Eds.)

Volume 116
High Latitude Rainforests and Associated Ecosystems of the West Coast of the Americas: Climate, Hydrology, Ecology and Conservation (1996)
R.G. Lawford, P. Alaback, and E. Fuentes (Eds.)

Volume 117
Global Change and Mediterranean-Type Ecosystems (1995)
J. Moreno and W.C. Oechel (Eds.)

Volume 118
Impact of Air Pollutants on Southern Pine Forests (1996)
S. Fox and R.A. Mickler (Eds.)

Volume 119
Freshwater Ecosystems of Alaska. Ecological Syntheses (1997)
A.M. Milner and M.W Oswood (Eds.)

Volume 120
Landscape Function and Disturbance in Arctic Tundra (1996)
J.F. Reynolds and J.D. Tenhunen (Eds.)

Volume 121
Biodiversity and Savanna Ecosystem Processes. A Global Perspective (1996)
O.T. Solbrig, E. Medina, and J.F. Silva (Eds.)

Volume 122
Biodiversity and Ecosystem Processes in Tropical Forests (1996)
G.H. Orians, R. Dirzo, and J.H. Cushman (Eds.)

Volume 123
Marine Benthic Vegetation. Recent Changes and the Effects of Eutrophication (1996)
W. Schramm and P.H. Nienhuis (Eds.)

Volume 124
Global Change and Arctic Terrestrial Ecosystems (1997)
W.C. Oechel et al. (Eds.)

Volume 125
Ecology and Conservation of Great Plains Vertebrates (1997)
F.L. Knopf and F.B. Sämson (Eds.)

Volume 126
The Central Amazon Floodplain: Ecology of a Pulsing System (1997)
W.J. Junk (Ed.)

Volume 127
Forest Decline and Ozone: A Comparison of Controlled Chamber and Field Experiments (1997)
H. Sandermann, A.R. Wellburn, and R.L. Heath (Eds.)

Volume 128
The Productivity and Sustainability of Southern Forest Ecosystems in a Changing Environment (1998)
R.A. Mickler and S. Fox (Eds.)

Volume 129
Pelagic Nutrient Cycles: Herbivores as Sources and Sinks (1997)
T. Andersen

Volume 130
Vertical Food Web Interactions: Evolutionary Patterns and Driving Forces (1997)
K. Dettner, G. Bauer, and W. Völkl (Eds.)

Volume 131
The Structuring Role of Submerged Macrophytes in Lakes (1998)
E. Jeppesen et al. (Eds.)

Volume 132
Vegetation of the Tropical Pacific Islands (1998)
D. Mueller-Dombois and F.R. Fosberg

Volume 133
Aquatic Humic Substances: Ecology and Biogeochemistry (1998)
D.O. Hessen and L.J. Tranvik (Eds.)

Volume 134
Oxidant Air Pollution Impacts in the Montane Forests of Southern California (1999)
P.R. Miller and J.R. McBride (Eds.)

Volume 135
Predation in Vertebrate Communities: The Bialowieża Primeval Forest as a Case Study (1998)
B. Jedrzejewska and W. Jedrzejewski

Volume 136
Landscape Disturbance and Biodiversity in Mediterranean-Type Ecosystems (1998)
P.W. Rundel, G. Montenegro, and F.M. Jaksic (Eds.)

Volume 137
Ecology of Mediterranean Evergreen Oak Forests (1999)
F. Rodà et al. (Eds.)

Volume 138
Fire, Climate Change and Carbon Cycling in the North American Boreal Forest (2000)
E.S. Kasischke and B. Stocks (Eds.)

Volume 139
Responses of Northern U.S. Forests to Environmental Change (2000)
R. Mickler, R.A. Birdsey, and J. Hom (Eds.)

Volume 140
Rainforest Ecosystems of East Kalimantan: El Niño, Drought, Fire and Human Impacts (2000)
E. Guhardja et al. (Eds.)

Volume 141
Activity Patterns in Small Mammals: An Ecological Approach (2000)
S. Halle and N.C. Stenseth (Eds.)

Volume 142
Carbon and Nitrogen Cycling in European Forest Ecosystems (2000)
E.-D. Schulze (Ed.)

Volume 143
Global Climate Change and Human Impacts on Forest Ecosystems: Postglacial Development, Present Situation and Future Trends in Central Europe (2001)
J. Puhe and B. Ulrich

Volume 144
Coastal Marine Ecosystems of Latin America (2001)
U. Seeliger and B. Kjerfve (Eds.)

Volume 145
Ecology and Evolution of the Freshwater Mussels Unionoida (2001)
G. Bauer and K. Wächtler (Eds.)

Batch number: 10755284

Printed by Printforce, the Netherlands